Advances in e-Engineering and Digital Enterprise Technology – I

Advances in e-Engineering and Digital Enterprise Technology – I

Proceedings of the Fourth International Conference on

e-Engineering and Digital Enterprise Technology (e-ENGDET)

Leeds Metropolitan University, UK
1–3 September 2004

Edited by

K Cheng, D Webb, and **R Marsh**

Professional Engineering Publishing

Professional Engineering Publishing Limited
Bury St Edmunds and London, UK

First Published 2004

ISBN 1 86058 467 5

A CIP catalogue record for this book is available from the British Library.

Printed by The Cromwell Press, Trowbridge, Wiltshire, UK.

Acknowledgements

The organizational arrangements for the conference have been undertaken by staff within Innovation North, Faculty of Information and Technology, Leeds Metropolitan University, UK. The editors are especially grateful for the help given by Mr Jack Hunt, Dr Jeremy Toussaint and Ms Marion Baldwin.

The editors also wish to thank the many referees and the following members of the 4th e-ENGDET Conference Committee for their assistance with paper reviewing.

Professor P Barrar	University of Manchester, UK
Professor C R Boer	ITIA-CNR, Italy
Professor S Y Chou	National Taiwan University of Science and Technology, Taiwan
Dr A Crispin	Leeds Metropolitan University, UK
Professor B J Davies	International Journal of Advanced Manufacturing Technology, UK
Dr J Efstathiou	University of Oxford, UK
Professor K Feldmann	University of Erlangen-Nuremberg, Germany
Professor R Gonçalves	UNINOVA – CRI, Portugal
Professor M Gregory	University of Cambridge, UK
Professor X J Gu	Zhejiang University, China
Professor D K Harrison	Glasgow Caledonian University, UK
Professor H S Hu	University of Essex, UK
Professor G Huang	University of Hong Kong, Hong Kong
Professor T Ito	University of Tokushima, Japan
Professor P Y Jiang	Xian Jiaotong University, China
Professor W B Lee	Hong Kong Polytechnic University, Hong Kong
Professor P G Maropoulos	University of Durham, UK
Professor D T Pham	University of Cardiff, UK
Dr Y Qin	Strathclyde University, UK
Professor M Shpitalni	Technion-Israel Institute of Technology, Israel
Professor J Z Y Wang	University of Nevada at Las Vegas, USA
Professor P K Wright	University of California at Berkeley, USA
Dr Y Yusuf	University of Hull, UK
Professor L Zheng	Tsinghua University, China

Sponsored by

leeds metropolitan university EPSRC The IEE I MECH E

Related Titles of Interest

Title	Editor/Author	ISBN
Advanced Manufacturing Processes, Systems, and Technologies (AMPST 99)	M K Khan, Y T Abdul-Hamid, C S Wright, and R Whalley	1 86058 230 3
IMechE Engineers' Data Book – Third Edition	C Matthews	1 86058 456 X
Computer-aided Production Engineering (CAPE 2003)	J A McGeough	1 86058 404 7
Advances in Manufacturing Technology XVII	Y Qin and N P Juster	1 86058 412 8
Cams for Industry	J Reeve	0 85298 361 X

For a full range of titles published by Professional Engineering Publishing (publishers to the Institution of Mechanical Engineers) contact:

Marketing Department
Professional Engineering Publishing
Northgate Avenue
Bury St Edmunds
Suffolk IP32 6BW
UK

Tel: +44 (0) 1284 763277: Fax: +44 (0) 1284 718692
E-mail: marketing@pepublishing.com
www.pepublishing.com

About the Editors

Professor Kai Cheng is professor of Precision Engineering at Leeds Metropolitan University. His current research interests focus on the design of precision machines and instruments, nano/micro engineering, e-design, and e-manufacturing. Professor Cheng has published about 160 papers in learned international journals and refereed conferences, authored/edited three books, and contributed five book chapters. His BEng (Hons) and MSc degrees were obtained from Harbin Institute of Technology, and he has a PhD from Liverpool John Moores University. He worked as a senior lecturer in Glasgow Caledonian University before joining with Leeds Metropolitan University. Professor Cheng is a member of the IMechE and the IEE. He is the leader of the Advanced Manufacturing Technology Research Group (AMTRG), which is currently working on a number of research projects funded by UK EPSRC, EU 6th Framework Program, KTP Programs, DTI, and industry. The group has a record of international excellence in the areas of precision engineering and Internet-based e-design/e-manufacturing, which can be found at the website: http://care.ies.lmu.ac.uk/

Professor David Webb is currently a Professor of Engineering Modelling and Simulation at Leeds Metropolitan University and Head of the Centre for Applied Research in Engineering. Professor Webb obtained a DPhil in Physics in 1975 from the University of York and his current research interests include computer-aided modelling and the simulation of structural components and manufacturing processes using the finite element method. He has published over a hundred journal and conference papers in the areas of manufacturing, biomechanics, crashworthiness, and the modelling of complex materials.

Dr Rodney Marsh was, until recently, Head of the School of Engineering at Leeds Metropolitan University. His industrial background comprised a traditional five-year apprenticeship as a mechanical engineer, followed by twelve years as a production engineer in the motor industry in Europe. Dr Marsh became a member of the Institution of Production Engineers (now IEE) in 1971 and a Chartered Engineer in the same year. In 1973 he became a member if the Institution of Mechanical Engineers. His academic specialism includes design for manufacture, flexibility in manufacture systems, lean manufacture, and supply chain management. Dr Marsh has been external examiner at twelve UK universities for courses at undergraduate and postgraduate level, as well as MPhil and PhD projects.

Contents

Web-based CAD/CAM/CAPP/PDM

Virtual and Collaborative Engineering

e-Business, e-Commerce, and e-Services

Computing Infrastructure for e-Engineering

Digital Factory/Enterprise

On-line Education and Industrial Training

Preface

e-Engineering and digital enterprise technology are becoming the catalysts and prime enablers for the most radical changes in industry since the industrial revolution. They will impose real time transparency on all manner of industrial operations/transactions such as:

- reducing barriers to entry;
- empowering customers;
- reinforcing the need for manufacturers providing commodity products to differentiate themselves;
- enabling enterprises to be agile and responsive throughout the product life cycle.

It is, therefore, with great pleasure, and some degree of timeliness, that we host the 4th International Conference on e-Engineering and Digital Enterprise Technology to be held in Leeds, UK, 1–3 September 2004.

The proceedings include 63 refereed papers from a dozen countries on different continents, thus making the conference truly international. Topics covered in the conference include:

- e-Design and e-Manufacturing
- e-Supply Chain Management, e-Logistics, and Mass Customization
- Web-based CAD/CAM/CAPP/PDM
- Virtual and Collaborative Engineering
- Web-based Modelling and Simulations
- Remote Control, Condition Monitoring, and Tele-robotics
- e-Business, e-Commerce, and e-Services
- Computing Infrastructure for e-Engineering
- Digital Factory/Enterprise
- On-line Education and Industrial Training

The conference will provide an excellent opportunity to meet formally and informally with friends and colleagues, old and new, from around the world.

We look forward to welcoming you to Leeds in September 2004.

Professor Kai Cheng, Conference Chairman
Professor Dave Webb, Conference Vice-Chairman

Keynote Paper

Digital enterprise technology in production networks

P G MAROPOULOS, D G BRAMALL, P CHAPMAN, W M CHEUNG, K R McKAY, and B C ROGERS
School of Engineering, University of Durham, UK

ABSTRACT

Enterprises organised as distributed production networks face challenges related to the planning and synchronisation of their manufacturing operations, and of controlling the quality, cost and production time of products under development. Digital enterprise technology (DET) is a framework for new technologies which can shorten product development and realisation, by estimating and therefore controlling quality, cost and delivery factors for products at an early stage in their lifecycle. DET can also be deployed to facilitate planning and synchronisation of work across an extended enterprise. Particularly for high value and high complexity products, these techniques represent an emergent synthesis of strategies for design, manufacture and product lifecycle management.

1 INTRODUCTION

Digital enterprise technology (DET) has been defined as "the collection of systems and methods for the digital modelling of the global product development and realization process, in the context of lifecycle management" [1]. Distributed enterprises can take advantage of emerging methods for product modelling, process planning, workflow management and process simulation to coordinate product development. Being able to estimate during design and control during the production phase the manufacturing schedule, quality, cost, manufacturing time, process risks and workflow allows the enterprise to innovate and remain successful.

DET is implemented as a synthesis of technologies and systems from five main technical areas: the design of the product, design of processes, design of factories, technology ensuring the conformance of the digital product, process and resource domain with the real physical artefacts and equipment, and the integration with enterprise-wide, back-office systems for resource availability information, logistics and communication.

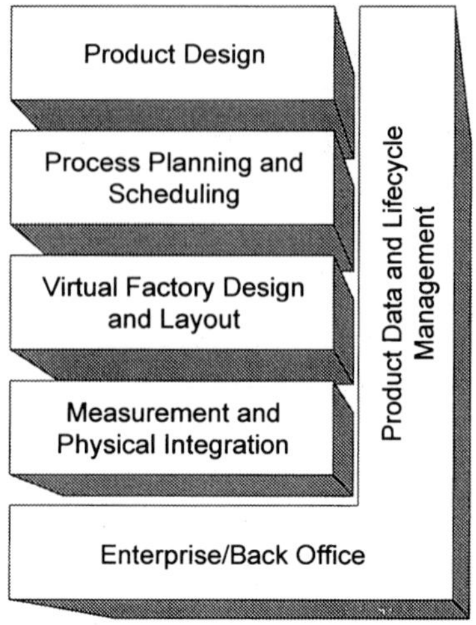

Fig. 1 The five technical areas of DET

Technology for the flexible deployment and web-based integration of these technical areas forms the backbone of DET. Internet methods, product related data repositories and standards for communication and the exchange of engineering data allow DET techniques to scale up and be deployed across an enterprise.

A *production network* [2] is made up of the functional units of the enterprise, its partners and potential partners. For a distributed production system, these entities are configured to form a virtual or extended enterprise. The formation and operation of these networks highlights a requirement for the development and deployment of two DET technical areas:

1.1 Early planning for production networks

Early planning allows a production network to be configured using objective metrics and then provides the ability to estimate and optimise costs, quality, manufacturing time frames and expected profitability. This should be possible even if a new product design is still at the conceptual design stage. The appropriate production units and partners must be chosen to make up the virtual enterprise for each venture, and the production costs, quality and delivery estimated given the resources available. Several iterations may be necessary before the final membership of a virtual enterprise is decided upon, or membership may deliberately be left flexible throughout the product lifecycle.

1.2 Workflow and production data management

The distributed production network needs technology for product lifecycle management to be deployed across all locations on the network in order to cooperate effectively. *Product data management* (PDM) systems are groupware technologies used for the storage, organisation and sharing of product related data. *Workflow management systems* (WFMS) coordinate the actions of a distributed team of designers and engineers, and together with PDM, provide a solid platform for the deployment of DET functionality in distributed and collaborative environments.

2 AGGREGATE PROCESS PLANNING

Aggregate process planning [3,4,5,6] is a novel, computer automated process planning (CAPP) technology which promotes production-oriented manufacturability assessment for discrete parts and assemblies, even in the conceptual stages of development, when it is not possible to fully describe the geometry and configuration of a product. The aggregate process plan is intended as a guide to explore the manufacturing and assembly options for a product, by providing an indication of cost, manufacturing time and quality. It provides the designer with the ability to envisage the production methods and resource requirements which would result from a particular design.

In particular, *resource aware* aggregate planning refers to the generation of *rough cut* manufacturing schedules in which the production schedule is constrained to only use selected and available network resources. This facilitates the technical evaluation of production options early in the design phase, during which the majority of lifecycle costs are decided. This technology can be applied to support strategic design decisions and evaluation of "what if" conceptual scenarios within the context of a production network. The results of these early plans provide useful indications of the areas in the enterprise that would benefit from further detailed analysis using other digital enterprise technology tools, such as factory simulation and logistics.

3 AGGREGATE MODELLING REQUIREMENTS

The core information requirements for aggregate planning centre around three key aspects of the manufacturing system; the product, the process that will be used to make it and the resources that will be deployed for production and integration.

3.1 Aggregate product model

The product model must support the transition of the design from conceptual design through to detailed design in a form that is suitable for integration with the process planning system and downstream detailed design tasks. An object-oriented, feature-based model provides the framework for modelling of the component parts and assemblies. Aggregate product models are a vital element of creating design representations that are conducive to rapid, automatic evaluation during the early design stages in terms of widely used manufacturability metrics such as *quality, cost and delivery* (QCD).

The two general requirements of an *aggregate product model* (APM) are as follows [4]:

The first requirement is that the APM representation should support the transition of design data from the uncertain, conceptual stage through to the detailed design stage. The model should therefore be able to:

- Represent incomplete data during conceptual design.
- Model uncertainty in feature dimensions during embodiment design.
- Add new information, including detailed design features.
- Model elements within standard sub-assemblies that may include full dimensional tolerances.

The second requirement is that an APM must represent the design intent in a way that is suitable for integration with aggregate process planning algorithms. It must therefore:

- Use a representation built on a generic software component model.
- Maintain the integrity of data objects representing features throughout the design stage.
- Represent assembly connectivity at feature level.
- Incorporate industry standards for quality in terms of dimensioning and tolerancing.

Within a feature-based aggregate product model, features need to encapsulate knowledge in order to facilitate design change that is consistent with design objectives and priorities. Features exist within a taxonomy of classes; therefore, a great deal of geometric, manufacturing and testing information can be inferred from such a classification. The other important requirement is that a feature model must support complex inter-relationships between multiple features (assembly and joint features), so the APM incorporates two classes of feature relations, *geometry feature relations* and *assembly feature relations* [4]. The information content of the above feature relations is enhanced by the specification of dimensions such as radius or length, parametric quantities and dimensional tolerances.

3.2 Aggregate Process Model

Aggregate process models are required for the early calculation of manufacturability metrics during product definition [3]. They derive the measures of manufacturability (QCD) by means of heuristic methods rather than computationally expensive, exhaustive calculations. The method used depends on the availability and completeness of data in the product model.

At their simplest, aggregate process models use an equation relating a manufacturability measure to product model drivers, such as a dimension, volume or parametric quantity. This equation can usually be derived from regression analysis of historical production data, or from repeated simulated manufacturing trials in a more detailed process planner. Higher level aggregate models make direct use of more complete data such as multiple dimensions, dimensional tolerances, and process based parameters, to select processes and derive estimates of manufacturability measures.

Core capability checks must also be built into the process model in order to eliminate feature-process combinations which are not feasible due to geometrical or material limitations. This eliminates invalid plans from the planning search space and speeds up optimisation.

3.3 Aggregate Resource Model

The term resource aware planning is used to indicate a dynamic inter-relationship between the planning entities (products and processes) and the enterprise resources, including humans, machines and consumables. Enterprise level resource data should ideally be available from an *enterprise resource planning* (ERP) or other back-office system, so as to conduct evaluations with respect to the real availability and existing commitments in the production network.

The resource aware, process planning procedure should construct schedules which are valid with respect to the available resources; never over-committing any resource by having it in more than one place or involved with more than one process at one time. Thus, the resource model captures the potential production capacity of the production network and the technical capability of the production resources that are distributed across the enterprise.

4 IMPLEMENTATION OF AGGREGATE PLANNING

Early technology demonstration systems such as CAPABLE/Space [6] have been developed to explore the possibilities provided by the kind of systems that are likely to become available in the medium term. The following operating requirements were established:

- The system must cope with complex products and large scale resource models.
- A process plan must be generated without precise product specifications or complete data.
- The planning procedure should identify the major manufacturing steps and assist the user in deciding upon the best allocation of components to factories in the network.
- Analysis of the resultant plan should provide sufficient data to be able to drive further refinement of the product or resource model.

4.1 Capable/SPACE

A prototype feature-based design and planning system was developed to test the aggregate planning hypothesis. Production schedules are generated using a multi-criteria optimisation approach involving process planning, process option generation, constraints, scheduling and optimisation.

4.1.1 Option generation from Feature-Process-Resource mapping

The planning engine uses a two stage process to intelligently allocate processes and resources as required to features in the product model. Thus, the system is capable of identifying (and evaluating) all possible manufacturing options, for each product feature, using the available enterprise resources.

A map is pre-calculated relating each feature to a list of theoretically feasible processing options and another map is constructed relating each process to a list of compatible and available resources. During the first stage of planning, the feature-to-process map is used to identify which processes are capable of producing a feature, regardless of that feature's dimensions and dimensional tolerances. The second stage uses the process-to-resource mapping to identify the particular resources available within the enterprise for realising the selected processes.

In most cases, there will be a vast number of options available in terms of manufacturing and assembling a product within the enterprises' production network. In order to intelligently explore this search space, the process plan is incrementally improved by selecting, at random, new manufacturing options for some features, and evaluating their fitness for purpose using the objective function of the simulated annealing [7] optimisation procedure.

4.1.2 Technological and Physical Constraints

A set of technological constraints are applied to ensure that the initial selection of processes and resources is valid, and that those processes and resources obtained through the mappings are capable of producing the features defined within the product model:

- Feature dimensions and tolerances along with surface finish and empirical dimensional ratios (such as drilling: hole length vs. diameter) may prohibit the use of a particular process.

- Physical constraints limit the feature and process combinations to capable resources; the constraints ensure that selected resources are capable based upon their operating parameters, such as bed size, tool, feed and spindle speed as well as the *statistical process control* (SPC) history in producing similar features.

4.1.3 The Hybrid Optimisation Algorithm

In order to evaluate the vast search space, produced as a result of the mappings, artificial intelligence derived optimization methods such as simulated annealing [7] and genetic algorithms [8] must be used. An early technology demonstrator of the aggregate planning system made use of a hybrid simulated annealing algorithm featuring a *greedy local search* [6]. As a result of this pilot implementation, four key principles were established for developing optimisation functions for process planning across production networks:

- The search space should be explored on the basis of a multi-criteria objective function that combines *quality, cost, delivery and knowledge* (QCD+K) metrics. To achieve this requires the network's collaborators to agree standards for the weighting given to each driver.
- The weightings should be used by the product development team to provide "what-if" scenario planning and target specific improvements to the plan.
- The optimisation requires a neighbourhood function to control the generation of new solutions from the current one. This can either result in a change to a single resource or to a process and resource simultaneously. The suggested neighbourhood function adapts to the energy of the overall solution such that in the early stages of optimisation large changes are forced by changing the process and resource selection simultaneously. In the latter stages, smaller changes are made by changing only the resources.
- To maximise the effectiveness of the optimisation functions, some local optimisation of sub-plans, using a greedy algorithm, is desirable. For example, multiple similar features at the same level in the bill-of-materials should often be manufactured using the same process and in the same set-up so there is no need to optimise each feature individually.

4.2 Advanced Scheduling

In addition to the above techniques, a new scheduling engine was designed for the successor to CAPABLE/Space. The engine was developed using the embedded *knowledge based system* (KBS) CLIPS [9] to take advantage of the Rete [10] algorithm, which matches data to rules using a special data structure called a *join network*. The engine is able to very quickly generate a sequence of tasks which never over-commits any of the resources.

The scheduling engine generates a valid sequence of operations from the process plan using a heuristic procedure. To decide which operation should be scheduled next, it needs no information other than the length of time an operation takes and the availability of the resources assigned to the operation. Aggregate process models provide the means needed for the expert system to estimate the duration of operations, calculating the resulting quality of the feature and the impact on the resources involved. The availability of resources matches with specially designed rules to give a list of candidate operations to be added next. The system selects an operation from the set, and repeats the cycle until there are no more operations. Other special expert system rules prevent more than one component of an already constructed sub-assembly from being operated on at the same time, so this new system does not rely on a hierarchal ordering of sub-assemblies.

The planning system is never prevented from producing a schedule due to a *resource waiting deadlock* [11], because tasks only seek to acquire resources at the time that they start, and so never hold resources while waiting to acquire more.

5 DET AND RESOURCE AWARE PROCESS PLANNING IN PRODUCTION NETWORKS

From the industrial perspective, the key challenge now that the technical barriers to the implementation of resource aware planning have been mostly overcome, is to demonstrate that DET systems and workflows can be implemented in production networks with a view to realising the following benefits for distributed design and planning systems [1]:

- Cooperative and distributed design and planning.
- Capturing and making use of design and process knowledge.
- Resource aware planning for multiple sites.

5.1 Planning strategies for production networks

Two distinct methodologies for utilising aggregate process planning methods in production networks have been identified and validated [6]. Termed *product distributed analysis* (PDA) and *resource distributed analysis* (RDA) the two techniques differ in the way that the product models and resource models are created and shared across the production network. Both techniques make use of the redundant nature of resources in the production network [2] and the modular nature of the product model.

5.1.1 Product Distributed Analysis

In the first method, product distributed analysis (PDA), a product model is electronically published, and the suppliers utilise this in generating process plans based upon their own factory models. The PDA configuration is used in two case studies [6] and is as follows:

- The main company posts the specified design or product module to the database and informs the supplier network of the models and specifications for download.
- A process planning analysis is then performed utilising the supplier's own resource model and the resultant planning information is uploaded back to the database.
- The main company collates the submitted process plans and assesses the response; a new *request for interest* (RFI) may be posted, if necessary.
- The main company selects the members of the production network based upon the submitted plans.

The advantage of PDA is that it exploits the more accurate knowledge that exists at the supplier level, leaving the details of process modelling and the final detailing of the product to local suppliers.

5.1.2 Resource Distributed Analysis

The second method, resource distributed analysis (RDA) utilises the aggregate planning procedure on a larger problem, with the system making use of the distributed factory models of the suppliers. The RDA procedure is as follows:

- Suppliers, who wish to become part of the main company's supply network, electronically submit a digital resource model of their production facilities and equipment, which is stored in the database.
- The main company downloads these supplier models, and with a given product model utilises the process planning optimisation software to create production routes, with which it can evaluate and rank the selected suppliers.
- The selected plans are uploaded into the database and the appropriate suppliers informed.
- Each supplier retrieves their respective process plans and decides to accept, reject or improve upon the results.

The advantage of RDA is a global optimisation over a single search space, rather than piecing together fragments of plans and schedules from many low level suppliers. Also, this method can be carried out without the involvement of any suppliers.

6 WORKFLOW AND PRODUCT DATA MANAGEMENT

Extended enterprises require support for business processes which involve many distributed locations and personnel. *Product data management* (PDM) systems, a variation on the groupware theme, facilitate communication and coordination among product development teams. They provide document storage and management, including revision control, and various levels of integration with CAD systems to support distributed product development as required by the DET framework.

A complementary capability increasingly provided alongside PDM is workflow management. A workflow is the set of tasks, resources and informational elements involved in the fulfilment of a business process [12]. Traditionally, business workflows were based around a template consisting of the ad hoc workflow: negotiation, acceptance, assignment and review [13]. A *work flow management system* (WFMS) [14] provides the ability to script a business process, providing a flowchart to show the current state of the process and an 'inbox' for users to receive notifications about tasks assigned to them.

Two competing philosophies of cooperative work have grown out of two different perspectives. *Computer supported cooperative work* (CSCW) (14) systems support the so called map approach, emphasising knowledgeable actors cooperating within a shared information space. On the other hand, WFMS supports the script approach, providing scripts and fairly inflexible templates for highly structured business processes. They give their users the ability to be notified when all the requirements for commencement of a task have been satisfied, sometimes called rendezvous [13].

Product development teams require WFMS in order to structure the design process. They also require a shared information environment because of the number and complexity of interactions between different elements in a complex product assembly. DET therefore requires both approaches to be integrated to succeed.

In addition to these requirements, the ability to distribute the elements of a task over the internet is important to an extended enterprise. CSCW and WFMS have grown up with the internet, and systems now commercially available can operate in a distributed manner over the internet and through *World Wide Web* (WWW) based portals. Figure 2 shows the relationship between core DET functions for product lifecycle management, and web-enabled client applications making use of data stored in PDM/ERP and made available over the internet.

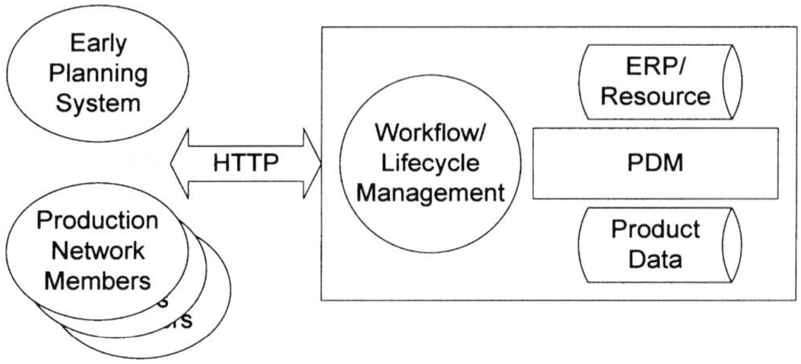

Fig. 2 Web-enabled DET functions

6.1 Future of Workflow Systems

The present generation of workflow systems could give way in time to more sophisticated approaches, involving less scripted transactions and a more loosely coupled relationship based on dynamically discovering partners, creating short lived networks, and deploying software agents to participate in such organizations. Aggregate planning and other DET technologies will play a crucial role in the decision making and planning required within such networks.

7 CONCLUSIONS

Digital enterprise technology provides a new framework for product lifecycle management across a distributed production network. Of the five key technology areas of DET, manufacturing planning/scheduling and PDM/Workflow are critical to any successful distributed manufacturing venture.

Aggregate design and process planning provide a computer automated process planning (CAPP) solution that can be used to evaluate quality, cost and delivery from the earliest stages of product development. Resource aware scheduling enables production networks to be optimised based on conceptual stage product designs, by evaluating manufacturing options in a digital context that reflects the real situation on the ground.

Product data management (PDM) and workflow management systems (WFMS) are essential to any distributed product development effort, because of the many interdependencies between different sections of product data, and the need to control the process of product design. World Wide Web enabled portals ease the deployment of such systems across a heterogeneous production network containing companies with varying IT capabilities

This research has identified a requirement and proposed a framework for a major future area of innovation in digital product development in distributed enterprises.

8 ACKNOWLEDGEMENTS

The authors wish to acknowledge the financial support of the EPSRC, UK Grants GR/N 11285 and GR/R 26757, and the generous help of the collaborating companies, EADS Astrium, Mabey & Johnson, ArvinMeritor Automotive and the LSC Group.

REFERENCES

1. Maropoulos, P. G., 2003, "Digital Enterprise Technology – Defining Perspectives and Research Priorities", *International Journal of Computer Integrated Manufacturing*, 16(7-8), 467-478.
2. Wiendahl, H. P., Lutz, S., 2002, "Production in Networks", *Annals of CIRP*, 51(2), 573-586.
3. Maropoulos, P. G., 1995, "A Novel Process Planning Architecture for Product Based Manufacture", *Proceedings of the IMechE Part B - Journal of Engineering Manufacture*, 209(4), 267-276.
4. Bradley, H. D., Maropoulos, P. G., 1998, "A Relation-Based Product Model for Computer Supported Early Design Assessment", *Journal of Materials Processing Technology*, 76(1-3), 88-95.
5. Maropoulos, P. G, Bramall, D. G., McKay, K. R., Rogers, B., Chapman, P., 2003, "An Aggregate Resource Model for the Provision of Dynamic "Resource Aware" Planning", *Proceedings of the IMechE Part B - Journal of Engineering Manufacture*, 217(10), 1471-1480.
6. Maropoulos, P.G., McKay, K., Bramall, D. G., Rogers B., Chapman, P., 2003, "Dynamic and Distributed Early Planning Assessment by a Hybrid Simulated Annealing and Greedy Algorithm", *Proceedings of the IMechE Part B – Journal of Engineering Manufacture*, 217(10), 1429-1442.
7. Kirkpatrick, S., Gelatt Jr., C. D., Vecci, M. P., 1983, "Optimization by Simulated Annealing". *Science*, 220(4598), 671-680.
8. Goldberg, D. E., 1989, "Genetic Algorithms in Search", *Optimization and Machine Learning*, Addison-Wesley, Readings, MA, USA.
9. Giarratano, J. C., Riley, G., 1993, "CLIPS Reference Manual Version 6.0: Volumes 1 & 2", *NASA Lyndon B. Johnson Space Center Information Center*, Information Systems Directorate, Software Technology Branch, USA.
10. Forgy, C. L., 1982, "Rete: A Fast Algorithm for the Many Pattern/Many Object Pattern Match Problem", *Artificial Intelligence*, 19(1), 17-37.
11. Coffman, E. G., Denning, P. J., 1973, "Operating System Theory", *Prentice-Hall*, ISBN 0-13-637868-4.
12. Basu, A., Blanning, R. W., 2000, "A Formal Approach to Workflow Analysis", *Information Systems Research*, 11(1), 17-20.
13. Fischer, L. (Ed.), 2004, "*The Workflow Handbook 2004*", Future Strategies Inc., Lighthouse Point, FL, USA, ISBN 0-9703509-6-1.
14. Schmidt, K., 1997, "Of Maps and Scripts: The Status of Formal Contracts in Cooperative Work", *In: Proceedings of GROUP97*, Phoenix, AR, USA, 138-147.

e-Design and e-Manufacturing

Internet-based, end-user design and e-manufacture

R J BATEMAN and **K CHENG**
Advanced Manufacturing Technology Research Group (AMTRG), Leeds Metropolitan University, UK

ABSTRACT

With the rapid growth of the Internet and developments in rapid prototyping technologies, there now exists the real possibility to create an innovative type of e-manufacturing which will allow customers to directly design items on their own PCs and then transmit the data via the internet to a 'high street' *rapid manufacturing* facility in the customers own locale, thus minimising problems associated with logistics and delivery.

The term *devolved manufacture* has been previously employed to describe this concept in e-manufacturing were the actual manufacturing phase is relocated to the customers end of the process rather than passing through an intermediate factory site.

After reviewing some of the issues and technical challenges involved with such a system, this paper discusses the issues involved with the development of such an internet based system and appropriate conclusions are then drawn about future development needs.

Keywords: e-manufacture, mass customisation, devolved manufacture

1 INTRODUCTION

New Technologies or methodologies generally create the opportunity to experiment with novel design, manufacturing or marketing strategies and processes. The concept of *mass customisation* gives such an opportunity to effectively integrate the customer into the design process. The term was first coined by Davis in 1987 [1] and then popularised by Pine in his 1993 book [2].

At its simplest, mass customisation aims to produce a range of customizable products by allowing the customer to select from a range of possible (pre-defined) combinations and thus customise the product to their own choosing. To be successful and economically viable, the associated production systems have to be set up to allow this wider choice to be produced efficiently without incurring the overheads associated with normal dedicated mass production techniques and machinery (Figure 1).

Manufacturing methodologies and technologies like concurrent engineering (CE), just in time (JIT), computer integrated manufacture (CIM), flexible manufacturing systems (FMS), have been closely allied to the introduction of mass customisation, but as Piller [3] states *"...the main distinctive principle of mass customization is a mechanism for interacting with the customer and obtaining specific information in order to define and translate the customer's needs and desires into a concrete product or service specification ...*". A major hurdle to be overcome in the path to mass customisation is the capturing of the 'choices' made by the customer.

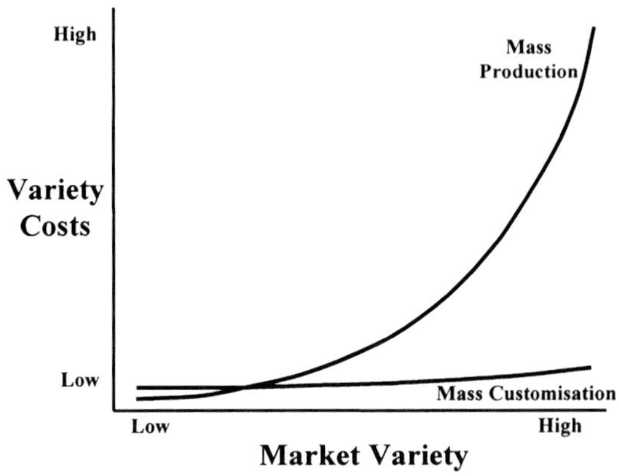

Fig. 1 Variety cost as a function of market variety [4]

The conventional process by which a customer chooses a product or service has inbuilt feedback mechanisms. The choice is usually made by selection of articles seen 'in-the-tin' in a store or perhaps after having seen images on television or in a catalogue. In all cases, the customer usually has a very clear idea of exactly what it is they are selecting. The extra choices made available in a mass customisation system requires a more complex feedback process – one which will interactively allow the customer to see or hear what the outcome of their specified choices will be.

The systems used to allow this process are generally known as configurators, co-design systems (or platforms), toolkits or choice boards. The use of these systems allow the customer to find what the options available are, make and receive feedback on their selections and often make payment. The web based system used by Dell Computers [5] allows customers to select from a range of possibilities to specify their own computer. From the manufacturers point of view, this information can then be used to automatically produce material and component lists, manufacturing schedules, delivery information, etc. It also has one further bonus in that the customer pays for the product up front – leading to radical changes in the manner in which working capital is employed.

2 TOWARDS MASS 'INDIVIDUALISATION'

2.1 Current developments

With the development of new technologies like 3D laser scanners, rapid prototyping and concept modelling machines, some companies have pushed the boundaries of the mass customisation concept towards what might be termed mass *individualisation* - i.e. where the product or service is created specifically for one individual only rather than allowing more than one customer to make the same choices resulting in identical products.

The well known sportswear company Adidas created the 'miAdidas' [6] project which uses laser scanning techniques to create a 3D model of a customers feet. This model is then used to create perfect fitting customised/individualised sports shoes for that individual. The service is primarily aimed at sports stars and professionals and is priced accordingly as a premium or value-added service.

Other companies have made use of the Internet to capture customers design data which they then use in a more traditional manner. eMachineShop [7] are one of a number of companies which will machine individual parts from customers own drawings – created by a downloadable free programme. ToyBuilders.com [8] will design and manufacture personalised action figures based on photographs supplied by the customer.

2.2 Devolved Manufacturing

Most, if not all implementations of the mass customisation philosophy are run along conventional manufacturing lines where the eventual destination of the customers 'design' is the company production facility (factory, print shop, fabrication shop, etc.).

This approach could be deemed 'factory-centric' – i.e. the process is set up with a conventional manufacturing model in mind and the sole aim of inducing customers orders into the existing (or planned) production facility. Once manufactured, the products are then shipped out to the customer, whatever their geographic location may be, in the usual manner. As such these products are then subject to the same manufacturing and logistical problems and limitations as exist in an equivalent mass production process.

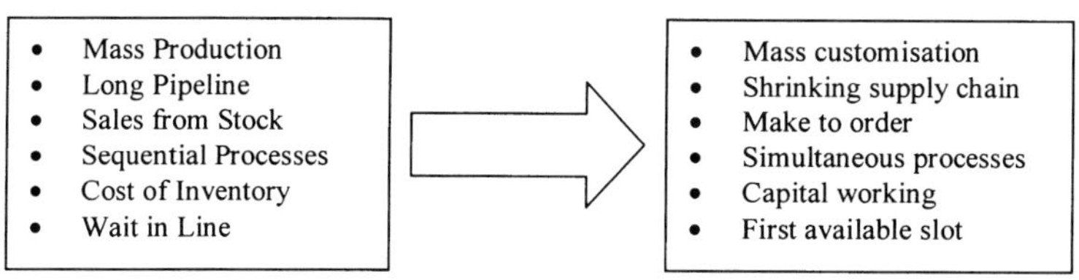

Fig. 2 Transformation by e-manufacturing [9]

Whilst these movements towards mass customisation and the development of new technologies, infrastructures and working methods are significant steps along the path to true *e-manufacture* (Figure 2) the limitations and problems of the conventional approach to manufacturing and distribution still exist.

The term devolved manufacturing has been previously used by the authors [10] to describe a novel approach utilising developments in rapid prototyping/manufacturing technologies and the Internet to *devolve* the manufacturing facilities closer to the customer – for example into the high street - thus logistical problems associated with long distance distribution.

Limited, but successful examples of this process of devolution of manufacturing facilities ever closer to the customer already exist. The photo-processing booth or shop is a familiar sight in most parts of the world. Instead of the undeveloped films being sent to a central processing laboratory, most film sizes can now be developed and printed in on the premises, using an automatic processing machine generally no bigger than a moderately sized photocopier.

The development of CNC (computer numerically controlled) type automated spectacle lens grinding machines allowed the 'Vision Express' company [11] to install these machines in their high street shops. A customer may now take their prescription into the high street shop and have their spectacles 'manufactured' on site, often in around 1 hour. This has dramatically reduced the time taken by for the manufacture of a customer's prescription spectacles from perhaps 2-3 weeks to a couple of hours by removing the need to send the order to a centralised manufacturing facility and then send the finished article back. From the customers' viewpoint, both of these situations represent major steps forward in time saving and convenience.

It is obvious that the maturing of the appropriate technologies must be seen as crucial enablers for both these cases. Add into the mix emerging developments with the Internet/WWW, rapid prototyping/manufacturing technologies and software applications, the potential of the devolved approach becomes very interesting.

3 MAKING IT HAPPEN

There now exists the very real possibility to create an innovative type of e-manufacturing facility which will allow customers to directly design items on their own PCs and then transmit the data via the internet to a 'high street' *rapid manufacturing* facility in the customers own locale, thus minimising problems associated with logistics and delivery. This section reviews the issues and technical challenges involved with such a system, and describes the development of such a prototype internet based system.

3.1 Challenges to be addressed

On the technology side, the challenges to be addressed with any new or developing technology are really 'the usual suspects' of technical capability and cost. The key areas to be integrated for the devolved manufacturing approach to become possible and economically viable are the Internet/WWW as communication medium, mass customisation methods and techniques to manage the system and rapid manufacture technologies to actually create the products

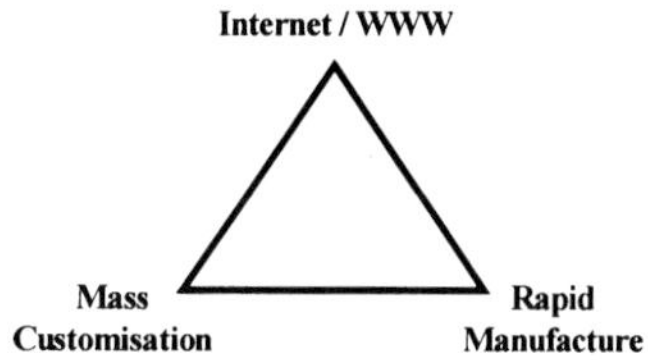

Fig. 3 – Key areas for integration

3.1.1 Technical capability and cost issues

Current generation rapid prototyping type machines essentially work either by additive or subtractive methods. Where the additive method is used, parts or models are usually built up layer by layer and the three-dimensional geometry of the part is constructed from CAD-generated solid or surface models. Using this laminar approach, highly complex parts can be built up quickly and automatically from a variety of materials. Subtractive methods start from the opposite end with a block or billet from which material is then removed either by traditional CNC type machining methods or perhaps by some other method like lasers or spark erosion.

The current cost involved with the various technologies are still regarded by many companies as high, but as would be expected with any developing technologies the costs have reduced significantly since the machines first appeared on the market. As more companies adopt this technology, the economies of scale will lead to inevitable reductions in cost of machines and materials.

Current generation rapid prototyping technologies are generally regarded as painfully slow when compared with standard mass manufacturing technologies like injection moulding, but productivity gains in the design, prototyping and functional testing phases of new product development process often offset the costs of introducing the new technology.

Current rapid prototyping technology extends only to single material machines. Where products or models need to be made with parts in different materials, either multiple machines are required or the machine needs to be paused, cleaned and then reloaded with different materials. To truly extend the manufacturing capability of this technology, machines with the capability to make products from several materials concurrently are needed although there have been recent developments to include a range of additives into the mix as a part is being created thus changing the properties different parts of the model to deal with different functional requirements.

There is some anecdotal evidence to suggest that there have been subtle changes in the manner in which rapid prototyping technologies are developing due perhaps to a result of the shift in perception of the technologies from being suitable only for creating models or prototypes to being fledgling 'proper' manufacturing technologies. Some manufacturers are looking closely at developing new material and small inexpensive devices aimed at very specific sections of various markets like healthcare.

3.2 System design issues

A prototype system is currently being developed to explore the technical and user-interface issues associated with devolved manufacturing approach.

3.2.1 Product design limitations

There are a number of issues to be addressed when creating a system which will have the capability to allow an end-user or customer to design pretty much anything they wish too. The technical limits of the machinery used will of course provide some constraints.. The maximum size of the designed object will be determined by the 'working space' of the machine – for an additive type machine, needing to fit within the appropriate cuboid specified by the machines dimensions or with subtractive system the size of the 'blank' used. Materials will also be pre-determined by the capability of the available system, and the appropriate costs involved would be relatively straightforward to calculate.

Given the almost total flexibility (within these limits) there are issues to be resolves regarding the what is the appropriate amount of 'freedom' that customer should be allowed. If the customer is to be seen as *co-designer*, does this infer some shared responsibility for the functioning of the finished object ? Normal design processes usually include some functional and safety testing to ensure finished products comply with the relevant product liability and safety laws. It is clear that great care must be taken in defining exactly what the service is that is being offered to the customer.

If the product is a functional item, whose responsibility is it to ensure that the item fulfils appropriate legal requirements and functions properly ? It is clear that the closer to a functioning item the customer wishes to design the greater the risk involved. It may be possible to assess the amount of design freedom appropriate for a given product in terms of it's intended functionality. All items could be placed on a 'risk' continuum with say, a purely decorative item categorized as low risk and a fully functioning item as high-risk at the other end. Some (admittedly quick and dirty) research suggests that it could be a promising qualitative approach but at present exactly how this assessment might be made on a quantitative basis remains unclear.

3.2.2 User interface issues

If the decision is made to allow specified design freedom to a customer, the problem of automatically codifying this 'freedom' in a coherent system remains. Taking a very simple example of a wine glass (on which, for the sake of simplicity all vertical lines are straight and the shape is symmetrical!) it quickly becomes apparent (Figure 4) that the number of parameters can quickly become high –

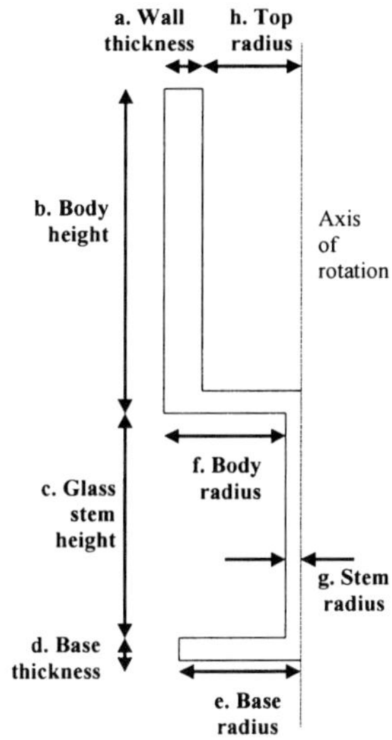

Fig. 4 – Wine glass parameters

and this is not including other parameters such as materials, addition of logos, text, etc.. If constraints are to be added onto the various parameters e.g. there are functional limits to the diameter of the base of the wine glass – too small and it will fall over, too big and it will become unusable then again the determination of these constraints . How such parameters and their appropriate constraints can be identified and codified and then translated into appropriate user-interface (what methods are best suited to allow customers to modify which parameters – e.g. numeric data entry into a table, slider controls or mouse driven 'drag and stretch') is the subject of ongoing research by the authors.

4 CONCLUSIONS

Trimmers [12] asserts that there are 11 current business models in use in e-commerce. These being E-shops, E-procurement, E-malls, E-auctions, Virtual Communities, Collaboration Platforms, Third Party Marketplaces, Value-chain Integrators, Value-chain Service Providers, Information brokerage, Trust & Other Services. It is imperative that the correct business model should be chosen for the devolved manufacture approach to succeed.

There are only really a small number likely scenarios for the development of the requisite infrastructure for devolved manufacture to become commonplace. These could include the expansion of an existing business which already has most of the infrastructure required (e.g. high street premises, communications and logistics networks, etc.) which would allow the large 'branded' organisation to expand and develop products or services specifically tied to their own brand and markets; sideways development by a general service chain (Tom Wohlers [13]) has suggested that photocopy shops would be ideally placed for this) or the development of a network of independent general purpose 'rapid manufacture' facilities (possibly by the franchise route). If/when such development takes place, it is reasonably certain that it will be through evolutionary not revolutionary changes.

One of the most difficult choices to make with mass customisation is deciding what level of customisation should be offered in order to make the company the most economically viable. It does not take much imagination to see that the consequences of offering too great a range of customisation for the target market will be wasted effort and resources, and too little choice for the chosen market will lead to lost opportunities. Figure 5 shows a simple decision matrix to determine one of four suggested strategies.

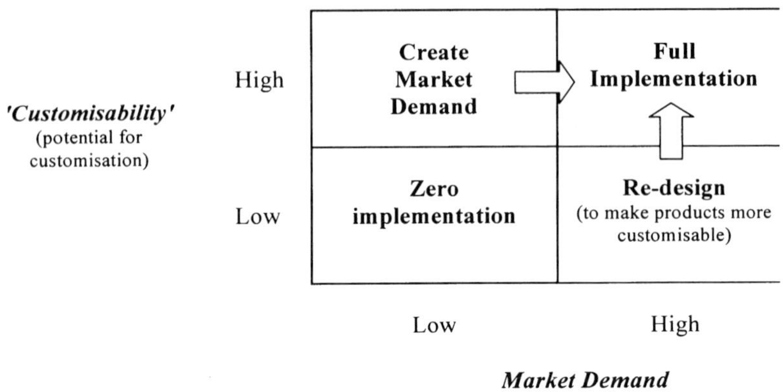

Figure 5 - Devolved Manufacturing Strategy Matrix

In cases where high customisability and high market demand exists, this would suggest strong commercially viable opportunities exist for the devolved manufacture approach. Where there is low potential for customisation and low market demand devolved manufacture would probably not be the appropriate strategy. Where there exists high market demand, but the product has low potential for customisation, the optimum strategy could be to consider re-design of the product to increase the customisability. In the final case of low market demand, high potential for customisation, the appropriate strategy would be to create a market by a high profile marketing campaign (once the market demand is created, this moves the product into the 'high/high' section).

There are obvious limitations with present technologies and infrastructures which would make the creation and implementation a fully working devolved manufacture approach difficult and uneconomic, and there seems really little choice but to wait until the technology 'catches up' with the approach.

5 SUMMARY

This paper examines the concept of a 'devolved' type of e-manufacturing which will allow customers to design and have objects manufactured in a local 'high street' facility, reviews the challenges involved in developing a such a system.

6 REFERENCES

1. Davis, S. (1987) *Future Perfect*, Reading: Addison-Wesley.
2. Pine, B.J. (1993) *Mass Customization*, Boston: Harvard Business School Press.
3. Piller, F., and Franke, N.: *Key Research Issues in User Interaction with Configuration Toolkits,* International Journal of Technology Management (IJTM), Vol. 26 (2003), No. 5/6, pp. 578-599
4. Anderson, D.M., (1998) *Agile Product Development for Mass Customisation*, McGraw Hill,
5. www.dell.co.uk, date last accessed 03/03/2004
6. miAdidas.com, date last accessed 03/03/2004
7. eMachineShop.com, date last accessed 03/03/2004
8. ToyBuilders.com, date last accessed 03/03/2004
9. *UK Manufacturing - we can make it better*, Manufacturing 2020 Foresight Report (URN 00/1380), 2001, www.foresight.gov.uk
10. Bateman, R and Cheng, K, (2002), *Devolved manufacturing*, International Journal of Concurrent Engineering, 10(4),291-298
11. www.visionexpress.co.uk, date last accessed 03/03/2004
12. Timmers, P., (1999) *Electronic Commerce – Strategies and Models for B2B* Trading, Wiley & Sons, NY.
13. Wohlers, T. (2004) *Rapid Prototyping, Tooling & Manufacturing State of the Industry Annual Worldwide Progress Report*, Wohler Associates Inc.

MATLAB-based e-design for aerostatic journal bearings

S-J CHEN, B-B YAO, and **P SHI**
School of Mechanical and Electrical Engineering, Harbin Institute of Technology, China

ABSTRACT

This paper presents an e-design approach by using Internet and MATLAB as tools for developing a web based aerostatic bearings design support system. Through an exemplar of remote design of the aerostatic journal bearings, the detailed implementation of the MATLAB-based web design support system is described. The implementation demonstrates how to use MATLAB in the Internet/Intranet environment for engineering design applications in particular.

Keywords: MATLAB Web Server; Remote Design; Aerostatic Journal Bearings

1 INTRODUCTION

With the development of computer and communication technology, Internet/Intranet has become a standard platform for developing new information systems. In this background, remote design has been a new design method for designers. It could not only use design resources (include product resource and intelligence resource) in large scales, but also the customers can participate in design, so it provides customized products and better services for customers.

This paper describes the detailed implementation of the MATLAB-based web application system through an example of remote design of the aerostatic journal bearings. The research presented demonstrates the usage of MATLAB in the Internet/Intranet environment for computing intensive engineering applications in particular.

2 DESIGN METHODS FOR AEROSTATIC JOURNAL BEARINGS

Generally, there are three ways to design aerostatic journal bearings, which are listed as follows:

(1) Manual design based on the empirical equations, experimental data and charts.
(2) Design by running some client programs according with the optimal design rules.
(3) Use numerical computing methods, such as finite difference, FEA (Finite Element Analysis), this method depends on professional software.

However, all the methods above can not let customers participate in design, also can not meet the needs of modern manufacturing industry completely, which is facing many new challenges including remote design and machining efficiently according to the customers' requirements. Inetrnet/Intranet-based remote design can meet these requirements, but most CGI (Common Gateway Interface) languages are not competent for complex numerical computing, such as matrix operation and non-linear equation's derivation, but all these problems can be solved easily by using MATLAB.

3 DEVELOPMENT OF REMOTE DESIGN SYSTEM BASED ON MATLAB WEB SERVER

3.1 The basic theory

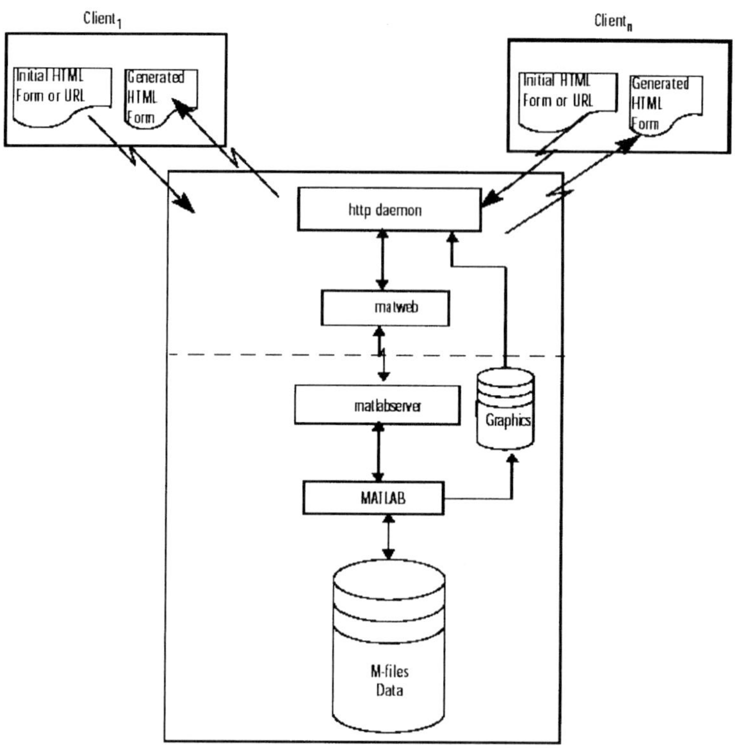

Figure.1 MATLAB on the Web

The basic theory of MATLAB Web Server is: users submit the data which they cares via the clients' browsers, the httpd deamon of the WWW(World Wide Web) sever could submit the data to matweb, matweb is the key CGI program to MATLAB Web Server [1]. Then matweb transmit the data to matlabserver which is running background, matlabserver take charges of the communications between web application and matweb, it is a multithread TCP/IP server, then matlabserver runs a matlab process and transmit the data to this matlab process for computing according to the matlab application program(the M-files), when finish computing, matlab export computing result to customers' browsers through matweb in HTML format. All this is accomplished under the collaboration of matweb and the matlab application program. Figure.1 shows a diagram illustrating how MATLAB operates over the Web.

3.2 Input document

The input documents programmed in the MATLAB environment is as follows:

```
<!-- STEP 1 Choose either the NT version or the Unix version of the form tag (depending on which platform the matweb client program will be run):-->
<!-- NT version: -->
<form action="/cgi-bin/matweb.exe" method="POST">
<!-- Unix version: -->
<form action="/cgi-bin/matweb" method="POST">
<!-- STEP 2 Create a hidden field naming your M-file. Replace MY_M_FILE with the name of main MATLAB function of your application.(An HTML input field of type "hidden" is commonly used to pass variables to a web server. It is not displayed by the browser.)-->
<input type="hidden" name="mlmfile" value="my_m_file">
<!-- STEP 3 Add all your other HTML form tags here. Replace MY_INPUT_VARIABLE_1 with the name of an input variable in your application. -->
<p>My input variable 1: <input type="text"
name="my_input_variable_1">
<!--Create additional input variables here.-->
<!-- STEP 4 Create a "submit" input tag for the user to click to send the input to your program.-->
Creating Input Documents
<p><input type="submit" name="Submit" value="Submit"></p>
<!-- STEP 5 Add the name of your main application function to the file
matweb.conf. See the matweb.conf file in the wsdemos directory .) -->
```

3.3 M-files

There are many M-files used in the system developed. The typical ones are listed below:

```
function retstr = mfile_template(instruct, outfile)
% STEP 1
% Initialize the return string.
```

```
retstr = char('');
% STEP 2
% Set working directory.
% The variables INSTRUCT.MLDIR and INSTRUCT.MLID are
provided
% automatically to all MATLAB Web Server applications that
use
% the matweb program.
cd(instruct.mldir);
% STEP 3
% Get the HTML form input variables
my_input_variable_1 = instruct.my_input_variable_1;
% STEP 4
% Perform your MATLAB computations, graphics file creations,
% etc., here:
% STEP 5
% Put variables that you want to put into your HTML output
document in an output structure. You create an HTML output
document from OUTPUT_TEMPLATE.HTML.
outstruct.my_output_variable_1 = More MATLAB computations
creating ...
scalars, matrices, cell arrays, graphics files, etc.;
% STEP 6
% Call the function HTMLREP with the output structure you
just
```



```
% created and the filename you created from
OUTPUT_TEMPLATE.HTML.
% Replace <OUTPUT_TEMPLATE.HTML> with the name of the HTML
output
% file you created using OUTPUT_TEMPLATE.HTML.
% This call fills the string RETSTR to return and optionally
% writes the output as a file if a valid filename is given
as the
% second argument to the present function.
templatefile = which('<OUTPUT_TEMPLATE.HTML>');
if (nargin == 1)
retstr = htmlrep(outstruct, templatefile);
elseif (nargin == 2)
retstr = htmlrep(outstruct, templatefile, outfile);
end
```

3.4 Output document

As described in section 3.2, in the system, there are also output documents programmed in the MATLAB environment. Some of them are listed as follows:

<!-- Modify this file to create your own HTML output document and save it as <MY_OUTPUT>.html, where <MY_OUTPUT> is replaced by a name that has meaning within the context of your application.-->
<!-- STEP 1
Display a MATLAB scalar or character string. Replace
<MY_OUTPUT_VARIABLE_1> in the following line with the name of the MATLAB variable you want to display. Change the other text to something meaningful within the context of your application. -->
My output variable 1 has been computed to be
$<my_output_variable_1>$
<!-- STEP 2 Put all your other HTML tags here -->

4. DESIGN CASE STUDIES

Based on the proposed architecture and scheme in section 3, a remote e-design support system has been developed by the authors. With the support of the system developed, the authors have carried out a number of design case studies on various aerostatic bearings as requested by industrial companies. The design results produced by the system are well received by the industry. These cases are also helpful for the evaluation and validation of the system. A number of snap shoot of the system screens are shown in Figures 2, 3, 4, 5. The system interface design is based on the requirements from human-computer interactions and the industry. Furthermore, as aerostatic bearings are special complex mechanical elements, their design requires specialists and computing intensive, the development of web-based e-design system for the bearings are very helpful for application-problem solving, achieving design agility and responsiveness [3].

(1) Design of aerostatic journal bearings

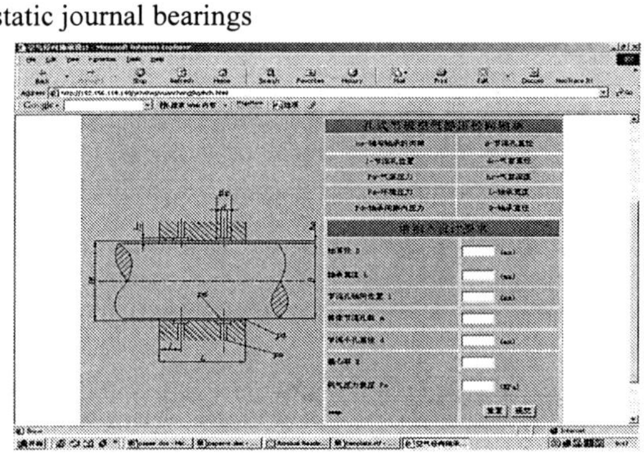

Fig.2 The input webpage

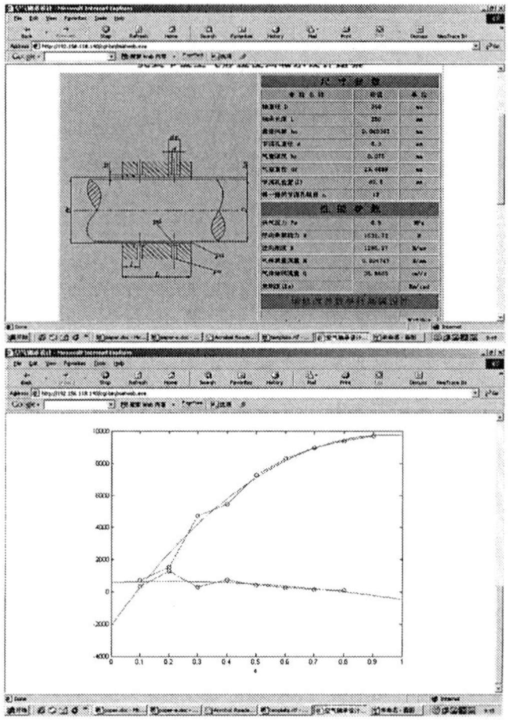

Fig.3 The output webpage

(2) Design of aerostatic end thrust bearings

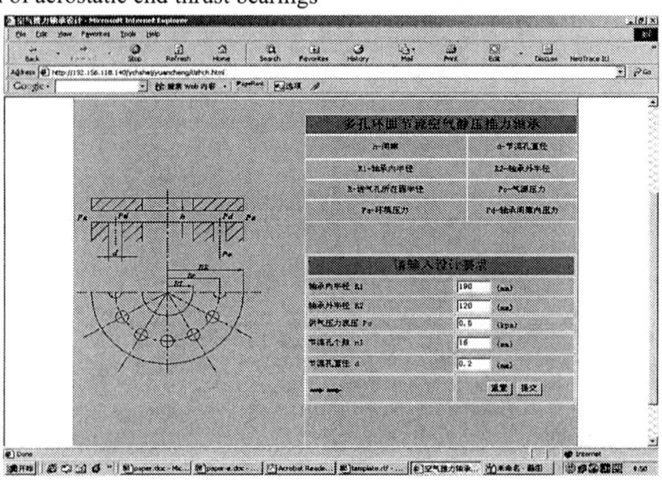

Fig.4 The input webpage

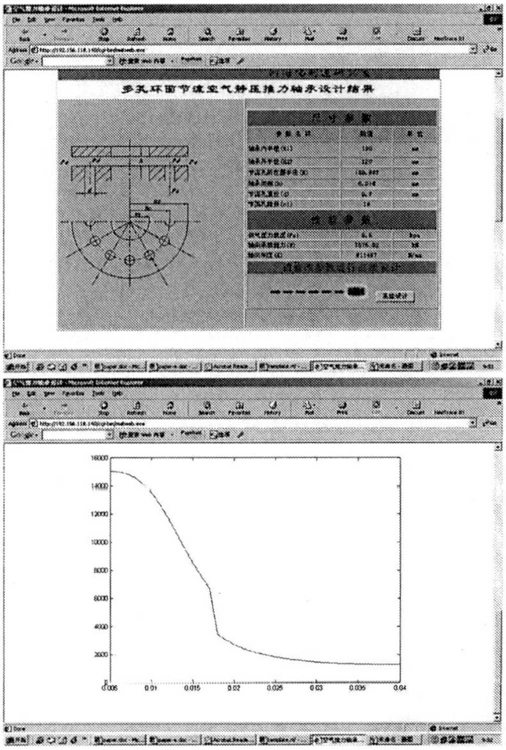

Fig.5 The output webpage

5. CONCLUSIONS

This paper described the detailed implementation of the MATLAB-based web application system. MATLAB Web Server is competent for the complex web computing. Also we can develop other remote support system, for example: online stimulation.

ACKNOWLEDGMENTS

This research is financially supported by the team development fund of the Cheung Kong Scholar Program (CKSP) at Harbin Institute of Technology.

REFERENCES

1 He Qiang and He Ying, 2002, "MATLAB Extended Program", Beijing, Tsinghua University Press.

2 The MathWorks, 2003, "MATLAB Web Server".

3 Cheng, K., Pan, P.Y. and Harrison, D.K., "Internet as a tool with application to agile manufacturing: a web-based engineering approach and its implementation issues", International Journal of Production Research, 38(12), 2000, 2743-2759.

Internet-based optimization of machining process

L DONG and **C FAN**
Engineering Training Centre, Shanghai Maritime University, China
Y ZHAO
Department of Mechanical Engineering, Harbin University of Science and Technology, China

ABSTRACT

This paper presents an optimum method of an Internet-based facility for the machining processes. A network architecture model is developed. The system complies with TCP/IP and is supported by Browser/Server model. The longest tool life is taken as objective function and the mathematics models of optimum machining parameters are established. A value analysis method is used to program the algorithm in order to get optimum machining parameters. Users input optimum machining parameters in the client and then the users will get the simulation curve of cutting force. The users can evaluate the optimum results and decide the machining parameters via Internet.

KEYWORDS: Internet-based, Optimization, Machining Process

NOTATION

t_0: auxiliary time of procedure
t_{ct}: time of tool exchange per workpiece
t_m: cutting time of procedure
C_{pr} : production cost
M : expense unit time of this procedure
C_t : expense per edge of tool used
T : tool life
T_{pr}: production time for single workpiece
l_w: cutting stroke of drill (mm)
f_{mmps}: feed speed of drill (mm/sec)
f : feed rate
B: depth of hole (mm)

Δ: feed of drill in advance
D: diameter of drill
n: rev of spindle
C_T: modify coefficient
m, n, q: indexes
T_p: tool life when productivity is the highest
η: transmission efficiency of machining tools
P_J: power of main motor
V: cutting speed
F_c: main cutting force
F_z : cutting force in feed direction
M: torque

1 INTRODUCTION

Globalization of manufacturing makes the competition between enterprises more and more intense these years. The competitive ability of enterprise can be improved by reducing cost, shortening the cycle of products development and enhancing the response ability to market. Metal cutting processes are very complex. Mnay cutting experiments need to be done when a new product is developed. The cycle of products development can be shortened and cost greatly reduced by optimization of machining process.
Many industries and research institutions are working towards gaining more and more speed and agility in manufacturing. The Center for Online Manufacturing Optimization (COMO) of the McMaster Manufacturing Research Institute (MMRI) has developed an Internet Based Manufacturing Design System which utilizes the Client/Server computing model of the Internet technology. The engineer can choose to test the optimized process by creating a physical prototype. The MMRI's Machining System Laboratory (MSL) has several state-of the-art CNC machines that are linked to the COMO system. By uploading the optimized process information to the lab, an actual physical part can be machined to the client's specifications(1). The Center for Nontraditional Manufacturing Research of University of Nebraska-Lincoln developed an Internet-based Real Time Remote Monitoring System (IRTRMS), this system was successfully developed using Data Acquisition System and Client/Server paradigm of the Internet technology(2). Cybercut is an experimental fabrication testbed for an Internet-accessible, computerized machining service(3),(4).
Internet-Based Optimization of Machining Processes presented in this paper is based on Browser/Server computing model of the Internet technology.

2 INTERNET ARCHITECTURE OF MACHINING PROCESSES OPTIMIZATION

The development of Internet-based manufacturing technology is rapid. But most of references are based on Client/Server Internet architecture which has some disadvantages.

2.1 Architecture of Client/Server

Client/Server is a kind of two layers architecture. The idea is to divide an application into two parts: foregrounding and background. The calculations in foregrounding are done by the Client and the calculations in background are done by the Server. A typical architecture of Client / Server is shown in Fig.1.

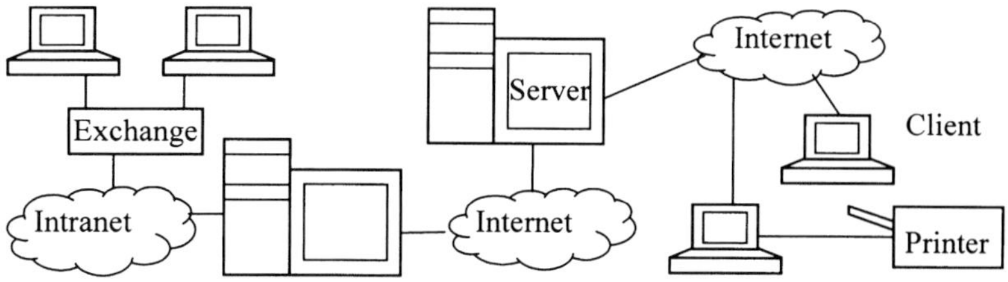

Fig. 1 Client/Server architecture

The application programs were formerly installed only in the Server, but in this architecture they are divided into two parts and installed in the Client and Server separately. The application programs that run in Client include: expression of information sent to users, mechanism of receiving input, data alternation and the effective test of data sent to the Server. The Client's function is independent, it does not depend on the Server, it provides graphic interface to users and exchange information with users directly. However, the Server provides query and management of database and calculation data on a large scale. Compared with Terminal/Mainframe, the efficiency of the system rises greatly and the efficiency of sharing resource enhances accordingly. Data stream through the net also decreases. There are some other advantages, such as alternate high efficiency, friendly user interface, the ratio of property and value high. But most of the calculations are done in the Client in this architecture with two layers, it requires that Client has better properties. The maintenance and upgrade of system is more, and the maintenance fee of software for the Client is high. The database in Server must keep connection with every activity of users, calculation resource is not made full and rational use, with the increase of users, system properties will become poor and users will wait for a long time to get results because the increased number of users.

2.2 Architecture of Browser/Server

With the development of the Internet and its relative technology, Internet has become a large distributed platform of calculation. Web browsers have good functions of striding over platform and multi-media, and people pay more attention on it. Browser/Server (B/S) is a kind of three layers architecture and consists of Browser, Web server and data server. The typical architecture of Browser/Server is shown in Fig.2.

Hypertext Transfer Protocol (HTTP) in the Browser/Server architecture is used to pass information. Users visit Server by Browser and send requests, the responds are replied to users after Server deals with the requests. Compared with traditional Client/Server (C/S), Browser/Server has features as following:

Different software must be installed according to the operation system used and applications of background visited in the Client for C/S. However, Web Browser only is installed as platform of user's applications for B/S, the development, maintenance and upgrade are all done in the Server.

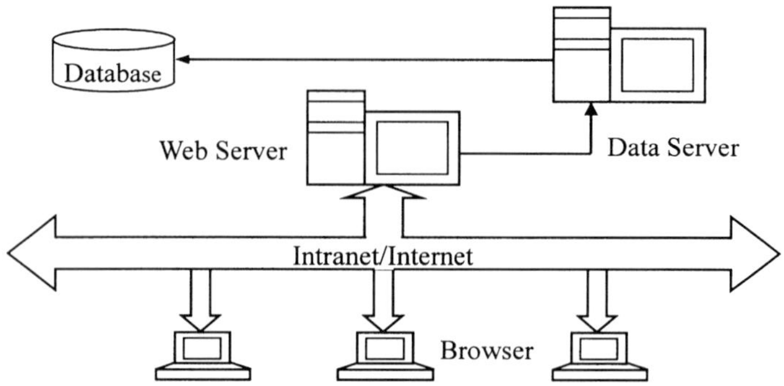

Fig. 2 Browser/Server architecture

Multi-Transfer Protocols are required for C/S, but only the TCP/IP is used for all system to communicate for B/S.
The edition needs be upgraded continually so that user's system is updated for C/S. but the upgrade and maintenance of system only is done in Web server for B/S, users may download when they use, it is assured that users use the newest edition every time.
From the above, it can be seen that B/S has many advantages, such as flexibility, convenience, low cost, strong function of exchange. This paper is based on B/S.

2.3 Internet-based architecture of simulation and optimization of machining processes

The architecture of Internet-based simulation and optimization of machining processes was constructed in this paper. It is shown in Fig.3. It is based on B/S model. There are no special application soft wares to be needed in the Client except for VRML.
Web Server should have a basic function, namely, multi-Web pages can be extracted by users, and user's requests are lined, so that the system of simulation and optimization of machining processes can be used by many users at the same time.

3 INTERNET-BASED OPTIMIZATION OF DRILLING PROCESS PARAMETERS

Drilling is one of the most important machining methods in metal cutting, metal cutting processes effect directly production cycle. The efficiency, quality and economics of drilling are affected by such things as machine tools, cutting parameters and the drills (cutting tools). How to choose a rational combination of these to achieve an ideal result is studied in this paper.

3.1 The common steps of machining parameters choice

Machining parameters (cutting speed V, feed rate f and depth of cut a_p) may be chosen for a certain production condition. The effect each of these on tool life is different. If tool life T is taken as an objective function, the effect of cutting speed on tool life is the biggest among them, next is feed rate, depth of cut is constant for certain drill, so feed rate f should be chosen firstly and the value should be as big as possible when the conditions and requirements of machining permit it, then cutting speed is chosen. So the optimization of drilling process means the best combination of cutting speed and feed rate for certain production.

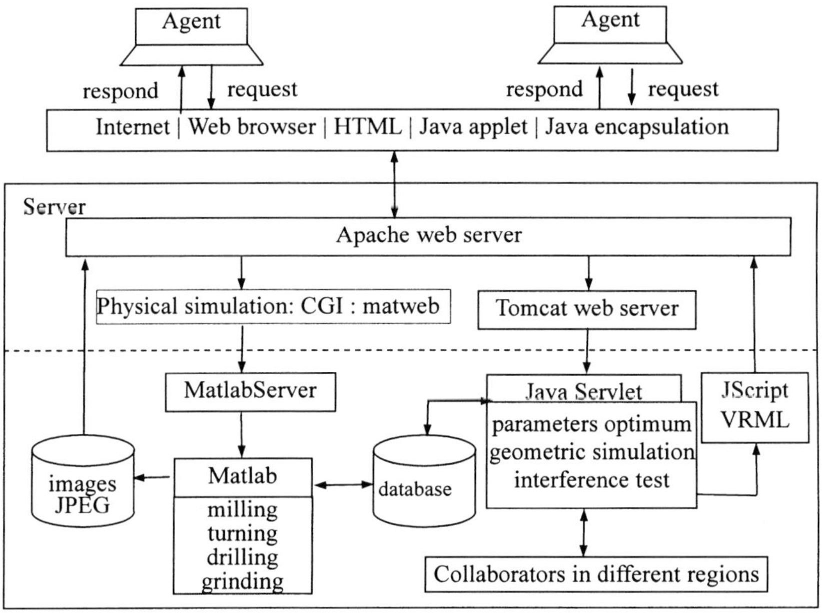

Fig. 3 The architecture of Internet-based simulation and optimization of machining processes

3.2 Establishment of objective functions of machining parameters optimization

Profit is the main goal of enterprises. Cutting parameters are often the main factors which effects productivity. The lowest cost or the highest productivity is regarded as the objective function in this paper. The independent variables are v *and* f. The average cost of every procedure includes man-hours and cost of tools. Assume that expense unit time of this procedure is M, the expense per edge of tool used is C_t, tool life is T, so the production cost C_{pr} is the objective function. Its model is Eqn. 1.

$$C_{pr} = t_{pr} M + \frac{t_m}{T} C_t \tag{1}$$

the basic model of production time for single workpiece is Eqn. 2. (unit: min/piece)

$$t_{pr} = t_m + t_o + t_{ct} \frac{t_m}{T} \quad (2)$$

where

t_0 — auxiliary time of procedure, includes workpiece load and unload, tool in and out, and break time of operator

t_{ct} — time of tool exchange per workpiece, includes tool load and unload, tool alignment and tool regrinding

t_m — cutting time of procedure, it is determined by Eqn. 3 for drilling

$$t_m = \frac{l_w}{f_{mmps} \cdot 60} = \frac{B + \Delta + D \cdot \cot\phi / 2}{fr \cdot \mathrm{n}} \quad (3)$$

where

l_w —— cutting stroke of drill (mm)

f_{mmps} —— feed speed of drill (mm/sec)

B —— depth of hole (mm)

Δ —— feed of drill in advance

We established formula of tool life and drilling parameters according to formula of cutting speed and tool life established by W.F. Taylor. it is shown in Eqn. 4.

$$T = \frac{C_T D^q}{n^{\frac{1}{m}} f_{mmps}^{\frac{1}{n}}} \quad (4)$$

where

D — diameter of drill

n — rev of spindle

f — feed rate of tool

C_T — modification coefficient

m, n, q are indexes

Production time of per piece t_{pr} is expressed by Eqn. 5.

$$t_{pr} = \frac{B + \Delta + D \cdot \cot\phi / 2}{fr \cdot \mathrm{n}} + t_o + t_{ct} \frac{B + \Delta + D \cdot \cot\phi / 2 n^{\frac{1}{m}-1} f_{mmps}^{\frac{1}{n}-1}}{C_T D^q} \quad (5)$$

We can see production time per piece t_{pr} is a function of tool life, the minimum T_p expresses the fewest man-hours from Fig.4, namely, the highest productivity.

Assume $\frac{dt_{pr}}{dT} = 0$, at this time, $T = T_p$

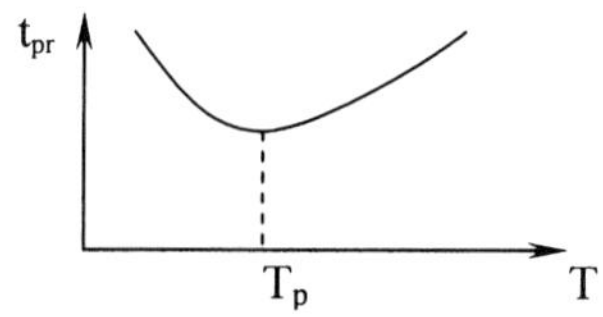

Fig. 4 Sketch map of relation between t_{pr} and T

T_p is the tool life when productivity is the highest.

3.3 Constrain functions

All kinds of constrain conditions must be considered if an optimum scheme is achieved for the objective function. The change range of cutting speed and feed rate will be limited in drilling. These constrains are shown in the following sub-sections.

3.3.1 Power constraint

Practical power required is less than effective power of machine tools, it is shown as Eqn. 6

$$F_c V - \eta P_J \leq 0 \tag{6}$$

where

η — transmission efficiency of machining tools, for driller Z3040X16/1, η=0.75
P_J — power of main motor (kw) for driller Z3040X16/1P_J=3
V — cutting speed (mm/s)
F_c— main cutting force (N)

3.3.2 Constraints of cutting and torque

Usually, there is the largest value of cutting force permitted in feed direction, which should be met in practical application. The cutting force F_z in feed direction is less than the biggest cutting force F_{zm} permitted, it is as Eqn. 7.

$$F_z - F_{zm} \leq 0 \tag{7}$$

The same situation, torque M is less than the biggest torque M_m permitted, it is as Eqn. 8.

$$M - M_m \leq 0 \tag{8}$$

3.3.3 Constrains of feed rate and spindle speed

The spindle rev of machine tools should be lower than the biggest spindle rev permitted by machine tools, it is as Eqn. 9.

$$n - n_m \leq 0 \tag{9}$$

Spindle speed of machine tools should be less than the biggest speed permitted of machine tools, it is as Eqn. 10.

$$f - f_m \leq 0 \tag{10}$$

3.3.4 Constrain of surface roughness

The roughness value of constrain should be lower than that required. If it is roughing, there is special requirements for the surface quality. General machining conditions can meet the constrain conditions.

3.4 Internet based implementation of machining parameters optimization

By above objective function and all kinds of constrain conditions, optimum theory are used to program so that the optimal drilling parameters (cutting speed *V* and feed rate *f*) can be gotten. Users input optimum machining parameters in the Client and get the simulation curve of cutting force. The users can evaluate the optimum results and decide the machining parameters via the Internet. The simulation of cutting force is described in the section below.

4 DRILLING PROCESS PHYSICS SIMULATION BASED ON MATLAB WEBSERVER

MatLab is a high property numerical calculation and visual mathematics software which was produced by MathWorks Ltd. It includes numerical analysis, matrix operation, signal processing and graph display. It also provides a convenient and friendly interface. By using the Web Matlab, people all over the world can transmit data to the Matlab Web Server through the Internet. With the aid of powerful calculation and graph display functions of Matlab, the calculation and graph results can be obtained.

The Web application in MatLab consists of two parts, one of them being MatLab Webserver, which actually is an executive application program (matlabserver.exe), providing an executive environment for MatLab application programs. The other is Web serve agent. It is also an executive program (matweb.exe). It will redirect all the requests from MatLab to matlabserver.exe and process them. As shown in Fig.3, MatLab Web application must be based on some standard Web serve environment. The browser users can request the documents in Web server by TCP/IP, while the MatLab Web serve agent sieves all requests. If the request is from MatLab Web, then it will be processed by MatLab Web serve program. Otherwise it will be processed by standard Web server. To set the MatLab Web application environment, the MatLab sever agent must be installed on the same computer with the Web server of the system. In this paper, Apache Web Server is used as the Web server of the system, which may be located in the same computer with MatLab Web serve program or not. During realization, the MatLab Web serve agent and MatLab Web serve program must be properly configured, which can be realized by matweb.conf and matlabserver.conf separately.

4.1 HTML documents

The HTML documents, which request MatLab Web serve, accomplish two functions. One is to receive the input from the browser user, including drill diameter, chisel angle, helix angle, point angle, workpiece material, feed rate per tooth, spindle rotation speed, etc. This is the same as the processing method of the common Web application. It can be realized by HTML form. The other function is to set a special flag and to redirect the request to MatLab Web. Its realization method is also the same as the calling method of the common CGI program. The second function can be realized by the following codes.

```
<form action="/cgi-bin/matweb.exe" method="POST"    target="outputwindow">
<input type="hidden" name="mlmfile" value= "drillsimu">
```

4.2 The setting of MatLab web application

The setting of MatLab Web application is mainly to set the MatLab Web serve agent, that is, as long as a MatLab Web application is added, a configuration must be added in the configuration file of the agent. For example, the filename of the file m in the HTML form mentioned above is drillsimu. Accordingly, we must add the following content into matweb.conf:

```
[drillsimu]
mlserver=our hostname
mldir=e:/pub/icons
```

where mlserver=our hostname indicates the IP address of the computer on which the MatLab Web serve program is located, while the instruction below is mainly the setting of the MatLab work directory and the global variables which may be used in the application.

4.3 Development of MatLab application program

MatLab application program has three functions.

(1) Receive input value of HTMLform;

(2) Call all kinds of functions of MatLab to do scientific calculation and drawing;

(3) Output the results and graphs to HTML page and return to client.

In Matlab application programs, the output results call HTML template (functemplate.htm), as shown in Fig.5.

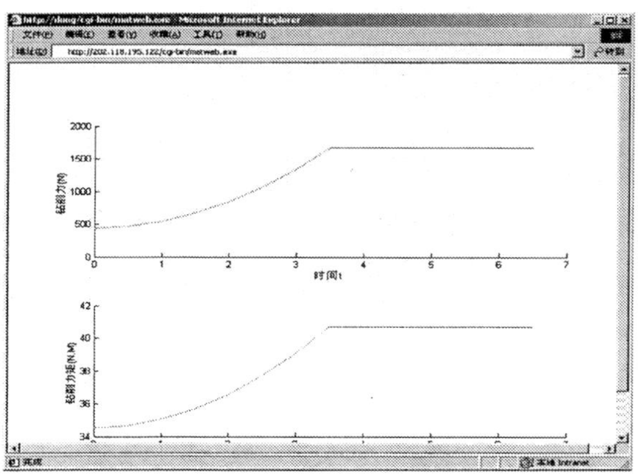

Fig.5 The HTML page of results display

5 CONCLUDING REMARKS

The optimization and simulation of machining processes is the key technology of Internet-based manufacturing. An Internet-based optimum method of machining processes is provided in this paper. A network architecture based on Browser/Server was developed and the users may view the optimized and simulated processes on their own computers, together with data showing the optimal machining parameters and the load forces on the tool. By making adjustments to the initial machining strategy and re-running the optimization and simulation, an optimal machining process can be determined.

ACKNOWLEDGMENTS

The authors would like to acknowledge the financial supports from Shanghai Education Committee, the Science and Technology Bureau of Harbin, and the Science and Technology Department of Heilongjiang Province Government for this research project.

REFERENCES

1 Bailey, T., Elbestawi, M.A., El-Wardany, T. and Viens, D., "Internet Based Implementation of Multi-Axis Milling Process Simulation and Optimization", *ICEM CIRP Int. Seminar*, Capri, Italy, 2000.

2 Kapoor, S. and Rajurkar, K.P., "Internet-Based Real Time Remote Monitoring System for Machining Processes", *Transactions of NAMRI/SME*, 1999,Vol. XXVII, pp.257-262.

3 Wright, P. and Dornfeld, D., "CyberCut: A networked Machining Service", *Transactions of NAMRI/SME*,1998, Vol.XXVI, pp.281-286.

4 Smith, C.S., "CyberCut: A World Wide Web Based Design- to-Fabrication", *Manufacturing Systems*, 1996, Vol. 15, Issue 6, pp.432-442.

5 Deitz, D., "Product Development on the Web", *Mechanical Engineering*, 1997, Issue 1, pp.66-69.

6 Lau, H., "The New Role of Intranet/Internet", *Technology for Manufacturing Engineering with Computers*, 1998, Vol. 14, pp.150-155.

atcPORTAL – a generic and extensible distributed agent-based portal for analytical target cascading in optimal design of complex systems

G Q HUANG
Department of Industrial and Manufacturing Systems Engineering, The University of Hong Kong, China

ABSTRACT

Analytical Target Cascading (ATC) has emerged as an integrated approach and methodology for optimal design of complex hierarchical systems with good convergence. It facilitates collaborative design problem solving and enables distributed computing to extend both the capability and capacity. ATC is very powerful yet requiring comprehensive understanding and efforts in setting up specific projects. This paper presents a generic and extensible information infrastructure called atcPortal in the form of a web portal where the ATC analyst defines the ATC problem using a special-purpose XML-based language called atcXML, conducts the ATC analysis, and obtains the analytical results. atcPortal not only reduces the overheads for the analyst to set up the projects on the side of practical applications, but also allows the researchers to experiment and extend the ATC methods under different system structures and/or different ATC strategies.

Keywords: Analytical Target Cascading, Target optimization, Web portal, Web services, XML, Hierarchical system modeling.

1 INTRODUCTION

Analytical Target Cascading (ATC) has recently emerged as a new approach to solving target optimization problems in complex engineering systems which can be represented hierarchically in terms of constituent components and modeled mathematically in terms of decision variables (Kim, Papoplouo et al 2000). The formulation addresses the problem of propagating design targets through a hierarchically partitioned product development process, so that the local targets are consistent with each other and with the overall top level targets. According to Kim (2000), ATC includes four key steps: (i) development of appropriate models, (ii) partitioning the system, (ii) formulating the target cascading problems for each element of the partition, and (iv) solving the partitioned problem through a coordination strategy to compute all stated targets. The researchers who originally proposed the ATC approach at the University of Michigan have studied and demonstrated the convergence with theoretical rigor (Michelena et al 2002) and through numerous industrial case studies (mainly in the field of vehicle design) (Lauca et al 2002; Kim et al 2002; Nystrom and Larsson et al 2003). Researchers in the same group have also applied ATC for building design (Choudhary et al 2003). The framework has been applied for thermal design and analysis of a fictional three-zone building. The building is decomposed into two levels. Different analysis tools are employed for computing the performance targets at each decomposed level of the building. Interestingly, most of the publications on ATC are somewhat related to the projects and the people who are working or used to work at the University of Michigan (e.g. Mahmoud et al 2002a; Mahmoud et al 2002b; Nystrom and Larsson et al 2003) with a few exceptions (e.g. Ge, Lu and Suh 2002).

ATC is a powerful method with several important characteristics. Firstly, its convergence has been demonstrated, both theoretically and through case studies. It has been reported that the top-level component normally converges after about ten iterations (Kim 2001). Several case studies have compared results from ATC and ATO (All At Once) and reported that the results are generally consistent with each other. This indicates that ATC is at least as effective as ATO in solving the optimal system design problems. The efficiency of convergence has been improved by much reduced complexity of sub-problems which converge much faster than when they are treated as one big system. In the meanwhile, the solution spaces are narrowed progressively after each cycle of cascading and backtracking (rebalancing). This also speeds up the convergence.

Secondly, ATC provides excellent flexibility for conducting scenario analysis. For example, we can set targets to extreme lower and upper bounds (normally 0 and ∞) for testing the boundaries of decision variables in minimization and maximization. We can also set several targets within a range to test the sensitivity and trend of different decision variables. In addition, it is easy to observe if some decision variables have taken their values at the upper or lower bounds specified in the associated constraints. Such decision variables should be highlighted for the designers to decide if it is necessary and feasible to change such extreme bounds to achieve better solutions.

Thirdly, ATC facilitates collaborative problem solving in system design. Sub-problems of different nature can be solved by using different (heterogeneous) models relatively independent of each other. The specialist users with relevant disciplines can simply focus on the issues that they are familiar with. All these are made possible by decoupling the systems and replacing their interactions by passing target values of their responses between them.

Fourthly, ATC enables distributed computing over the Internet to extend computational capacity. This is a significant advantage over the ATO approaches. For example, the dynamic programming developed for optimal supply chain configuration is limited to certain number (typically 50-100) of sub-problems /stages. With ATC, sub-problems can be solved at separate computers. Sub-problems are much simpler and therefore can be solved by less powerful computers. More complex sub-problems can be scheduled to be solved on more powerful computers.

Fifthly, ATC supports optimization by multiple criteria. The objective function is generic in the sense that the discrepancies accumulate between the all responses (criteria) concerned and their desired targets in absolute terms. In fact, responses can be freely added or removed from the objective function to conduct scenario analysis.

Sixthly, ATC is not limited to its original basic form. In fact, it has been widely extended by both the researchers who originally proposed ATC to suit particular requirements of specific problems.

Although gradually gaining its popularity in the vehicle design engineering community due to the powerful features mentioned above, ATC has not yet found its wide applications in optimal design of complex engineering and industrial systems as it deserves. A simple literature search on the Internet and in the university libraries has only resulted in about two dozens of publications most of which are related to projects or researchers at the University of Michigan as mentioned previously. It is not clear why this is the case. Even after understanding how ATC generally works, the overhead of setting up an ATC analysis is quite large for non-experienced analysts. It is quite straightforward to define a complex hierarchical system using the ATC convention. It is by no means a trivial task to define the communications between the different local optimization systems and the protocol regarding cycles of cascading and backtracking down and up the hierarchy. These overheads may become the possible hindrances why ATC has not yet been used by the researchers and practitioners as it deserves.

The research reported in this paper is aimed at overcoming such major hindrances by proposing a generic and extensible computational infrastructure / framework called atcPortal. When designing and developing atcPortal, we adopt the latest Internet and web technologies. Unlike Chi Language (Fellini et al (2001), we propose to take advantage of the popular XML to create a new but easy to use language called atcXML, for not only modeling the hierarchical systems but also defining ATC optimization problems for all system components. atcXML also acts as a blackboard mechanism for presenting output results. A distributed computational multi-agent environment is created from the atcXML definition file. A Java-based program, atcEngine, has been developed to coordinate the distributed problem solving between individual components represented by TopAgents.

When we devise the atcPortal infrastructure, several jargons are created: Targent (design-time system component) and TopAgent (run-time system component). We added "n" in the word "Target" to reflect the extensive use of the concept of agents in developing this portal. We blended "T" and "r" with the word "agent" to reflect the fact that this portal is especially developed for supporting Target Cascading. Also, the name of TopAgents reflects the fact the individual agents are for solving Target Optimization problems.

The rest of this paper is organized as follows. Section 2 outlines the overview of the proposed atcPortal. Sections 3, 4 and 5 present some details about the three main atcPortal components, namely atcXML, atcEngine and smoUDDI, respectively. Because of limited space, it is not possible to include a complete case study in the paper. A prototype atcPortal website is under construction for presenting further information about the atcPortal project and case studies (http:www.digiprise.org/atcportal).

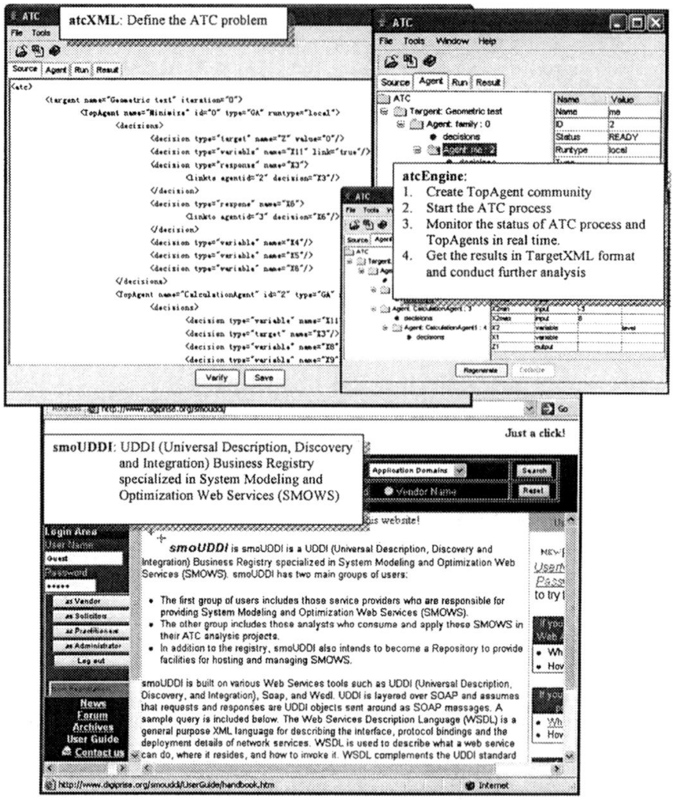

Figure 1 atcPortal overview

$$P_0 : Min \| z - 0 \| + \| x_{11} - x_{11,p1}^L \| + \| x_{11} - x_{11,p2}^L \| + \| x_3 - x_3^L \| + \| x_6 - x_6^L \| + \varepsilon_1 + \varepsilon_2 + \varepsilon_3$$
$$where\ z = x_1^2 + x_2^2;\ x_1^2 = x_3^2 + x_4^{-2} + x_5^2;\ x_2^2 = x_5^2 + x_6^2 + x_7^2$$
$$\| x_{11} - x_{11,p1}^L \| + \| x_{11} - x_{11,p2}^L \| \le \varepsilon_1;\ \| x_3 - x_3^L \| \le \varepsilon_2;\ \| x_6 - x_6^L \| \le \varepsilon_3;$$
$$g_1 : \frac{x_3^{-2} + x_4^2}{x_5^2} \le 1;\ g_2 : \frac{x_5^2 + x_6^{-2}}{x_7^2} \le 1;\ x_1, x_2, x_3, x_4, x_5, x_6, x_7, x_{11} \ge 0$$

$$P_1 : Min \| x_3 - x_3^U \| + \| x_{11} - x_{11}^U \| + \varepsilon_4 + \varepsilon_5$$
$$where\ x_3^2 = x_8^2 + x_9^{-2} + x_{10}^{-2} + x_{11}^2$$
$$\| x_8 - x_8^L \| \le \varepsilon_4;\ \| x_{10} - x_{10}^L \| \le \varepsilon_5;$$
$$g_3 : \frac{x_8^2 + x_9^2}{x_{11}^2} \le 1;\ g_4 : \frac{x_8^{-2} + x_{10}^2}{x_{11}^2} \le 1;$$
$$x_3, x_8, x_9, x_{10}, x_{11} \ge 0$$

$$P_2 : Min \| x_6 - x_6^U \| + \| x_{11} - x_{11}^U \|$$
$$where\ x_6^2 = x_{11}^2 + x_{12}^2 + x_{13}^2 + x_{14}^2$$
$$g_3 : \frac{x_{11}^2 + x_{12}^2}{x_{13}^2} \le 1;\ g_4 : \frac{x_{11}^{-2} + x_{12}^2}{x_{14}^2} \le 1;$$
$$x_6, x_{11}, x_{12}, x_{13}, x_{14} \ge 0$$

$$P_3 : Min \| x_8 - x_8^U \|$$
$$x_8^2 = x_{15}^2 - 2x_{15} + 1;$$
$$x_{15} \ge 0$$

$$P_4 : Min \| x_{10} - x_{10}^U \|$$
$$x_{10}^2 = x_{16}^2 - x_{16} + 1;\ x_{16} \ge 0$$

Figure 2 Example of system hierarchy and ATC sub-problems.

2 *atcPORTAL*: AN EXTENSIBLE AND GENERIC AGENT-BASED ATC PORTAL

atcPortal is a web portal designed and developed specifically for supporting and facilitating Analytical Target Cascading for optimal design of complex hierarchical systems. At present, only rudimentary and standard ATC facilities and mechanism are supported. Figure 1 presents a brief overview.

atcPortal involves two main phases: Definition and Execution. The Definition phase is heavily dependent on the ATC analysts. The output from this Definition phase is an XML-based definition file in the atcXML standard. atcXML is a special XML-based language devised for completely defining an ATC project / problem including hierarchical system modelling and optimization problems associated with individual system components. The Definition phase includes several tasks. Firstly, the system analyst defines and models the hierarchical system. This involves defining the system components, their key parameters and hierarchical relationships. Secondly, the system analyst defines target optimization problems for all the constituent system components. This normally involves setting up the objective function, establishing equality and inequality constraints over the decision variables and their domains. Thirdly, the system analyst chooses appropriate optimization solution methods and convergence criteria for all the system components to solve their target optimization problems. This stage usually involves specifying the locations of the optimization problem solvers and their associated parameters.

The second Execution phase is mainly managed by the atcPortal itself through a atcEngine. atcEngine is a multi-agent system where individual agents solve their local problems and are coordinated by a specific ATC strategy to achieve the overall optimum. atcEngine does not allow the ATC analyst to change ATC parameters anymore. Instead, the ATC analyst is able to inquire about the progress and status of the ATC application with the help of its user interface. atcEngine accomplishes several tasks. It firstly generates and sets up the multi-agent community based on the atcXML definition. Then, atcEngine starts its process with the top-level component (agent), cascades down and backtrack up the system hierarchy recursively until the top-level agent converges (assume that all sub-level agents converge). During this process, the analyst is able to interrogate the status of the ATC process without being able to change anything except for stoping the process.

atcPortal provides three key components equipped with appropriate user interfaces to support the three ATC stages. They are atcXML, smoUDDI, and atcEngine. atcXML is a special XML-based language devised for defining an ATC project / problem including hierarchical system modelling and optimization problems associated with individual system components. atcXML and its editor facilities fully supports the first stage of hierarchical system modeling and partially supports the second stage of optimization problem formulation.

Both definition and execution phases rely on the computational resources that must be made available for system modeling and optimization over the Internet. atcPortal provides a directory called smoUDDI where such resources are available. smoUDDI is a UDDI (Universal Description, Discovery and Integration) Business Registry specialized in System Modeling and Optimization Web Services (SMOWS). The three atcPortal components are constructed with a few important concepts which will be discussed in detail in the following three sections respectively.

```
<?xml version="1.0" encoding="UTF-8"?>
<ATC project="Geometric Test" team="GQ" date="14/2/2004">
  <Targent id="p0" name="p0" runat="local">
    <Optimizer  name="SGA P0" class="simpleGA0">
      <param name="population" value="1000"/>
      <param name="evolution" value="500"/>
    </Optimizer>
    <Decisions>
      <decision type="dependent" data="real" name="x1" lower="0"/>
      <decision type="dependent" data="real" name="x2" lower="0"/>
      <decision type="independent" data="real" name="x3" lower="0""/>
      <decision type="independent" data="real" name="x4" lower="0"/>
      <decision type="independent" data="real" name="x5" lower="0"/>
      <decision type="independent" data="real" name="x6" lower="0"/>
      <decision type="independent" data="real" name="x7" lower ="0"/>
      <decision type="independent" name="x11"/>
      <decision type="dependent" data="real" name="z" target="0" lower="0"/>
    </Decisions>
    <Links>
      <link decision="x11">
        <with decision="x11" of="p1"/>
        <with decision="x11" of="p2"/>
      </link>
      <link decision="x3" with="x3" of="p1"/>
      <link decision="x6" with="x3" of="p2"/>
    </Links>
    <Converge>
      <criterion on="or" decision="objective" type="ratio" value="0.1"/>
      <criterion on="or" decision="objective" type="deviation" value="0.01"/>
    </Converge>
    <Targent id="p1" name="p1" runat="local">
      <Optimizer  name="SGA P1" class="simpleGA1">
        <param name="population" value="1000"/>
        <param name="evolution" value="500"/>
      </Optimizer>
      <Decisions>
        <decision type="dependent" name="x8" lower="0"/>
        <decision type="independent" name="x9" lower="0"/>
        <decision type="dependent" name="x10" lower="0"/>
        <decision type="independent" name="x11" lower="0"/>
        <decision type="dependent" name="x3" lower="0"/>
      </Decisions>
      <Links>
        <link decision="x8" with="x8" of="p3"/>
        <link decision="x10" with="x10" of="p4"/>
      </Links>
      <Converge>
        <criterion on="or" decision="objective" type="ratio" value="0.1"/>
        <criterion on="or" decision="objective" type="deviation" value="0.01"/>
      </Converge>
      <Targent id="p3" name="p3" runat="local">
        <Optimizer name="SGA P3" class="simpleGA3">
          <param name="population" value="1000"/>
          <param name="evolution" value="500"/>
        </Optimizer>
        <Decisions>
          <decision type="dependent" name="x8" lower="0"/>
          <decision type="independent" data="discrete" name="x15"/>
            <member value="1.0"/>
            <member value="2.0"/>
            <member value="2.8"/>
          </decision>
        </Decisions>
        <Converge>
          <criterion on="or" decision="objective" type="ratio" value="0.1"/>
          <criterion on="or" decision="objective" type="deviation" value="0.01"/>
        </Converge>
      </Targent>
      <Targent id="p4" name="p4" runat="localhost:4446">
        <Optimizer  name="SGA P4" class="simpleGA4">
          <param name="population" value="1000"/>
          <param name="evolution" value="500"/>
        </Optimizer>
        <Decisions>
          <decision type="dependent" name="x10" lower="0"/>
          <decision type="independent" name="x16" lower="0"/>
        </Decisions>
        <Converge>
          <criterion on="or" decision="objective" type="ratio" value="0.1"/>
          <criterion on="or" decision="objective" type="deviation" value="0.01"/>
        </Converge>
      </Targent>
    </Targent>
    <Targent id="p2" name="p2" runat="local"/>
      <Optimizer name="SGA P2" runat="localhost:8011/o1"/>
        <param name="population" value="1000"/>
        <param name="evolution" value="500"/>
      </Optimizer>
      <Decisions>
        <decision type="independent" name="x11" lower="0"/>
        <decision type="independent" name="x12" lower="0"/>
        <decision type="independent" name="x13" lower="0"/>
        <decision type="independent" name="x14I" lower="0"/>
        <decision type="dependent" name="x6" lower="0"/>
      </Decisions>
      <Converge>
        <criterion on="and" decision="objective" type="ratio" value="0.1"/>
        <criterion on="or" decision="objective" type="deviation" value="0.01"/>
      </Converge>
    </Targent>
  </Targent>
<ATC>
```

Table 1 atcXML definition file for the case study shown in Figure 2.

3 *atcXML*: ATC DEFINITION LANGUAGE

atcXML is the special-purpose language devised in atcPortal for defining an ATC project. The ATC definition encompasses the following aspects:

- General project description information
- Defining the system hierarchical structure
- Defining system characteristics
- Defining ATC optimization models

In order to facilitate the discussion and to save space, a simple system shown in Figure 1 is used throughout this paper. This system is abstracted and modified from an example presented by Kim et al (2000). The example was slightly modified to create three levels in the hierarchy. Figure 2 shows the system hierarchy and Table 1 list the 105 lines of atcXML definition file. An ATC project is defined in the atcXML file between the pair of brackets <ATC …> … </ATC>. General project information are defined by the attributes of the "ATC" tag.

3.1 Hierarchical System Modelling

System modeling includes four general aspects: composition – what components constitute a system, configuration – how system components are related to each other, characteristics – by what system components are described, and constraints – how characteristics of system components are related to each other.

ATC applies predominantly to hierarchical systems. The composition of a system is relatively simple. That is, the system consists of a set of components which in turn consist of lower-level components. This recursive composition repeats until bottom-level elementary components which do not have further child components.

In atcXML, each system component is defined by the "Targent" tag – the pair of brackets "<Targent …> … </Targent>". Each Targent is defined with several key attributes, including "id" and "name". The "id" attribute is unique to the Targent and must not be the same between any two Targents. There are five pairs of the brackets, defining the five Targents corresponding to the five components of the hierarchical system shown in Figure 1.

The hierarchical relationships between system components are implied by the nesting relationships between the Targents.

3.2 Defining System Characteristics

Each system component (i.e. Targent) is described by a set of characteristics or decisions. atcXML distinguishes between three types of characteristics: input parameters whose values must be specified, independent decision variables and output dependent decision variables. All three types of decisions are defined in atcXML using one "decision" tag with a "type" attribute for differentiating them. The "data" attribute is for defining the data type of the decision, e.g. integer, real, discrete, etc. If it is of the continuous types (type="real" or type="integer"), then the domain may be specified by the lower and upper limits. If an output is discrete, its domain needs to be defined through a membership (Lines 62-64).

3.3 Defining Common and Cascading Decision Variables

Because of hierarchical decomposition of the system used in ATC, the relationships between the parent and children components between the two adjacent levels and the relationships between the sibling components of the same parent are described by linking variables shared between them.

Let us first consider the case of defining linking variables between sibling Targents at the same level. Individual sibling Targents first define their own variables independent of each other. If they share a particular common variable, then a dummy decision variable is created in their parent Targent. This dummy variable is linked to the common variables, as defined in Lines 23-26 for variable x11. Common variables may have different names. atcEngine will treat these three variables as the single one and build up the objective functions for the Targents (TopAgents) involved in due course.

The case of linking variables between the parent and child Targents at two adjacent levels is simpler to handle because such relationship is one-one. A few attributes are added in the "decision" tag, as in Lines 27-28 in Table 1.

3.4 Defining Response Targets

Target values may be defined for one or more output variables (responses) for Targents. As the name ATC implies, this is mandatory for the top-level Targent while optional for other Targents. This is easily done by explicitly defining the "target" attribute as in Line 20 for decision "z". atcEngine will adjust the objective function to include the term |z-Target| when creating the TopAgents.

3.5 Defining Targent Optimizers

In ATC, individual system components formulate their own target optimization problems and solve these problems with certain optimization methods. The target optimization problems are said to be formulated in the preceding section when system characteristics are fully defined. This section is only concerned with the solution methods.

At present, optimization solution methods are assumed to be reflected in the computational programs for individual system components. Such computational programs are in turn treated as blackbox web services – Optimizers – that are interoperable on the Internet.

This assumption has simplified the atcXML definition to a large extent. For example, Targents (TopAgents as their run-time components) in the case study use the Optimizers that are based on Genetic Algorithm. Five of such Optimizers have been developed and deployed as web services. It is assumed that the provider of an Optimizer register necessary information in order for this Optimizer to be called on the Internet. Such information may include the Internet location, input and output data templates, and parameters for setting up the optimization.

The "Optimizer" tag is used for defining the Targent's Optimizer, as in Lines 4-7. The Optimizer "simpleGA" in the above definition is an optimization web service built with simple / basic GA (Genetic Algorithm) and registered in smoUDDI. Parameters defined are specifically associated the optimization method. They are passed to the Optimizer when it is called upon. In this case, "Population", "Generation" and "Penalty" (on inequality constraints) are the three parameters for simpleGA. The convergence and/or stop criteria may be explicitly defined as parameters or derived from the parameters defined. For example, simpleGA will stop after a maximum generation of 5000 if other convergence criteria do not become valid.

3.6 Defining Targent Convergence Criteria

A Targent is said to converge if and only if its local solution is consistent with the solution obtained from the results backtracked from its children. The meaning of "consistent" refers to the satisfaction of the convergence criteria defined explicitly by the pair of <Converge> … </Converge> brackets as in Lines 30-33. The Targent converges when all compulsory criteria with attribute "and" must be met while any of the criteria with "or" attribute is satisfied.

The criterion defined in Line 31 can be interpreted as $| f_{backtrack} - f_{local} | / f_{local} \leq value$ where value=0.1 and criterion defined in Line 32 as $| f_{backtrack} - f_{local} | \leq value$ where value=0.01. These two criteria are defined for the overall objective function. Similar criteria may be defined for linking variables, in addition to the overall objective function of the Targent.

It is important to note the complete difference between the convergence of the Targent and its Optimizer. Assume the Optimizer is based on a simple GA (Genetic Algorithm) that is asked to converge into the best solution after a given number of generations. The convergence criteria of the Optimizer are only implicitly defined in atcXML through the Optimizer's parameters. The Optimizer must interpret them as designed.

4 *atcENGINE*: ATCPORTAL ENGINE

atcEngine in the atcPortal is the central mechanism that creates, starts, monitors and coordinates the participating TopAgents to accomplish the computational ATC process as defined in the atcXML file. atcEngine is basically a multi-agents system (MAS) consisting of a community of TopAgents. TopAgents are run-time implementation of Targents as defined in atcXML. The TopAgent community Organizer is responsible for creating TopAgents from Targents as instances of the TopAgent Java class. TopAgents are presently implemented in Java and each TopAgent will run in a separate thread. TopAgents are supposed to be "thin" in the sense that they provide containers for the corresponding Optimizers which are supposed to be "fat" and consuming substantial computational resources. Therefore, TopAgents are not designed to run in a distributed computational environment but Optimizers are.

In order to reflect the distributed and collaborative nature of ATC problem solving, atcEngine has been designed to maximize the autonomy of individual TopAgents in managing their own internal affairs and minimizing the central control over them. After their creation, TopAgents reason and maintain their own states and they reason and take their own actions. They interact with each other by exchanging messages.

Following the object-oriented convention, we discuss the TopAgent model in terms of properties and methods.

4.1 TopAgent Properties

A TopAgent maintains a set of properties. During the process, all attributes defined for Targents in atcXML are converted into TopAgent properties. In addition, the hierarchical relationships between system components (i.e. Targents in the atcXML file) are also captured as TopAgent properties. Connections to the remote Optimizers defined as Web services in atcXML are verified with smoUDDI and necessary information are deployed within the corresponding TopAgents.

In addition to the above design-time properties, each TopAgent has run-time properties, e.g. status property, inbox and outbox of messages, etc.

4.2 TopAgent status

The status of a TopAgent is defined by the following possible values:

- Idle: A TopAgent is not asked to do anything yet.
- Optimizing: A TopAgent has called its Optimizer to solve the local ATC optimization problem and is waiting for the result.
- Cascading: A TopAgent has cascaded downwards its children TopAgents and is waiting for their feedback.
- Backtracking: A TopAgent has converged and backtracked upwards its parent TopAgent.
- Succeeding: All TopAgents reach the “Succeed” status when the top-level TopAgent converges.
- Failing: A TopAgent fails when it does not have a solution to its local target optimization problem or any of its children backtracks with “failure”.

4.3 Messages and Message Box

Each TopAgent maintains a message box. When a TopAgent needs to interact with another TopAgent, it first prepares a message and then puts it in the message box of the receiving TopAgent. The TopAgent’s Messenger continuously process the incoming messages.

There are three types of messages: Cascade, Backtrack, and Optimize. The Cascade and Backtrack messages are exchanged between adjacent TopAgents in the system hierarchy. The purpose of a Cascade message is to cascade the values generated by the parental TopAgent for the hierarchical linking decision variables to its children. Therefore, it will take the following form:

```
<Cascade ID=Unique From=”p0” To=”p1”>
   <Decision name=”x3”>2.00</Decision>
   <Decision name=”x11”>1.00</Decision>
...
</Cascade>
```

Likewise, the purpose of a Backtrack message is to submit the values achieved upon the TopAgent convergence to its parent for the hierarchical linking decision variables to its parent. Therefore, it will take the following form:

```
<Backtrack ID=Unique From=”p1” To=”p0”>
   <Decision name=”x3”>2.01</Decision>
   <Decision name=”x11”>1.01</Decision>
...
</Backtrack>
```

The Optimize messages are internal in the sense that they are exchanged between the TopAgent and its Optimizer as web service. The Optimize message is in fact a pair of Request and Response messages sent and received by the TopAgent to and from its Optimizer. The purpose of the Request message is to starts up the remote web service (Optimizer) and passes the necessary decision parameters and variables to the Optimizer web service. The purpose of the Response message is to return all the resulting data from the Optimizer web service to the TopAgent. Both request and response messages basically follow the SOAP (Simple Object Access Protocol) Standard. The main body of a typical request message contains the location of the web service and parameters to be passed to this web service. For example a Request message for TopAgent "p1" looks like:

<Request to=" http://147.8.86.5/wsdl/demoGeo/p1">
<Decision name="x3:uvalue">1.12</Decision>
<Decision name="x11:uvalue">2.21</Decision>
<Decision name="x8:lvalue">0.98</Decision>
<Decision name="x8:old">0.85</Decision>
...
</Request>

The response message from the Optimizer of the "p1" TopAgent contains the values obtained from the Optimizer upon the given parameters:

<Response from=" http://147.8.86.5/wsdl/demoGeo/p0">
<Decision name="x3"></Decision>
<Decision name="x11"></Decision>
<Decision name="x8"></Decision>
<Decision name="x9"></Decision>
<Decision name="x10"></Decision>
<Decision name="objective"></Decision>
<Decision name="et"></Decision>
<Decision name="ey"></Decision>
<Decision name="er"></Decision>
...
</Response>

TopAgents process the incoming messages, update their properties and determine what should be done next. These tasks are accomplished by their built-in methods.

4.4 TopAgent Methods

The TopAgent Java class includes several main methods that accomplish certain special tasks. Messenger is the method mainly responsible for processing the incoming messages in the message box. It identifies and calls the appropriate method (action) depending on the type of the message and current status of the TopAgent.

The first method called by the messenger is the property manager method. It reads a message and updates the properties according to the message type and body information.

The Cascade and Backtrack methods are triggered bythe messenger method if the incoming message type is Cascade and Backtrack respectively. How Cascade and Backtrack methods perform the tasks are governed by the chosen ATC control strategy. There are four ATC strategies as discussed in Park et al (2000). At present, only Strategy IV is used because it has been used in most of the reported case studies. In Strategy IV of ATC, the Cascade method and the Backtrack method follow the procedure outlined in Figure 3.

The Cascade method starts with triggering the Optimize method and ends with sending Cascade messages to all its children if any; or otherwise a Backtrack message to its parent.

If all the children have backtracked, the Backtrack method triggers the Converge method. The Converge method is responsible (1) for assessing the values of the different terms included in the TopAgent's objective function from the children's feedback data, and (2) for checking if TopAgent meets the specified convergence criteria. The TopAgent ends its Backtrack method by preparing and sending a Backtrack message up to its parent if it converges; or otherwise triggering the Optimize method again.

The Converge method is responsible for testing if the specified convergence criteria have been satisfied. If the TopAgent does not have any children, then it converges when its Optimizer converges. If it does have children, then all the "and" criteria are first checked and then each "or" criterion is checked. This method changes the status of the TopAgent.

The Optimize method also changes the status of the TopAgent by temporarily setting it to "Optimizing". This status will be switched to "Cascading" after the Optimizer returns the response message. The Optimize method is triggered on two occasions: (1) within the Cascade method, and (2) within the Backtrack method after the TopAgent fails to converge upon rebalancing.

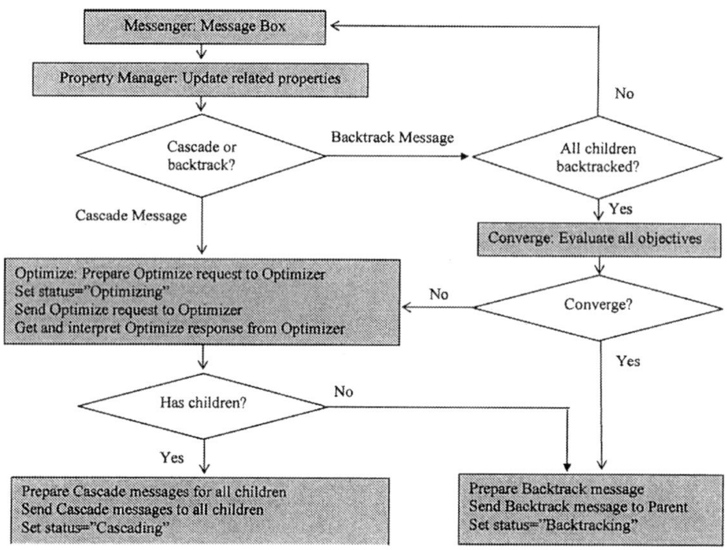

Figure 3 Information/work flow of a single TopAgent's Cascade and Backtrack methods.

5 TOWARDS SMOUDDI: ATCPORTAL'S UDDI DIRECTORY OF SMOWS

When formulating the ATC problem mathematically, Kim et al (2000) distinguishes between optimal design models and analysis models. Optimal design models call analysis models to evaluate responses from corresponding system components. Analysis models take lower level decision variables and parameters, as well as responses from lower level components in the system hierarchy, and return responses for optimization problems of upper level system components. Technically, analysis models are equivalent to equality constraints defined for an optimization problem.

smoUDDI is a UDDI (Universal Description, Discovery and Integration) Business Registry specialized in System Modeling and Optimization Web Services (SMOWS). smoUDDI has two main groups of users. The first group of users includes those service providers who are responsible for providing System Modeling and Optimization Web Services (SMOWS). The other group includes those analysts who consume and apply these SMOWS in their ATC analysis projects. In addition to the registry, smoUDDI also intends to become a Repository to provide facilities for hosting and managing SMOWS.

smoUDDI is built on various Web Services tools such as UddI (Universal Description, Discovery, and Integration), Soap, and Wsdl. UDDI is layered over SOAP and assumes that requests and responses are UDDI objects sent around as SOAP messages. A sample query is included below. The Web Services Description Language (WSDL) is a general purpose XML language for describing the interface, protocol bindings and the deployment details of network services. WSDL is used to describe what a web service can do, where it resides, and how to invoke it. WSDL complements the UDDI standard by providing a uniform way of describing the abstract interface and protocol bindings of arbitrary network services. WSDL assumes SOAP/HTTP/MIME as the remote object invocation mechanism. UDDI registries describe numerous aspects of web services, including the binding details of the service. WSDL fits into the subset of a UDDI service description.

6 CONCLUDING DISCUSSIONS

This paper has proposed and discussed a prototype web-based multi-agent system called atcPortal for specifically supporting optimal design of complex systems using Analytical Target Cascading. Its practical contribution is that ATC analyses can be easily set up and conducted with the help of this atcPortal, especially after it is fully developed. Its scientific contributions include at least the following two aspects. Firstly, the latest web technologies have been employed for the development and implementation of the atcPortal. The design optimization is truly distributed and collaborative on the Internet. Secondly, the atcPortal provides a basis for extending the ATC methods to incorporate and experiment with other strategies and for other types of system structures (e.g. hierarchical structures with limited number of shared components, and hierarchical structures with alternative branches). Discussions on these extensions will be reported separately in the near future.

Finally, although many of the limitations that this prototype atcPortal system suffers at present can be easily improved, it is important to point out a fundamental limitation on the balance between the generality and extensibility as set out in the title of the paper. Optimizers as web services collected in the smoUDDI are supposed to be generally reusable regardless if they are used in ATC analyses or for other applications. At present, they contain ATC-specific elements. In order to overcome this limitation, it is necessary to introduce two tiers of web services: one for Optimizers which are allowed to contain ATC-specific elements to enhance their extensibility and the other for Response / Analysis Models without ATC-specific elements to improve their generality. This is the area requiring substantial research and development efforts.

ACKNOWLEDGMENTS

Authors are most grateful to Hong Kong University Committee on Research and Conference Grants and William Mong Engineering Fund for financial supports that made this research possible.

REFERENCES

1. Choudhary, R., Malkawi A., Papalambros, P.Y., "A Hierarchical Design Optimization Framework for Building Performance Analysis" Proceedings of the Eight International Building Performance Simulation Association (IBPSA) Conference, Eindhoven, Netherlands, August 11th-14th , 2003.
2. Etman, L.F.P., Kokkolaras, M., Papalambros, P.Y., Hofkamp, A.T., Rooda, J.E., "Coordination Specification for Analytical Target Cascading Using the Chi Language," Proceedings of the 9th AIAA/ISSMO Symposium on Multidisciplinary Analysis and Optimization, Work in Progress, AIAA-2002-5637, Atlanta, Georgia, USA, September 4-6, 2002.
3. Fellini, R., Kim, H.M., Kokkolaras, M., Michelena, N., and Papalambros, P.Y, 'Target Cascading for Design of Product Families," Proceedings of the 4th Congress on Structural and Multidisciplinary Optimization, Dalian, China, June 4-8, 2001.
4. Ge, Ping, Lu, Stephen C.-Y. and Suh, Nam P., 2002, "An Axiomatic Approach for Target Cascading of Parametric Design of Engineering Systems", Annals of the CIRP, Vol. 51/1/2002, pp. 111-114.
5. Kim, H. M., Kokkolaras, M., Louca, L., Delagrammatikas, G., Michelena, N., Filipi, Z., Papalambros, P., and Assanis, D., "Target Cascading in Vehicle Redesign: A Class VI Truck Study," International Journal of Vehicle Design, Vol. 29, No. 3, pp. 199-225, 2002.
6. Kim, H.M., Michelena, N. F., Papalambros, P. Y., and T. Jiang, "Target Cascading in Optimal System Design," Proc. 2000 ASME Design Automation Conference, DAC paper 14253, Baltimore, September 10-13, 2000.
7. Kim, H.M., Rideout, D.G., Papalambros, P.Y., and Stein, J.L., " Analytical Target Cascading in Automotive Vehicle Design," Proceedings of the 2001 ASME Design Automation Conference, DAC-21079, Pittsburgh, Pennsylvania, USA, September 9-12, 2001.
8. Kim, H.M., Target Cascading in Optimal System Design, Ph.D. Dissertation, Department of Mechanical Engineering, University of Michigan, Ann Arbor, Michigan, USA, 2001.
9. Kokkolaras, M., Fellini, R., Kim, H.M., Michelena, N., and Papalambros, P., "Extension of the Target Cascading Formulation to the Design of Product Families," to appear in the Journal of Structural and Multidisciplinary Optimization, 2002.

10. Louca, L.S., Kokkolaras, M., Delagrammatikas, G.J., Michelena, N.F., Filipi, Z.S, Papalambros, P.Y., and Assanis, D.N., "Analytical Target Cascading for the Design of an Advanced Technology Heavy Truck, "Proceedings ASME International Mechanical Engineering Congress and Exposition, New Orleans, Louisiana, November 17-22, 2002.
11. Mahmoud, H.A., Kabamba, P.T., Ulsoy, A.G., and Brusher, G.A., Ranking Subsystem Targets According to their Influence on System Performance, In: Proc. American Control Conf., Denver, CO, June 2003.
12. Mahmoud, H.A., Kabamba, P.T., Ulsoy, A.G., and Brusher, G.A., Target Reduction and Balancing Using System Norms, In: Proc. American Control Conf., Anchorage, AK, May 2002.
13. Michelena, N., Kim, H. M., and P. Y. Papalambros, "A System Partitioning and Optimization Approach to Target Cascading," Proc. 12th Int. Conf. on Engineering Design, (U. Lindemann et al, eds.), Vol. 2, Munich, 1999, pp. 1109-1112.
14. Michelena, N., Park, H., Papalambros, P., "Convergence Properties of Analytical Target Cascading," Proceedings of the 9th AIAA/ISSMO Symposium on Multidisciplinary Analysis and Optimization, AIAA-2002-5506, Atlanta, Georgia, USA, September 4-6, 2002. To appear in AIAA Journal.
15. Nyström, Mattias., Larsson, Tobias., Karlsson, Lennart., Kokkolaras, M., Papalambros, P.Y. "Linking Analytical Target Cascading to Engineering Information Systems for Simulation-Based Optimal Vehicle Design," ICED 03, 14th International Conference on Engineering Design, Research for Practice - innovative products, processes and organisations, 19-21 August 2003, Stockholm, Sweden
16. Rideout, D.G., Stein, J.L., Ferris, J.B. "Target Cascading: A Design Process for Achieving Vehicle Ride and Handling Targets," Proc. IMECE 2001 Symposium on Advanced Automotive Technologies, Paper DSC-9-

e^2-MES – an e-service-driven networked manufacturing platform for extended enterprises

P JIANG, G ZHOU, G ZHAO, Y ZHANG, and **H SUN**
CAD/CAM Institute, Xi'an Jiaotong University, China

ABSTRACT

In this paper, a kind of e-service-driven networked manufacturing model is put forward. Referring to this model, some key enabling techniques, like e formalizing a machining equipment, encapsulating manufacturing resources concerning the principle of core capability protection, generating a tree that consists of enterprises and suppliers based on products to be manufactured, configuring a manufacturing executive sub-system inside an enterprise, scheduling the manufacturing processes, tracking the manufacturing information flow in the range of extended enterprise, etc, are discussed in detail. Furthermore, a testbed called as e^2-MES is developed with Java web solution. The running example from the industry indicates that the above idea is feasible and practical.

1 INTRODUCTION

Currently, the new requirements of global markets, customized products, low manufacturing cost and fast product delivery, etc, together with the rapid progresses of information, Internet and e-Commerce technologies, are driving traditional enterprises innovation. Extended enterprise, as a new organization mode of enterprises and implementation mode of networked manufacturing, is becoming more popular in academic and industrial areas. The concept of extended enterprises can be described as a way of seeking the external manufacturing resources of enterprises to form the strategic alliance. Its focus is based on the contracts. Starting from the point of collaboration, it can fully take advantage of the core capabilities of each allied enterprise and finally arrive at the "win-win" goal by connecting the key value units of each allied enterprise [1].

According to the dynamic construction of extended enterprises and their scheduling and control mechanisms, many researches and practices are undertaken in recent year [2-4]. However, referring to these results, there exist some shortcomings such as only coming down to the theoretic research and short of practice, lacking the whole information tracking and managing, interaction and share across the enterprises, to badly hinder the extended enterprises from further studying and practising.

To deal with the above problems, we propose a kind of e-Service-driven networked manufacturing model for extended enterprises. The concept of it will be described in next section. Through taking advantage of this model, the whole manufacturing process can be controlled well and the goals of cost, quality and production cycle tracking of manufacturing process at the level of extended enterprises can be reached.

2 PHILOSOPHY TO AN E-SERVICE-DRIVEN NETWORKED MANUFACTURING

Before describing the implementing architecture of e-service-driven networked manufacturing platform, we have to define some basic concepts used in this paper.

2.1 Basic Concept of e-Service

Services is defined as a kind of procedure-related activity sequences applied in different application areas whose core working mechanism is that referring to the requirements of service receivers and recurring to the specific media, the service providers supply the service receivers with a series of value-added activities having rewords or non-rewords such as information and material flow, problem resolving, procedure implementation and spot operation. Its key factors mainly include service providers, service receivers, service carriers, service media and service contents, etc.

E-Service is defined as a new kind of service mode which is introduced into the networked manufacturing environment and makes the traditional service processes have the networked, information, digital and electronic-payment natures. The concept of e-Service extends that of the traditional. Besides the basic characteristic of traditional services, its implementation also mainly relies on digital processing of information, e-Commerce and networked technologies. Figure 1 illustrates a general e-Service model.

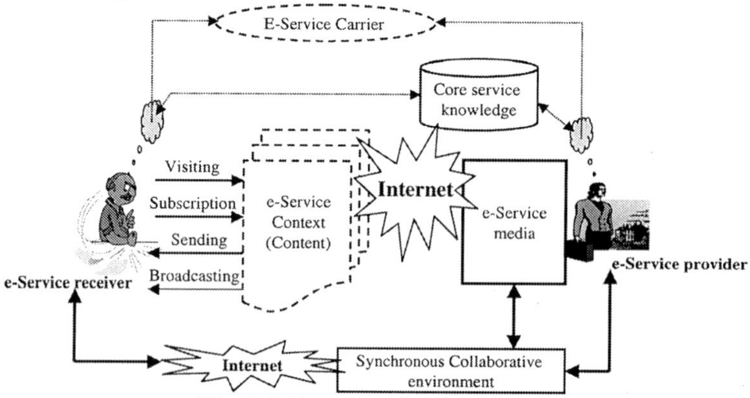

Fig. 1 A General e-Service Model

Generally, there exist three types of e-Service modes, that is, active services, passive services and mixed services. The active services refer to that the service providers create the service contents while the passive services to that the service receivers build the service contents with the help of the service media provided by the service providers; the mixed services refer to that the service contents are created by both service providers and receivers based on the collaborative working mode.

2.2 Basic Concept of e-Service-driven Networked Manufacturing

As mentioned above, the traditional networked manufacturing mode is short of information tracking and managing, interacting and sharing across enterprises in a "transparent" manner for customers, suppliers and manufacturers. We propose an e-Service-driven networked manufacturing model whose core points are to construct a ASP-based networked manufacturing service platform, called as e^2-MES (Manufacturing Execution System for Extended Enterprises), through taking advantage of information and Internet technologies and adopting the scheme of service processes corresponding to manufacturing processes. On the basis of e-formalization, encapsulation and integration for the bottom-level manufacturing resources, the customers, suppliers and manufacturers can collaboratively participle in the manufacturing activities in a "transparent" manner via this platform. At the same time, by means of plug and play mode, ASP-based enabling tools as the service tools such as web-CAX/PDM/ERP can be integrated into this platform easily.

3 A WEB-BASED IMPLEMENTATION FOR E^2-MES

3.1 Implementing Architecture

According to the constructing mechanism and working logic of extended enterprises, we propose a novel implementing architecture for e^2-MES shown in figure 2, which is constructed by adopting three-tire networked information infrastructure (Browser/Server/Database) and whose serving objects mainly include the customers submitting manufacturing tasks and the manufacturers/suppliers dispersing different remote sites. Here, it must stated that the manufacturing task is defined as a description in which planning, assigning, ordering, machining or assembling operations for the parts/assemblies/products which can be decomposed into a series of sub-manufacturing tasks.

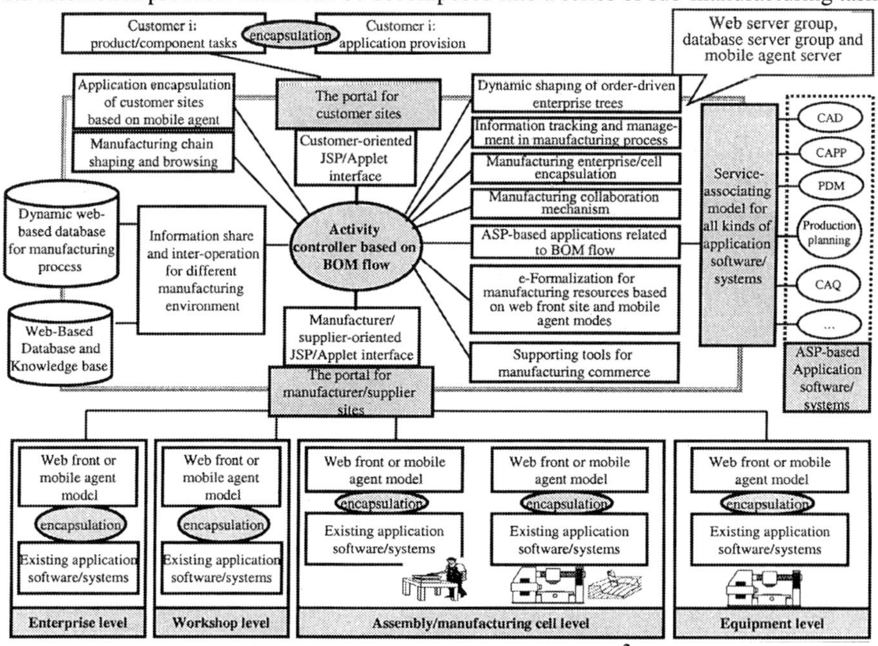

Fig. 2 An Implementing Architecture for e^2-MES

Starting from the point of viewing the structure, this platform mainly consists of a series of functional modules, namely the customer-oriented portal, the manufacturer/supplier-oriented portal, the BOM-flow-based activity controller, the web-based database and knowledge base, and a series of ASP-based enabling tool, etc.

From the perspective of viewing the running logic step by step, this platform can be described as follows:

- First, through adopting the order-processing tool based on the product BOMs provided by this platform, the customers input the corresponding product-related parameters such as the batch, delivery time, quality and cost of product to be manufactured and submit them to the platform to form the product-related manufacturing tasks.
- Then, the customers adopt the bid tool provided by the platform to select an optimal enterprise as the core enterprise corresponding to the submitted manufacturing tasks.
- Third, the core enterprise analyzes the product manufacturing nature and correlates assemblies/parts to different types, that is, outsourcing, and in-house and out-house machining, and further form the rough production plans, stock plans and bid documents based on the BOMs of the submitted products. Here, in-house machining assemblies/parts correspond to the core enterprise, outsourcing assemblies/parts to the stock branch of core enterprise, and out-house machining assemblies/parts to the leaguers of core enterprise.
- Fourth, according to out-house machining assemblies/parts, the core enterprise adopts the direct assignment or bid mode to find out the correspondent leaguers to form the first-level extended enterprise. At the same time, each allied enterprise also divide the responsible manufacturing tasks into three different types of outsourcing, in-house machining and out-house machining and assign them to other enterprises to form the next-level extended enterprise. This operation is repeated until each leaguer's responsible manufacturing tasks cannot be further divided and finally a multi-level extended enterprise can be dynamically constituted. According to the above results, a structure tree to represent the above extended enterprise whose root node corresponds to the core enterprise and branch and leaf nodes to the suppliers and leaguers of core enterprise and multi-level sub-suppliers and sub-leaguers derived from their up-suppliers and up-leaguers is dynamic shaped and called as an enterprise tree. In addition, the whole tree corresponds to the product-based order while the leaf nodes to the different level product plans and stock plans.
- Fifth, concerning the rough production plans and the information of manufacturing tasks to be manufactured, the leaguers corresponding to the leaf nodes prepare the process plans and further form the detailed production plans. Based on these production plans, the leaguers can configure their manufacturing resources inside and allocate these manufacturing tasks to their workshops, manufacturing cells and facilities.
- Finally, according to the manufacturing task allocation results and under the precondition of e-formalizing facilities, together with encapsulating the core capabilities of manufacturing cells, workshops and enterprises, the real-time production scheduling and control for manufacturing cells can be realized. During the scheduling procedure, the process-related manufacturing data can be collected and transmitted up to corresponding workshops and enterprises by adopting the e-formalization templates. Under this case, the scheduling and control results for workshops, enterprises and even extended enterprises can be concluded through taking advantage of the scheduling results of manufacturing cells.

Based on the above description, e^2-MES is taken as a manufacturing service platform for extended enterprises whose serving objects mainly focus on task-driven collaboration and control, information interaction and share across enterprises to implement the manufacturing tasks, and finally arrive at the "win-win" goal.

3.2 Some Key Enabling Techniques

3.2.1 e-Formalization for facilities as a Basis to Up-streaming Manufacturing data

In e[2]-MES, it is a key point to e-formalizing facilities because the real-time up-transmission of manufacturing data produced by the bottom level facilities such as process sequence, running status, processing quality and audio and video data is the basis of rightly scheduling and controlling ranging from manufacturing cells to extended enterprises, and also provide the information source for information tracking and managing of extended enterprises.

E-formalization of a facility deals with two aspects, that is, enhancing data collection capability, especially in the down-to-up stream, at the bottom level and attaching a web server in the front of the facilities. In this paper, we don't discuss much more technical details concerning the first aspect. Instead of this, the last aspect is emphasized on in depth.

Generally speaking, the core of e-formalization for a facility is to set up a web server in the front-end of this facility and to use it to open a bi-directional channel in between a variety of down-to-up-stream and up-to-down-stream data that are concerned with products to be manufactured, facility running procedure, high-level scheduling, etc. Through the web server, several service functions are attached and make the facility have the e-natures. As shown in figure 3, a CNC lathe can be e-formalized based on Javá solution. Here, e-service based on a "process" takes an important role.

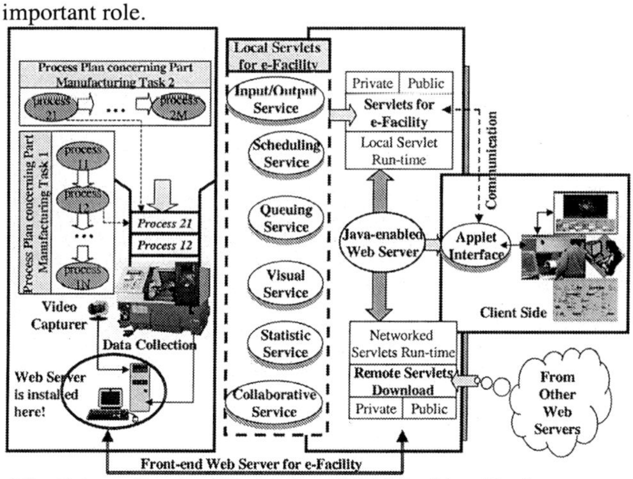

Fig. 3 An e-Service Model for CNC Machine Tools

From this reference model of CNC machine tools, we learnt that the e-formalization of a CNC lathe is concerned with three steps. Step 1 deals with a setup of some necessary hardware units like video capturer, sensors, data sampling board, etc. The purpose to do this is to capture the on-site machining video stream that makes people have the confidence to know current progress, to sample facility-running status data that diagnose potential machine faults, and to collect product-machining-related quality data. Step 2 is to install a Java-enabled web server in the front-end of the machine tool and to develop some server-side programs and scripts like servlets, servlet-applet pairs, Java server-side pages, HTML pages, etc. These programs and scripts will enable the whole e-service functions with which the e-formalization is reached. The correspondent e-service functions include input/output, queuing, scheduling, statistic, visualizing, and collaborating, etc, and are implemented mainly by means of local servlets and servlet-applet pairs and subordinately in terms of networked servlets downloaded from other servers. Step 3 is to integrated all together under the framework of facility e-formalization. About the details of the above e-service functions, literature [5] and [6] can be referred to.

3.2.2 Order-driven Manufacturing Task Allocation and Networked Manufacturing System Configuration

In e^2-MES, order-driven manufacturing task allocation realizes the following objects:

- The optimal selection of core enterprises;
- The classification of manufacturing tasks to form the outsourcing, in-house machining and out-machining types;
- The optimal selection of multi-level leaguers and optimal configuration of manufacturing resources based on different types of manufacturing tasks;
- The dynamic formation of the multi-level enterprise tree and extended enterprises based on the results of manufacturing task allocation;
- The reallocation of manufacturing tasks inside an enterprise and the dynamic formation of networked execution systems.

According to the above objects, several key technologies/tools should be considered and developed, which include the description and establishment of data and manufacturing resources, the model for manufacturing capabilities and the classification for manufacturing tasks based on TQCSE rules, the strategies/models for manufacturing task allocation, the optimal selection of leaguers and shape of enterprise trees based on the combination of computer-oriented automatic evaluating bids and expert-oriented manual evaluating bids, and the formation of networked manufacturing execution systems, etc.

Concerning the above requirements, two kinds of strategies/models are proposed to realize the manufacturing task allocation, that is, direct assignment and bid. Here, the definitions and running logic of the two strategies/models are not discussed in details; the interested readers can attach the reference [7] and [8]. Figure 4 illustrates a bid-driven manufacturing task allocation model.

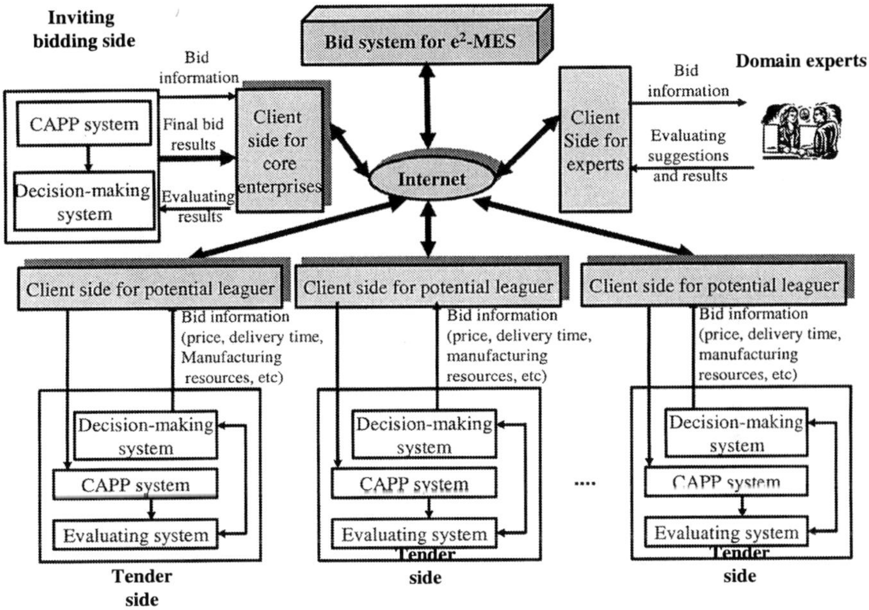

Fig. 4 Bid-driven Manufacturing Task Allocation Model

As shown in figure 4, during the bid process, the cost evaluation of a manufacturing task is the basis of deciding the bidding price and becoming one of key factors for potential leaguers winning a bid. Under this case, we present the following formula to calculate the cost of a manufacturing task.

$$C_{total} = E_m W_m + \sum_{i=1}^{n} T_i^{process} C_{unit}^{process} \tag{1}$$

Here, C_{total} represents the total cost of a manufacturing task expended in lubricating process; E_m represents the price per unit weight (KG) of the rough of products/assemblies/parts; W_m represents the total weight of the rough of products/assemblies/parts; $T_i^{process}$ represents the man-hour of a process; $C_{unit}^{process}$ represents the cost of a man-hour, which includes the depreciation fees of facilities, wages of workers, and power expenditure, etc.

Generally, the man-hour calculation of processes mainly relies on their process rules. The concrete calculating method is not discussed in details in this paper; anyone who is interested in it can consult the reference [7].

After the manufacturing tasks have been allocated to the correspondent enterprises and reallocated inside each enterprise to its workshops, manufacturing cells and facilities, the extended enterprises will be dynamically constituted and the networked manufacturing execution systems can be built finally.

3.2.3 MC-based Production Scheduling

As mentioned in section 2, based on the reverse orientation reasoning rules, the production scheduling for the whole extended enterprise can be derived from the MC (manufacturing cell)-based production scheduling. The concrete reasoning philosophy can be consulted in reference [9] and [10]. This paper mainly concerns with production scheduling implementation for manufacturing cells.

As we know, after the manufacturing tasks are allocated to a manufacturing cell, they in fact represent a series of process sequences. Under this case, the objects of production scheduling are to arrange these processes to different facilities at the right time to satisfy the initial requirements such as delivery time and lubricating quality and furthermore, during the lubricating process, rearrange the rest of processes well in real time under the condition of some facilities going wrong. In order to meet the above requirements, some optimal scheduling algorithms are adopted and the typical ones mainly include NN-based scheduling algorithm, heuristic rule-based algorithm and genetic algorithm, and so on. In this paper, we adopt genetic algorithm to implement the optimized scheduling for manufacturing cells. Here, we do not discuss the algorithmic model and scheduling mechanism, which was realized and presented in our previous literature [9], the interested reader can consult it. Figure 5 shows a Gantt chart from GA-based scheduling for a manufacturing cell.

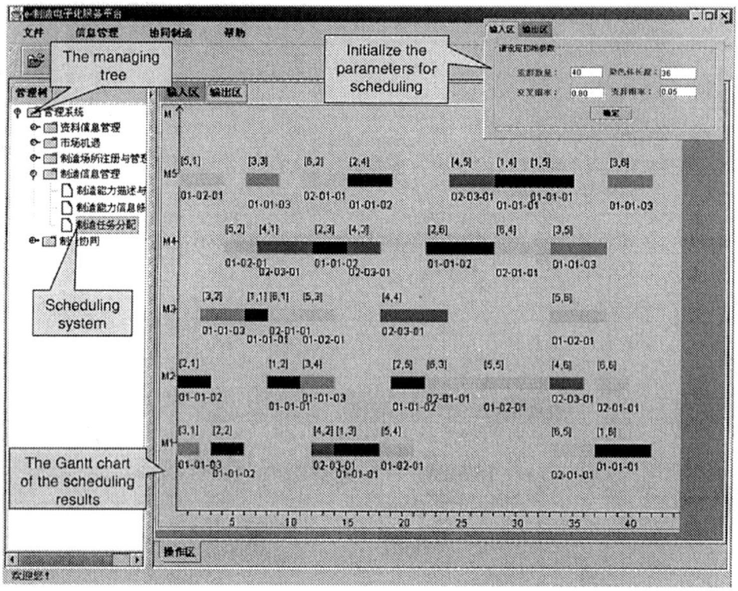

Fig. 5 A Gantt Chart from GA-based Scheduling for Manufacturing Cell

3.2.4 Manufacturing Process Tracking

It is the key point to track the manufacturing information in real time across enterprises in an extended enterprise, which also becomes the bottleneck in current networked manufacturing systems. To deal with this problem, we propose an idea of BOM-based manufacturing information tracking and managing and build an information tracking and managing tool for the whole manufacturing process across enterprises.

According to the manufacturing information evolvement process shown in figure 6, we divide the manufacturing information into four types, that is, design information corresponding to design BOM, process information to process BOM, manufacturing information to manufacturing BOM and commerce information to commerce BOM. Through querying, gathering and tracking these types of information and implementing its transmission by means of down-to-up-stream and up-to-down-stream modes, we can easily realize the whole information management of e^2-MES and support the manufacturing process correctly execution. Figure 7 illustrates a running interface of BOM-based manufacturing information tracking and managing, which is realized by adopting Java Web solution, especially JSP+HTML technologies.

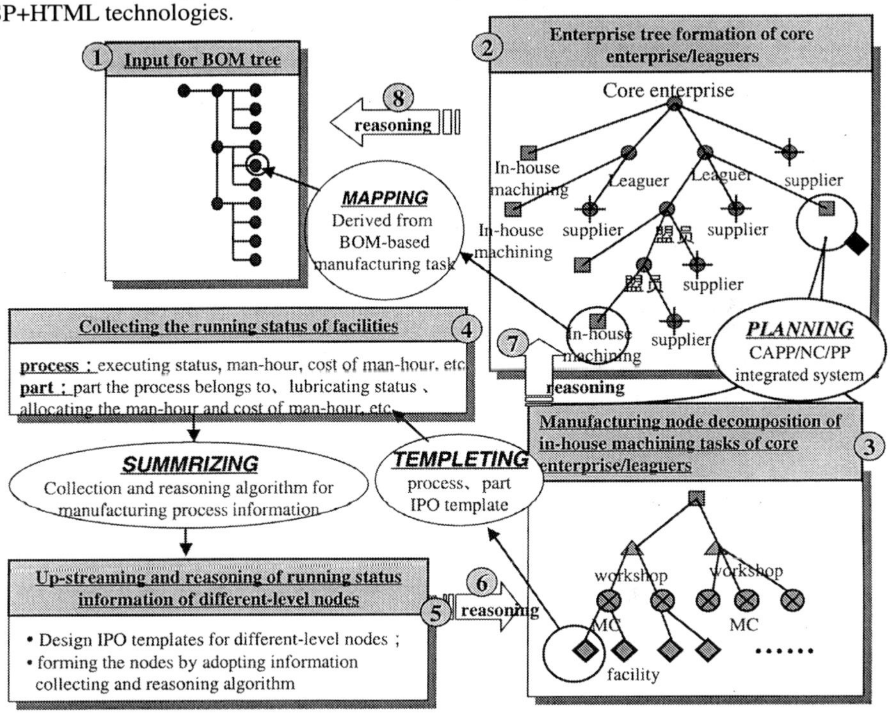

Fig. 6 Manufacturing information evolvement procedure in e^2-MES

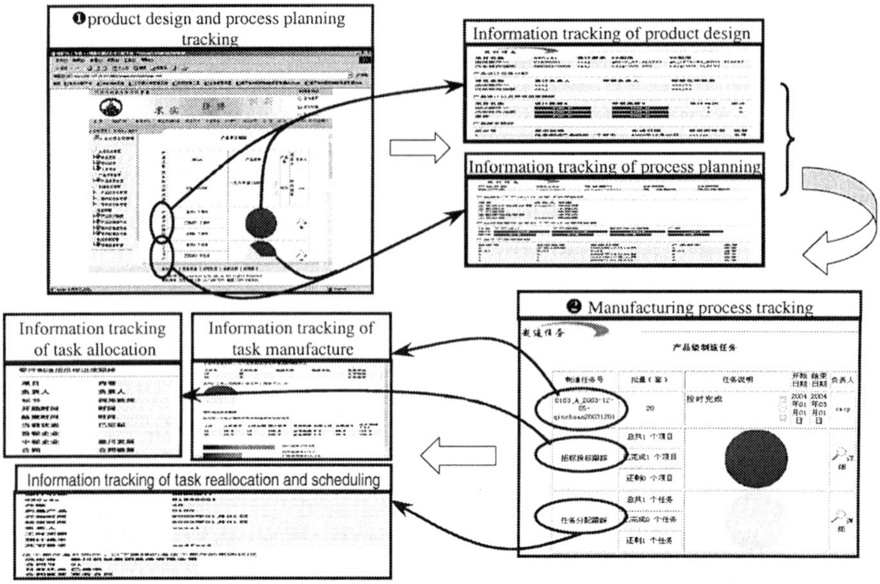

Fig. 7 Running interface of BOM-based manufacturing information tracking

3.2.5 Collaboration Support in Manufacturing

Because legacy collaborative tools like NetMeeting for information share in the form of pseudo B/S architecture and cannot meet the requirements in e^2-MES such as the real-time information interaction, display of complicated engineering drawings and documents and supporting ASP nature, we still develop a novel synchronously collaborative tool in the real B/S mode. This tool is based on Java media framework (JMF) and consists of a series of general "applet-servlet" pairs which integrate sound and video, and include whiteboard and online chat to support manufacturing collaborations. Here, the whiteboard can be used for displaying VRML geometric models besides general images and for marking some comments on line. Because of the platform-independent natures of java and B/S mode, this collaborative tool overcomes the shortages of legacy collaborative tools and can communicate and interact with one another by downloading to any operation system. Figure 8 shows the collaboration mechanism based on process plan review for a valve part.

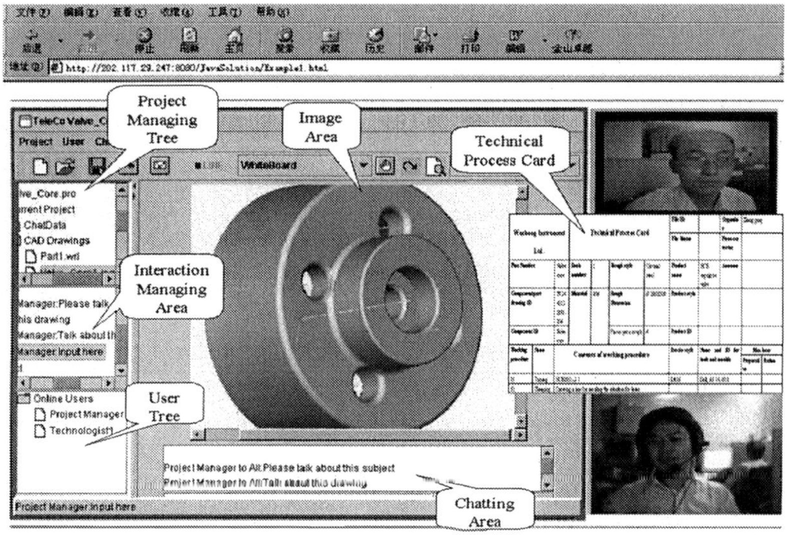

Fig. 8 Collaborative Support to Process Plan Review

4 CASE STUDY

Based on the above ideas, we develop a prototype of e^2-MES by adopting Java web solution. Figure 9 is the snapshot of the whole running process of this platform, which is described step by step as follows.

- The customer submits his product-based order to this platform;
- The customer select the optimal enterprise as the core enterprise by taking advantage of the bid tool provided by the platform;
- According to its own capabilities, the core enterprise decomposes the product-based manufacturing task into three different types, that is, outsourcing, in-house machining and out-house machining sub-manufacturing tasks;
- The core enterprise allocate these sub-manufacturing tasks to corresponding enterprises or suppliers to form an enterprise tree by adopting direct assignment or bid strategy/model and finally constitute an extended enterprise;
- Each leaguer reallocates its correspondent manufacturing tasks to its inner workshops, manufacturing cells and facilities based on production plans and process plans and finally form the networked manufacturing execution system. In the meanwhile, based on GA-based Gantt chart, the dynamic schedule for manufacturing cells can be realized; based on the e-formalization templates, the bottom level information is transmitted up and collected at the up level such as enterprise-level or extended-enterprise-level. Under this case, the information tracking, transmitting and scheduling across enterprises can be realized well.
- Finally, the product-based manufacturing task is implemented.

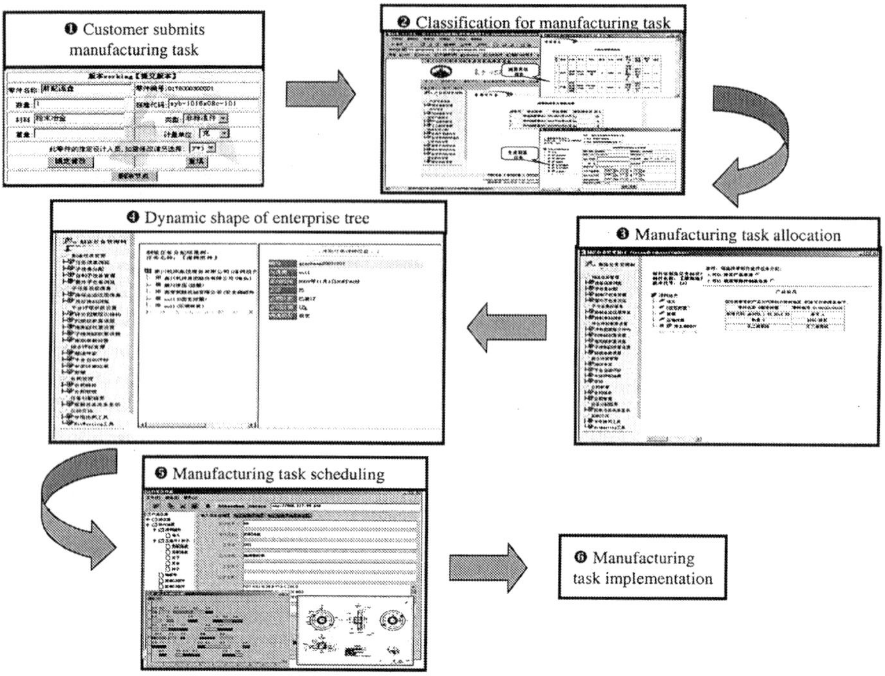

Fig. 9 A Run Example Using e^2-MES

5 CONCLUSIONS

On the basis of analyzing the current researches for networked manufacturing modes and extended enterprises, this paper presents and builds a prototype of e-service-driven networked manufacturing platform called as e^2-MES. Using this platform, we can efficiently and effectively realize the collaboration and control, and the information interaction and share across the enterprises in an extended enterprise to implement the correspondent manufacturing tasks and arrive at the "win-win" goal. Furthermore, e^2-MES is an ASP-based networked manufacturing platform, whose theories, architecture, running logic, service procedure and implementing methods, etc. will provide a new reference for studying and developing networked manufacturing platform as well as pave a new way for the implementation of networked manufacturing mode.

ACKNOWLEDGEMENTS

The research is under the supports of: national 863 High-Tech Research and Development Program of China (Project No.: 2001AA415230), and University Key Teacher Foundation of China Ministry of Education. We hereby thank them for the financial aids.

REFERENCES

[1] http://www.informationbuilders.com/The Extended Enterprise.htm.

[2] W.B. Lee, H.C.W. Lau (1999). Multi-agent modeling of dispersed manufacturing networks. Expert Systems with Applications, Vol. 16, pp297-306.

[3] Cheng Tao (2000). Research on some key issues in distributed networked manufacturing theory and technology, Doctoral Dissertation, Huazhong University of Science of Technology, 2000.

[4] YANG Shuzi, WU Bo, etc. (2000). Network Manufacturing and Enterprise Integration. China Mechnical Engineering, 2000, Vol. 11, No. 1-2, 45-48.

[5] Pingyu Jiang (2003). Approach to an Implementing Methodology and Application for e-Manufacturing Systems. Chinese Journal of Computer Integrated Manufacturing Systems, in Press.

[6] Pingyu Jiang, Yingfeng Zhang (2003). Visualized Part Manufacturing via an Online e-Service Platform on Web. Int. Journal of Concurrent Engineering: Research and Applications, Vol.10, No. 4, 267-277.

[7] Guanghui zhou (2003). Research on Modelling and Scheduling Method for e-Service Platform Oriented Networked Manufacturing Chain. Doctorate Dissertation, Xi'an Jiaotong University.

[8] Guanghui Zhou and Pingyu Jiang (2003). Dynamic Formation of Task-Driven e-Manufacturing Federation. International Conference on Agile Manufacturing, December 4-6, 2003. Beijing, China, pp99-104.

[9] Yingfeng Zhang, Pingyu Jiang, and Guanghui Zhou (2003). GA-Driven Part e-Manufacturing Scheduling via an Online e-Service Platform, International Journal of Integrated Manufacturing Systems, Vol.14, No.7, pp575-585.

[10] Pingyu Jiang and Guanghui Zhou, etc. (2004). Monograph: e-Service Theories and Implementing Methods for Networked Manufacturing, The Science Press, in Press.

Implementation perspectives of design re-use through global manufacturing

L LI, D K HARRISON, and **J S LYNN**
School of Engineering, Glasgow Caledonian University, UK
K CHENG
School of Engineering, Leeds Metropolitan University, Leeds

ABSTRACT

With the force of drastic global competition today, companies and manufacturers have to promote their development processes to provide customers with products which are more sensitive to their needs. The costs and time to develop new products will potentially influence the companies' strategy in market demands. Importing the ability to reduce the development cycle and refine the infrastructure of product design and manufacturing will approach the goals. The technologies in modern engineering and manufacturing have been changed drastically and are still developing. The engineers and designers have been challenged with increasing demands from customers who wish their products had the rich features but keeping low costs. Reducing cost is an essential and effective way for wealth generation and companies' survival in the scale of economies. Importing the ability to find all existing similar components for a given desired component is the key to cost reduction. A new version of a solution-making search agent system could be developed and implemented to achieve the aims by automatically retrieving existing parts that have similar shape and features to the given part. The proposed system aims to provide an on-line intelligent support for determining the closest matching parts and the related manufacturing data will be analyzed and indexed in a backup database.
This paper investigates the feasibility and makes a plan to support developing new solution-making system which could be deployed to achieve agile design and manufacturing purposes.

Key words: Data retrieval, Similarity Search, 3D Search Agent for Engineering, Web-based Design.

1 INTRODUCTION

Nowadays, the volume of information available on the World Wide Web is rapidly increasing. The trend of the information era offers the potential massive volume of information, which can be obtained and reused easily. The Internet attracts companies and manufacturer's eyes and changed the development's priority from the traditional orientation, by reducing time-to-market and the cost of acquiring the latest technical information, which has become necessary for leading companies to beat their competitors. The Internet provides a vast, easy to access and versatile platform to designers, manufacturers and customers. Furthermore,

CAD/CAM technologies are increasingly playing an important role in the manipulation of manufacturing data and design products. In order to increase their potential customers and win business, companies use CAD/CAM to produce sensitive designs more quickly and cheaply than their competitors.

While creating a new product, designers have to access a vast variety of combinations such as price, quality and manufacturing operations. Even creating the simplest components will derive a significant cost that is often quoted as a large amount of money per part per annum [1]. The collaborative of reusing existing design and manufacturing could be more practical and push forward these aims to come true, involving changing the development cycle of product with keeping low cost, time-saving and improving product capability. With these good points of reusing designs, how to find the desired parts to take advantage of are concerned by designers wherever and whenever they are available. Many companies have established item catalogues to control the indexing of common parts [2], however, this approach is often ineffective and time-consuming especially for surveying and gathering non-standard items. The usability could be improved via developing intelligent support systems to help in the processes of locating existing designs and manufacturing data. The Internet can be viewed as the infrastructure or a common workspace to support a possible intelligent solution-making system for data sharing between engineers, designers and researchers, moreover Internet is already sufficiently enhanced to offer the necessary technologies to support companies showing off their parts catalogue and available CAD graphical models. Thus, often Manufacturers like to post their 3D computer-generated models of their product through the Internet as a means of advertising their production capabilities.

As the number of 3D models available on the Internet grows, finding out the information the user desires on the web is quite difficult, although the existing search engines such as yahoo, AltaVista, HotBot, and many other applications have made navigating the web easier. These services provide users with access to large databases containing keywords across references to URLs (Uniform Resource Locator). Unfortunately, these traditional text-based services are not always effective for 3D data, users have to spend a lot of time on the Internet to download and filter information finding new items of interest. The fact motivates us to develop an intelligent search agent system for collecting and filtering models, which supports queries based on 2D, 3D sketches /models, and text keywords. The heart of a search engine is the search agent, which collects the information to be placed in the database. Intelligent agents can transform passive search, retrieval engines into active, personal assistants. Franklin and Graesser [3] propose the autonomous agent as software, robotic entities that are capable of independent action in open, unpredictable environments. The proposed autonomous intelligent agent system could be the solution to simply the complexity and reduce the volume of information retrieved to meet maximum demand from users. The proposed system will combine a search agent and content analyzer, which lets the user organize, re-use and discover new patterns for the users own personal information.

This paper proposes the integration of the Internet and the development of product as a means of reducing the development cycle. The new engineering tool will be Internet enabled and make use of a number of advanced methodologies and web-based technologies such as intelligent agent, database integration and efficient, secure programming languages, which contribute to the industry design and manufacturing purposes. The new methodologies and knowledge in various research areas are involved in the integration and the popular technologies are critically analyzed, discussed and determined in order to make benefits to the progress of developing new powerful tools.

2 DATA RETRIEVAL AND SIMILARITY SEARCH

2.1 Information Retrieval in Modern Engineering

Information retrieval has been a very active research area recently [4]. The most well known examples are text based search engines that have become the part of people's daily lives.

For decades, people have been fantasizing about how to create interesting 3D models in computer graphics. Some of the most forward thinkers have been using CAD/CAM techniques to develop products for the advantage of improving quality, reducing cost and lead time etc, as well as facilitating rending 3D models on the Internet to attract potential customers. Motivated by the fact, designers and users take advantage of using existing designs and obtain them from a large accumulation of 3D models over the World Wide Web. In the future, the challenge to users will shift from "How can we create and publish them over the web?" to "How can we find them out?" The research will follow the trends of data retrieval in matching, classification and retrieval of 3D models. The retrieval based on the shape similarity has been involved in many fields such as geometry, computer vision, and molecular biology etc [5] [6]. A good example of using a neural network system is to extract features from bit-mapped 2D images to identify parts' code [7]. According to these content-capable search engine systems can return similar images ranged on color, edge detection, and shape similarities, the geometric features can be extracted from 2D and 3D CAD models and indexed as the identification stored in the database. There is much work on classifying the similarity of 2D models and this could be extended to 3D surface models by observing various 2D views [8]. The challenge of our prototype system is to build efficient modular interfaces for specifying 3D queries and analyze the shape features for similarity search. An appropriate and effective algorithm is the core to implement the task of shape matching.

2.2 Similarity Search and Pattern Recognition

Matching is the central problem in graphics analysis and pattern recognition. Cybendo [9] presents an approach, which is concerned with the normalization of a 3D object's representation and then extracting geometric features. When the objects are available after presentation, they need to be normalized to an orientation before rasterizing into voxels, which enables a further voxel - voxel comparison. This voxel - voxel approach is computationally accurate and intensive. After being normalized, the work is involved in computing translational, rotational and second order 3D moment invariants of an object such as: surface area, centroid and smallest rectangle axis align. These invariants are used to extract features and make data searching.

A good edge-matching algorithm, Chamfer matching is an edge matching technique, first proposed by Barrow [10]. The edge points of a model were transformed by a set of transformation equations to the edge points of a similar one distorted somehow. Chamfer matching attempts to find the best matching for the edge points from the two models by minimizing a generalized distance between them. Borgefors [11] applies edge detection to create a feature map on binary format for distance transform. Distance transform (DT) converts the edge of an object into a DT model in which each bin denotes the distance to the nearest edge bin. Based on their method, the estimates of rotation, scaling and translation between the model and target model can be achieved and matching can be finally realized for partial shapes.

Some research work uses graphic histograms as a way to analyze the similarity of an image. The histogram of 2D or 3D graphics records the frequency distribution of levels involved. It is already seen that values in a graphic are vectors, typically with three components. It might be attempted to perform an operation such as histogram equalization on each of the three components separately.

Ankerst [12] uses the benefits of shape histograms to analyze the similarity of 3D objects. The work of the research is to store histograms of how many surfaces resident at different radii from center of mass. It suggests three techniques for decomposing the space. Shell, sector and combination of the former two, based on a sphere located at the center of the mass of the solid 3D models, are used to decompose the space for calculating shape histogram. Histograms are calculated on the decomposed surfaces.

Similar to Borgefors's and Ankerst's approach, Funkhouser [13] considers feature recognition based on the measures for dealing with distance and angles between random points on the object's surface. The effectiveness of the adopted method was evaluated by a database of over 100 models with the complex variety of shapes such as cars, boats, pens, phones etc. Accuracy has been tested for shapes recognition and nearly 70 percent was achieved.

These approaches present a similar method to the theory of the histogram, using the vector space for queries. They are useful based on checking an object's orientation invariance properties and efficiently improving calculation.

3 PROTOTYPE SYSTEM OVERVIEW

3.1 The Architecture of Proposed System

For the prototype system, the shape-based search agent allows users to upload a model file and retrieve similar images based on shape and color similarities.

The architecture of the prototype system is shown in Figure 1. The development phases are in four steps: user's query, data indexing, features matching and parts retrieval. The system can be performed to test its capability offline, and then be enabled to go live for user query on the Internet. Figure 1 presents an overview of the system architecture and highlights its main features of functionality.

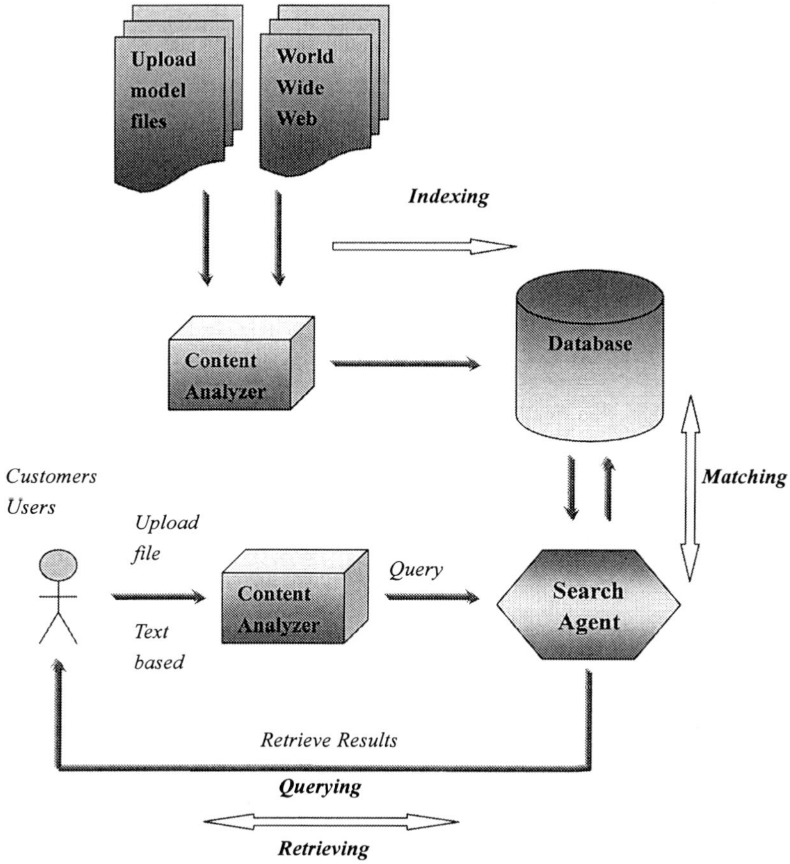

Figure 1 – The Architecture of the Proposed System

(1) Indexing: it needs to fill 3D models within a database for storage of models by surfing over the World Wide Web. The 3D data of these models presents the capability of products, and will be indexed into a database as its own features for future comparison.
(2) Querying: The proposed system provides an interface to allow users to search interactively for 3D models. In order to retrieve models efficiently, the system supports query methods based on typing keywords, uploading 2D, 3D models, and then the entry will be analyzed so that it will submit to an intelligent search agent for matching later on.
(3) Matching: For each user query, the agent will immediately query the database on the web server using the data generated by the analysis program. Database is responsible for matching by its indices that are already stored in database.
(4) Retrieving: The system can return the resulting models that are most similar to the query one. Due to the bandwidth usability, the proposed system should query, analyze and return resulting 3D models within the tolerance of user.

3.2 Parts Modelling Format

As mentioned, the Internet will be used as an infrastructure for the proposed system on account of the power of the Internet that shares text, value and data but also supports delivery of high quality and interactive 3D graphical content. Thus, much work has been done for enterprises and numerous modelling systems have been developed to build 3D detailed solid models that focus on designing and testing parts using CAD/CAM technologies. The part and shape modelling systems e.g. IGES, ACIS, STEP etc, always section 3D models into consecutive 2D slices and then scan these layers using different storage formats concerning 3D, since the data exchange becomes the key factor in concurrently processing data among different modelling systems. Although the standard and specification of 3D models exists, the data exchange is still an unstable factor for CAD systems. This fact motivates prototype system to adopt a compatible output format for representation in all commercial CAD systems rather than translation between different formats. The choice of shape presentation is STL (STereo Lithography) format. STL provide the openness in a multi-system environment, and allows a CAD system to process external data and provide a good level of integration. Besides, all CAD systems have already included the module for translating shapes to STL format. STL is a format designed for rapid prototyping, which is intended to represent the parts or shapes by a number of triangles. As presented in [14], STL files can include solids that are described by their mesh structures.

3.3 Web Publishing Overview

Interfaces design is a part of the suite of system development. It describes the way to implement a system not only for interaction between human and system, but also taking over the collaborative of separate modules that deals with distinct jobs in the system. First, the appropriate user interfaces designed could smoothly navigate users to take an easy tour over the web-capable system Second, different web techniques, programming languages and methodologies adopted would affect the system performance due to the immediate issues involved using related methodologies e.g. user requirement, database connection/query, code portability, system security, maintenance and upgrade in the future.

The database management system (DBMS) is used for indexing 2D, 3D models. The popular relational database systems are Microsoft SQL Server, Microsoft Access, Oracle and MySQL. They are Windows OS compatible, support relational database concepts, queries and updates. The application development software could be the collaborative of using Java Server Pages, Servlet, HTML, Vrml, and Java etc. They are platform compatible and provide capable APIs so as to allow modular web-based application development and support accessing to the DBMS.

The architecture for the web publishing overview applying these techniques and methods has been investigated and initialized. Figure 2 gives the idea of data flow diagram that shows the end users interaction, database accessing and the combination of various web techniques and backup applications (analyzer, search agent).

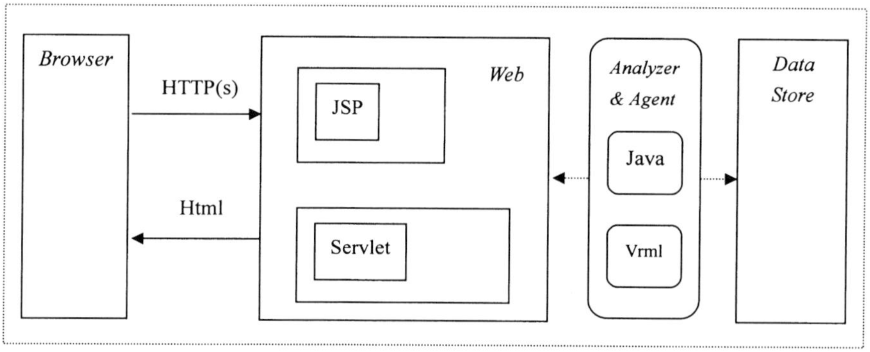

Figure 2 – Web Publishing Overview

4 IMPLEMENTATION OF THE APPROACH

In order to assess the feasibility of a similarity search based on 2D, 3D content, an analysis system has been built that involves a graphics histogram program for features extraction. The users can select samples and find the similar ones in the left panel showing a thumbnail view of each model. The current system screenshots are shown in Figures 3-4. The set of test files are all acquired from the Internet.

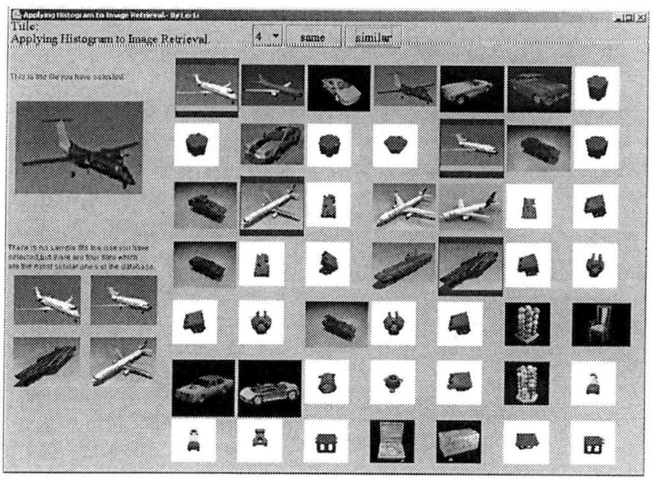

Figure 3 – Samples of plane model retrieved

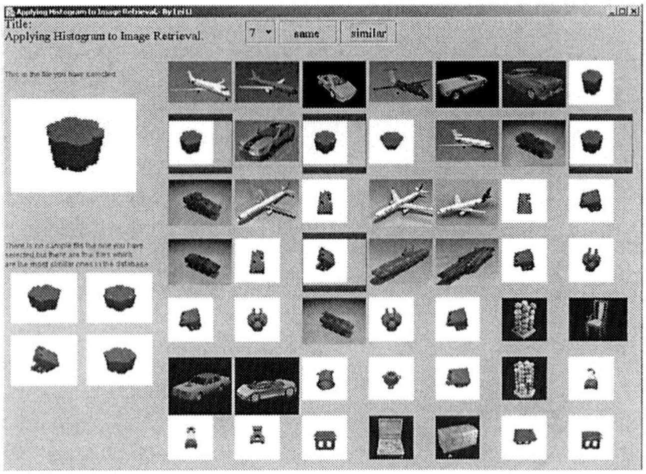

Figure 4 – Samples of part model retrieved

Using available technologies, the prototype here has been developed using Java and it is convenient to use Java Applet to facilitate web publishing. Java provides a number of networking capabilities to make it easy for developing networking and web-based tools. Java takes advantage of independence from platforms so as to makes its code portable running over different systems [15].

5 CONCLUSIONS AND FUTURE WORK

Due to the initial imagination of providing an intelligent web capable system, this paper preliminarily analyzes the feasibilities and implementation perspectives of a new version content-based retrieval system in order to promote agile design and manufacturing purposes. This paper has presented the details of the proposed system for similarity search in engineering and the implementation of the prototype system based on a similarity recognition method.

Whilst the initial performance of the test system proved it practicable, future work will focus on the more extensive experiments continued for 3D shapes matching through using different methods as the literatures suggests as a wide range of active research areas. Developing a formal method of evaluating the search engine performance is obviously vital, which can generate similarity data diagrams from the results retrieved. The users can assess the system performance online or off line and provide advice and criticisms. The modification of the proposed system could be done based on the assessment from users.

REFERENCES

1 Berchtold.S and Kriegel.H.P, 1997, "Similarity Search in CAD Database Systems", pp.562-570.

2 Sharma.S.C, 1978, "A critical study of the classification and coding systems in manufacturing companies", Lehigh University, US.

3 Franklin.S and Graesser.A, 1996, "Is it an Agent, or just a Program? : A Taxonomy for Autonomous Agents", Springer-Verlag, Berlin, Germany.

4 Dura.R, Hart.P and Stork.D, 2001, "Pattern Classification", seconded. J. Wiley, New York.

5 Arman.F and Aggarwal.J, 1993, "Model-based object recognition in dense-range images—A review", pp. 25-43.

6 Loncaric.S, 1998, "A survey of shape analysis techniques: Pattern Recognition", pp. 987–996.

7 Kaperthi.S, 1991, "A neural network system for shape-based classification and coding of rotational parts", *International Journal of Production Research*, vol.29, No.9, pp.1775-1785

8 Capparrelli.F and Rocket.P.I, 1999, "Novel Approach to Nearest Neighbour Search in High Dimensional Spaces for 3D-Object Recognition", *7th International Conference on Image Processing and its Applications*, pp.14-16.

9 Cyberko.G, Bhasin.A and Regli.W.C, 1997, "Pattern Recognition of 3D CAD Objects", *Smart Engineering Systems Design*, vol.1, pp.4-8.

10 Barrow.H.G, Tenenbaum.J.M, Bolles.R.C and Wolf.H.C, 1977, "Parametric correspondence and chamfer matching", *Proceedings of the Fifth International Joint Conference on Artificial Intelligence*, Cambridge, MA, pp.658-663.

11 Borgefors.G, 1988, "Hierarchical chamfer matching: A parametric edge matching algorithm", *IEEE Trans. on PAMI*, vol. 10, No. 6, pp. 849-865.

12 Ankerst.M, Kriegel,H.P and Seidl.T, 1999, "3D Shape Histograms for Similarity Search and Classification in Spatial Database", *SSD'99*, Springer, China.

13 Funkhouser.T, Osada.R, Chazelle.B and Dobkin.D, 2001, "Matching 3D Models with shape Distributions", *International Conference on Shape Modeling*, Institute of Electrical and Electronics Engineers, Italy.

14 Standard Template Library Programmer's Guide, http://www.sgi.com/tech/stl/, [accessed 12/11/2003].

15 Morelli.R, 2003, "Java Java Java Object-Oriented Problem Solving", Upper Saddle River, N.J.: Prentice Hall, US.

Applying Internet technologies to manufacturing SMEs in the UK North East region – barriers and opportunities

R MARASINI, K IONS, I HAMILTON, and **M AHMAD**
B2B Manufacturing Centre, University of Teesside, UK

ABSTRACT

Internet technologies are increasingly being adopted by UK businesses to facilitate collaboration, trade, learn, manage company business processes and deliver services. Large companies are reaping the benefits of such technologies, but few manufacturing small to medium enterprises (SMEs), which account for 97.8% of manufacturing companies in the North East (NE), have the skills, facilities and finance to benefit from its application. It is essential that barriers to innovation be removed to enable SMEs who are low investors in technology to benefit from technology adoption. In collaboration with the University of Teesside and Warwick University and through ONE North East RDA and ERDF funding, the Business-to-Business Manufacturing Centre (B2B-MC) was set up to help improve the regional economy by assisting manufacturing SMEs to apply internet technologies to enable product innovation, learning, improved manufacturing techniques and the delivery of products against ever shortening timescales. This paper reviews the applications and effectiveness of current internet-based initiatives available to SMEs in the NE region and describes the approach developed by the B2B-MC to support technology adoption. The major barriers and opportunities of internet technology adoption and challenges to support SMEs in the region are also presented.

1 BACKGROUND

Internet technologies are increasingly being applied by UK business to facilitate collaboration, trade, learn, manage company business processes and deliver services. More and more companies, particularly OEMs, are using E-Business as a means, primarily, of decreasing costs, improving sales and marketing and managing people. Large companies are reaping the benefits of internet technologies but few manufacturing small to medium enterprises (SMEs), which account for 97.8% of manufacturing companies in the North East (ONS, 2003) have the skills, facilities and finance to benefit from its application.

The North East has a business population of just fewer than 100,000, which accounts for about 10% of all UK businesses (Small Business Service, 2003). In all UK regions, SMEs (0 to 249 employees) make up over 99% of the business population. The North East has an SME base which accounts for 54% of employment and 45% of turnover (Small Business Service, 2003). ONE North East (2002) recognises the regions continued dependence upon traditional manufacturing, currently representing almost a fifth of the region's workforce and contributing a greater proportion of regional GDP than any other region bar one. The DTI (2001) predicts that over the next ten years, SMEs will be the main drivers of economic growth, product innovation and job creation. It is essential that barriers to innovation be removed to enable SMEs who are low investors in technology to benefit from technology adoption. The Tees Valley Joint Strategy Unit (JSU) recognises the need for radical action to accelerate the growth of the electronic, knowledge based economy in the North East to ensure the levels of growth required by the region, and that the development of e-commerce is necessary to achieve this.

In collaboration with the University of Teesside and Warwick University and through ONE North East RDA and ERDF funding, the Business-to-Business Manufacturing Centre (B2B-MC) was set up to support SMEs in the North East (NE) region of the UK. The objective of the centre is to help improve the regional economy by assisting manufacturing SMEs to apply internet technologies to support product innovation, learning, improved manufacturing techniques and the delivery of products against ever shortening timescales. The centre aims to promote adoption of such technology through targeted support and the development and demonstration of technology demonstrators. This paper reviews the applications and effectiveness of current internet-based initiatives available to SMEs in the North East region and describes the approach developed by the B2B-MC to support technology adoption. The major barriers and opportunities of internet technology adoption and a brief introduction to the proposed technology demonstrators will also be presented.

2 B2B MANUFACTURING CENTRE AND OBJECTIVES

Warwick Manufacturing Centre in partnership with the DTI established the National Business to Business Centre in 2002 to support businesses to adopt web technologies that could positively affect their performance (National B2B Centre, 2004). The B2B Manufacturing Centre is establishing a B2B Hub for NE region, which will transfer established and recognised international B2B best practice pioneered by the National B2B Centre to manufacturing SMEs in the North East of England. The hub employs experts in B2B applications who work with regional business clusters, centres of excellence and the region's Business Links to encourage accelerated B2B development within the region. The project aims to improve regional manufacturing effectiveness at SME level, by supporting and expanding the effective use of ICT. Through its activities the project will also contribute to the achievement of two of ONE North East's (2002) 6 key objectives: "Establishing a new Entrepreneurial Culture and "Recognising Universities and Colleges at the Heart of the Region's Economy". The project will deliver a comprehensive range of bespoke solutions to SMEs and will demonstrate documented business benefits. Specific quantifiable objectives in support of the aim, which will be developed and delivered during the projects lifetime, will be to:

- Develop 7 Technology Demonstrators (Collaborative Product Development, E-trading, Supply chain, Digital manufacturing, Logistics and Distribution, Data Exchange and Visualisation).
- Assist 70 companies in the adoption of B2B Technologies.
- Create 26 jobs, safeguard 50 jobs and establish 6 new business incubators.
- Deliver 110 learning opportunities and achieve 3000 website 'hits'.
- Develop and share B2B best practices.

3 OPPORTUNITIES IN NORTH EAST AND CURRENT INITIATIVES

3.1 Opportunities

The North East has a strong industrial base that includes diverse manufacturing and process industries. The Tees Valley Joint Strategy Unit (2000a) identifies 15 active sectors in the Tees Valley - manufacturing (chemicals, offshore, engineering, iron and steel, electronics, food and drink, textiles and clothing), construction, and services (business and financial services, public administration, the voluntary and community sector, health and care, tourism and leisure, retail and freight support). The JSU emphasises that further investment in IT and support is required in fabrication and engineering, business and financial services, and the delivery of services for public administration. In addition to the SMEs in offshore engineering, as in many other industries, are not in a position to make necessary investment and require individual support and guidance to ensure that their efforts are effective. Supply chain development has been emphasised for food and drink industry, electronics sector and adoption of new technologies in textiles and clothing sector. ICT development and support in the NE region will provide employment, business survival and competitive advantage among other UK regions.

The regional ICT strategy board (2003) believes supply, demand and skills as 3 key areas to address in order that the region develops as an E-Competent region quickly and effectively. The supply means ensuring the right infrastructure exists; the demand means ensuring there are enough users for the infrastructure; and skills means people, business and organizations need to know how to use the tools provided by the supply and demand programs. These, in turn, should provide access to global market opportunities, access to information, services, employment and education opportunities, exploitation of knowledge for all, ICT skilled force and underpin collaborative working and cluster development and enhance the potential for individuals and communities.

3.2 Initiatives in the NE region

The UK Government has given priority to take advantage of the 'broadband Revolution' and wants every citizen to be able to take advantage of the Internet and Broadband by 2005. Regional differences in SME online access appear to persist in the UK (Dixon et al. 2002).

The North East Regional Portal n-e-life.com (www.n-e-life.com) is a major initiative in North East, which seeks to be the electronic definition of the region. The Portal is joint venture between ONE NorthEast and the private sector. The Portal allows businesses and lifestyle organisations to promote their products and services. In line with existing government policy it seeks, with UK online (http://www.ukonline.gov.uk) to show how the Internet and web-based technologies can be brought to individuals and businesses in the North East of England. It has a major role in helping businesses to embrace the Internet and to date 66,000 businesses are present on the Regional Portal.

There are following other main initiatives in the North East region that are aiming to remove the barriers to innovation and enable SMEs who are low investors in technology to benefit from the adoption of ICTs without financial risk.

3.2.1 RCAE Regional Computer Aided Engineering Project

The RCAE project is a joint venture between Durham, Northumbria and Teesside Universities supported by the ERDF. The partners provide specialist assistance in the provision of, initial exploration, analysis and implementation of CAE and related solutions.

3.2.2 RCID - Newcastle University - engineering design, analysis

The RCID provides advice and assistance on engineering design issues. The Centre has been jointly funded through ERDF, the University of Newcastle, and contributions from regional growth-orientated businesses.

3.2.3 Centre for Design Research (CfDR) - University of Northumbria

CfDR is an academic and professional partnership within Northumbria University to provide a research-led design service to industry, commerce and the public sector in industrial design, interaction design, conceptual design research, innovation management, medical product design, medical design research, new media and interface design, design learning

3.2.4 Centre for Rapid Product Development – University of Northumbria

The Centre was established by the University of Northumbria with assistance from the ERDF, to provide an integrated design, analysis, test, prototype production and manufacturing service in-house and raise awareness of Rapid product development.

3.2.5 Agility Group - Durham University

The Agility Group defines itself as engaging a company in "Manufacturing Knowledge & Technology Transfer" The Group is composed of three strands: Process and Factory Improvement, Business Strategy and Logistics, Digital enterprise Technology.

3.2.6 Productivity North East (PNE)– Durham University

PNE is an employer-driven network, which represents industry within the region. It organises seminars, conferences, company visits, a benchmarking club and other networking events. PNE gives access to best practices available in Manufacturing and Quality; Health and Safety; Engineering and Maintenance, etc.

3.2.7 Manufacturing Challenge - general manufacturing support including cluster clubs

The Manufacturing Challenge offers assistance to manufacturers to enable them to be more competitive achieved through a network of business mentors sharing knowledge and best practice; informative events; Partnership links with business support organisations and educational institutions

3.2.8 RTC North - technology transfer, business networks, houses RCME

RTC North helps clients to exploit new opportunities and improve competitiveness through the application of technology, and knowledge. RTC North manages the Regional Centre for Manufacturing Excellence (RCME). The RCME offers free services for diagnosing, analysing and finding solutions to any manufacturing related issue.

3.2.9 Cleveland Innovation

Cleveland Innovation provides product innovation support. It has industrial, product and engineering designers. Cleveland Innovation is a key member of the North East Technology Support (NETS) partnership, which provides European-funded technology support to SMEs.

3.2.10 AMAP - Sunderland University - CAD design & CAE training, ergonomics

AMAP is part of the School of Computing & Technology within the University of Sunderland. AMAP provides problem-solving solutions to manufacturers, offering training or advice, or working in partnership-based research.

3.2.11 Digital Factory - Bishop Auckland College - RP, CAD

Digital Factory assists SMEs with the selection of suitable CAD technology, new product development, rapid prototyping and reverse engineering.

3.2.12 Tees Valley Engineering Partnership

TVEP is a partnership of senior executives from across the engineering industry, and the educational support organisations, which are working to address the potential threats and opportunities for industry.

It is obvious that there are several initiatives in NE region that are targeted to support businesses in manufacturing sector. Close co-ordination between the initiatives is very crucial to minimise overlap of scope and better delivery of the objectives in the region.

4 B2B-MC METHODOLOGIES

The B2B-MC has adopted various state-of -the-art techniques to evaluate the IT status of SMEs, identify requirements and solutions, build and facilitate partnerships and design and implement solutions. The methodology used to support SMEs is as following:

4.1 Contact with SMEs

SMEs profiled by NE-Life were contacted via a variety of means including snail mail, e-mails, e-shots and telephone calls. An ongoing on-site visit programme is proving successful - 29% of SMEs who have been visited thus far have accepted the assistance of the B2B-MC. In addition to this, a series of awareness events to showcase the application of internet technology to a particular problem have been planned. One such event, (real-time collaborative product development) revealed that out of 1400 SMEs mailed about 5% of companies responded and attended the event. Feedback from SMEs regarding such events has thus far been very encouraging.

4.2 Supporting SMEs

The B2B-MC has adapted a model to support SMEs, which is presented in Figure 1. The methodology follows the six-sigma approach that began in the 1980's at Motorola. Two main six Sigma models (Six Sigma, 2004): Define, Measure, Analyse, Improve and Control (DMAIC) and Design, Measure, Analyse, Develop and Verify (DMADV) are utilised. The DMAIC model, which is used for improving existing business processes is practised by the SME and the DMADV, which is used to develop new products, services and processes, is adapted by B2B-MC team. Depending upon the SME requirement, either model is used to support SMEs.

4.2.1 Initial Visit

During an initial visit, centre specialists elicit data about company processes and identify areas for improvement. The output of this process is an initial visit report.

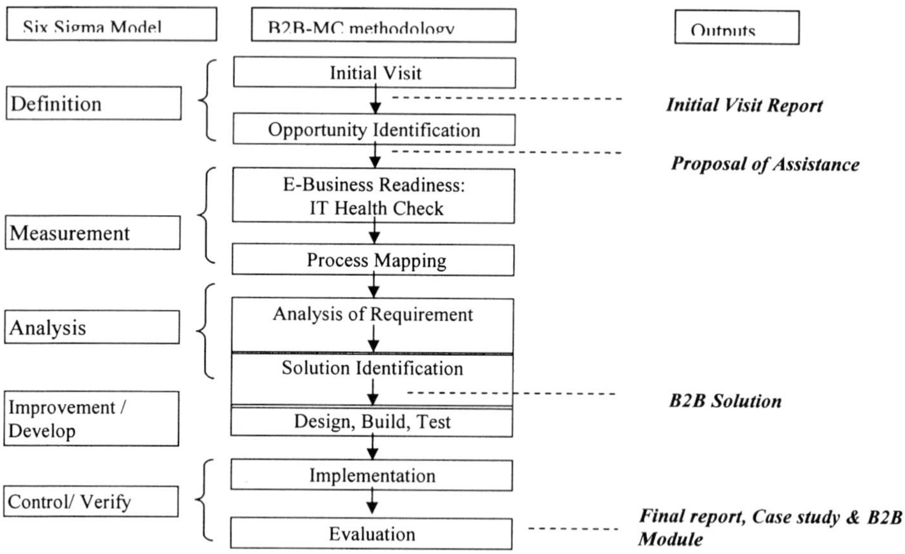

Figure 1: Methodology used to support SMEs (adapted from Business plan, B2B-MC, 2003)

4.2.2 Process mapping, IT health check and analysis

This involves, the submission of a proposal of assistance to the SME which if agreed to is followed by business process mapping and an IT health check. The objective of the process mapping is to track existing process, bottlenecks and areas where ICT tools can be used to improve the business performance. The IT health check developed by Jones (2000) as part of a European Social Funded (ESF) project operated by the University of Teesside, is carried out by completing a standard comprehensive questionnaire. The results obtained from analysis provide the classification of IT level of the company and guidelines for improvements and progression to the next level in the hierarchy. The results of analysis also assist the B2B-MC team to select appropriate solutions for the company. Figure 2 shows the 5 IT capability levels that can be identified by the IT health-check. The model is based on the Capability Maturity Model (CMM) developed by Carnegie Mellon University.

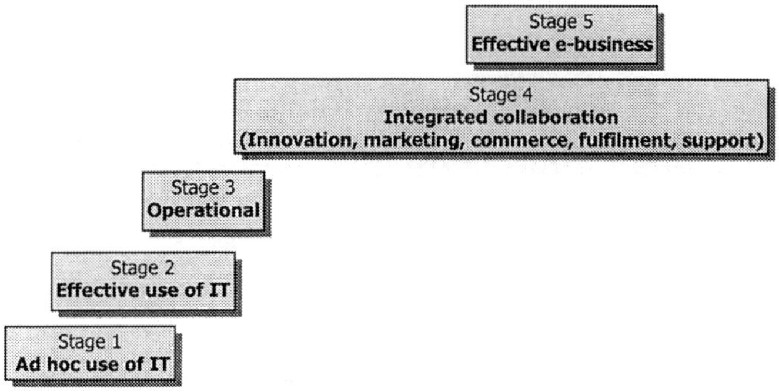

Figure 2: Capabilities of SMEs (Adapted from Paulk *et. al.* 1995)

4.2.3 Design, Build and Test

In agreement with the SME and the solution provider, the process improvement program is designed and IT system if appropriate is developed and tested.

4.2.4 Implementation

Upon successful testing, agreed by the solution provider and the company, the IT system is implemented in the company.

4.2.5 Evaluation

In the final phase, the IT implementation and its performance are evaluated and further actions are suggested if appropriate. The evaluation includes the analysis of number of jobs created and /or saved, benefit to the turnover and environmental performance in terms of using less paper, less travel etc. The project is recorded as a best practice example or success factors for future use. A final report is also produced highlighting the success and best practice guidelines. Course modules are developed based on case studies, for undergraduate and postgraduate courses.

4.3 Awareness events, demonstrators, and learning events

Awareness events are organised to showcase the application of new technologies. The main aim of these events is to build the confidence of SMEs. It is anticipated that the events and support provided to SMEs will build a ladder of IT development enabling them to move up the adoption ladder.

5 FINDINGS IN NORTH EAST REGION

This section presents the findings of the North East region. 17 companies have been profiled to-date through semi-structured interviews and then cross-tabulating their IT capabilities. Figure 3 shows the status of the companies in using ICTs and Figure 4 the size bands of companies interviewed.

5.1 Adoption of ICTs

The DTI (2003, p.14) suggest, "There is clear evidence of an emerging digital divide between smaller and larger businesses". However, the cluster of manufacturing SMEs interviewed, despite including a majority of micro businesses seem to be bucking this trend - at least in terms of internet connectivity and adoption of basic ICTs. For example, internet access is 100% compared to 87% for all NE businesses and 90% for all UK businesses. Additionally, broadband access, having a website and using external email all compare favourably with national and regional figures. The most probable explanation for this is that it is that those companies who have thus far responded to the marketing activities of the B2B project already have a positive attitude to ICT adoption and are keen to move up the adoption ladder.

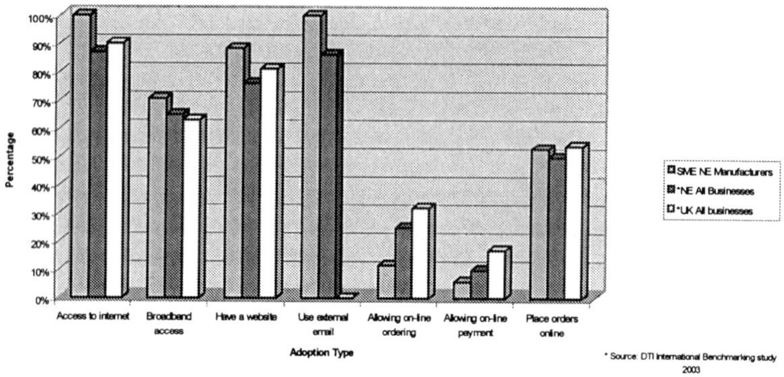

Fig. 3 Adoption of ICTs by SMEs in NE

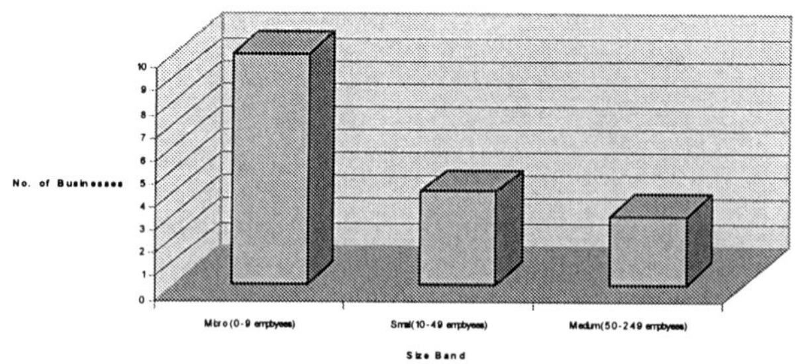

Fig. 4: Size bands of SMEs in contact with B2B-MC

In terms of electronic trading, it is clear that in line with all UK and NE businesses (DTI, 2003) NE manufacturing SMEs are more likely to place orders online than enable ordering and/or payments for their products or services. The most commonly cited reason SME manufacturers gave for not enabling online ordering was that their products or services were simply not suitable. There are clearly two challenges here; firstly, how to engage the significant number of SMEs who do not respond to ICT adoption initiatives and secondly how to encourage SMEs move up the adoption ladder.

5.2 Barriers to e-Business and internet technologies

There are barriers to promote internet technologies and e-business activities for innovation and manufacturing growth in the NE region, some of the barriers are similar to those highlighted for UK SMEs in Dixon et al. 2002. The NE has a relatively low level of R&D activity, technology transfer and innovation and the overall level of expenditure on R&D is low in the region as a whole (Tees Valley Joint Strategy Unit, 2000b). There is also a lack of awareness of the potential of ICT, the belief that ecommerce is not applicable to SMEs type of products or services and fears over set up costs and security (ONS, 2003). Many SMEs are reluctant to introduce technology, especially the Internet, (Tees Valley Joint strategy Unit, 2000c) and consequently the use of the web by SMEs in the region is still relatively undeveloped. It is difficult to convince SMEs to find innovative uses for ICTs for several reasons:

5.2.1 Skills

The region has a very weak skills base, not only in terms of hi-tech skills but in basic skills with major problems in basic education (numeracy and literacy) skills, as well as computer and IT literacy (Tees Valley Joint Strategy Unit, 2000b). There is therefore, often the lack of staff to implement ICT solutions effectively.

5.2.2 A lack of awareness of the potential of ICT

There is a belief that e-commerce is not applicable to SMEs type of products or services.

5.2.3 High initial set-up costs

The belief that there is considerable financial risk in setting up ICT solutions- even simple websites. The smaller the internal resource base the less likely the business is to use ICT.

5.2.4 Security/privacy issues

Despite the fact that the "B2B" SME community, is a large potential user of electronic money systems there is some suspicion about the robustness of such systems.

5.2.5 Fear of IT and reluctance to change

Many SMEs were found to have fear of using ICT systems and some sort of reluctance to use new technologies thinking that it will be very difficult to adapt in their business.

5.2.6 Trust and quality of support to the companies

There are several initiatives in the region with overlapping objectives, thus leading to competition between various organisations and projects. This has led to a negative perception amongst some SMEs who site "initiative overload" as a barrier to engagement with technology transfer initiatives. Our main concern lies in building the trust and confidence upon the quality of the service provided by the B2B-MC. Liaison between regional bodies, the DTI and the region's universities is crucial to achieve this.

5.2.7 Ownership of data

The problems of ownership of data and exposing the business to some other commercial organisation are seen to be a major concern among SMEs.

5.2.8 Commitment of time by the SMEs,

Many IT systems not suitable to SMEs and many SMEs run their business in adhoc basis, there is lack of commitment for time that SMEs spent in research and development activities or to investigate their business potential.

6 CONCLUSIONS

This paper has presented a methodology applied by the B2B-MC methodology to support manufacturing SMEs in the North East region of UK. The UK North-East scenario was presented and the barriers and opportunities for the adoption of B2B internet technologies were discussed. The main challenge is to engage the significant number of SMEs who do not respond to ICT adoption initiatives and to build the trust between the concerned parties. Increased awareness of the available internet technologies and development of low cost solutions tailored to the need of SME is very crucial for the competitive business benefit in the NE region and in the UK.

ACKNOWLEDGEMENTS

The authors would like to acknowledge the funding from ONE North East RDA and ERDF and support from the B2B Manufacturing centre staff in University of Teesside and Warwick Manufacturing Group, Warwick University.

REFERENCES

1. Business Plan (2003), B2B Manufacturing Centre, University of Teesside.
2. Dixon, T., Thompson, B. and McAllister, P. (2002) “The value of ICT for SMEs in UK: A critical literature review”, Report for Small Business Research Programme.
3. DTI (2002) “Financing the enterprise Society: Financial Services for Small and Mid-sized Enterprises in 2010”
4. DTI (2003) “Business in the Information Age – International Benchmarking Study”
5. Jones, A. (2000) “IT health check”, School of Computing, Uinversity of Teesside
6. National B2B Centre (2004) www.nb2bc.co.uk/ accessed 15 April 2004
7. Office for National statistics (2003) “Commerce, Energy and Industry – Size Analysis of United Kingdom Businesses”
8. ONE North East (2002) “Realising our potential: The Regional Economic Strategy for the North East of England”
9. ONE North East (2002) 2Manufacturing in the Regions: North East”
10. Paulk, C.M., Weber, C.V.,Curtis, B. and Chrissis, M.B. (1995) “The Capability Maturity Model: Guidelines for Improving the Software Process. Addison-Wesley Publishing Company”
11. Six Sigma (2004) “Six Sigma DMAIC roadmap” http://www.isixsigma.com/library/content/c020617a.asp , accessed 15 April 2004.
12. Small Business Service (2003) “Press release - Small and Medium Enterprises (SME) Statistics for the Regions 2001” http://www.sbs.gov.uk/press/news119.php, accessed 15 April 2004.
13. Tees Valley Joint Strategy Unit (2000a) “Technical Report 4: Sectoral analyses and strategies”
14. Tees Valley Joint Strategy Unit (2000b) “Technical Report 5: Social exclusion in the Tees Valley”
15. Tees Valley Joint Strategy Unit (2000c) “Technical Report 6: Enhancing the competitiveness of the Tees Valley”.
16. Tees Valley Joint Strategy Unit (2004), “Developing a Local Electronic Economy Project in the Tees Valley”, http://www.teesvalley-jsu.gov.uk/leep.htm, accessed April 15, 2004.
17. The Regional ICT Strategy Board (2003) “Towards an eRegion v.3.0. Strategic electronic development of the North East of England”

Components of design for the superplastic forming process

G McCORMICK, P BROOMHEAD, J KULON, and **D J MYNORS**
Design and Systems Engineering Department, Brunel University, Uxbridge, UK

ABSTRACT

The superplastic forming process is relatively new, and often under-exploited as designers tend to design to the limitations of the older longer established processes. To achieve the full potential of superplastic forming, designers must have access to easily available design tools to enable them to exploit the process. This paper examines the potential of elements of a design tool being developed to support designers in the adoption of superplastic forming. The design tool utilises web based technologies and has the potential to link complex representations of the behaviour of the workpiece being formed with design rules and process simulation packages.

1 INTRODUCTION

Metal forming is a collection of technologies with a long history and a great deal of accumulated experience. As a result the scope and limitations of most processes are well understood. There are some formalised design rules, such as those for sheet metal forming (1), hot forging (2) and die design (3). There is equally a great deal of experience in choosing the appropriate material and process for a component, and in designing the component within the process and material capabilities, and in using the full scope of these process capabilities. It is only this accumulated experience which makes some of the choices seem obvious. The true difficulty and subtlety of the designer's job becomes more apparent when a new product is envisaged: the design of the product cannot be considered independently of the choice of material and process, and vice versa.

Traditionally designs and technology have co-evolved incrementally. New technology is enough of a risk by itself, so it is initially used only to produce better versions of an old design. Only later is the design modified step by step to take greater advantage of the technology, while the technology itself evolves and produces further new possibilities. A good example is the history of the motor car. This was originally known as the horseless carriage, and it was literally a carriage stuck on top of an engine. Then the engine and carriage were integrated with a chassis, but the carriage legacy could still be seen in the trunk strapped to the back. Then that too was integrated, but it still remained as a partitioned rear luggage space. It was only much later that hatchbacks with integrated seating, luggage space and rear access developed. The same progression can be seen with signalling. Even as late as the 1950s cars commonly had semaphore style signals which were extended and withdrawn electrically in imitation of hand signals.

Superplasticity was first described in 1912 (4) and has been heavily researched since the 1960s. It arises when the material's strain hardening exponent *m* is greater than about 0.3. Under these conditions necking is limited, and when m=1 there is no necking and the material behaves like a perfectly Newtonian viscous fluid. However superplasticity is not in fact a new invention: ancient arsenitic bronzes (2500 BC) may have been superplastic and Damascus steel for sword making (350 BC) may have had superplastic properties. Superplasticity is not confined to metals, and it is possible that some parts of the earth's deep layers are superplastic (5).

Superplastic alloys are now commercially available that allow strains of several hundred percent to be achieved within certain controlled ranges of temperature and strain rate. Superplastic forming (SPF) is a relatively new metal forming process (5) that takes advantage of the high strains which can be achieved with these alloys. It has allowed the creation of complex 3D profiled parts in near net shape from sheet metal, with excellent finish and dimensional accuracy, and without joins. Typical benefits are increased strength and reduced waste, component count and weight. Also, because stresses are low and most dies are one-sided, SPF die production is relatively cheap and easy, which in turn may allow reduced time to market.

Examples of SPF application include:

- Complex body panels and wings with an excellent surface finish for low volume car production.
- Titanium heat exchangers for undersea oilfield use.
- Hollow fan blades in advanced gas turbines.
- Sets of kitchen pans.
- The Eurofighter foreplane.
- An aluminium enclosure for mobile phone base stations.

As SPF is a new technology design rules are not well established. The process is still evolving and looking at it from an historical perspective, we suggest that it is often not being used to its full potential. In the next section we indicate design options which may often be underused with SPF. In section 3 the limitations of current design resources are indicated, and in section 4 a design tool which addresses the specific needs of design with SPF is described. This design tool offers a modular framework which gives scope for progressive development and enhancement. Some potential enhancements to the tool are discussed in section 5, and conclusions are summarised in section 6.

2 UNREALISED POTENTIAL OF SPF

SPF is a new technology. Would it be too strong to say that relative to the full potential of this technology we are metaphorically clamping a carriage on top of an engine, or imitating hand signals? In the nature of things we cannot say how designs will evolve before designers have actually innovated. Moreover it would be invidious to criticise specific designs, but we can readily list some specific attributes of SPF that are often not used to their full potential. A fuller list of suggestions is available via the design tool described in section 4.

- SPF allows strains of several hundred percent, but strains typically used are of the order of 100% and often less.
- With SPF it may be possible to reduce thickness and radii to a degree that would be uncomfortable with conventional designs.
- Complex 3D shapes are readily achieved with SPF but are not easy to design.
- Extra strengthening features such as ribs and domes could often be added with next to no extra production cost.
- Re-entrant shapes and trapped parts are possible but uncommon.
- Laminated and multi-sheet SPF parts give light-weightedness, impact resistance, thermal control and fine local control of properties, but at present are used in a very small range of products.
- Useful variations of product thickness can be achieved with pre-forming or with some of the more sophisticated SPF methods.

This is not to denigrate the work of designers: some SPF techniques are not straightforward and there are specific difficulties with design for SPF. For instance it is not practical simply to specify limits on radii: the allowable radius depends on the local gauge, but that in turn depends on the shape of the die and the pressure/temperature profile used. Finite element simulation may be used to check designs, but the full constitutive equations may need to take account of dynamic grain growth, so the simulation can be rather specialised and very slow, which limits the amount of design exploration which is practical. Furthermore there are many variations on the SPF process, and many closely linked parameter choices.

3 LIMITATIONS OF AVAILABLE RESOURCES

Consideration of the current state of SPF designs led the authors to examine the issues that need to be addressed in order to design with SPF, and other issues which need to be tackled to make design with SPF easier.

It was found that the number of superplastic materials known to science is large, but there are a smaller but still substantial number of commercial alloys. Unfortunately standard databases and datasheets do not detail superplastic parameters. Some relevant properties can be found from scientific papers, but it is not always clear whether the alloy described has had the same treatment as readily available alloys of the same name. Superplastic parameters depend strongly on the specific treatment given to the alloy, and may vary from batch to batch. There is therefore a need for develop a database of SPF material properties. This should ideally store both generic and batch specific data.

Many SPF products are built using a small range of processes. In fact we encountered not only a wide range of processes, but an equally wide range of tooling and control options which can be applied to these processes.

It is common to check SPF designs using finite element analysis. Packages typically used include ABAQUS (6) and MARC (7). These are general purpose packages. There is not yet any truly specialised SPF package, nor is there any equivalent of the very quick approximate calculations available in specialised sheet metal packages, such as Fastform (8) and Autoform Onestep (9). It is possible that sheet metal packages will be enhanced to include basic SPF, (Pam-stamp 2G (10) offers such a facility) but considering the wide variety of SPF processes we encountered there seems to be a gap in the market.

While the difficulty of using SPF to its full potential can be reduced by addressing each of these issues separately, we found that the three issues of design, material choice, and process choice are in reality so closely linked that if possible they should be considered jointly. Separate design steps may require separate types of calculations, but designers need to go backwards and forwards through these steps or round and round them. An integrated tool is called for.

4 THE DESIGN TOOL

In order to promote the use of SPF to its full capability a prototype web based design tool has been developed. This is available at http://www.brunel.ac.uk/research/forming/Spfguide/. A very simple and easily understandable framework has been devised. The design processes is represented as a series of steps, and each step is implemented using a module of software. Different modules can then be developed or enhanced according to need and opportunity. This design tool has the potential to link complex representations of the behaviour of the workpiece being formed with design rules and process simulation packages.

The design/material choice/process choice interaction is one where everything seems to depend on everything else. For instance with some materials back pressure should be used to avoid cavitation, while other materials which cavitate less readily may allow a simpler process but require a more expensive die material, yet have a different maximum allowable strain. In the design tool this interaction is dealt with by iteration. Choices are narrowed then explored in greater detail, after which the user may backtrack or cycle through the steps again as long as desired. Control remains in the hand of the user. The design cycle modelled in the tool is shown in figure 1 below. The thick solid lines show the designer's path through the process. Dashed lines show key information flows.

At present the design revision stage is implemented as a checklist, which prompts the designer to compare the proposed design with options and possibilities that are available via SPF but not always considered. Extra detail about some of the suggestions is available via hyperlinks.

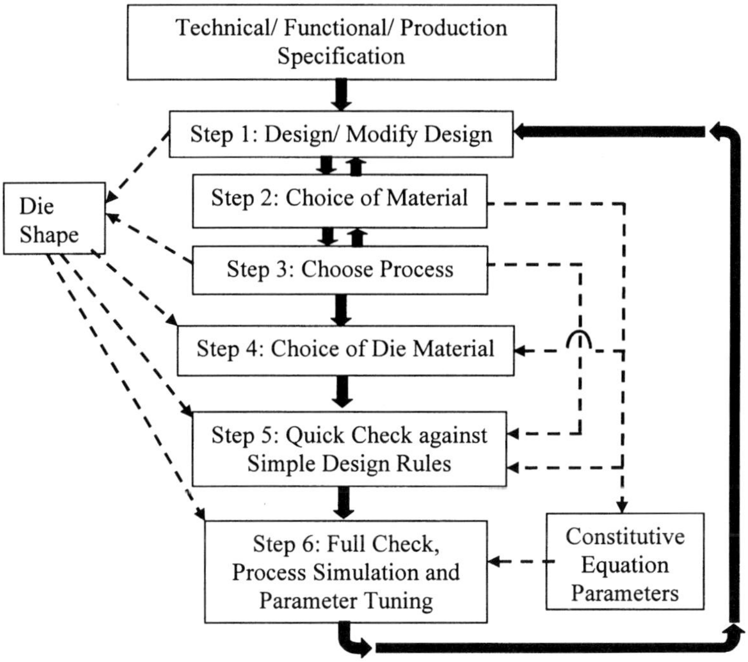

Fig. 1 The SPF design cycle, as implemented in the design tool.

Material choice is made in two stages. In the first stage material capabilities are specified. Requirement could for instance be for a cryogenic material with high strength to weight ratio. A database of SPF materials is examined and a shortlist shown to the user. The user selects one material from the shortlist, and is given a summary of the chosen material's properties. The material choice can be revised before moving on to the next stage. An example of this is shown in figure 2 below.

Although there are two or three commonly used SPF processes, many variations exist. The user is given a list of processes and the opportunity (via hyperlinks) to see explanations of each, and their particular benefits. A process must be chosen before moving on to the next stage, though of course the choice may be revised later.

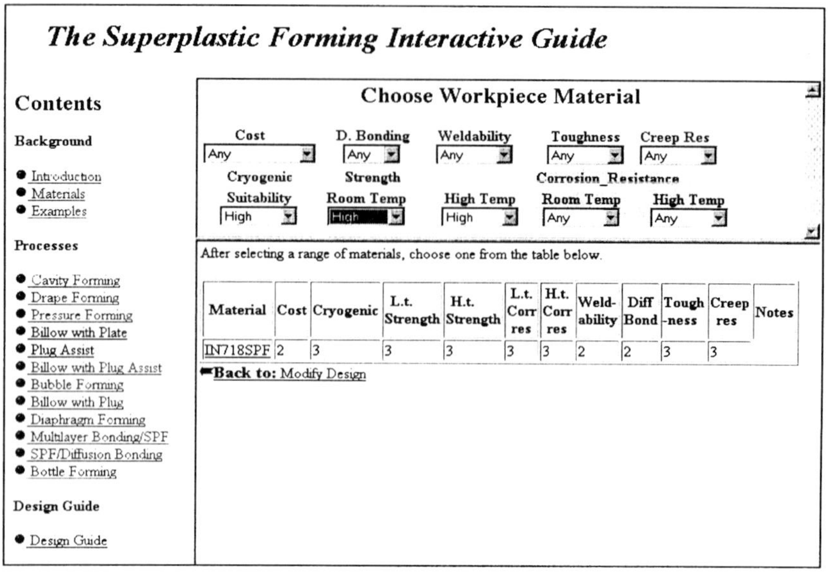

Fig. 2 Design tool, choice of workpiece material

The choice of die material depends on the workpiece material (for instance graphite would be a dangerous choice with titanium) and also on the degree of precision required and the production quantity. The user is given a choice of the three cheapest materials which can meet the requirements, (see figure 3).

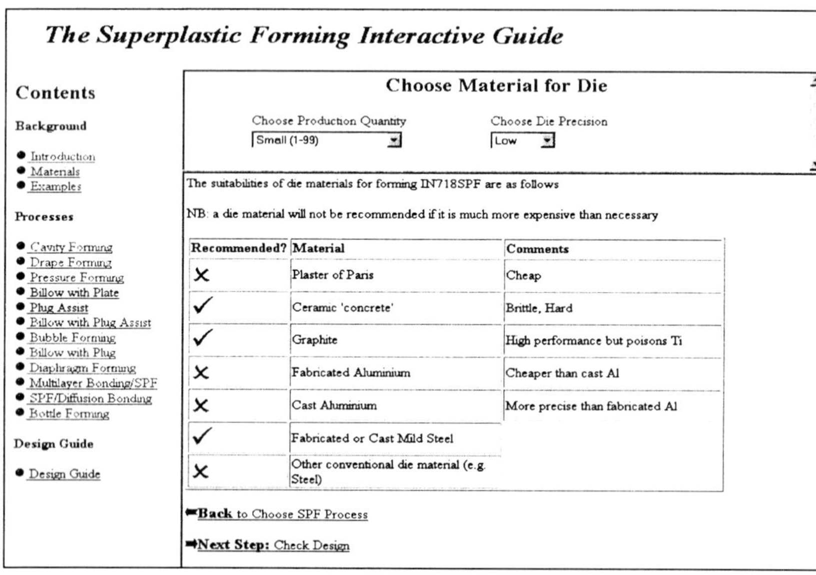

Fig. 3 Design tool, choice of die material

At this stage some approximate rules appropriate to the workpiece material may be used to check the design. A particular issue with SPF is that the thickness profile of the product depends strongly on the shape of the product, as well as the process and parameters chosen, but most of the design rules are relative to the expected local gauge of the product. An estimate must be made of the final thickness profile before checking the design rules.

If the simple check looks promising a more time consuming finite element simulation may be undertaken. The parameters for this simulation are made available from the database of material properties. At present the simulation must be conducted externally to the design tool.

5 THE DESIGN TOOL'S POTENTIAL.

Although basic at present, the above design tool captures essential features of design for SPF, and has the potential to link complex representations of the behaviour of the workpiece being formed with design rules and process simulation packages. Functionality can be enhanced module by module.

5.1 Database Security

We are aware that material properties data, batch specific data, constitutive equations, and design rules for specific alloys may be subject to commercial confidentiality. There is no difficulty with partitioning such data into publicly and privately accessible portions, as the data the rules and the shell have been kept quite distinct. The database may also store alternative versions of the constitutive equations as some versions are more accurate than others within a limited range, and others are more general but more approximate.

5.2 Quick Check of Design Rules

The design rules in the database are at present based on (11). Design rules for specific alloys are required. Although design rules are usually derived from experience, there is also potential, with SPF, for developing design rules by systematic exploration of design parameters using finite element simulation. At present calculations are simple and approximate. It will be possible to increase the accuracy of these calculations without sacrificing speed by importing data in STEP format (12) and using a geometric approach to estimate final product thickness profiles. Alternatively it should be possible to derive approximate profiles by assuming perfect viscous behaviour, perfect contact friction, and using a membrane approximation. Such calculations might be performed by the web server, or by a machine linked to the web server.

An area which has yet to be explored is the incorporation of design rules specific to a type of product.

5.3 Finite Element Simulation.

The database, which at present stores generic constitutive equation parameters, could quite easily store alternative equations to suit different circumstances or different simulation packages (some packages require the equations to be monotonic, some do not). It would be very simple for the data sent to the design tool to indicate the package to be used and for the design tool to choose the most appropriate equations and communicate them back using extended markup language (XML).

Since the quick check is envisaged to be based on data in STEP format it would be possible to go further, and optionally perform the finite element calculations under the direction of the design tool, using the most appropriate mix of software and computing power, including grid based parallel computing networks.

5.4 Process Optimisation

Process optimisation attempts to improve products by adjusting process parameters. Again, since data would be available in STEP format this optimisation could usefully be controlled by the design tool. A wide range of possibilities can be envisaged.

The simplest option is to optimise process temperatures and the pressure profile used. Process temperatures are usually fixed, and many of the finite element packages used for SPF already have enhancements which will predict effective pressure profiles.

The next step beyond this might be to optimise slightly more complex processes such as billow with plate, or plug assist. Here parameters such as the plug or plate profiles could be chosen to achieve desired product thickness profiles. This optimisation would depend on either having a very efficient approximate method, or access to a fast (possibly parallel) finite element server.

At a similar level of complexity, but requiring very accurate constitutive models, would be optimisation of the less common but potentially very effective non isothermal or pulsed pressure SPF processes.

5.5 Design Review and Design Optimisation

The design review module implemented in the design tool is currently in a checklist format. For specific product areas the formatted data could be analysed and critiqued automatically using artificial intelligence techniques.

At a more challenging level, part design could be optimised to meet a functional specification subject to SPF design rules. Even more useful would be joint optimisation of part design, process choice and process parameters. Use of parallel processing would almost certainly be essential, and specialised modules for particular sorts of design might be appropriate. This would be several steps beyond current routine practice. However all the individual elements of such a scheme already exist. For example:

- The database technology required is already routine: only the data presents a problem.
- Some general-purpose finite element packages can effectively simulate SPF, given sufficient understanding and care.
- XML, STEP and network communication technologies are now widely available.
- Grids of fairly ordinary computers linked in parallel are already available. With the right software they can be hundreds of times faster than any desktop PC or workstation.
- Metaheuristic optimization methods such as simulated annealing have for years been the norm for designing sets of complex circuit boards where the number of possibilities to consider is in the billions of billions (13).

6 CONCLUSIONS AND SUMMARY

Superplastic forming (SPF) is a relatively new metal forming technology which can allow the creation of complex 3D profiled parts with excellent finish and dimensional accuracy, and without joins. Many uses of SPF are somewhat conservative: the full potential of this technology is rarely exploited.

Designing with the full potential of SPF must take into account not only the many varieties of SPF process and tooling, but the way that design materials and process cannot be considered independently. To aid this process a prototype web based design tool has been built. This web tool has a simple and easily understood structure. Each part of the structure of the tool is capable of enhancement, so that it is potentially capable of linking complex representations of the behaviour of the workpiece being formed with design rules and process simulation packages.

Full exploitation of this tool will require an enhanced database of SPF alloy properties, sets of alloy-specific design rules, and a rapid method of calculating expected product thickness profiles. The tool also has the potential to coordinate sophisticated design reviews, and design or process optimization. Integration of such capabilities would greatly enhance the abilities of designers to derive benefit from SPF. The basic technologies for these enhancements already exist and it can confidently be expected that integration of these options can be realised. This integration will prove a significant step to fuller exploitation of the potential of SPF.

REFERENCES

1 Lundh, H., Bustad, P.A., Carlsson, B., Engberg, G., Gustafsson, L. and Lidgren, R. (1998) "Sheet Steel Forming Handbook: Size Shearing and Plastic Forming", SSAB Tunnplat AB, Borlange Sweden

2 Sheridan, S.A. (1972) "Forging Design Handbook", ed P.U. Unterweiser. American Society for Metals

3 Paquin, J.R. and Crowley, R.E. (1987) "Die Design Fundamentals", (2nd Edition) Industrial Press Inc, New York

4 Bengough, G.D. (1912) "A Study of the Properties of Alloys at High temperatures", *J. Inst. Metals*, Vol. 7, pp 123 - 174

5 Nieh, T.G., Wadsworth, J. and Sherby, O.D. (1997) "Superplasticity in Metals and Ceramics", Cambridge University Press

6 Hibbit, Karlsson and Sorensen Inc. (2004) "ABAQUS User's Manual", http://www.abaqus.com.

7 MSC Software Corporation (2003) "MSC.Marc 2003 User's Guide", http://www.marc.com

8 http://www.forming.com/fastform.htm/brochures/fastform.pdf (2004)

9 http://www.autoform.com/products/product_onestep.html (2004)

10 http://www.esi-group.com/SimulationSoftware/Stamping_process/features.html/view (2004)

11 Padmanabhan, K.A. and Davies, G.J. (1980) "Superplasticity", Mechanical and Structural Aspects, Environmental Effects, Fundamentals and Applications. Springer Verlag (Berlin, Heidelberg, New York) p267.

12 ISO10303, "Industrial Automation Systems and Integration", ISO, Geneva, Switzerland. (Multiple documents, 1999-2004).

13 Sechen, C. and Sangiovanni-Vincentelli, A. (April 1985) "The Timberwolf Placement and Routing Package", *IEEE J. of Solid-State Circuits*, Vol SC-20, No 2, pp 510 – 522.

Building a web-based coupling quotation system in mechanical engineering

M RIAZ and **J LU**
School of Computing and Engineering, University of Huddersfield, UK

ABSTRACT

Traditional quotation systems in most industries are paper-based system, which is expensive and time consuming. An economical secure and reliable system is required with the highest performance. This research is to reach a solution to expedite this process and aims to make a positive difference to the mechanical and manufacturing industry. A web-based database system has been created for ordering and generating non-standard 2D designs for flexible couplings via a series of predefined options and variable dimensional information. Flexible couplings measure output torque, transmit power, compensate for end movement and also protect equipment from overload. Evolutionary Prototyping Lifecycle Model - EPLM, Database Management systems - DBMS, and Client-Sever systems - CSS are implemented in this investigation. It is found that the system eliminated the attention of the drawing office for all couplings. The system would assist some industrial sales and engineering teams worldwide, by providing customers with up to date information and subsequently offering them the highest quality of service. It concluded that a reduction in lead-time (the amount of time, defined by the supplier that is required to meet a customer request or demand [5]) and engineering designs costs have been achieved from this investigation.

1 INTRODUCTION

All non-standard product designs require the attention of the engineering drawing department. For example, at Textron Power Transmission, the time taken for each non-standard drawing is approximately 2-3 hrs. However due to the increasing demand for non-standard components a drawing could take up to 2-3 Weeks. A software solution is required to expedite this process.

In engineering design, couplings connect two pieces of rotating equipment, whilst allowing some degree of misalignment or end movement [5]. Figs 1a and 1b show a photograph and a drawing of a coupling. Couplings measure output torque, transmit power, compensate for end movement and also protect equipment from overload.

a.

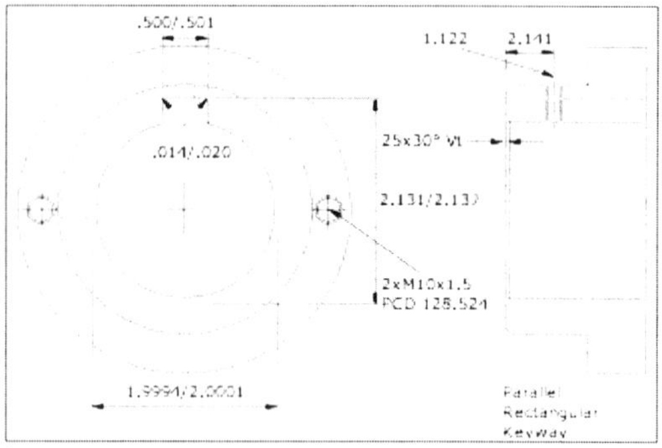

b.

Figure 1 Photograph (a) and drawing (b) of a Coupling [8-9]

To develop a database system to configure standard/non-standard couplings, generate drawings and place orders would alleviate the attention of the sales department and eliminate the use of the engineering drawing department. This research may assist industries worldwide by making a positive difference to the supply of drive solutions, and by providing customers with up to date information and subsequently offering them the highest quality of service. A reduction in lead-time and engineering design costs are some of the benefits from this study.

Most database systems especially those related to mechanical engineering are rarely extended to generate drawings. This is probably because research is usually carried out in how to design a database system and implement it as a knowledge base system for mechanical engineer designers. An example of this is in a paper published by Ron C.J.A.M. Broens and Marc J.de Vries [6]. In this paper the authors discuss the design of an intelligent database for use as a knowledge management system accessed by designers through the Internet. So far this has been viewed as an innovative project as very limited literature has been reported relating to extending database systems to generate drawings up to date.

2 METHODOLOGY

In this research, the following methodologies are employed, i.e. Evolutionary Prototyping Lifecycle Model, Database systems, and Client-Sever systems.

2.1 Evolutionary Prototyping Lifecycle Model

Neufelder reports that in a government survey of software development projects, sixty percent of all errors made were in the requirements phase and only fifteen percent of these errors were actually caught in that phase [4]. As software reliability is directly related to error content, a software development life cycle model that focuses on ways to reduce errors at the requirements phase will inevitably increase the likelihood of more reliable software. Prototyping allows the software developer to express an aspect of the design in the form of working code to the client. Prototyping is a way of disambiguating functionality that may otherwise be difficult to express in words. Schneider describes the prototype developer as the "explainer" and the receiver of this explanation as the "listener" [7]. An evolutionary prototyping approach is useful when requirements are changing rapidly, when the customer is reluctant to commit to a set of requirements, or when no one fully understands the application area. A prototyping approach focuses on gathering a correct and consistent set of requirements [3]. This approach lends particular strength to building quality software by means of the ongoing clarification of existing requirements and the discovery of previously missing or unknown requirements. By clarifying the requirements through enhanced understanding, prototyping has been adopted as a technique in this research to improve the accuracy of the requirements.

In the design stages of the system factors such as cost, compatibility, reliability and security issues need to be carefully considered. Economical secure and reliable applications are required with the highest performance. Some degree of functional decomposition is required and is the process whereby a system is broken down along functional lines into a number of smaller sub-systems [1]. This process has been applied to this system to partition the requirements into smaller sub-systems as shown in the work of Kyu-Yeul Lee et al [1].

2.2 Database Technologies Available

In the mainstream market some popular vendors are Oracle, IBM, Microsoft Sybase and Computer Associates. Traditional industry surveys show that Oracle leads the field with IBM second and Microsoft trailing [2]. As far as the above vendors are concerned their products are everyman databases apparently aiming to do everything for everybody. Despite offering mainstream database products Computer Associates and Sybase are niche market vendors. Some of the products offered in these niches can be used directly to replace a mainstream database.

According to the Bloor Research Report [2], the conventional fat-client view of the client/server model is regarded as wasteful in resources, more expensive to manage and has availability problems. The latest client hardware i.e. TV set-top boxes, mobile phones, personal digital assistants (PDAs), cannot fit the fat client model. In the classic two-tier client/server database architecture the client is usually an application while the server is the database system e.g. SQL Server, Oracle.

N-tier databases are most commonly implemented as a three-tier model. They extract the business rules and place them in an independent middle layer. All data is passed through the middle layer, thus the server and client never communicate directly. The middle tier provides a layer of abstraction. Client applications are written to communicate with the middle tier and do not depend upon the type of server that actually stores the data. The back-end server could be changed e.g. from SQL Server to Oracle, and only the middle layer would require modification.

A database management system (DBMS) gives the user access to their data and helps them transform the data into information. These systems allow users to create, update, and extract information from their databases. The main advantages to a computerised database system are speed, accuracy, and accessibility.

Microsoft SQL Server 2000 is capable of supplying the database services needed by large systems. Large servers may have thousands of users connected to an instance of SQL Server 2000 at the same time. SQL Server 2000 has full protection for these environments, with safeguards that prevent problems, such as multiple users trying to update the same data simultaneously. With the increase in e-commerce and client server applications SQL Server allows compatibility with ASP. ASP technology was used in implementation of this system.

2.3 Client and Server systems

In order to communicate over TCP, a client program and a server program need to establish a connection with one another. Each program binds a socket to its end of the connection. A socket is an end-point of a two-way communication link between two programs running on a network. To communicate, the client/server reads from and writes to the socket bound to the connection. The server has a socket that is bound to a specific port number and is in a waiting state, listening to the socket for a client connection request. To make a connection request, the client needs to know the hostname of the computer on which the server is running and the port number to which the server is connected (fig. 2).

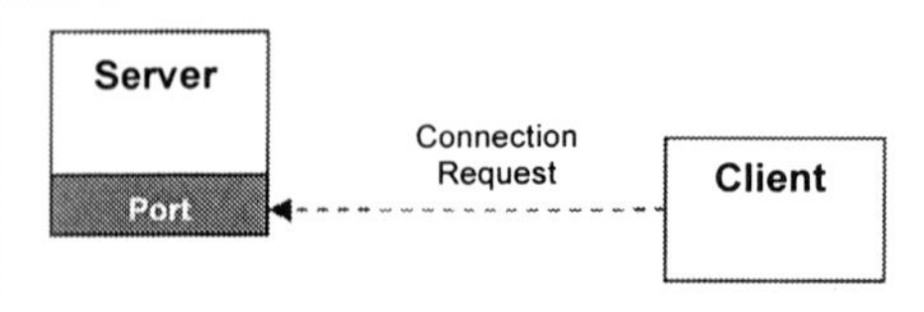

Figure 2 Client Connection Request

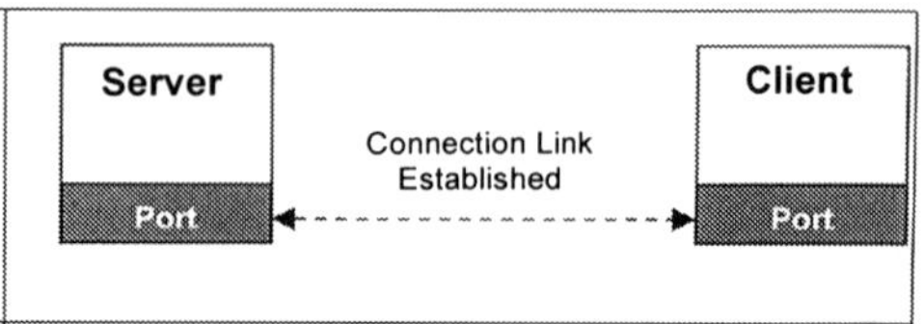

Figure 3 Communication established

When the server accepts a client connection request, it creates a new socket bound to a different port. New sockets and consequently different port numbers are required in order for the server to continue to listen to the original socket for connection requests and to tend to the needs of the connected client. If the client connection request is accepted, a socket is successfully created and a communication link is established between the client and the server (figure 3). The client and server can now communicate by writing to or reading from their sockets. The java.net package in the Java platform provides classes that implement the two-way connection between the client program and the server program on a network. The classes allow Java programs to communicate over a network in a platform-independent fashion.

3 SYSTEM DESIGN

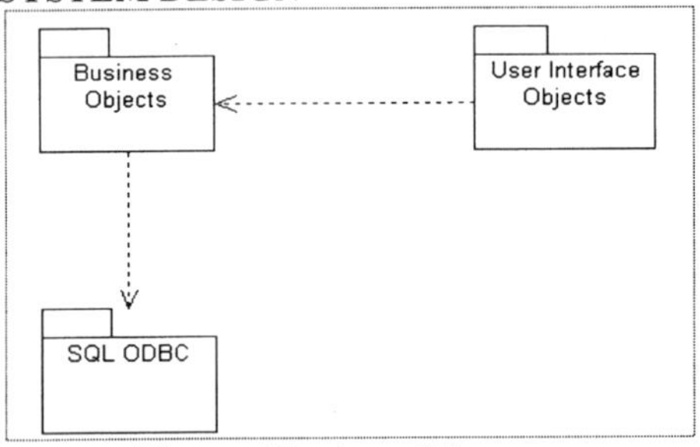

Figure 4 Package Diagram

UML has been used in the software requirements phase to establish the specification of the system. Structured methods use functional decomposition in which the overall system was mapped as a function and broken down into sub functions. In UML classes are grouped together into higher-level units and expressed as package diagrams as shown in fig 4.

The dependencies have also been included to the relations between the packages i.e. that there is dependency between the classes that are contained in the packages. The above packages are composed of classes. Each package has a set of classes that are similar in some way. The public classes are visible in the package i.e. these are visible to other classes and other packages. The package structure shows that the SQL classes are only visible to business objects and not to the User interface objects.

UML Use Case Diagrams were used in the requirements phase to describe the functionality of the system. Figure 5 below shows the available functionality for the role of a customer. Customers can login/logout of the system, create orders, view their own orders and download PDF documents. The diagram also depicts the multi-tiered database architecture whereby the web server is the middle tier.

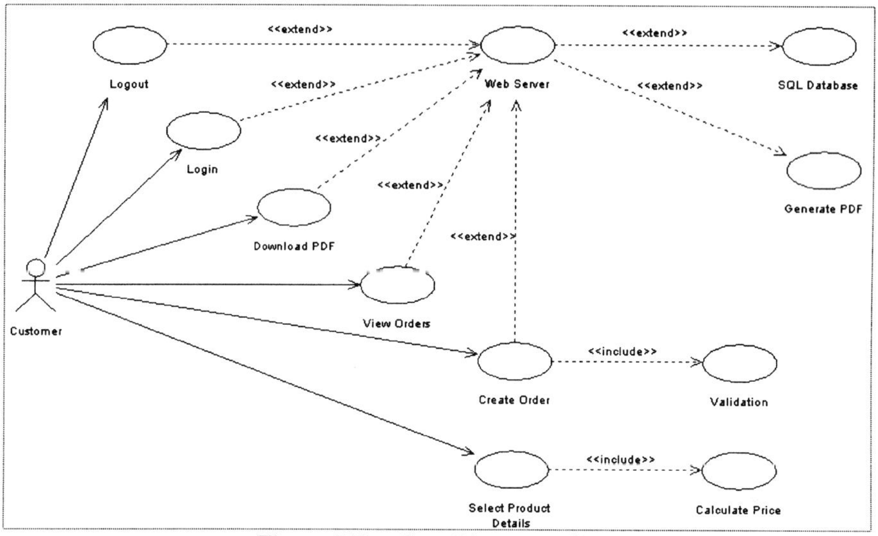

Figure 5 Use Case Diagram - Customer

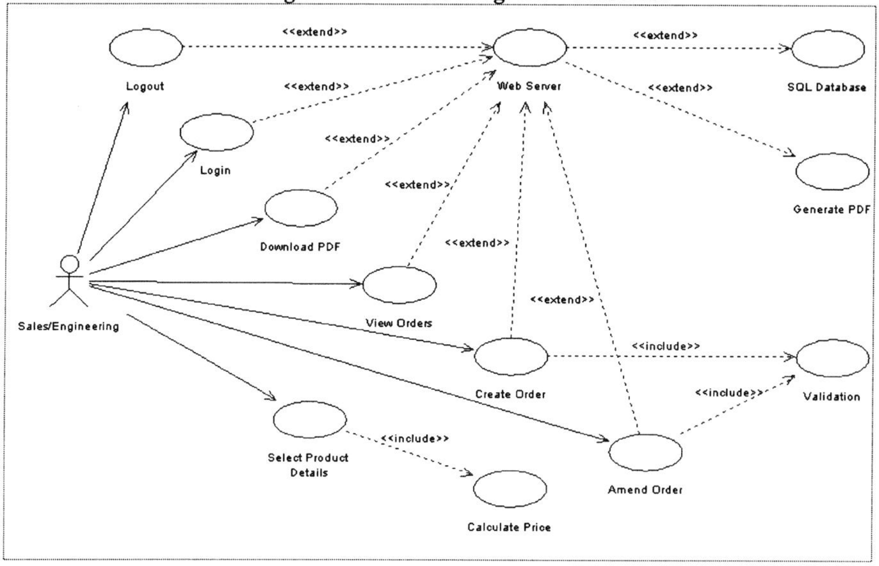

Figure 6 Use Case Sales/Engineering Personnel

The Use Case diagram *(fig.6)* shows the functionality for all Sales/Engineering Personnel. Sales/Engineering Users have almost the same functionality of the customer with the exception that they can view/amend any order. The diagram also depicts the multi-tiered database architecture whereby the web server acts as the middle tier.

3.1 Entity Relationship Diagram

Database Normalisation is the process used by database designers to efficiently organise and reduce data redundancies in a database. This process helps eliminate data anomalies where insertions, deletions, updates destroy the integrity of the database. The results from the normalisation carried out in the design stages of the database are shown below *(fig. 7)*.

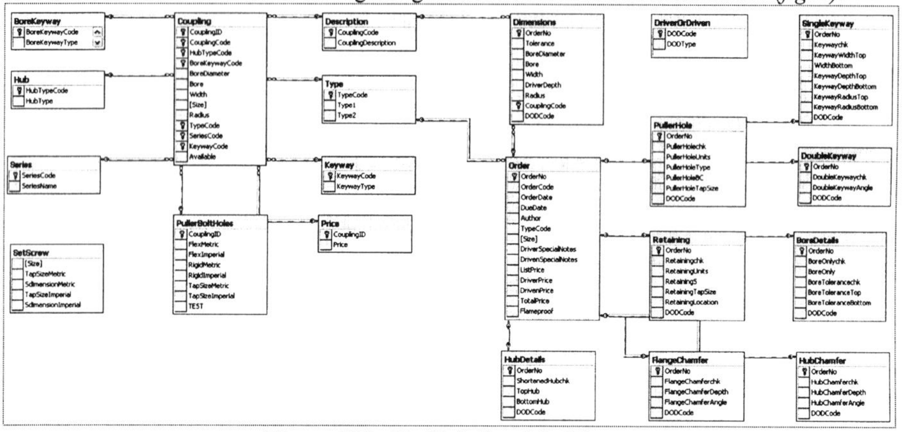

Figure 7 Entity Relationship Diagram

4. IMPLEMENTATION AND RESULTS

4.1 Generating Drawings

The drawings that need to be generated dynamically are not required to scale, as they need to fit the standard A4 paper size. A technique was designed to generate the drawings for standard/non-standard couplings. Firstly 23 different drawings were created using adobe illustrator. The drawings were all created using the same properties such as width and height (5166mm x 3081mm) and background layer transparency (transparent). To preserve the vectorised format the drawings were saved as WMF Images. Vector graphics perform more efficiently on the web because they are based on mathematical computations, rather than the pixel-by-pixel information used by bitmaps. For example, if the two circles in figure 8 were separate vector graphics being displayed in a browser both images would have the same file size.

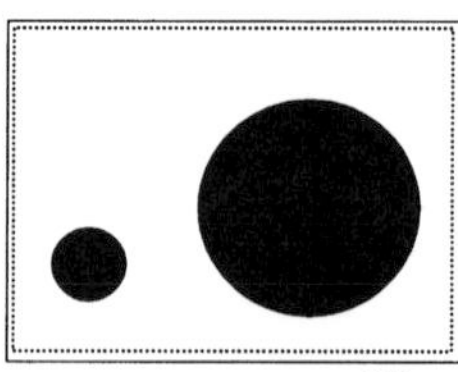

The only difference between the two circles is the radius, which can be adjusted by a mathematical calculation. Vector graphics use mathematics in this way to manipulate images.

Figure 8 Vector Images

By superimposing these images on each other, any standard/non-standard coupling drawing can be formed i.e. the drawing shown in figure 1b. This technique can be explained in a simplistic manner by using smiley faces as an example (see figs. 9-10). First 7 drawings *(fig.9)* need to be created, each with the same properties i.e. width, height layer transparency.

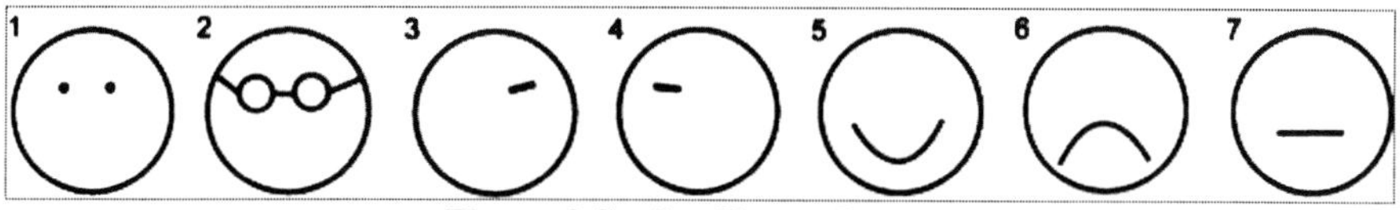

Figure 9 Smiley Vector Images

To create a drawing of a face with a smile and left eyewink, images 1, 4 and 5 are superimposed on each other to give the following result.

Figure 10 smile and left eyewink

The total number of unique drawings that can be created from superimposing the images on each other in this example is 49 (i.e. 7 x 7).

4.2 Interface Design

The interface design *(fig. 11)* is sophisticated, to display all information on one screen. Factors such as maintaining a corporate look and user friendliness were considered when designing the interface.

Tabs were used in the interface design to accommodate for all the information to be integrated onto one screen. The rectangle on the top right of the design is to be used for displaying coupling images. Directly below this is a table that will dynamically display product information i.e. width, depth radius. The table will be implemented as a separate web page embedded within an IFrame to allow the table to reload its data without refreshing or changing the main screen. The IFrame will act as a wrapper around the web page containing the table.

To the left of the images area and table lay tab panes. Although the uppermost tab pane only has one tab it has been designed to look like a tab pane to maintain a consistent look. The first tab on both the middle and bottom tab pane is to be used for the input and configuration of standard couplings. The middle tab pane will be used for driver couplings whereas the bottom pane will be used for driven couplings. These panes will also have a second tab used for the input and configuration of non-standard couplings. The second tab will drop down vertically on the middle tab pane and cover the bottom tab pane. This is because there are considerably more options for non-standard couplings. However the second tab on the bottom tab pane cannot drop down vertically as part of it would disappear off the screen therefore it has been designed to open up horizontally and masks the table to its right.

The small icons to the right of both tab panes will be used to display the current drawing configuration of the corresponding coupling in the images area. An add/update button will be present on the screen depending on whether an order is being placed or amended. The menu button to the left will redirect the user back to the main menu. The name of the current user logged in shall be displayed on the far left of the screen. If an order is being amended then the name of the author will be displayed beneath the current user logged in.

4.3 Final Interface

Above designed systems have been implemented by using Java Client-Server Sockets, Applets, Active Server Pages, SQL, JavaScript, VBScript, HTML, DHTML,CSS with the following tools SQL Analyzer, Macromedia Dreamweaver MX, and Netbeans IDE.

The final release of the interface is shown in figure 12. The interface does not differ from the original design as the design was carefully considered and planned well. As there are no tab panes in ASP The tabs have been created by using tables and altering the z-index of each table. For example when the initial form is loaded the z-index of the options tab is set to zero and the z-index of the standard tab is set to one. The z-index of the highest table is seen. When the options tab is selected the z-index values are altered to place the options tab in front and the standard tab behind it.

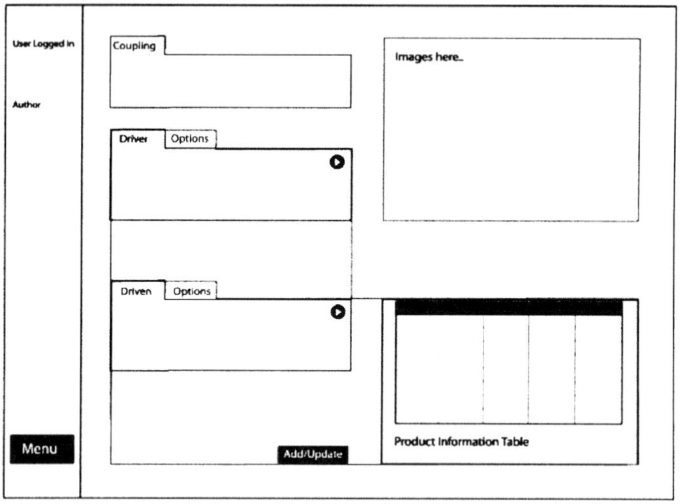

Figure 11 Interface Design

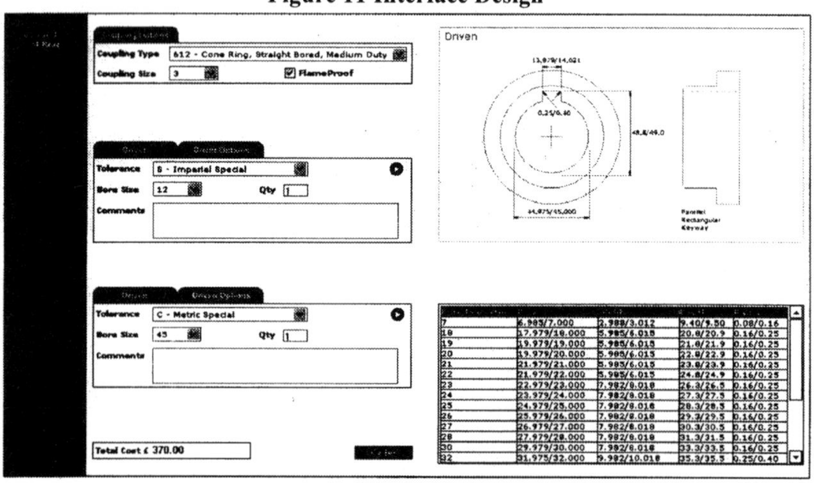

Figure 12 Final Interface

The view screen *(fig.13)* allows a user to open an order by clicking on the view cell of the row corresponding to the order. Similarly a PDF document can be downloaded by clicking on the PDF drawing cell. The rows are highlighted on the mouse over event for each row to make it easier to read the information from the table. Additional search functionality has also been implemented which allows a user to search for an order via the order number. The radio buttons at the top of the screen allow previous records to be displayed depending on preference i.e. last 6 months, last 12 months, all.

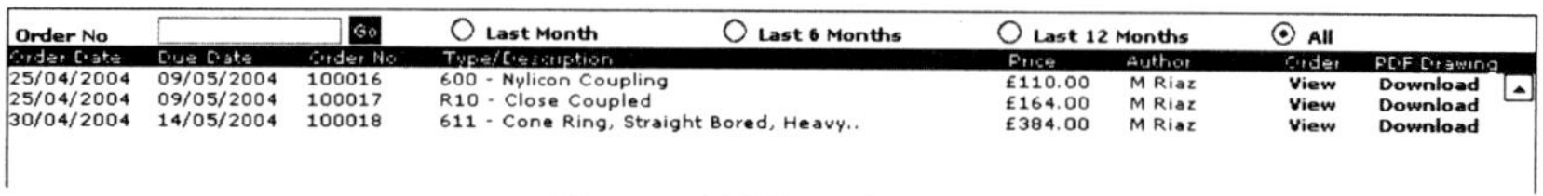

Figure 13 View Screen

4.4 Testing

The initial system tests concentrated on small subsections of the application. When the tests were completed on all subsections, the overall system and its integration between systems was tested. This approach facilitates the finding of errors. This is because it would be difficult to isolate the cause of an error if the system was tested as a whole.

Furthermore it is important to note that errors in software are not just concerned with programming logic or system failures. The worst defects are those that cause the software to fail to meet its requirements. If the software does not have the required capabilities then it is deemed as useless.

The acceptance testing method was used for this project. Acceptance testing allows the tester to prove that the system meets all the user requirements or business logic. It is a form of validation testing and proves that the system works. The testing was performed by end-users i.e. sales/engineering personnel. The user manipulated data through the system developed without relying on the knowledge of the requirements but using their knowledge of the business logic. Acceptance testing is the final stage of the testing process and uses real data rather than simulated data and confirms whether or not the system is ready to 'go-live'.

5 DISCUSSION AND CONCLUSIONS

The table III shows the number of non-standard couplings drawings requests made to the engineering drawing department over the past five years.

Table III Non-standard couplings drawings requests 1999 - 2003

Year	Number of Non-Standard Drawings
1999	67
2000	81
2001	119
2002	133
2003	152

The figures show that from 1999 to 2003 the number of non-standard drawings has almost doubled. It is clear that the request for the number of drawings is increasing on average at about 23% per year.

The graph below *(fig.14)* forecasts the number of man-hours David Brown Textron will save by using the system developed for non-standard coupling drawings in the next 5 years based on the average rate of increase per year. The number of man-hours is calculated by multiplying the number of non-standard drawings per year by the average time taken per drawing. **Man hrs = number of non-standard drawings x avg time taken/drawing**

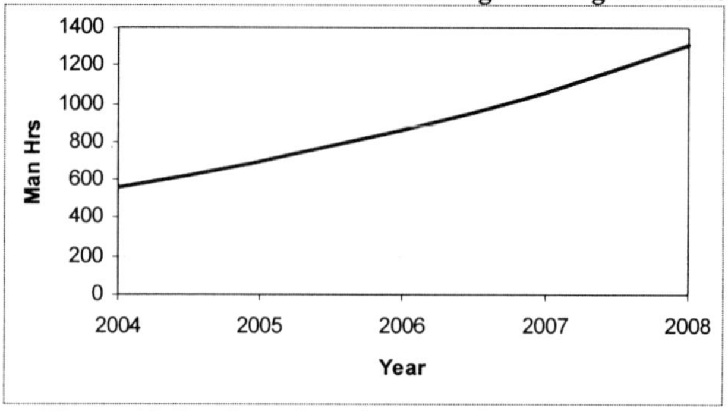

Figure 14 Graph to show forecasts from 2004 - 2008

The system has been developed for the engineering industry to allow for the integration of other products with minimum modification required to database tables and classes. A reduction in lead-time and engineering designs costs could be significantly improved, according to benefits shown in table IV.

Table IV Benefits of developed system

Portability	Any user with a web browser can access a web application
Centralised Data	Data is centralised on SQL Server making it easy to backup and secure
Low maintenance	Implementation as a web application increases the maintainability and adaptability of applications
Low Cost	No additional overheads are required to use the system
User friendly	Familiar and attractive interfaces, using colour and graphics, encourage its use

REFERENCES

1 Kyu-Yeul Lee , Won-Joon Lee and Myung-Il Roh, (2004) "Development of a semantic product modeling system for initial hull structure in shipbuilding", *Robotics and Computer-Integrated Manufacturing,* Volume 20, Issue 3, June 2004, Pages 211-223

2 Bloor Research Report 2001, Bloor Research

3 Carter, Antón. (2001) "Evolving Beyond Requirements Creep: A Risk-Based Evolutionary Prototyping Model"

4 Neufelder, Ann Marie (1993) "Ensuring Software Reliability" Marcel Dekker

5 Riley C, Distribution Manager (2003) "Textron Power Transmission"

6 Ron C.J.A.M. Broens, Marc J.de Vries, (2003) "Classifying technological knowledge for presentation to mechanical engineer designers", Faculty of Technology Management, Eindhoven University of Technology, The Netherlands

7 Schneider, K (1996) "Prototypes as Assets, not Toys – Why and How to Extract Knowledge from Prototypes", (Proceedings of ISCE-18), Institute of Electrical and Electronics Engineers, Inc.

8 Textron Power Transmission, 2004

9 M.Riaz, 2004, Internal Report, University of Huddersfield, School of Computing & Engineering

A study on universal electronic table of product commonality-orientated design and its system development

R LI and **T ZOU**
Centre of Enterprise Information Engineering Technology, Zhejiang Institute of Science and Technology, Hangzhou, China
S ZHANG
Hangzhou Cigarette Factory, China

ABSTRACT

Competition in manufacturing has been focused on the ability of enterprise to satisfy unpredictable needs of customer since years ago. Under this background, it is very important for manufacturing enterprises to obtain competition advantage to improve all kinds of product commonality during designing a customized product. In this paper a universal electronic table of product commonality-oriented design method is presented through making use of the 40 GB data for designing 3-serial industrial turbines in Hangzhou Turbines Ltd (HTL). The design data have been cumulated in a variety of cases throughout the past 30 years. Furthermore, the development of a prototype system is described and the system applications are explored.

1 UNIVERSAL ELECTRONIC TABLE

Universal electronic table (UET) is a sheet or sheets of computer data table with all kinds of information of components and parts (CAP), such as the related quantity, name, trademark, finish size and so on; meanwhile, it also attaches all kinds of necessary views to operate those computer data table, such as comprehensive evaluation view, aided design view etc.

Generally speaking, UET possesses the following characteristics:
1) UET can show the exact use quantity of each kind of CAP in enterprise, as well as which products have already made used of them. Moreover, the upper levels till product, basic information or drawing of each CAP, can also be easily known through UET.
2) UET can show each product has used which CAP, as well as each CAP's use quantity, upper CAP and drawing etc.
3) UET can easily show all kinds of taxes of CAP, such as according to the use quantity and so on; of course, it also provides many kinds of evaluation method for these taxes.
4) UET can easily find out whichever CAP's detailed information, as well as its place in all kinds of taxes, stock etc.

2 A UET PROTOTYPE SYSTEM DEVELOPED FOR HTL (UPSDFH)

The object model of UPSDFH is shown in Fig. 1. Here, all objective entities are in the form of objects, including entering original data, data statistics, results disposition, overall evaluation,

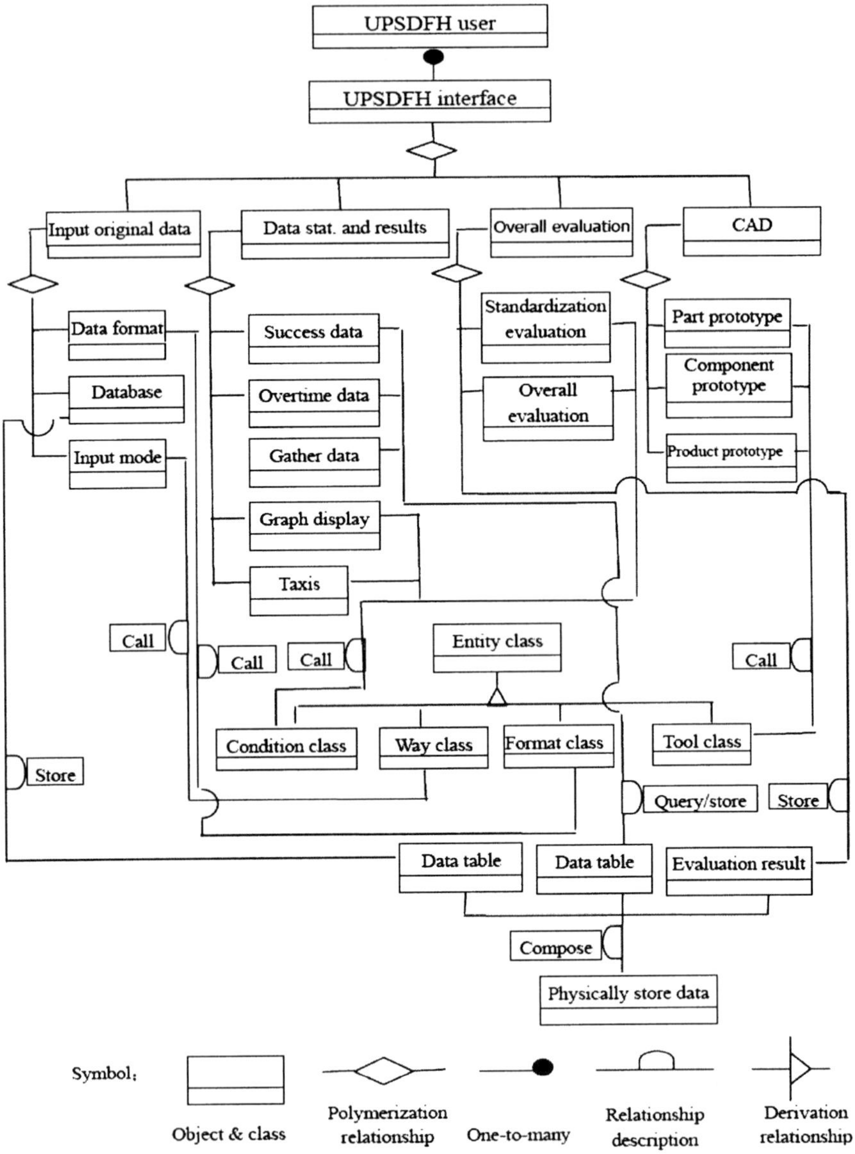

Fig. 1 Object model of the UPSDFH

and computer aided design, etc. Object sets with same contribution and behaviour are shown in class, such as condition class, way class, format class and tool class etc. From Fig. 1 it can be found that this model describes static, structure properties of UPSDFH, meanwhile it is mapping of entity object and relationship between objects.

Generally speaking, UPSDFH mainly consists of six modules which are shown in Fig.2.

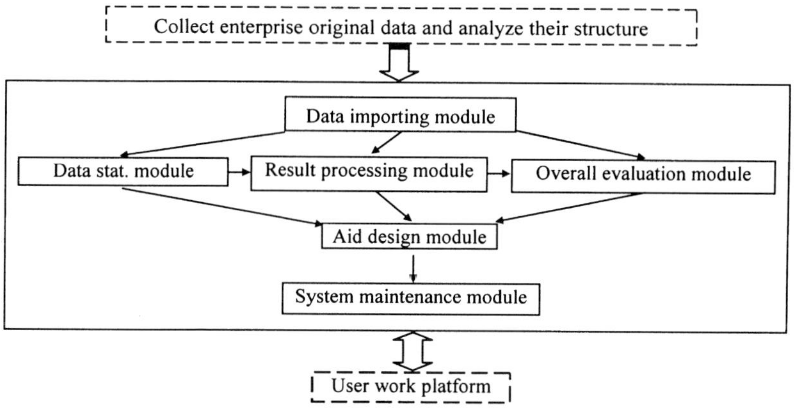

Fig. 2 System structure scheme of the UPSDFH

1) Data importing module
For example, applying the system in a new enterprise, the first step should be done is to build the database of system through importing all the related data accumulated in the enterprise up to now. Of course, data importing module is also necessary for enterprise to renew an old system's database.

2) Data statistics module
For example, CAP stat., stat. result retrieval (such as stat. result table, detail result table, success data table and time-out data table etc.), stat. result displaying (for example, product total quantity, and CAP total quantity) and so on.

3) Results processing module
For example, merging stat. result (such as final stat. result, detail stat. result), ranking stat. result (such as taxis of total quantity of product, taxis of total quantity of CAP), displaying stat. result sequencing of (such as in the form of bar, graph and so forth), retrieving specified product (such as retrieving all CAP of a product, retrieval of classification product, and displaying result in the form of figure), as well as retrieving a specified CAP and so forth.

4) Overall evaluation module
Usually include two parts: evaluating the commonality of CAP and evaluating the commonality degree of a product. Here, the system also provides functions, such as option of evaluation index, relative weight coefficient of each index, as well as related retrieval according to a selection of evaluation index value or a specified CAP.

5) Computer aided design module
Include all kinds of retrieval or stat. of CAP according to its name or code. For example, after inputting "steel wire" in the retrieval window of CAP name, the system will tell designer all kinds of information about "steel wire", such as all the steel wire which have been used before and whose detail information (such as quantity, type, price, basic figure), all the higher level CAP's information of these steel wire and whose sequence, the commonality evaluation index value of these steel wire, and so forth.

6) System maintenance module
Include printing result(such as retrieval result, figure displaying result), previewing result, adding/deleting record, printer setting up, exporting all records in a window in the form of text format etc..

3 APPLICATIONS ILLUSTRATION: FREQUENCY DRAWING OF PRODUCT OR PART

Running UPSDFH, it is easy to get all kinds of information related to the commonality of CAP, such as which kinds of part have been used in a product and how much of a kind of part have been used in a special product.

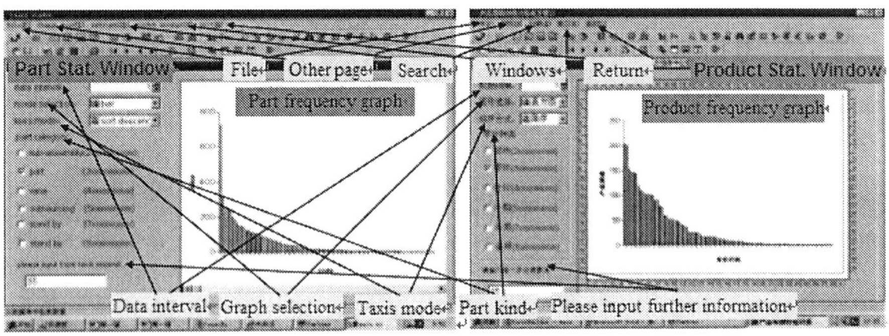

Fig.3 Part frequency graph **Fig.4 Product frequency graph**

Fig.3 shows the "Part Stat. Window" of UPSDFH, as well as the "Part frequency graph" of block shape part and stand in HTL. It is easy to see from Fig.3 that most of block shape part and stand have been used only one time up to now.

Fig.4 shows the "Product Stat. Window", as well as the "Product frequency graph" of block shape part and stand in HTL. It is easy to see from Fig.4 that most of block shape part and stand have been used only in on product up to now. Another interesting aspect worth to be mentioned is that a kind of pipe of heat transition in HTL, whose code is 37030376046, has been used 4,376 pieces, even though only used in one product.

4 CONCLUSIONS

It can be easily known through running UPSDFH that, usually one fourth kind of CAP contribute to 80 percent of CAP quantity used in today's enterprise. In other words, the so called PARETO law is valid. It goes without saying that designer should make the most of those belong to one fourth kind in order to improve the commonality of product.

ACKNOWLEDGEMENTS

This research work is financially supported by the Natural Science Foundation of China (Grant No. 70272056) and the National High Technology 863 Programs in China (Grant No. 2002AA414510).

REFERENCES

1 David, M. and Anderson, P. E., "Implementing Mass Customization", Agility & Global Competition, 1998, 2 (2): 36-49.

2 Li, Renwang, "Research on Some Theory and Method Problems of Mass Customization", Doctor's Degree Dissertation of Zhejiang University, 1999 (In Chinese).

Research on information models to support fixture design fixture assembly processes

S LI
College of Mechanical Manufacturing Science and Engineering, Southwest University of Science and Technology, Sichuan and College of Mechanical Manufacturing Science and Engineering, Sichuan University, Chengdu, China

ABSTRACT

Sharing of information is an important issue in Networks Manufacturing. The use of information models is an approach to support the intensive share of information required in Intelligent Manufacturing through the manufacturing processes. This paper explores the definition of the structure of information models for supporting the related Fixture Design and Fixture Assembly Planning (FD&FAP). This research presents the view that the Information Models required to support FD&FAP is captured according process requirements of the workpiece and the information required of FD. The information required of the workpiece process is the representation of the information about a workpiece machining, and the information required of FD&FAP captures the information about the element of fixture locating, clamping design and assembly processes. The key problem discussed is the structure of these information models that would consider not only the information support for the functionality of FD&FAP, but also their interaction in a shared environment.

Keywords: Information Models; FD&FAP; Networks Manufacturing.

1. INTRODUCTION

Nowadays, industrial production is under great pressure due to market globalization, shortened life of products, increased product diversity, greater demands of the clients on quality, and shorter delivery times. In order to face those requirements and to successfully compete in the market, enterprises must be able to produce faster with more quality and flexibility. As a consequence, they must use new technologies for production, revise their organizational structure, and be aware of the consequences of decisions taken in the early stages of product design.

The research of fixture related can be classified in two main areas: Fixture Design (FD) and Fixture Assembly Processes (FAP). The main goals of FD are to reduce the cost of Fixture Design and to simplify the operations required for performing them. The Assembly Planning area focuses on Fixture Assembly Processes (FAP) and Assembly Sequence Generation.

The traditional FD approaches provide guidelines on how to design fixture, or propose a method to assess the fixture characteristic. However, they don't consider the assembly processes by which the fixture will be assembled. On the other hand, there is no way of sharing with the designer the information that is considered during FD&FAP and which could help in the decision taking process. This loss of information can be avoided by creating a link between the FD&FAP.

Fixture Assembly Processes are like assembly process. In order to obtain an integrated assembly process, Rampersad [1] proposed a division of the variables in the assembly model into a fixture process and assembly system. Each of these variables consists of three levels with relations between them and with other systems. In the case of a change in one or more of the variables, other variables undergo a corresponding change. The properties of the assembly process can be regarded as a function of the properties of the fixture and the of fixture assembly.

Because of the interaction of the different areas of the enterprise, a key characteristic of Networks Manufacturing is the intensive information interchange. This information includes sharing process design and fixture design data between the areas, and different software application in all the stages of the fixture design activity. Information Modelling allows the uses of that information in a shared environment. There are several papers in the literature [2~6] in which is proposed the use of standard information models to support the product realization process.

In pursuing information modelling strategies to support FD&FAP, this work has utilized a number of modelling techniques which are explained below in order to define the FD&FAP requirements in the different stages of the fixture design activity.
The result of this investigation is the definition of the data models described in the paper. The work is currently in progress to build an object-oriented database using these models which will be tested using information about the fixture design of large electrical machines, but which will be applicable in many different fixtures.

2. INFORMATION RELATED TO FD&FAP

The research reported in this paper is based on the concepts of the research developed in the College of Mechanical Manufacturing Science and Engineering, Southwest University of Science Technology, China and is funded by the project [7]. The use of an Information Model of the workpiece and an Information Model of the fixture were proposed in order to solve the problem of information modelling in a Networks Manufacturing Environment.

An Information Models of the workpiece is a computer representation of workpiece data that contains detailed information about a workpiece or a family of workpieces, so it can support the applications interacting in the fixture design activity. The Information Model of the workpiece must allow those applications to have access to the information and also to be a repository for the data they create.

The Information Models of thefixture describes and captures the information about the FD&FAP. In the project the structure of the Information Models of thefixture was divided in tow functional levels in order to capture the function of the fixture: fixture resources and fixture assembly resources. For the level of the fixture resources the graphics library, locating design and clamping design is modelled. For the level of the fixture assembly resources the Processes, Resources and Strategies are modelled.

In this research, an integrated approach is followed, considering the effects on the complete system caused by any change in it. This approach will allow the system to be a support for the designer working in a Networks Manufacturing environment. In an integrated approach, the structures of workpiece information, assembly processes, resources and strategies are considered simultaneously. When any of their aspect is modified, everything changes.

In the information structure that is proposed in this research, the designer will have access to the information about the resources and processes that are available in order to consider them from the early stages of design. This means that the fixture design is made with a complete awareness of the facilities available. Moreover, the design of the assembly system is considered in the design of the fixture.

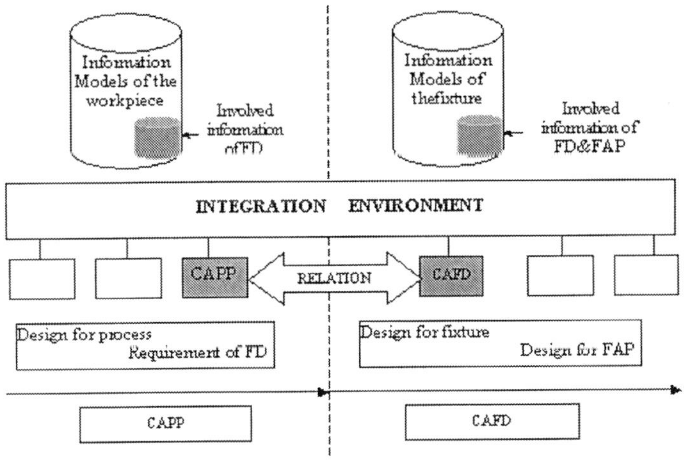

Figure 1 Information Models of the workpiece and Information Models of thefixture interactions

The approach followed in this research is shown in Figure 1. As may be seen from the diagram, the cylinders inside the Information Models of the workpiece and the Information Models of the fixture represent the information structure related to fixture design defined in order to support the FD&FAP related activities during the fixture design related activities. In the applications level of the diagram, the CAPP and the CAFD are illustrated with an arrow between them, representing the interaction explored in the research. This interaction will allow the designer to take into account the main APP considerations when doing the CAPP during the design stages of the fixture design related activities. Therefore, only some details need to be defined in the post-design stages of the fixture design activities.

3. METHODOLOGY USED FOR THE DEFINITION OF THE INFORMATION MODELS AND THE DEFINITION OF FUNCTIONS

In order to define the structure of the information models in the enterprise view level, an IDEF-0 model of the functions performed by FA&FAP in the fixture design activities are constructed. Once the model of the functions is complete, the following step is to build a model for capturing the processes and sequence of activities in an IDEF-3 model.

The information given in the elaboration diagrams of the IDEF-3 method was used to build the Information view level of the models. The domain analysis for building the computational view of the system was done using Booch-UML methodology. In the following stages of the research, an Object-Oriented database will be used to build the computational view of the system and to implement the information models.

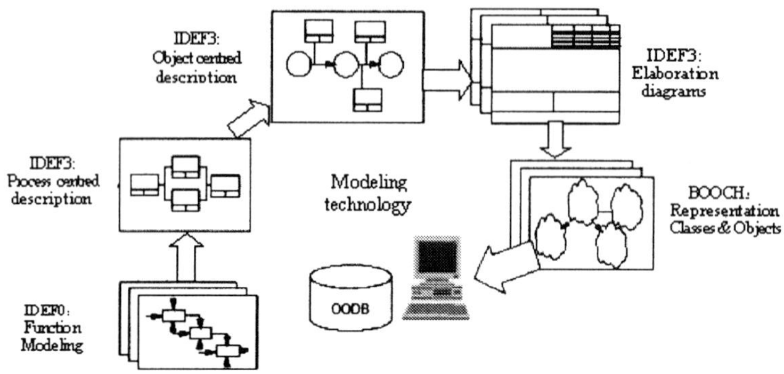

Figure 2 Methodology used for the definition of the Information Models

The sequence of utilised methods for defining the structure and contents of the Information Model is shown in Figure 2.

Once the structure of the Information Model is defined by using the mentioned methods, it is necessary to include the detailed information requirements imposed by the functions of the FA&FAP applications. The basic functions of those applications are shown in Table 1.

Table 1 Basic function of FD&FAP

Functions of Fixture Design	Functions of Fixture Assembly Process
• Identify geometric characteristics that affect fixture (i.e. dimension, tolerance, feature, etc.) • Consider process characteristics of the workpiece (including machining feature, machining precision, etc.) • Identify fixture assembly sequence (including processes required, time spent, reorientations needed, etc.) • Insertion Analysis (location error, locating error, clamping error, etc.) • Classify the fixture elements into groups with reference to the graphics library • Identify possible problems for fixture design (fixture element, clamping device, fixture configuration, etc.)	• A precedence diagram for the fixture assembly sequence • Assess the time required for each operation and for the complete assembly process • Identify the tools required for manipulating the part • Identify resources available and select equipment • Identify recommended strategies • Change the planning according to the actual situation of the process and to balance the lines • Support the creation of the documentation for the operators • Recognise possible tangling with other parts, alternative ways of fitting the parts, etc

These functions, together with the IDEF-0 and IDEF-3 representations, provide the basis for building the UML-Booch diagrams that constitute the basis for the creation of the computational level of the system. Some of these diagrams are shown below.

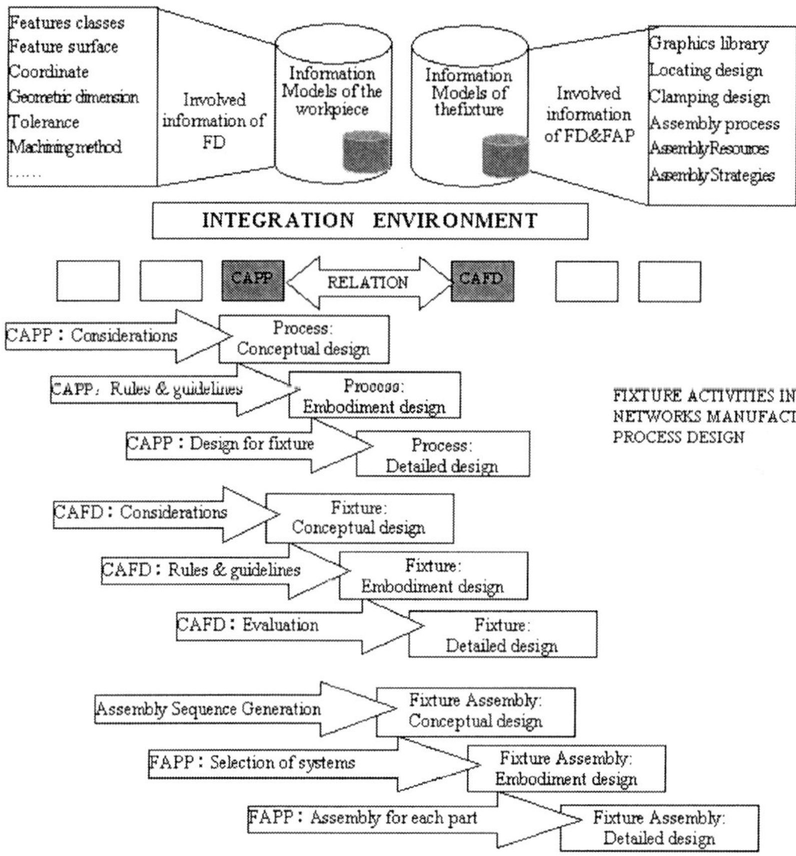

Figure 3 Structure of the information models and stages of the fixture design activities in which FD&FAP are used

4. INFORMATION MODELS OF FD&FAP

4.1 Fixture design activities based on Networks Manufacturing

In order to support FD&FAP related applications through the fixture design stages, the basic internal structures of the information Model of the workpiece and Information Model of the fixture are divided in the areas shown in Figure 3.

The Information Model of the workpiece will represent the data levels of the workpiece process information. The Information Model of the fixture will be able to model both the function and behaviour of the fixture Resources, assembly Processes, and Strategies in the

FD&FAP levels. The instantiation will be done in the Generic and Variant Levels. The modelling views that will be analysed include the specification, functionality, process information, standards and regulations, etc. The information related in the FD&FAP is based on those basic structures.

In Figure 3, the related activities of the FD&FAP in Networks Manufacturing are shown. In the early stages of the FD&FAP, the basic considerations, rules and guidelines of the fixture design are applied, and in detailed design stage, the evaluation method of structured fixture design is used in order to optimise the fixture design. Simultaneously to the design of the fixture, the Assembly System and the Fixture Assembly Processes are designed. This allows the processes designer to take into account the FAP considerations since the early stages of the fixture design, influencing the design of the fixture system and fixture assembly processes that will be needed. As a result, the work that has to be done in the fixture design stages in FAP is reduced and simplified.

The internal relations between the levels and modelling views of the Information Models of the fixture shown in Figure 3 are represented in Figure 4 and Figure 6 using class diagrams.

4.2 Process information model of the workpiece

The Information Models of the workpiece process is a computer representation of the manufacturing process information of the workpiece. The manufacturing process information of the workpiece shown in Figure 4 includes workpiece features, Surface features, Processes features, basic dimension, size- tolerance, Location tolerance, etc.

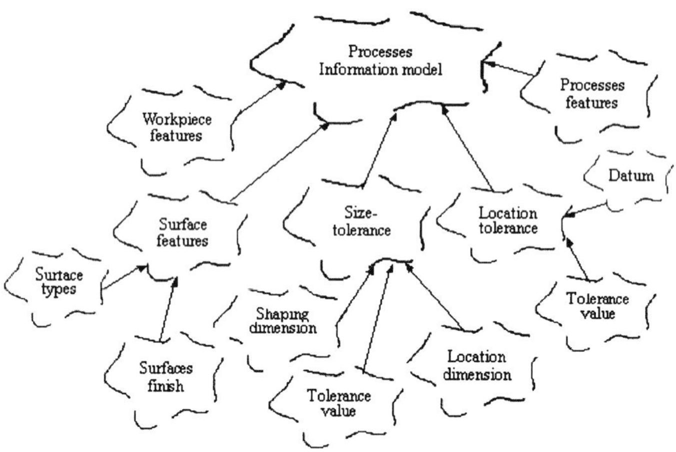

Figure 4 Processes information model of the Workpiece

4.3 Class diagram of FD&FAP

In Figure 5, some of the interactions between the information contained in the Information Models of the workpiece and in the Information Models of the fixture are represented. The represented interactions consider the following cases:

- The Process of the workpiece is designed according to the existing facilities.
- New fixture is designed for machining the workpiece.
- The assembly process of the fixture is designed at the same time.

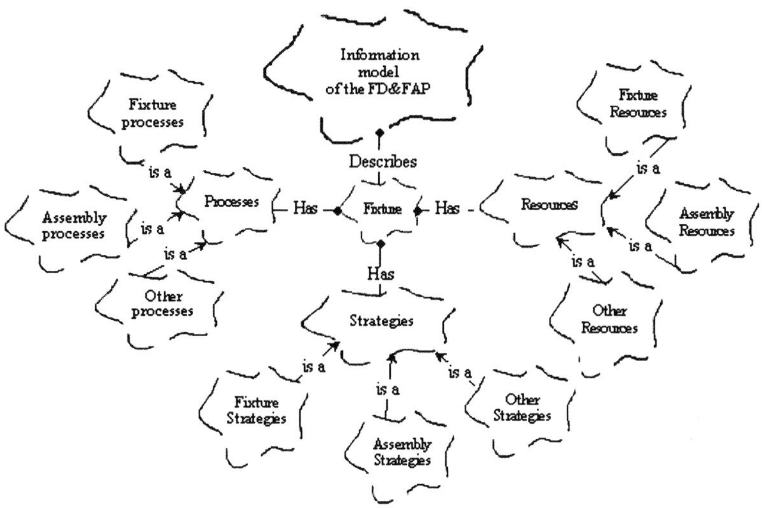

Figure 5 Class diagram for Information Models of The fixture using Booch notation

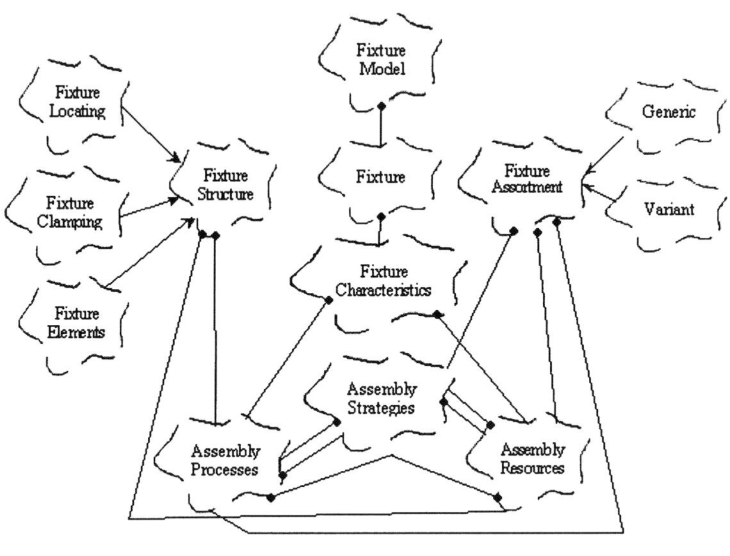

Figure 6 Interaction between assembly-related information in the D&FAP

The Assembly Processes, Assembly Resources and Assembly Strategies classes involve a number of subclasses that represent the detailed information. In Figure 6, some of these subclasses are shown and the relationships between the main classes are illustrated.

In the proposed system, these interactions will be done through the FD&FAP applications by direct interaction between the Information Models of the workpiece and the Information Model of the fixture. These models are a source and repository for the information needed and generated by the applications.

5. CONCLUSIONS

In this research, a new approach to the interaction of fixture-related information for the support of the FD&FAP activities in Networks Manufacturing environments using Information Models of the workpiece and Information Models of the fixture has been presented. The following contributions are made in this work:

- The fixture design related information structures needed within Information Models of the workpiece and Information Models of the fixture have been identified.
- A preliminary description of the interactions between the fixture assembly-related information contained in the Information Models of the workpiece and the Information Models of thefixture has been presented.

The author will pursue further work to explore how FD&FAP applications should interact with the information contained in the Information Models in order to provide effective support to the designer through the fixture design activities. The Information Models of the workpiece and the Information Models of the fixture will be populated with industrial information and their interaction with the FD&FAP will be tested.

REFERENCES

1 Rampersad, H. K. (1995) "An Integral Assembly Model", *Journal of Intelligent Manufacturing*, 6, 41-51.

2 Jjurrens, K. K., Algeo, M. E. A. and Fowler, J. E. (1996) "Beyond product design data: data standards for manufacturing resources, in Rapid Response Manufacturing: Contemporary Methodologies, Tools, and Techniques", Dong J. (ed.), Chapman & Hall, New York.

3 Kjellberg, T. and Bohlin, M. (1996) "Design of a manufacturing resource information system", *Annals of the CIRP, 45(1)*, 149-152.

4 Molina, A., Ellis, T. I. A., Young, R. I. M. and Bell, R. (1995) "Modelling manufacturing capabilities to support concurrent engineering", *Concurrent Engineering: Research and Applications, 3(1)*, 51-60.

5 Giachetti, R. E. (1999) "A standard manufacturing information model to support design for manufacturing in virtual enterprises", *Journal of Intelligent Manufacturing, 10*, 49-60.

6 Young, R. I. M. and Bhell, R. (1995) "Product and Manufacturing Models: Information Tools for the Designer", Report from the Department of Manufacturing Engineering, Loughborough University, UK.

7 Li Shuangyue, Shen Liangqiong and Liou Jianbo (2003) "Reasoning approach for supporting position and workpiece deformation in the automation fixture design", Proceedings of the Third International Conference on Electronic Commerce Engineering (ICeCE2003), Hangzhou, China, 24-26 October 2003.

‘Subdivision projection method’ for detecting gouging interference in three-axis NC machining

Y S YANG, H R SHEN, W DONG, and **J ZHANG**
Department of Mechanical Engineering, Shanghai Maritime University, People’s Republic of China

ABSTRACT

The paper presents a “Subdivision Projection Method” for detecting the gouging interference in 3-axis NC machining. Firstly, the machining surface and the partial cutter under the cylinder are triangularly subdivided, respectively obtaining two sets of triangular plane pieces. Secondly, it is assumed that the cutter is located at an interference-free initial position above the machining surface, imaged that the cutter moves along the axis vector towards the part surface, and “halts” at the first contact point between the cutter and the part surface. Moreover, the solution of this first contact point can be simplified as the seeking of an intersection point between two sets of triangular pieces. In order to greatly decrease the interference-measuring time, the concept of “Interference Characteristic Point” is introduced, and the method of primary-detection and detailed-detection together with the “containing boxes” technology are adopted. The aim of “Subdivision Projection Method” is to provide a general algorithm for inference avoidance in 3-axis NC machining.

KEYWORDS

CAM; NC machining; cutter interference avoidance; “Subdivision Projection Method”

NOTATION

$x_1^i, y_1^i, z_1^i, x_2^i, y_2^i, z_2^i, \ldots$ ——Coordinate value of the triangle’s top points in the subdivision of ball-end cutter.

P_1^i, P_2^i, P_3^i ——The triangle's top points in the subdivision of ball-end cutter.

a, b, c, d ——The coefficient of the equation on plane $\Delta P_1 P_2 P_3$.

L—— The distance between plane $\Delta P_1 P_2 P_3$ and the half-sphere center.

ε—— The subdivision precision.

R—— Radius of a ball-end cutter.

n—— The subdivision times.

e—— The subdivision allowable tolerance.

α—— The angle between the adjacent partition points.

N—— The division node number of cylinder boundary circle.

p_1, p_2—— Two adjoined CC points.

o_1, o_2—— The corresponding CL points of p_1 and p_2.

$T1$—— The triangular pieces of the cutter.

$E1_t$—— The side of triangular pieces of the cutter.

$V1_t$—— Top points of triangular pieces of the cutter.

$T2$—— The triangular pieces of the machining surfaces.

$E2_t$—— The side of triangular pieces of the machining surfaces.

$V2_t$—— Top points of triangular pieces of the machining surfaces.

t—— The mark of the triangular pieces of the cutter.

Δ_1—— Step-length in the process of primary-detection.

Δ_2—— Step-length in the process of detailed-detection.

$Z_{\inf}$—— The maximal interference's quality.

$Z_l^{'}$—— Coordinate value at interference-free CL point.

Z_l—— Coordinate value at initial CL point.

1 INTRODUCTION

Manufacturing enterprises always choose the method of NC machining on the basis of machining efficiency and relevant expenses, and the majority of complex surfaces on parts only need 3-axis NC machines to meet their machining requirements. Therefore, 3-axis NC machining still has a wide range of application now. Since the cutter interference of a non-ball cutter is more complex in NC machining programming, many methods of avoiding interference of the ball-end cutter are not suitable for the non-ball cutter. Up to now, no suitable way of solving the problem of interference of the non-ball cutter has been found (1),(2),(3),(4). Hence, the general algorithms for avoiding the interference of the 3-axis ball-end cutter, torus-end cutter and flat–end cutter NC machining remain a problem in the practical machining.

The algorithms for tool path generation and relevant interference checking are pivotal problems in 3-axis NC machining programming, and have been widely studied in various countries (1),(2),(3),(4). Article (4) is fruitful and presents a method that the first cutter contact points (CC points) between the cutter and the parts is decided by the projection of the cutter towards the machining surfaces. Then the cutter location points (CL points) in NC machining are calculated by the offset of the CC points. Finally the interference-free tool paths are generated. Because the solution for the intersection between the cuter and the machining surfaces is a complex geometric question, including, for example, the calculating question of the non-linear equation, there is an obvious shortage of this method. Otherwise, the CL tool-path generated by this method is incapable of deciding what location undercut will appear on machining surfaces, and it is not convenient to subsequently remove the undercut volume by using a smaller ball-end cutter. Paper (5) proposes a method for gouging checking, by triangularly subdividing the machining surface and the cutter. Thus, a complex problem of interference detection has been transformed to a simple solution of the intersection between two sets of triangular plane pieces. However, the rare use of this method is due to the huge calculating work on subdivisions and intersection. The authors of this paper present a general algorithm i.e. the "Subdivision Projection Method" for detecting interference in the 3-axis NC machining, based on the two above-mentioned methods. The "Subdivision Project Method" focuses on solving interference-checking problems of non-global surface cutters (torus-end cutter and flat-end cutter), on providing a general algorithm for checking the interference in 3-axis NC machining, and on supplying a type of practical tool in controlling the machining quality of the web-based CAD/CAM/CAPP system.

2 MAIN CONTENTS OF THE "SUBDIVISION PROJECTION METHOD"

The main contents of "the Subdivision Projection Method" are as follows. Firstly, the machining surface and the partial cutter's surface under the cylinder are triangularly subdivided, so that two sets of triangular plane pieces are obtained. Secondly, an algorithm for tool path generation that can calculate the data of the original CC points and the original CL points in NC machining is adopted. Thirdly, the projection method is applied to check the interference areas on the machining surface. That is, it is assumed that the cutter is located at an interference-free initial position with above the machining surface, imaged that the cutter moves along the axis vector towards the part surface, and "halts" at the first contact points between the cutter and the part surface, namely, the demanded CC points. The CL points in NC machining can be calculated by the offset of the CC points, and it is certain that these CL points will not gouge machining surface. Moreover, whether the cutter contacts the machining surface is judged by the solution of the intersection points between two sets of triangular plane pieces on the cutter body and the machining surface. Finally, through a comparison between the CL points and the initial points, it is possible to judge whether the cutter interferences on part surface occur, and obtain the value of the interference. In a word, after the above-mentioned process, a complex problem of interference detection has been transformed to a simple solution of the first contact points in the process in which the cutter moves in the projection direction towards the part surface. Moreover, the determination of these first contact points is simplified as the seeking of the intersection points between two sets of triangular pieces.

Axis Z in 3-axis NC machining usually is defined as fixed axis, and it is supposed to be the direction of cutter movement. To generate the tool path in NC machining, first of all, it is necessary that a guiding track is determined and a driving program is weaved. Each point on the guiding track is defined as the initial location before the cutter moves towards the machining surface. When the driving program makes the cutter at the initial location move in the projection direction, a set of interference-free points are generated. For improvement of the interference detection efficiency, the concept of "interference characteristic point" is introduced to decrease the total area of interference detection, and the method of primary-detection and detailed-detection together with the "containing box" technology are applied to restrict the distance of the cutter motion in the projection process.

3 TRIANGULARLY SUBDIVISION OF THE CUTTER AND THE MACHINING SURFACE

3.1 Partly subdivision of the cutter

The types of cutters often used in the complex surface NC machining are the ball-end cutter, torus-end cutter and flat-end cutter. The outline of these cutters is composed of line and arc, and considered as the collection of plane (circle plane of cutter bottom), cylinder, sphere and torus. When the cutters move in the direction of axis Z, the surfaces of the cutter under the cylinder, i.e. the circle plane on the bottom of the flat-end cutter, the half-sphere on the bottom of the ball-end cutter, the torus and the circle plane on the bottom of the torus-end cutter, first contact the machining surface. Hence, partial triangular subdivision of the cutter's surfaces under the cylinder is necessary. Generally speaking, if only the controlling precision of the triangular subdivision is given, this triangular method for the cutter shape will meet the quality control requirements in NC machining.

3.1.1 Ball-end cutter

In the case of triangular subdivision of the half-sphere on the ball-end cutter, since the sphere is a regular surface with the equal curvature, the subdivided result should be as proportioned as possible. Moreover, the quantity of subdivided triangle should be as small as possible. Thus, a subdivision method with a triangle of equal sides is adopted in this paper. As shown in Fig. 1, six points with equal distance between each other are taken on the maximal circumference of the half-sphere. Then these six points are linked in sequence and six triangles are formed after connecting them with the center point of the sphere on the bottom. Afterwards, each triangle denoting one sixth of the half-sphere is divided again, and a new triangle is formed by linking the three center points of the arc line which correspond to each side of the above-mentioned triangle. Thus, one big triangle is subdivided into four small triangles. This process will carry on again and again, till the plane pieces of the triangle satisfy the subdivided allowed precision. The calculation is as follows.

$$P_0^{'}(0,0,-R), \quad P_1^{'}(R,0,0) \quad P_2^{'}(\frac{R}{2},\frac{\sqrt{3}R}{2},0);$$

$$P_1^i(x_1^i,y_1^i,z_1^i), \quad P_2^i(x_2^i,y_2^i,z_2^i), \quad P_3^i(x_3^i,y_3^i,z_3^i);$$

$$P_1^{i+1}(x_1^{i+1},y_1^{i+1},z_1^{i+1}), \quad P_2^{i+1}(x_2^{i+1},y_2^{i+1},z_2^{i+1}), \quad P_3^{i+1}(x_3^{i+1},y_3^{i+1},z_3^{i+1});$$

Assuming that the equation of $\Delta P_1 P_2 P_3$ is $ax+by+cz+d=0$

where $$a = (y_3 - y_1)(z_2 - z_1) - (y_2 - y_1)(z_3 - z_1)$$

$$b = (x_2 - x_1)(z_3 - z_1) - (x_3 - x_1)(y_3 - y_1)$$

$$c = (x_3 - x_1)(y_2 - y_1) - (x_2 - x_1)(y_3 - y_1)$$

$$d = -ax_1 - by_1 - cz_1$$

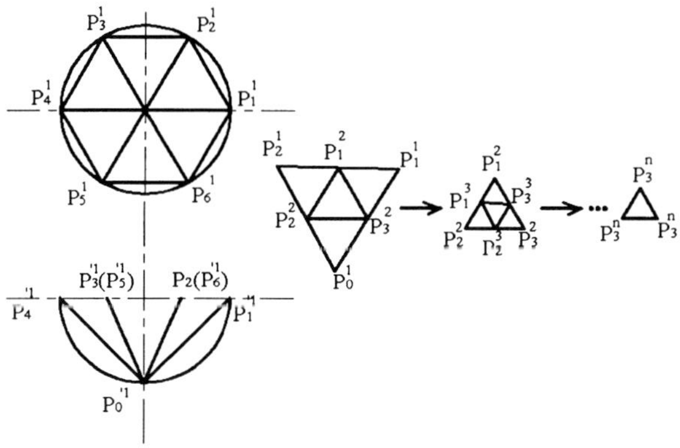

Fig. 1 Triangular subdivision of the ball-end cutter

Then the distance between plane $\Delta P_1 P_2 P_3$ and the half-sphere center is obtained by

$$L = \frac{|d|}{\sqrt{a^2 + b^2 + c^2}} \tag{1}$$

Given the subdivided precision ε, then the subdivided times n is:

$$\varepsilon \geq R - L \rightarrow n \tag{2}$$

3.1.2 Torus-end cutter

It is necessary that the torus on the torus-end cutter is triangulated because it may first contact the machining surface. Considering that the difference between the two lengths of the boundary circle on the torus is not long, the torus is triangularly subdivided by applying the chordal deviation-controlled method, as shown in with Fig. 2. Firstly, the boundary circle on the cylinder of the cutter is divided by the chordal deviation-controlled method. Given the subdivided allowable tolerance e, the angle between the adjacent partition points α can be expressed as

$$\alpha = 2\cos^{-1}(1 - \frac{e}{R}) \tag{3}$$

The division node number of the cylinder boundary circle N is an integer of less than $\frac{2\pi}{\alpha}$. Then, the division nodes on the boundary circle of the cylinder are defined as the initial value of the partition points in the circumferential direction on the torus of the cutter. The triangular subdivision on the torus may also carry through by using the chordal deviation-controlled method. Afterwards, its partition points are again defined as the initial value for the next step and this procedure continues until the boundary circle on the bottom plane of the cutter. Similarly, this method can also be used in the triangular division of the circular plane on the bottom of the cutter.

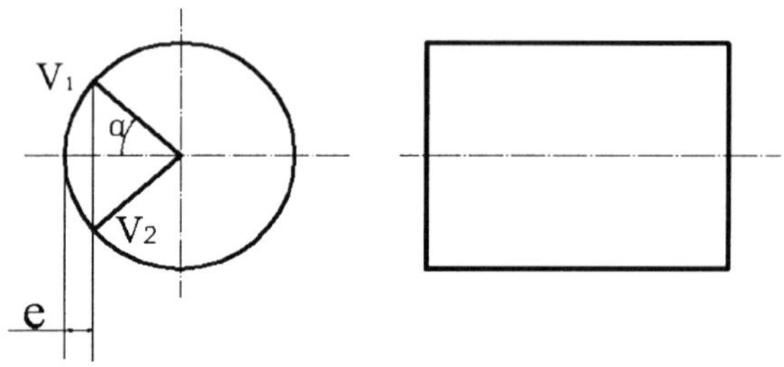

Fig.2 Precision control of flat-end cutter or torus-end cutter

The triangular division of the circular plane of the flat-end cutter can employ the same method.

3.2 Triangular subdivision of the machining surface

Firstly, NURBS(Non-Uniform Rational B-Spline) surface modeling is used in the machining surface. Then the machining surface is triangularly subdivided. When subdivided allowable tolerance is given, the grid step-length of this subdivision can be estimated by a definite formula (which has been stroked out in this paper)

4 MEASURES FOR IMPROVEMENT OF EFFICIENCY

4.1 The determining of the possible interference areas

Obviously, it is a great waste of time to detect every CL point on the tool paths. So the concept of "Interference Characteristic Point" is introduced to improve the efficiency of interference, by detecting only some possible interference areas (2). Supposing p_1 and p_2 are two adjoined CC points on the same or adjoined CC tool-paths, and the corresponding CL points are o_1 and o_2 respectively.

If the following relations can satisfy,

$$\begin{array}{c}(o_1p_1 g p_1o_2 \geq 0 \quad \text{and} \quad o_2p_2 g p_2o_1 \geq 0) \\ \text{and} \quad (p_2p_1 g p_2p_2 \geq 0 \quad \text{or} \quad p_1p_2 g p_1p_1 \geq 0)\end{array} \tag{4}$$

Then CC points are called as the "Interference Characteristic Points". It can be inferred that except themselves, other interference CC points only exist on their adjoined areas. Because the interference areas of the machining surface are always partial, according to these points, the location and the scope of these areas can be estimated quickly. And the calculation of the interference is largely reduced. Moreover, the detection of the "Interference Characteristic Point" is accomplished at the same time when the CC data are formed. In other words, when a new CC point is formed, formula (4) may be used to detect its adjoined points on the same and anterior CC tool-paths, and the corresponding markers are given.

4.2 Application of the "containing box"

Here are two definitions:

(1) The "containing box" of the surface is the minimal cylinder which includes the surface on the projection circle of the cutter on *XY* plane (the cylinder is parallel to *Z* axis).

(2) The "containing box" of the cutter is the maximum cylinder which includes the cutter body first contacting the machining surface. In other words, the "containing box" of the flat-end cutter is the circle plane on its bottom, one of the ball-end cutter is the minimal cylinder which includes its half-sphere, and one of the torus-end cutter is the minimal cylinder which includes its torus.

To apply the "containing box" technology, the first step is to move the "containing box" of the cutter to the first contact location of the "containing box" of the surface. By applying this technology, it is certain to shorten the cutter's moving distance, to eliminate some empty detection during the moving process of the cutter and to improve the efficiency of interference detection. During the second step of the projection of the cutter towards the machining surface, the step-length is first made longer to search the initial location where the cutter first contact the machining surface. Then a shorter step-length is made to move the cutter upwards or downwards to identify a more exact location.

5 THE PRACTICAL ALGORITHM

5.1 The types of intersection between the boundary triangles

The interference checking of the cutter is simplified as the seeking of the intersection points between the two sets of triangular pieces of the cutter and the machining surfaces. When the cutter's triangle first contacts the surface's triangle, the following theorem must be obeyed.

Theorem (6): if the cutter first contacts the surface's triangle on a point outside of the triangle, the cutter should contact the side of the triangle before or at the same time as it contact the inside of the triangle.

It can be concluded from the above theorem, that if the contact point between the cutter and the machining surface is inside or on the sides of the triangular pieces, then this point is the first contact point between the cutter and the machining surface; If the contact point is outside the triangular pieces, then the cutter should move farther in the projection direction for the cutter to contact the machining surface, and the cutter must first contact one side of the surface's triangle before or at the same time as it contact any other point of this triangle.

This shows, the types of intersection between two boundary triangles as follows (in Fig. 3).

(1) The two triangles are on the same plane (Fig. 3(1)).

(2) The intersection of the two triangles is a point which is on a side of the triangles (Fig. 3(2)).

(3) The two points of the intersectional line is the intersectional points of the two triangles. Moreover, two top points of one triangle are on the two sides of the other triangle respectively (Fig. 3(3)).

(4) Two top points of the two triangles are featured with superposition (Fig. 3(4)).

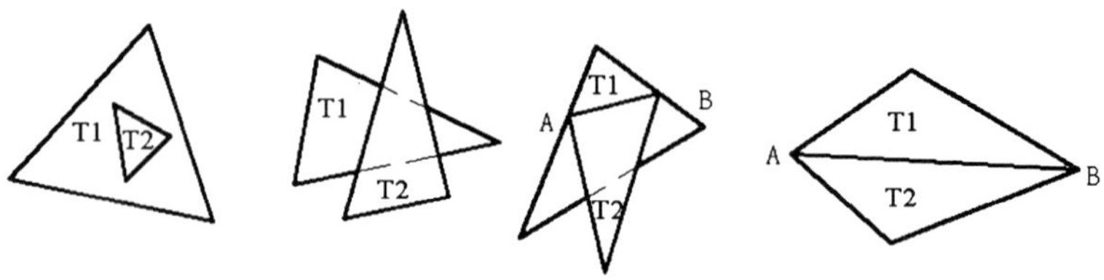

(1) (2) (3) (4)

Fig. 3 Types of intersection between two boundary triangles

According to the intersectional types of the afore-mentioned two boundary triangles, an algorithm for the interference detection in 3-axis NC machining can be designed. Supposing that the triangular pieces of the cutter are *T1*(side $E1_i$, top points $V1_i$, i=1,2,3), the triangular pieces of the machining surface is *T2*(side $E2_i$, top points $V2_i$, i=1,2,3), and both of the triangular pieces have the assured directions. Moreover, supposing the positive direction of triangles is that an outward movement of their normal vector, the negative direction is its inward movement, and the direction between the two is defined as on the triangular pieces.

5.2 The practical algorithm

Step1. An algorithm is applied to generate the original tool-paths, and to calculate the corresponding original CC points and original CL points.

Step2. Formula (4) is used to quickly detect the initial CC points and CL points on the whole parts surface, and to identify the possible interference areas.

Step3. A guiding path on one interference-free plane above the parts surface serves as the initial location of the cutter before projection. Moreover, this guiding path corresponds to the initial CL tool-path.

Step4. The initial locations on the guiding path are taken to form the "containing box" of the cutter and the parts surface. Then the "containing box" of the cutter is moved along the projection vector to contact the "containing box" of the parts surface.

Step5. The first interference checking point is reached. When the primary-detection is done, the cutter is moved by step-length □$_l$ each time(depending on the radius of the cutter), and the top points of every triangle($V1_1$, $V1_2$, $V1_3$) is calculated .

Step6. The intersection point of two sets of triangular pieces in the "containing boxes" is calculated.

(1) A triangular piece of the cutter is taken, and marked as *t*.

(2) Whether piece *t* intersects the triangle in *T2* is judged. If the triangle is just connected or

intersected, then the serial number of the triangle and its type of intersection are recorded, and the procedure turns to Step8, otherwise the procedure turns to the next step.

(3) Whether the triangle taken in *T2* is the last one is judged. If not, then next triangle is taken and the procedure turns to Step 6(2).

(4) Whether piece *t* is the last one is judged. If not, the next one is taken and the procedure turns to Step 6(1).

Step7. If two sets of triangular pieces in the "containing boxes" *T1* and *T2* are not intersectant, then the procedure turns to Step 5.

Step8. If *T1* and *T2* are just contacted or intersected, then the accurate calculation for the first contact point of the cutter and the machining surface is taken. When the detailed-detection is done, the cutter is moved against the projection vector by a step-length □$_2$ (depending on the interference-checking precision).

Step9. The intersectional calculation in Step 6 is repeated. If pieces *T1* and *T2* are not intersectant, then the procedure turns to the next step; otherwise to Step 8.

Step10. The center point of the intersectional and non-intersectional location between the cutter and the parts surface serves as the first contact point, and the interference-free CL points can be obtained.

Step11. If the coordinate value of the interference-free CL point is greater than that of the initial CL point, then this point is the interference point, and the serial number of the interference point is recorded; otherwise it is the interference-free point.

Step12. For the decided interference points, the CL points near them are chosen, and the procedure turns to Step 5. The procedure will continue until all adjoined CL points are the interference-free points.

Step13. For all detected interference CL points, the initial CL points are replaced by interference-free CL points. Afterwards, the undercut volume is removed by using a smaller ball-end cutter. Furthermore, the maximal interference's quantity of every CL point Z_{inf} can be calculated by the coordinate value at the interference-free CL point $Z_l^{'}$ and the coordinate value at the initial CL point Z_l.

$$Z_{\text{inf}} = Z_l^{'} - Z_l \tag{5}$$

Step14. The results are output. End.

By fully using the interface resources provided by system CADDS, the functional modules of NC machining programming have been developed, and the computer simulation of the cutter interference detection has been realized. Finally, an example of generating the interference-free tool path on a complex surface of a mould is accomplished to verify the correctness of the "Subdivision Projection Method".

6 CONCLUSIONS

(1) The concept of "Interference Characteristic Point" is introduced to decrease the total areas of interference detection and improve the efficiency of the "Subdivision Projection Method".

(2) After the building of the "containing boxes" of the cutter and the parts surface, and the adoption of the method of primary-detection and detailed-detection, the efficiency of the "Subdivision Projection Method" is farther improved.

(3) The "Subdivision Projection Method" transforms the complex problem of the cutter interference detection into a simple problem of the judgment of the intersection between two simple bounded triangular planes. This will provide a general algorithm for avoiding cutter interference in the 3-axis ball-end cutter, torus-end cutter and flat-end cutter NC machining. Moreover, this algorithm is steady-going and easy to realize.

REFERENCES

1 Choi, B.K., Lee, C.S., et al., 'Compounds surface modeling and machining', *Computer-Aided Design*, Vol 20, No.3, pp127-136, 1988.

2 Choi, B.K. and Jun, C.S., 'Ball-end cutter interference avoidance in NC machining of sculptured surfaces', *Computer-Aided Design*, Vol 21, No. 6, pp371-378, 1989.

3 Tang, K. and Charles, C.C., 'Offsetting surfacing boundaries and 3-axis gouge-free surface machining', *CAD*, Vol 27, No. 12, pp915-925, 1995.

4 Allan Hansen and Farhad Arbab, 'Fixed axis tool positioning with built-in global interference checking for NC path generation', *IEEE journal of Robotics and Automation*, Vol 4, No. 6, pp610-621, 1988.

5 Zhang, H.M., 'Study on the design of complex surfaces and NC machining technology', *Doctoral dissertation in Zhejiang University*, Hangzhou, China, 1995.

6 Yang, Y.S., 'The study on the methods for avoiding cutter interference in multi-axis NC machining of complex surface', *Doctoral dissertation in Nanjing University of Aeronautics & Astronautic*, Nanjing, China, 1998.

7 Bobrow, J. E., 'NC machining tool path generation from CSG part representations', *Computer-Aided Design*, Vol 17, No.2, pp69-76, 1985.

8 Hsu, P.L. and Yang, W.T., 'Real-time 3D simulation of 3-axis milling using isometric projection', *Computer-Aided Design*, Vol 25, No.12, pp769-775, 1993.

9 Hwang, J.S., 'Interference-free tool path generation in the NC machining of parametric Compound Surface', *Computer-Aided Design*, Vol 24, No.12, pp667-675, 1992.

e-Supply Chain Management, e-Logistics, and Mass Customization

Table retrieve on the Internet for supply-chain management supplier searching

J CHEN, M H WU, and **P BARBER**
School of Arts, Design, and Technology, University of Derby, UK

ABSTRACT

This paper highlights the ongoing research worked aimed at developing an E-based supplier selection Expert System, which is a vital area in the ongoing development of E-manufacturing. The paper outlines the approaches to data retrieving and analyses from web-based tables for supplier information.

The research has produced an intelligent search system using the Internet to achieve the information retrieving functions. This system is capable of comparing captured information from the Internet to identify the correct suppliers.

1 INTRODUCTION

Purchasing is one of the most important processes within business. One of the key activities within the purchasing process is supplier selection [1]. In today's highly competitive environment, the right suppliers are not only supply low cost, high quality products, but also delivery the right items at the right time to the right place. On the other hand, the World Wide Web (WWW) makes worldwide purchasing a reality that every enterprise, from small business to global cooperating, has the opportunity to get the information from worldwide potential suppliers quickly and directly. Thus, these enterprises have more possibilities to find their best suppliers than the others, which exclude the WWW. There is a large amount of information available on the WWW, much of this information is formatted into table format to be used in simplify human-computer interactivities. Hence, how to find the useful information within these formatted data on the Internet could be an essential issue. There have several papers reported with different table detection approaches [2][3][4]. But all these previous methods adopted on a small group of tables, or highly specific list of Web pages [5].

The main aim of this paper describes the new methodology for table detection, table cells retrieving and analyze of online suppliers by the HTML files list of web search engines searched, (online suppliers would be assumed only by capacitors and resistors suppliers, and Google search engine will be used), and finally developed the table retrieving system, the advantages of this methodology are:

- Retrieve wider range of data and table formats via Search engine.
- Keep in touch most important suppliers, which selected by market, not some specific domain names.

The paper is organized as follows. The first section describes the retrieve methods on the web search engines, followed by studying existing table formats on the different tables and generally present approach convert different tables into the pre-defined standard tables on section two. Section three, data mining from standard tables and results report. Finally, in section four, future work would be concluded.

2 RETRIEVE URL METHODS ON THE WEB SEARCH ENGINES

The first factor for table retrieve system could be obtained the Uniform Resource Locators (URL) of suppliers. Because this paper mainly describes how to retrieve useful information from web tables, existing web search engine will be adopted. In this paper, URL resources of online suppliers returned by the search engine Google will be analyzed. Because Google can only return a maximum at 100 URLs at once, the following expression theoretically can return all URLs Google can find. The character of Google search engine is:
http://www.google.com/search?q=keyword 1+...+keyword κ & num=β & start=γ.

μ denote number of keywords. β ($1\leq \beta\leq 100$) denote number of URL returned by search engine. γ ($0\leq \gamma\leq\alpha^* \beta$), here ($\alpha \subseteq N$) denote the start number of each search. So the following expression could present the URL list R is:

$$\mathbf{R}=\sum_{\mu=\delta^*\beta+\gamma}^{(\delta+1)^*\beta+\gamma-1} \mu \quad (0\leq \delta\leq\infty).$$

With expression above, every single URL by keywords could be automatically retrieved. It is possible that not every URL, which search engine returns, is capable for the keywords, for example, a URL links to an image or a PDF file. Moreover, search engines might return duplicate URLs to make the system analyses the same table twice.

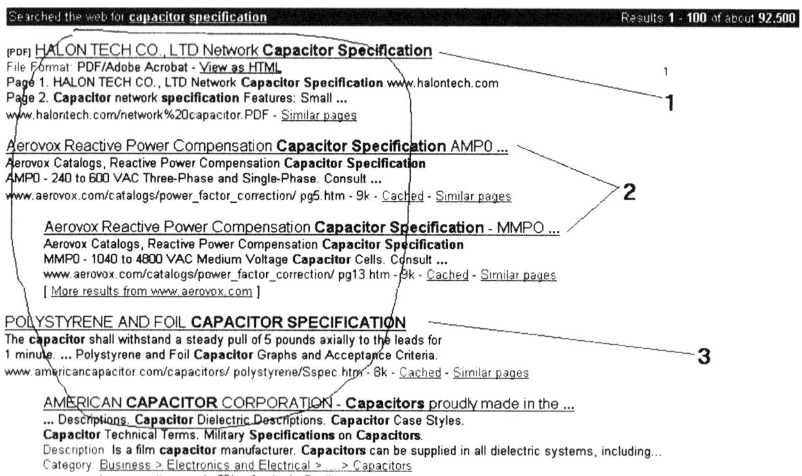

Fig. 1 Search results list

Fig. 1 above shows the random five results by key words "Capacitor specification", the list can be divided into three different groups. First could be the other format of file, like Adobe PDF file, Image file etc, this kind of URL should be ignored because this paper involves in Web format file (HTML). Second group is different Web pages within the same domain name. The third is single web page in the single domain name. This paper focuses the last two groups because web tables can be found easily within these two forms.

3 STUDY EXISTING TABLE FORMATS ON THE INTERNET

There are many table formats on the Internet (Figure 2). But basically, in the square areas (red lines highlighted), the cells could be divided into columns and rows. And these different HTML tables could be grouped into two main different layouts.

- Regular HTML table;

This type of tables named standard table because amount of cells of each column and each row is equal.

- Irregular HTML table

Irregular tables have unequal amount of cells in each column or each row, or have unequal amount of cells in both columns and rows. This type of tables can be divided into two disparate groups, column-irregular table, row-irregular table.

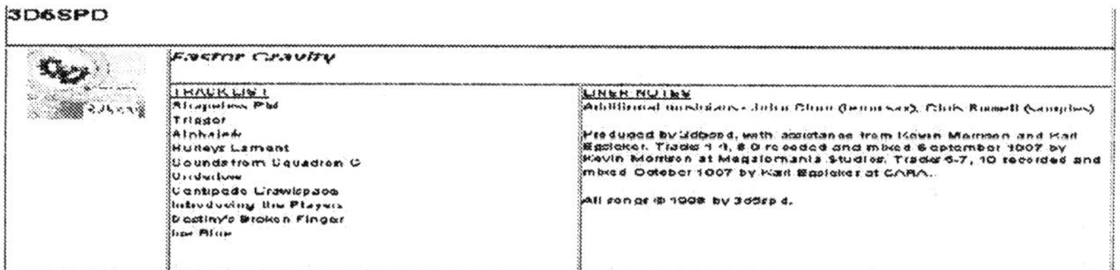

Description	Qty	ID	Digi-Key	Newark	Mouser
0.22 µF 50V polypropylene film and foil capacitor	2	C1	P3224	–	–
Alt. 0.47 µF 100V+ metalized polypropylene film capacitor	2	C1			1429-1474
Long-life electrolytic capacitor, radial	2 or 4	C2/C3			See below
Power indicator LED, red diffused, 1.8V, 1-5 mA, 3mm	1	LED	HLMP1700	90F5862	512-HLMP-1700

Fig. 2 Formats of tables

3.1 Identifying regular HTML tables

Since regular HTML table has the equal cells of each column and each row, it is quite easy to retrieve information of each cell from web tables and group them into database for further analyze through following group algorithm.

```
Input: T= source code of web pages.
Begin :
        Read (T)
        T1=Delimit T( <table, </table)
          T2=Delimit T1( <tr, </tr)
 For R=1 to count_num(T2)
              T3=Delimit T2(<td, </td)
                For S=1 to count_num(T3)
                 cells=T3(s)
                  N=database_insert(cells,S,R)
                End for
End for
```

Fig. 3 Group Regular Table Algorithm

3.2 Identifying irregular HTML tables

Irregular HTML tables have different amounts of cells in their columns or rows or both. Approaches to identifying irregular HTML tables can convert irregular HTML tables into regular HTML tables. That is to say, irregular HTML tables need to place extra cells into the corresponding columns or rows until the amount of cells in each column and row are even. And there are two algorithms for the two different types of irregular tables.

One is column-irregular table:

```
Input: T= source code of web pages
Begin :
     Read(T)
     T1=Delimit T(<table, </table)
     T2=Delimit T1<tr, </tr)
     For R=1 to count_num(T2)
         T3=Delimit T2(<td,</td)
           For S=1 to count_num(T3)
             If (checked "col=)
                D=num_of_col
             cell=T3(s)
            For C=1 to D
             array_insert(T3,cell)
           End for
         End if
     N=database_insert(S,R)
     End for
  End for
End
End
```

Fig. 4 Identify column-irregular table algorithm

The other is row-irregular table:

```
Input: T= source code of web pages
Begin :
    Read(T)
     T1=Delimit T(<table, </table)
    T2=Delimit T1<tr,</tr)
    For R=1 to count_num(T2)
        T3=Delimit T2(<td,</td)
          For S=1 to count_num(T3)
            If (checked "row=)
              D=num_of_row
             cell=T3(s)
           For C=1 to D
  Insert cells(cell, row position(s));
```

Fig. 5 Identify row-irregular table algorithm

The basic approach of converting tables is to looking for a specific HTML tag within a HTML page. Variations within tables can be caused by spanning cells using <*rowspan*> and <*colspan*>, the amount of if A tag named *colspan* makes a column-irregular table, *colspan=n* means that it fills in *n-1* cells into the specific column, makes the cells of it equals to other column, and a tag named *rowspan* makes a row-irregular table, *rowspan=n* means that in the following *n-1* rows, each row should fill the specific cell in once. After convert irregular tables into regular tables through these algorithms, useful data should be filter out from the converted tables.

4 DATA MINING AND ANALYSIS

Not all data from the converted tables contains the useful information, unrelated data should be filtered. In this paper, the tables should be related with capacitor or resistor, so a database or a dictionary file should be built as a filter to remove the unexpected tables. Because the ultimate objective of the system is to analysis the specifications of electronic items, then, provide the right items to the purchasing department. The other key component of the system is analysis the data from selected tables. This work is undergoing.

5 CONCLUSION AND FUTURE WORK

In this paper, presents three algorithms, which detect regular, column-irregular, row-irregular tables. And an algorithms for how to get result from the search engine "Google" automatically.

Future work should be analysis the data, which retrieved from selected tables, and return the useful information with, pre-define formats to purchasing. The future work is be studying.

REFERENCES

1 Vokurka, R. J., Choobineh, J. and Vadi, L. (1996) "A prototype expert system for the evaluation and selection of potential suppliers", *International Journal of Operations & Production Management*. Vol. 16, no .12, pp.106-127.

2 Chen, H. H., Tsai, S. C. and Tsai, J. H. (2000) "Mining Tables from Large Scale HTML Texts", *18th International Conference on Computational Linguistics*, Germany.

3 Lerman, K., Knoblock, C. and Minton S. (2001) "Automatic Data Extraction from Lists and Tables in Web Sources", *ATEM-01 International Joint Conference on Artificial Intelligence*, Seattle.

4 Hurst, M. (2001) "Layout and Language: Challenges for Table Understanding on the Web", *1st International Workshop on Web Document Analysis*, USA.

5 Wang, Y. and Hu, J. (2002) "A Machine Learning Based Approach for Table Detection on The Web", *International World Wide Web Conference*: Proceedings of the Eleventh International Conference on World Wide Web 2002, pp. 242-250.

6 Penn, G., Hu, J., Luo, H. and McDonald R. (2001) "Flexible Web Document Analysis for Delivery to Narrow-Bandwidth Devices", *International Conference on Document Analysis and Recognition*, USA.

A mixed integer programming model for integrating MRP and job shop scheduling

K J CHEN and **P JI**
Department of Industrial Systems Engineering, The Hong Kong Polytechnic University, Kowloon, Hong Kong

ABSTRACT

This paper presents a Mixed Integer Programming (MIP) model which succeeds in a systematic integration of the production planning and job shop scheduling problems. The proposed model explicitly considers capacity constraints, operation sequences, lead times and due dates in a dynamic demand environment. The objective of the model is to seek the minimum both production flow time and penalty costs including tardiness and earliness costs. The output of the model is operation schedules with order starting time and finish time. The integrated model is verified with a commercial software package and tested with a small example. The model will be further improved and meta-heuristics will be used to solve the model more efficiently.

1 INTRODUCTION

Traditional Material Requirements Planning (MRP) follows a top-down hierarchical approach. It starts with the generation of specific planned order releases for all final products, subassemblies and components. These order releases are subsequently translated into a set of tasks and due dates, and a detailed job shop scheduling problem is solved to meet these due dates. However, MRP ignores capacity constraints, assumes that lead times are fixed, and does not consider operation sequences of items (1-4). This creates many problems on the shop floor for later production, for example, there is no guarantee that a feasible production schedule exists for the generated production plan. When infeasibilities are found, a great deal of adjustments should be made to the production plan. The unreliable planning process drives planners to lengthen planned lead times to try to improve due date performance. However, longer planned lead times normally cause higher forecast error, longer queues, and even more unreliable planned lead times. Unquestionably, MRP and job shop scheduling are closely interrelated and they should be integrated together to generate realistic operations schedules for the shop floor.

Studies on MRP and scheduling integration have focused primarily on the development of heuristic approaches (3, 5-7). There appears to be scant research on presenting exact mathematical formulations and methods to settle the issue. Lasserre (8) proposed a decomposition approach to solve the integration problem. Dillenberger et al. (9) established an MIP model for resource allocation and multi-period production planning and scheduling. However, these authors did not consider the precedence relationships among the items. More recently, Kolish (10) introduced an MIP model to address the problem in which different customer orders need the same part types. A major limitation of this work is that it considered only a product with two levels, while a real product usually has many levels.

In the general planning and scheduling problem, the most common objective is the minimization of the makespan. Due to the growing interests in JIT production strategy in industry, planning and scheduling with earliness-tardiness penalties has received tremendous attentions (11-16). Wang (13) presented the mathematical description of earliness-tardiness production systems and developed a heuristic algorithm for it. However, only mass and one-of-a-kind product manufacturing systems, usually single-machine and flow shop environment, were studied. Although Kim and Kim (14) explored a short-term production problem for products having multi-level structures, with the objective of minimizing the weighted sum of tardiness and earliness of the items, their research was based on group technology (GT) assumption which simplified the integrated problem.

This paper presents a Mixed Integer Programming (MIP) model which succeeds in a systematic integration of the production planning and job shop scheduling problems. The proposed model explicitly considers capacity constraints, operation sequences, lead times and due dates in a dynamic demand environment. The objective of the model is to minimize both production flow time and penalty costs including tardiness and earliness costs. The output of the model is operation schedules with order starting time and finish time. The integrated model is verified with a commercial software package.

This paper is organized as follows: Section 2 introduces the problem under investigation. In Section 3, an MRP and scheduling integration model is developed. A small example to test the model is presented in Section 4. Finally, Section 5 concludes the paper with a summary and recommendations for future research.

2 PROBLEM DESCRIPTION

We consider a production planning and scheduling problem for products with multi-level structures. In such a structure, items (final products, subassemblies and components) have precedence constraints among them, that is, before processing parent items, their child items should be completed. (Here, a child item represents a lower level item that belongs to a parent item.) Each of these items requires various operations on eligible machines which are continuously available for production. Therefore, constraints of the time capacity should be considered. It is assumed that demands of items and their due dates are known in advance and similarly for processing time of operations.

Our problem is to find an optimal schedule for the demands that both production flow time and penalty costs including tardiness and earliness costs are minimized. Minimizing production flow time is equivalent to minimizing idle time or maximizing machine utilization. In addition, production flow time is chosen as the objective to be reduced because it is able to reflect two focuses in job shops: manufacturing lead times and WIP (work-in-process) inventory level. Another objective of the problem in this paper is to find a schedule with all jobs completed as closed to their due date as possible, which is on the fact that either early or late delivery of an order result in an increase in the production costs.

3 MODEL FORMULATION

In order to formulate the integrated model, the following parameters and variables are introduced.

Parameters:

D_i	Demand index ($i = 1, \ldots, n$)
M_k	Machine index ($k = 1, \ldots, m$)
P_i	Final product of demand D_i
Q_i	Quantity of demand D_i
N_{ip}	Number of item p needed for one final product P_i
t_{ipk}	Processing time required by item p of demand D_i on machine M_k ($p = 1,\ldots, P_i$)
r_i	Ready time of machine M_k
DD_i	Due date of demand D_i
I	Cost of idle time per hour
TC	Cost of tardy demands per day per job
EC	Cost of early demands per day per job
α	A large positive number
$A(p)$	Set of child items of item p
B	Set of item p, where $A(p) = \varnothing$

Variables:

C_{max}	Production makespan
S_{ipk}	Production start time of item p of demand D_i on machine M_k
C_i	Production completion time of demand D_i
L_i	Number of tardy days (real number) for demand D_i
E_i	Number of early days (real number) for demand D_i
L_i^I	Number of tardy days (integer) for demand D_i
E_i^I	Number of early days (integer) for demand D_i
X_{ipjqk}	1 if item p of demand D_i precedes item q of demand D_j on machine M_k; and 0 otherwise

Now, we have the following formulation:

$$\text{MIN}\left\{ I(mC_{\max} - \sum_{i=1}^{n}\sum_{p=1}^{P_i} t_{ipk} \cdot N_{ip} \cdot Q_i - \sum_{k=1}^{m} r_k) + \sum_{i=1}^{n}\left[TC \times L_i^I + EC \times E_i^I\right]\right\} \quad (1)$$

subject to:

$$C_i \le C_{\max} \qquad \forall i. \qquad (2)$$

$$S_{ipk} \ge r_k \qquad p \in B,\ \forall i, k. \qquad (3)$$

$$S_{imj} - S_{ink} \ge t_{ink} \cdot N_{in} \cdot Q_i \qquad n \in A(m),\ \forall i, j, k. \qquad (4)$$

$$S_{iP_ik} + t_{iP_ik} \cdot Q_i = C_i \qquad \forall i, k. \qquad (5)$$

$$S_{ink} \ge S_{lmk} + t_{lmk} \cdot N_{lm} \cdot Q_l - \alpha(X_{inlmk}) \qquad \forall i, k, l, m, n. \qquad (6)$$

$$X_{inlmk} + X_{lmink} = 1 \qquad \forall i, k, l, m, n. \qquad (7)$$

$$\frac{C_i}{8} - DD_i \le L_i \qquad \forall i \qquad (8)$$

$$DD_i - \frac{C_i}{8} \le E_i \qquad \forall i \qquad (9)$$

$$L_i^I \ge L_i \qquad \forall i \qquad (10)$$

$$E_i^I \ge E_i - 0.99 \qquad \forall i \qquad (11)$$

The objective (1) is to minimize the production flow time, job tardiness and earliness in order to minimize the machine idle cost and the penalty costs. Constraints (2) show that the completion time of any demand has to be less than or equal to production makespan. Constraints (3) ensure that the start time of the components should be greater than or equal to the machine ready time. Precedence constraints among the items are satisfied in constraints (4). The completion time of any demand is given by the start time of the final products plus the processing time of the final products. The expression is specified in constraints (5). Constraints (6) and (7) require that no two operations can be processed simultaneously on the same machine. Constraints (8) and (9) define tardiness and earliness of demands, respectively. Meanwhile, unit conversion is conducted in order to transform all the units into days. Expressions (10) and (11) are to convert the value of tardiness and earliness to an integer when the penalty costs are in the unit of days.

4 NUMERICAL RESULTS

Due to space limitation, only a small example is presented to verify the proposed model. The three-level product structure is shown in Fig. 1. Two machines, with 8 hours available per day, are eligible to process the items (Table 1). The ready time of M1 and M2 are Hour 5 and Hour 3, respectively. There are two demands, one requiring 10 Product C1 with due date Day 3 and the other requiring 25 Product C2 with due date Day 4. The following data are useful for calculating the costs.

- Cost of idle time at \$50 per hour.
- Cost of tardiness at \$250 per day per demand.
- Cost of earliness at \$50 per day per demand.

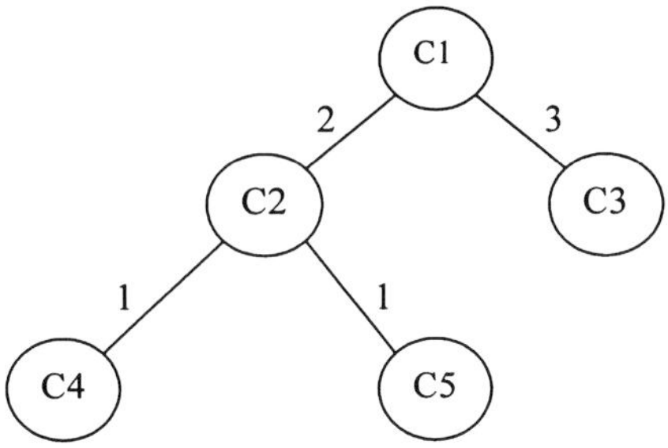

Fig. 1 Three-level product structure

Table 1: Machine processing time for the items

Items	Machine Number	Processing Time (hours)
C1	M1	0.7
C2	M1	0.5
C3	M2	0.5
C4	M2	0.1
C5	M2	0.2

The example was solved using the software CPLEX on a personal computer with Pentium 333 MHz CPU. The important data extracted from the results is summarized in Table 2. The result is graphically represented in the Gantt chart as shown in Fig. 2 (For convenience, item p of demand D_i is denoted ip).

Table 2 (a): Results of the example

Number of constraints	Total Number of variables	No. of integer variables	Makespan (hour)	CPU time (sec.)
62	45	30	38.5	0.77

Table 2 (b): Results of the example

Number of tardiness	Number of earliness	Tardiness penalty	Earliness penalty	Total cost
2	0	500	0	1050

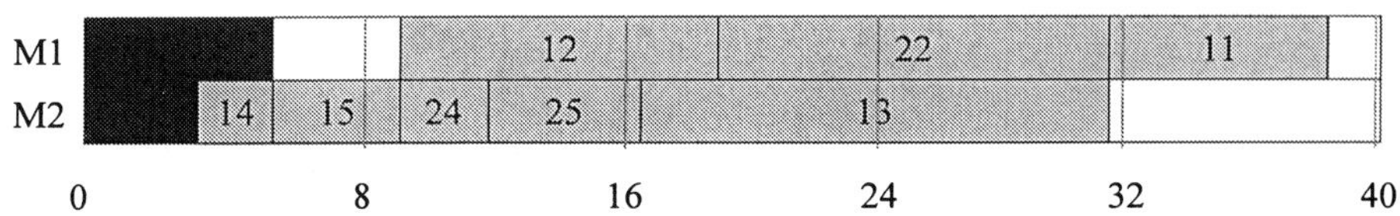

Fig. 2 Output in the form of Gantt Chart

The results show that the developed model can find optimal schedules with operation starting time and finish time, which is more realistic for the shop floor.

5 CONCLUSIONS

An MIP Model for integrating MRP and job shop scheduling, with the objective of minimizing both production flow time and penalty costs including tardiness and earliness costs, has been formulated. The proposed model explicitly considers capacity constraints, operation sequences, lead times and due dates in a dynamic demand environment and generates realistic operations schedules for the shop floor. However, the integrated problem is strongly NP-hard, and it is very difficult to solve practical scale problems in a reasonable amount of time. Therefore, the model will be further improved and meta-heuristics will be used to solve the model more efficiently.

REFERENCES

1. Billington, P. J., Mcclain, J. D., and Thomas, L. J. (1983), "Mathematical programming approaches to capacity-constrained MRP systems: review, formulation and problem reduction," Management Science, 29(10), 1126-1141.
2. Sum, C. C., and Hill, A. V. (1993), "A new framework for manufacturing planning and control systems," Decision Sciences, 24(4), 739-760.
3. Taal, M., and Wortmann, J. C. (1997), "Integrating MRP and finite capacity planning," Production Planning and Control, 8(3), 245-254.
4. VoB, S., and Woodruff, D. L. (2003), Introduction to Computational Optimization Models for Production Planning in a Supply Chain, Springer, Berlin.
5. Bahl, A. C., and Ritzman, L. P. (1984), "An integrated model for master scheduling, lot sizing and capacity requirements planning," Journal of the Operational Research Society, 35(5), 389-399.
6. Miltenburg, J., and Sinnamon, G. (1989), "Scheduling mixed-model multi-level just-in-time production systems," International Journal of Production Research, 27(9), 1487-1509.
7. Riane, F., Artiba, A., and Iassinovski, S. (2001), "An integrated production planning and scheduling system for hybrid flowshop organizations," International Journal of Production Economics, 74(1-3), 33-48.
8. Lasserre, J. B. (1992), "An integrated model for job-shop planning and scheduling," Management Science, 38(8), 1201-1211.
9. Dillenberger, C., Escudero, L. F., Wollensak, A., and Zhang, W. (1994), "On practical resource allocation for production planning and scheduling with period overlapping setups," European Journal of Operational Research, 75(2), 275-286.
10. Kolish, R. (2000), "Integration of assembly and fabrication for make-to-order production," International Journal of Production Economics, 38(3), 287-306.
11. Baker, R., and Scudder, G. D. (1990), "Sequencing with earliness and tardiness penalties: a review," Operation Research, 38(1), 22-36.
12. De, P., Ghosh, J. B., and Wells, C. E. (1993), "On general solution for a class of early/tardy problems," Computer and Operations Research, 20(2), 141-149.
13. Wang, D. W. (1995), "Earliness/tardiness production planning approaches for manufacturing systems", Computers and Industrial Engineering, 28(3), 425-436.
14. Kim, J. U., and Kim, Y. D. (1996), "Simulated annealing and genetic algorithms for scheduling products with multi-level product structure," Computers and Operations Research, 23(9), 857-868.
15. Wang, D. W., Chen, X. Z., and Li, Y. (1996), "Experimental push/pull production planning and control system," Production Planning and Control, 7 (3), 236-241.
16. Gordon, V., Proth, J. M., and Chu, C. B. (2002), "A survey of the state-of-the-art of common due date assignment and scheduling research," European Journal of Operational Research, 139 (1), 1-25.

Buyer-oriented collaboration and stimulation of electronics SCM

Z JIANG
School of Mechanical Engineering, Xi'an Jiaotong University, and Department of Industrial Engineering, Zhengzhou Institute of Aeronautics, People's Republic of China
X FENG and **J SHI**
Department of Industrial Engineering, Zhengzhou Institute of Aeronautics, People's Republic of China
X ZONG
School of Mechanical Engineering, Xi'an Jiaotong University, People's Republic of China
H WANG
Wolfson School of Mechanical and Manufacturing Engineering, Loughborough University, UK

ABSTRACT

This paper studies the coordination and stimulation in buyer-oriented E-supply chain management under the return contract. When the supplier in such supply chain gives the buyer a certain rebate rate to unsold products, the supply chain management can be coordinated. The buyer always has the motivation to join the E-supply chain management, but the supplier's motivation depends on the rate of commission and coordination mechanism. When the rate of commission is lowered to a certain level, the supplier will join the E-supply chain management even without any extra incentive. However, when the rate of commission is increased to a certain higher level, the supplier will not join the E-supply chain management even if the buyer offers the highest subsidy. When the situation is between the above two cases, i.e., if the buyer gives a certain amount of subsidy to the supplier, the whole supply chain still can be coordinated. The paper gives explanation to the failure of E-supply chain management and proposes a practical and operable plan of coordination and stimulation in E-supply chain management on the basis of return contract.

1 INTRODUCTION

The kernel issue of supply chain management (SCM) is collaboration and stimulation[1,2]. Contract (i.e. Return Contract, revenue sharing contract, backups contract, option contract, etc) is the most prevalent coordination and stimulation mechanism in traditional SCM. The collaboration and stimulation have been well studied in SCM. However, its researching object is based on traditional SCM. With the development and popularization of Internet, and the application of E-Commerce in SCM, the electronic supply chain management (e-SCM) is formed[3,4,5]. According to the different structure of E-Commerce, e-SCM could divided into three kinds: buyer-biased e-SCM, neutral e-SCM and seller-biased e-SCM, as Table 1.

* Project supported by the University Young Backbone Teacher Program of Henan Province, and the Tackle-Key-Program of S&T Committee of Henan Province of China (0424260002, 0424260016, 0424220192), and the National High-Technology R&D Program for CIMS (2002AA415250, 2003AA4Z3300), China.

Table.1 The classification and its characteristics of e-SCM

Classes	Excellency	Shortcoming	Samples
Buyer-biased e-SCM	Buyer win the lowest price through collecting seller info and inviting public bidding.	Seller faced with vehement market price; and paying for commission; and not enough supplier.	FreeMarkets FOB Covisint
Neutral e-SCM	Win benefits of seller and buyer through collecting info and inviting public bidding of seller and buyer each other.	The seller and buyer will unwillingness joining if there is not enough actors; and all paying for commission each other.	E-steel Chemdee FastParts
Seller-biased e-SCM	seller win the maximal price through collecting buyer info and inviting public bidding.	Buyer faced with vehement market price; and paying for commission; and not enough buyers.	E-Chemicals Telekom Deutsche

E-SCM is differ from the traditional SCM in that it has self characteristic in coordination and stimulation with depressing consumedly the transaction cost of buyer and seller, and paying the definite commission and leaguer fee for e-commence serve provider. This paper studies the coordination and stimulation in buyer-oriented E-supply chain management under the return contract. When the supplier in such supply chain gives the buyer a certain rebate rate to unsold products, the supply chain management can be coordinated.

2 MODEL EXPLANATION

Set c as the unit production costs of the supplier, set w_1 and w_2 respectively as the unit selling prices of traditional supply chain and e-supply chain. As the researching object is the buyer-oriented e-supply chain, and the buyer needn't to pay commission but the supplier needs to pay certain and invariable fee of membership (S) and a certain amount of commission that calculated according to the amount of money of the trade, and set ρ as the rate of commission. Set Q_1 and Q_2 respectively as the order quantity of the buyer in the traditional supply chain and in the e-supply chain. Set b_1 and b_2 respectively as the returning refund proportions that the supplier pays for the goods that the buyer couldn't sell out in traditional supply chain and in the e-supply chain. Set r as the retailing unit price of the distributor. Set A_s and A_b respectively as unit trading fee of the supplier and the buyer in the traditional supply chain. In e-supply chain, as the trading fee is reduced enormously with e-commerce and derived simply, set the trading cost of the supplier and the buyer as ***zero***. Set x as the amount of customer's requirement (CR). x is a continuous and random variable that is not negative, the density function is $f(x)$ and it's distributing function is $F(x)$.

In this model, the order of incident is as follows:

(1) The supplier participates the purchasing competitive bidding, and the buyer chooses the person who win the bidding form it, and get order from him;
(2) The supplier organizes the production and sends the goods to the buyer;
(3) The demand happens and the unmet demand is backlogged. The unsold goods are returned the supplier at a discount.

And the hypothesis is as follows:

(1) Neglecting the lose of reputation by short supply;
(2) The supplier's production capacity is unrestricted, and the relationship is positive proportion within the cost and the quantity of production.

(3) F(x) is successive and differentiable, and it has a contradictorily function. By way of seeking the obvious result, and supposing x accords with the equably distributing in block [0, D], so $f(x) = \frac{1}{D}$,

(4) For ensuring the profit of the supplier and the buyer in the traditional supply chain, that r-Ab>w1>As+c;

(5) In e-commerce, the supplier is produced by competition. So the price is lower than that in the traditional supply chain, that w1>w2>c;

(6) The retailing price is invariable, no matter where the buyer gets the goods, from traditional supply chain or e-commerce network-station.

(7) The Common Knowledge is the price, cost and demand of goods, and there is not anisomerous information problem between the supplier and the buyer.

3 COLLABORATION OF TRADITIONAL SCM

To comparing with the e-SCM and researching the stimulation when the manufacturer and the buyer join the e-supply chain. At firstly, studying the stimulation of traditional supply chain. In traditional supply chain, supplier must afford as production cost c and exchange fee A_s, and the distributor must afford the purchasing cost w_l and exchange fee A_b.

3.1 Integrated Supply Chain

In the integrated supply chain, supplier and distributor allocate products by ierarchy mechanism, so there is no exchange fee, that also means A_s=A_b=0. The whole supply chain's expected profit is as:

$$\Pi(Q) = -cQ + \int_0^Q rxf(x)dx + \int_Q^\infty rQf(x)dx \qquad (1)$$

According to literature [8], the production quantity optimistically of supply chain can be expressed as follows:

$$Q_T^* = F^{-1}(\frac{r-c}{r}) \qquad (2)$$

3.2 Distributive Supply Chain

In distributing SCM, the supplier and distributor are independent main-body, and their trading by market mechanism should afford the homologous fee, hat also means A_s=A_b≠0. Considering the trading fee, the distributor's profit (Π_T^B) is as:

$$\Pi_T^B = -w_1Q_1 - A_bQ_1 + \int_0^{Q_1} rxf(x)dx + \int_{Q_1}^\infty rQ_1f(x)dx \qquad (3)$$

So, according to literature [9], the economic order quantity of buyer is as:

$$Q_1^* = F^{-1}(\frac{r-w_1-A_b}{r}) \qquad (4)$$

Accordingly, if the order quantity is Q1*, the supplier's profit is as:

$$\Pi_T^S(Q_1^*) = -cQ_1^* - A_sQ_1^* + w_1Q_1^* \qquad (5)$$

Comparing (2) with (4), as w_l>c, $\frac{r-w_1-A_b}{r} < \frac{r-c}{r}$, and according to assumption 4:

$$Q_1^* = F^{-1}(\frac{r-w_1-A_b}{r}) < Q_T^* = F^{-1}(\frac{r-c}{r}).$$

Therefore:

Deduction 1: In a traditional supply chain, the economic order quantify of the buyer is less than the optimistic production quantity of the supply chain, in other words $Q_1^* < Q_T^*$.

The above-mentioned phenomenon is known as Double Marginalization [10]. Which means, when the trade price of supplier is higher than its marginal cost, the buyer's economic order quantity will be less than the optimum production quantity of whole supply chain. In order to controlling this phenomenon, the supply chain management must afford a collaborative mechanism, which is used by the buyer so that their economic order quantity can be equal to the whole supply chain's production. This paper studies the collaboration under the return contract.

According to [9], in the whole given back items and the rate of drawback of each unit of product is b_1, the buyer's economic order quantity q* is as:

$$q^* = F^{-1}(\frac{r - w_1 - A_b}{r - b_1})$$

Assuming $q^* = Q_T^*$, that is $F^{-1}(\frac{r - w_1 - A_b}{r - b_1}) = F^{-1}(\frac{r - c}{r})$

According to the assumption 4, gained $\frac{r - w_1 - A_b}{r - b_1} = \frac{r - c}{r}$, the consequence is $b_1 = \frac{(w_1 + A_b - c)r}{r - c}$. And the second deduction is as follows:

Deduction 2: The refund rate $\{ b_1 = \frac{(w_1 + A_b - c)r}{r - c} \}$ that the supplier provided can be coordinated to the traditional SCM under the strategy of surplus product return.

4 COLLABORATION AND STIMULATION OF E-SCM

In the e-SCM, suppliers have to pay the trade commission (ρw_2) to the website for each unit of agreement, in addition the firm use cost charged generally by the year. Because the e-business greatly reduced the trade cost. So, the assumption of its trade cost is zero.

4.1 The collaboration of the e-SCM

The e-SCM is the same as the traditional SCM, and the optimum production quantity of integrated supply chain is expressed as follows:

$$Q_E^* = F^{-1}(\frac{r - c}{r}) \qquad (6)$$

In distributive SCM, the supplier and the distributor respectively maximize their own profits. The distributor's profit (Π_E^B) is expressed as follows:

$$\Pi_E^B = -w_2 Q_2 + \int_0^{Q_2} rxf(x)dx + \int_{Q_2}^{\infty} rQ_2 f(x)dx \qquad (7)$$

And also according to [8], the optimum order quantity of the distributor (Q_2^*) can be expressed as follows:

$$Q_2^* = F^{-1}(\frac{r-w_2}{r}) \qquad (8)$$

Correspondingly, when the order quantity equals Q_2^*, the supplier's profit (Π_E^S) is as:

$$\Pi_E^S(Q_2^*) = (w_2 - c - \rho w_2)Q_2^* - s \qquad (9)$$

Comparing (6) with (8), we can find the third deduction:

Deduction 3: In e-SCM, the optimum order quantity of the distributor is less than the optimum production quantity in the supply chain, that is $Q_2^* < Q_E^*$.

This means that there is a "Double Marginalization" phenomenon in the e-SCM as well. So it is possible to coordinate under the return contract. The reasoning process is similar to that of the second deduction. So we come to:

Deduction 4: In e-SCM, the refund rate $\{ b_2 = \frac{(w_2 - c)r}{r - c} \}$ that the supplier provided for the surplus product can be coordinated to the e-SCM.

4.2 The stimulation of the e-SCM

Although the whole e-SCM can be more excellent through adopting the coordinating mechanism, but according to the stimulation compatibility, manufacturer and the distributor have motivation to engage in the e-SCM unless the profits of both them are respectively more than that in the traditional SCM.

4.2.1 The stimulation to the distributor(or buyer)

In the traditional SCM,

$$\Pi_T^B(Q^*) = -(w_1 + A_b)Q^* + \int_0^{Q^*}[rx + b_1(Q^* - x)]f(x)dx + \int_{Q_1^*}^{\infty} rQ^* f(x)dx$$

and also, in the e-SCM,

$$\Pi_E^B(Q^*) = -w_2 Q^* + \int_0^{Q^*}[rx + b_2(Q^* - x)]f(x)dx + \int_{Q_1^*}^{\infty} rQ^* f(x)dx$$

Assuming $\Delta\Pi^B = \Delta\Pi_E^B(Q^*) - \Pi_T^B(Q^*)$, introduce Q^*, b_1 and b_2. The consequence is as:

$$\Delta\Pi^B = \frac{1}{2}(w_1 + A_b - w_2)Q^* > 0$$

Hence, the fifth deduction can be expressed as:

$$\Pi_T^B(Q^*) = -(w_1 + A_b)Q^* + \int_0^{Q^*}[rx + b_1(Q^* - x)]f(x)dx + \int_{Q_1^*}^{\infty} rQ^* f(x)dx$$

Deduction 5: In e-SCM, the profit of the distributor is always more than that in the traditional SCM after coordinating. So the rational distributor should engage in the e-supply chain spontaneously without additional encouragement.

4.2.2 The stimulation to the supplier

In traditional SCM, the profit of the supplier is expressed as follows:

$$\Pi_T^S(Q^*) = (w_1 - c - A_s)Q^* - \int_0^{Q^*} b_1(Q^* - x)f(x)dx$$

In e-SCM, the profit of the supplier is expressed as follows:

$$\Pi_E^S(Q^*) = (w_2 - c - \rho w_2)Q^* - S - \int_0^{Q^*} b_2(Q^* - x)f(x)dx$$

Assume $\Delta\Pi^S = \Delta\Pi_E^S(Q^*) - \Pi_T^S(Q^*)$, Introducing Q^*, b_1 and b_2, the consequence is as:

$$\Delta\Pi^S = \frac{1}{2}(w_2 - w_1 + 2A_s - 2\rho w_2 + A_b)Q^* - S$$

Introducing Q^*, and assuming the equation above equals zero, the consequence can be expressed:

$$\rho^* = \frac{2A_s + A_b + w_2 - w_1}{2w_2} - \frac{rS}{(r-c)Dw_2}$$

If $\rho \geq \rho^*$, $\Delta\Pi^S \leq 0$, that is, the profit that the supplier gained in e-SCM is less than that in the traditional SCM. So, the rational supplier refuses to engage in e-SCM. And if $\rho < \rho^*$, $\Delta\Pi^S > 0$, the supplier will engage in it spontaneously.

So, the deduction can be expressed as follows:

Deduction 6: Under the coordinating mechanism of the surplus return, if the commission $\rho \geq \rho^*$, the supplier will not engage in e-SCM. And if the commission $\rho < \rho^*$, the supplier will engage in it spontaneously.

According to Deduction 5, if $\rho \geq \rho^*$, the supplier refuses to engage in e-SCM, but if the distributor will give the supplier some bonus from his own profit, the supplier has possibly motivation to engage in it. So it is helpful to make the coordination of the supply chain continuous. In underside, the maximum unit bonus that the distributor is able to provide will be deduced.

In e-SCM, the good quantity which the distributor able to sell is:

$$Q_E^E = Q^* - \int_0^{Q^*} (Q^* - x)f(x)dx = Q^* - \frac{Q^{*2}}{2D}$$

Assuming the commission rate under this condition is ρ_{Max}, the incremental profit per unit that the distributor gained is $\frac{\Delta\Pi^B}{Q_E^E} = \frac{r(w_1 + A_b - w_2)}{r+c}$, and the maximum unit bonus that the distributor is able to pay is $(\rho_{Max} - \rho^*)w_2 = \frac{\Delta\Pi^B}{Q_E^E}$, so we can come to

$\rho_{Max} = \rho^* + \frac{r(w_1 + A_b - w_2)}{(r+c)w_2}$. If $\rho > \rho_{Max}$, the maximum bonus paid by the supplier is not enough to profit the supplier from engaging in the e-SC, so the latter refuses to engage in it. Therefore the seventh deduction is as follows:

Deduction7: if the commission rate $\rho^* \leq \rho \leq \rho_{Max}$, the buyer gives a certain amount of unit bonus ($\rho - \rho^*$) to the supplier, the supplier will be stimulated to engage the e-SC and the whole supply chain will be improved, also the maximum unit bonus that the buyer gives is $\frac{r(w_1 + A_b - w_2)}{r+c}$; If the $\rho > \rho_{Max}$, the supplier will not engage the e-SC spontaneously, even if the buyer offers the maximum unit bonus.

5 CONCLUSIONS AND FURTHER RESEARCH DIRECTIONS

This paper studies the coordination and stimulation in buyer-oriented E-supply chain under the return contract and gets following conclusions: when the supplier in such supply chain gives the buyer a certain rebate rate to unsold products, the supply chain can be coordinated. The buyer always has the motivation to join the E-supply chain, but the supplier's motivation depends on the rate of commission and coordination mechanism. When the rate of commission is lowered to a certain level ($\rho<\rho^*$), the supplier will join the E-supply chain even without any extra incentive. However, when the rate of commission is increased to a certain higher level ($\rho>\rho_{Max}$), the supplier will not join the E-supply chain even if the buyer offers the highest subsidy. Then the E-supply chain is dismissed because of being short of participants, when the situation is between the above two cases ($\rho^*\leq\rho\leq\rho_{Max}$), as the buyer gives a certain amount of subsidy to the supplier, the whole supply chain can still be coordinated. The paper gives explanation to the failure of E-supply chain, and proposes a practical and operable mechanism of coordination in E-supply chain on the basis of return contract. According to the mechanism, we can learn how to define the rational rate of commission of E-commerce. We will do further research on how to make use of other contract to carry on the coordination and stimulation in e-supply chain management.

ACKNOWLEDGEMENTS

This work was supported by the University Young Backbone Teacher Program of Henan Province, and the Tackle-Key-Program of S&T Committee of Henan Province of China (0424260002, 0424260016), and the National High-Technology R&D Program for CIMS (2002AA415250), China.

REFERENCES

1 Chen F., "Information Sharing and Supply Chain Coordination" [R], Columbia University, 2002.

2 Cachon G, Lariviere M., "Contracting to Assure Supply: How to Share Demand Forecasting in a Supply Chain" [J], *Management Science*, 2001, 47(5): 629-646.

3 Kunshan Ying, "e-SCM analyzing" [J], *Policy and Management*, 2001, (1): 20-22.

4 Lancioni R A, Smith M F, Oliva T A., "The Role d the Internet in Supply Chain Management" [J], *Industrial Marketing Management*, 2000, 29(1): 45-56.

5 Ning Feng, Boyi Xu,."e-Supply chain Management" [J], *IE&Management*, 2001, 6(6): 67-71

6 Corbett C J, Tang C S., "Designing Supply Chain Contracts: Contracts Type and Information Asymmetric" [A], In: Tayur S, Ganeshan R. *Quantitative Models for Supply Chain* [M]. Boston (MA): Kluwer Academic Publishers, 1998. 269-297.

7 Barnes-Shuster D, Bass Y. "Coordination and Flexibility in Supply Contracts with Options" [R], University of Chicago, 2000.

8 Cachon G P., "Supply Chain Coordination with Contracts" [R], Warton School of Business, University of Pennsylvania, 2001.

9 Pastemack B A., "Optimal Pricing and Return Policies for Perishable Cowanodities" [J], *Marketing Science*, 1985, 4(2): 166-172.

10 Spongier J J., "Vertical Integration and Antitrust Policy" [J], *Journal of Political Economy*, 2002, 58: 347-352.

The implementation and evaluation of a 'micro' supply chain management system in a medium-sized UK manufacturer

A JONES and **J ROBINSON**
Production Planning, Saftronics Limited, Leeds, UK
B O'TOOLE and **D WEBB**
School of Engineering, Leeds Metropolitan University, UK

ABSTRACT

Companies often focus efforts to improve business efficiencies on internal systems and significant benefits can be achieved by so doing. However, the magnitude of the realised benefit is restricted if only processes within the organisation are targeted for improvement.

This paper describes the implementation of a bespoke web-based supply chain management system / online collaboration system. The main drivers behind the development of the system were late deliveries from the supplier and also sporadic and inaccurate delivery schedule information from the supplier.

The implementation approach is discussed along with the problems that were encountered en route and the measures that were undertaken to ensure that implementation was successful. The importance of metrics is stressed as these are essential if any improvements are to be quantified which in turn enables the justification of such projects in financial terms which is compounded by the 'soft' benefits that result from the use of such systems.

1 BACKGROUND

Manufacturing is vital to the economies of Europe, "more so than ever" according to the Director of Economic Affairs of the Union of European Industrial & Employers Confederations (UNICE). Considering employment alone manufacturing generates 4.5 times as many secondary jobs as retail and so is clearly important in the context of lowering unemployment. Despite the importance of manufacturing, employment in the sector fell by nearly 20% across much of Europe between 1970 and 1990 with the worst affected being the UK. Over the same time span import penetration doubled in the UK and these trends have continued coupled with major job losses across the sector as whole.

These job losses have, however, contributed to the almost 50% increase in UK productivity that manufacturing experienced between 1985 and 1994. However, existing standards of value adding capability within British industry still remain at a low level and if the UK manufacturing industry is to be competitive globally then the fat must continue to be trimmed and the strategic benefits that can be facilitated by recent advances in technology, for instance, the feasibility of supply chain planning and online collaboration, need to be built. Further, effective implementation and management of these 'devices for efficiency' for manufacturers are considered to be critical to long-term survival and competitive advantage.

In the manufacturing industry in particular it is especially important that we in the UK tend towards the 'crest rider' end of the spectrum of approach to technology as labour costs are extremely high relative to those in developing economies and this will be the only way that UK manufacturers will be able to be profitable and successful in the longer term.

One UK manufacturer keen to apply new ideas and approaches to solving old (supply chain) problems is Saftronics. Saftronics employs approximately 130 people and assembles motor control centres (MCCs) and control panels (CPs) the majority of which are supplied to the UK water industry. These MCCs & CPs consist of large bespoke fabricated and painted metal enclosures containing various low voltage electrical switchgear and components as specified by the customer.

2 METHODOLOGY ADOPTED

As part of a KTP (Knowledge Transfer Partnership) project various aspects of Saftronics' supply chain as well as internal planning and production processes were looked into with a view to unearthing improvement opportunities. The project stages included:

1. Identify key problems and opportunities for improvement
2. Propose solution and specify requirements
3. State expected business benefits
4. Develop and test the solution
5. Pilot the solution and Full roll-out of the solution
6. Review and evaluation

3 IDENTIFY KEY PROBLEMS AND OPPORTUNITIES FOR IMPROVEMENT

Production of any description was unable to begin until painted metalwork was delivered to Saftronics. The fact that production could not commence until the metalwork had been received coupled with the fact that metalwork deliveries were by default often significantly late was a real body blow to Saftronics' hopes of meeting customer delivery dates. However this metalwork delivery tardiness was only one dimension of the metalwork issue, the other, equally important in the view of the author (AJ), was that there was no certainty as to what was going to be delivered and when. Stores personnel coined this problem saying that it was 'like Christmas every day'. This resulted in the situation where planning and scheduling of production could only be undertaken with any hope of reliability *after* deliveries had been physically received by Saftronics. These issues of routine lateness and uncertainty combined to give rise to some of the serious operational problems that senior management had identified, namely that:

- Delivery dates of orders to Saftronics' customers were routinely after schedule.
- When Saftronics did deliver an order on time it was usually achieved at considerable additional expense by the use of over time or outsourcing.
- 'Planning' of production tended to be reactive and short term undertaken with reference to what metalwork was in the building rather than what should be produced, this reactive 'hand-to-mouth' approach of production control led to excessive dependence on expediting systems not to mention unnecessary stress.

In brief, digging into the processes deeper revealed there were a number of contributory factors which combined to present the various problems to Saftronics, for instance:

- Delivery performance of metalwork items was not measured or tracked in any real way.
- The dates that metalwork was wanted to be delivered to Saftronics tended to be imposed rather than arrived at by consultation and with respect to available capacity.
- There was considerable scope for metalwork orders to be 'lost' in no man's land, beyond the control of Saftronics, between fabrication and the paint shop.
- The fact that items of metalwork were not uniquely identified meant that there was no formal checking in process for items of metalwork that were received.

4 PROPOSE SOLUTION AND SPECIFY REQUIREMENTS

As mentioned already the set of problems to be addressed were substantially interlinked. There were four main areas that, if properly addressed would greatly reduce many of the headaches that Saftronics had been experiencing, these were:

- Measurement of performance
- Realism in relation to targets and plans
- Traceability of items of metalwork
- Visibility of job status and targets both internally and in the supply chain

4.1 Measuring Performance

If the adage 'you cannot control what you cannot measure' is to be believed then it was fundamentally important that metalwork delivery performance be monitored against each job's project plan and that this data be kept and monitored routinely.

4.2 Realistic Plans

In favour of attempting to impose unachievable delivery dates on the fabricator it is better to consult with them and ask them to give a delivery date that they are willing to commit to. The focus here being on eliciting workable dates such that plans will be met and will start to be seen as being credible.

4.3 Traceability

Metalwork items would be uniquely identifiable so as to enable a formal parts checking in process to be followed by Goods In at Saftronics.

4.4 Visibility

Saftronics should give the fabricator advance warning of orders i.e. as soon as the company wins an order they should alert the supplier about the job, its size and when designs would be scheduled for release to the supplier. This would mean that the supplier is able to make informed planning and resourcing decisions based on an anticipated order schedule. For Saftronics, the information required for workable production plans would be enhanced by the supplier alerting Saftronics of any problems i.e. advance warning of late deliveries to Saftronics.

These suggestions, while they are of significant value, they do not guarantee relevancy, timeliness or visibility. To these ends a database system accessible by both Saftronics and the metalwork supplier would be required to replace the previous spreadsheet system. Both parties having easy a access to a centralised database would ensure that each would be working to the same set of information and that this information could be up-to-date (live / real-time).

4.5 'Micro' Supply Chain Management System

So as to make the information easily accessible to both parties it was decided that the system would be deployed via the world wide web and access to the system would be controlled by username and password. The micro SCM system would provide the following functionality via a web interface:

- Allow jobs with end customer delivery dates to be defined and allow the bill of material (BOM), from the metalwork point of view, for a job to be entered.
- The system needs to be date oriented rather than activity oriented i.e. the emphasis must be on what date Saftronics will receive metalwork parts rather than at what stage in the fabrication process a job is with the fabricator.
- Easy reporting of metalwork parts that need to be chased.
- Quick tracking and reporting of parts that were delivered late... and why.
- The content and functionality made available to system users is dependent upon their login level, for instance only Stores members are able to check in metalwork parts, only the supplier is able to update the 'revised delivery date' information.
- Provide the fabricator with information to assist decision-making regarding rough cut capacity requirements.
- Provide Saftronics with a more source of reliable and up to date information to enable production planning to be undertaken in a meaningful way.

4.6 Sticks and... Carrots!

To clarify the ground rules and to define the context within which the micro SCM system would be used an agreement was drawn up between Saftronics and the fabricator, the main points of interest consisted of:

- Fabricator to update the online schedule with 'revised delivery dates' as appropriate on at least a daily basis.
- Fabricator to take control of the paint process as well as the production of the metalwork. For this extra service (and responsibility) Saftronics would pay a management fee to the fabricator. This removes scope for problems falling between the cracks (i.e. finger-pointing between the fabricator and the painters) and therefore increases visibility and eases monitoring of any slippage as ownership lies only with one party.
- Supplier receives additional 'motivation' for achieving delivery dates they agree to i.e. a percentage of the agreed fee for a job is forgone if the agreed delivery dates are missed.

- Deliveries are signed for by Stores and this delivery date is recorded on the metalwork logging system (a part of the micro SCM system).

5 STATEMENT OF EXPECTED BUSINESS BENEFITS

The aims of the proposed solution were to increase the on-time delivery of metalwork to Saftronics and, where this could not be achieved, to improve the quality of information (i.e. accuracy and timeliness) coming from the fabricator regarding revised delivery dates. The achievement of this would mean that production 'planning' could be attempted and Saftronics would be in a better position to meet customer deadlines and therefore win repeat business with less Sales effort.

6 DEVELOP THE TEST SOLUTION: THE MICRO SCM SYSTEM

After achieving buy-in from Saftronics senior management and also the managing director of the fabricator the contract was agreed and signed off (as per 4.6). This was a green flag for the process translating functional requirements (and user expectations!) into a robust and effective 'micro' SCM' prototype system (based around an MS Access 2000 database).

The web interface to the database comprised of a suite of ASP pages (active server pages) written in VBScript such that the system could be used by anyone with a browser (and a valid username and password). In respect of the users of the micro SCM system the following categories were defined:

- Engineers – able to enquire about the delivery status parts of any project, view performance measurement charts etc.
- Planning – as for Engineers and also able to define new projects on the micro SCM system.
- CAD – as for Engineers and also able to explode a project into its composite parts.
- Fabricator – as for Engineers and also able to update the 'revised delivery dates' for metalwork parts.
- Stores – as for Engineers and also able to book in received metalwork parts.
- Admin – as all of the above and with a additional functionality to administer the database / website.

The main screens of interest to Saftronics from a planning and operational point of view are the 'Delivery Status of Metalwork Parts' page (Fig. 1.) and the 'Detailed Delivery Status' page (Fig. 2.) The key points of the 'Delivery Status of Metalwork Parts' page (Fig. 1.) are:

- lists all the live jobs – live from Saftronics point of view.
- gives high level information about each job including its size, who the Engineer responsible for the job is and key dates, namely:
    - 'Planned Order' – the date the order is due to be released to the fabricator.
    - 'Actual Order' – the date the metalwork order was released to the fabricator.
    - 'Planned Delivery' – the agreed delivery date of the parts comprising the job.
    - 'Revised Delivery' – updated by the fabricator this date reflects any expected change to the planned delivery.

- o 'Actual Delivery' – the date that the final part for the job was received at Saftronics.
- the report can be sorted by the above dates.
- flag indicating whether the job has been fully received or not.
- late ordered and late received jobs are highlighted with yellow backgrounds.
- the report page can be filtered by Engineer.
- hyperlinks are provided to the 'Detailed Delivery Status' page for each live job.

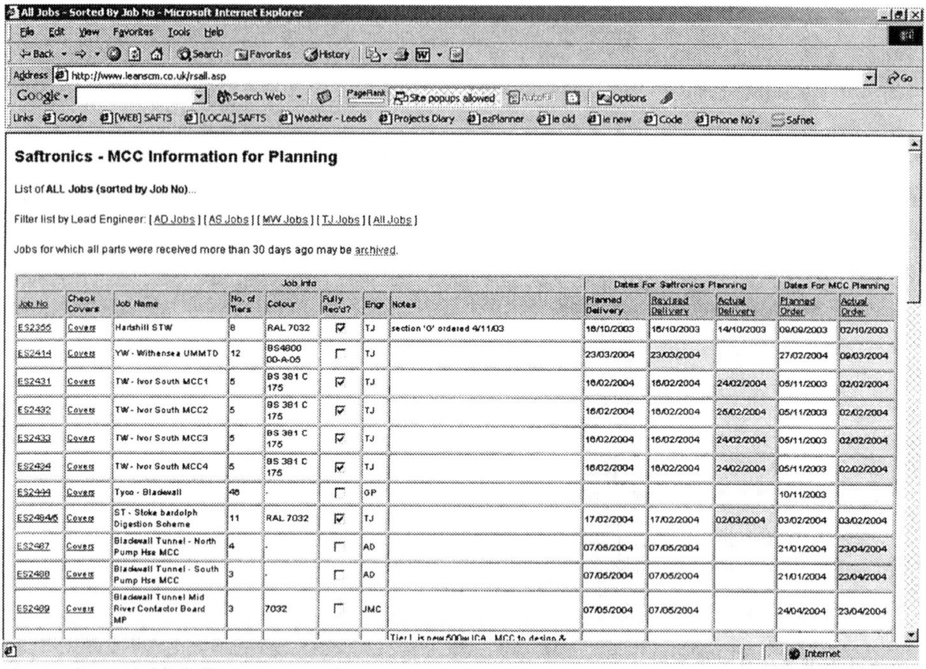

Job Info								Dates For Saftronics Planning			Dates For MCC Planning	
Job No	Check Covers	Job Name	No. of Tiers	Colour	Fully Rec'd?	Engr	Notes	Planned Delivery	Revised Delivery	Actual Delivery	Planned Order	Actual Order
ES2356	Covers	Hartshill STW	8	RAL 7032	☑	TJ	section 'G' ordered 4/11/03	16/10/2003	16/10/2003	14/10/2003	09/09/2003	02/10/2003
ES2414	Covers	YW - Withensea UMMTD	12	BS4800 00-A-05	☐	TJ		23/03/2004	23/03/2004		27/02/2004	09/03/2004
ES2431	Covers	TW - Ivor South MCC1	5	BS 381 C 175	☑	TJ		16/02/2004	16/02/2004	24/02/2004	05/11/2003	02/02/2004
ES2432	Covers	TW - Ivor South MCC2	5	BS 381 C 175	☑	TJ		16/02/2004	16/02/2004	26/02/2004	05/11/2003	02/02/2004
ES2433	Covers	TW - Ivor South MCC3	5	BS 381 C 175	☑	TJ		16/02/2004	16/02/2004	24/02/2004	05/11/2003	02/02/2004
ES2434	Covers	TW - Ivor South MCC4	5	BS 381 C 175	☑	TJ		16/02/2004	16/02/2004	24/02/2004	05/11/2003	02/02/2004
ES2444	Covers	Tyco - Bladewall	48	-	☐	GP					10/11/2003	
ES2464/5	Covers	ST - Stoke bardolph Digestion Scheme	11	RAL 7032	☑	TJ		17/02/2004	17/02/2004	02/03/2004	03/02/2004	03/02/2004
ES2487	Covers	Bladewall Tunnel - North Pump Hse MCC	4	-	☐	AD		07/05/2004	07/05/2004		21/01/2004	23/04/2004
ES2488	Covers	Bladewall Tunnel - South Pump Hse MCC	3	-	☐	AD		07/05/2004	07/05/2004		21/01/2004	23/04/2004
ES2489	Covers	Bladewall Tunnel Mid River Contactor Board MP	3	7032	☐	JMC		07/05/2004	07/05/2004		24/04/2004	23/04/2004
							Tier L is new 500w ICA MCC to design &					

Fig. 1 Roles of Each Category of User at Each Stage in the Planning, Design, Manufacture and Receipt of Metalwork Parts

The key points of the 'Detailed Delivery Status' page (Fig. 2.) are:

- lists every metalwork part that has been defined for the selected job.
- for each part it displays various details as follows:
  - o 'Part ID' – the part's ID code – unique to the job.
  - o 'Part Type' – the category of metalwork part i.e. door, back plate, tier etc.
  - o 'Order Date' – the date the part was ordered from the fabricator.
  - o 'Planned Delivery Date' – the part's initially agreed date for delivery to Saftronics.
  - o 'Revised Date' – defaults to the 'Planned Delivery Date' but can be changed by the fabricator should the need arise.
  - o 'Received Date' – the date that the part was received at Saftronics. If this field is blank this indicates that the part has not yet been received and if it has a yellow background this indicates that the part was received late.
  - o 'Notes' – a field displaying any notes added by CAD, the fabricator or Stores.

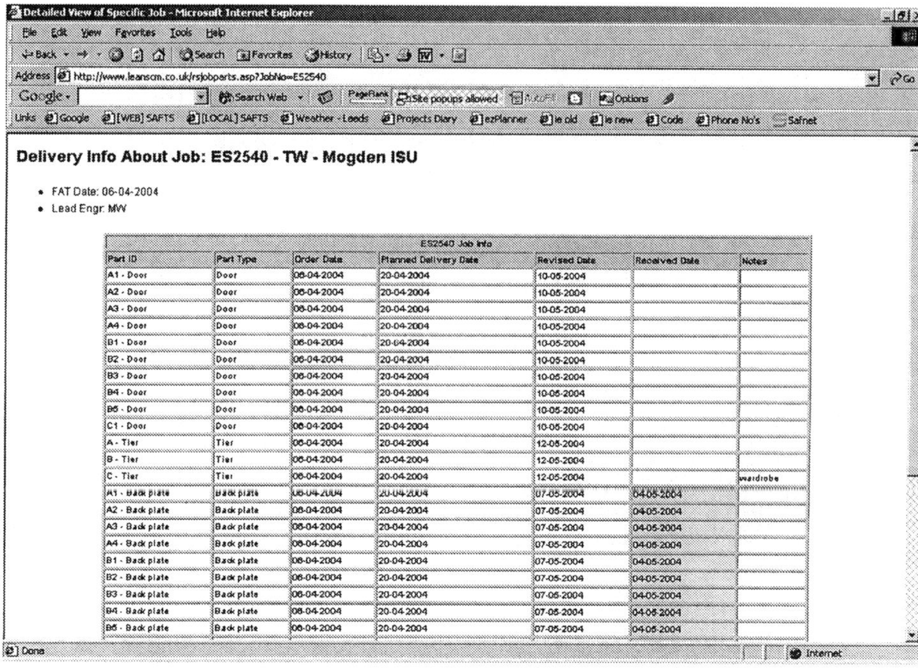

Delivery Info About Job: ES2540 - TW - Mogden ISU

- FAT Date: 06-04-2004
- Lead Engr: MW

ES2540 Job Info						
Part ID	Part Type	Order Date	Planned Delivery Date	Revised Date	Received Date	Notes
A1 - Door	Door	06-04-2004	20-04-2004	10-05-2004		
A2 - Door	Door	06-04-2004	20-04-2004	10-05-2004		
A3 - Door	Door	06-04-2004	20-04-2004	10-05-2004		
A4 - Door	Door	06-04-2004	20-04-2004	10-05-2004		
B1 - Door	Door	06-04-2004	20-04-2004	10-05-2004		
B2 - Door	Door	06-04-2004	20-04-2004	10-05-2004		
B3 - Door	Door	06-04-2004	20-04-2004	10-05-2004		
B4 - Door	Door	06-04-2004	20-04-2004	10-05-2004		
B5 - Door	Door	06-04-2004	20-04-2004	10-05-2004		
C1 - Door	Door	06-04-2004	20-04-2004	10-05-2004		
A - Tier	Tier	06-04-2004	20-04-2004	12-05-2004		
B - Tier	Tier	06-04-2004	20-04-2004	12-05-2004		
C - Tier	Tier	06-04-2004	20-04-2004	12-05-2004		wardrobe
A1 - Back plate	Back plate	06-04-2004	20-04-2004	07-05-2004	04-05-2004	
A2 - Back plate	Back plate	06-04-2004	20-04-2004	07-05-2004	04-05-2004	
A3 - Back plate	Back plate	06-04-2004	20-04-2004	07-05-2004	04-05-2004	
A4 - Back plate	Back plate	06-04-2004	20-04-2004	07-05-2004	04-05-2004	
B1 - Back plate	Back plate	06-04-2004	20-04-2004	07-05-2004	04-05-2004	
B2 - Back plate	Back plate	06-04-2004	20-04-2004	07-05-2004	04-05-2004	
B3 - Back plate	Back plate	06-04-2004	20-04-2004	07-05-2004	04-05-2004	
B4 - Back plate	Back plate	06-04-2004	20-04-2004	07-05-2004	04-05-2004	
B5 - Back plate	Back plate	06-04-2004	20-04-2004	07-05-2004	04-05-2004	

Fig. 2 Roles of Each Category of User at Each Stage in the Planning, Design, Manufacture and Receipt of Metalwork Parts

The prototype database consisted of approximately 20 ASP files containing over 8,000 lines of VBScript. Testing was done by the developer prior to piloting to ensure that all scripts and required functionality were working correctly. It should be noted that the development process was undertaken in close contact with the stakeholders of the system so as to facilitate ownership and take-up of the system when it was rolled out formally. The required domain name, bandwidth and disk space on a web server able to run the ASP scripts were kindly made available to the project by me-too.net Ltd.

7 PILOT AND FULL ROLL-OUT OF THE MICRO SCM SYSTEM

While much 'selling' of the micro SCM system had been undertaken by this stage this process needed to be stepped up a gear as key personnel were now required to use the system rather than just nod their heads to the virtues of a theoretical system – a massive difference. It should be born in mind also that the author (a KTP Associate) had no line authority over anyone in the company and was therefore wholly reliant upon the goodwill of the people he was dealing with – this represented a useful opportunity to develop diplomacy and persuasion skills.

A stakeholder analysis was performed and potential winners and losers were identified. More attention was paid to the potential losers at this stage as they were considered to present the biggest problem in respect of a smooth implementation, members of this category were:

- CAD – required to enter extra details about the jobs they were drawing.
- Stores – required to spend (potentially) more time checking metalwork parts in.

In the case of CAD the offered stance was the acknowledgement that yes they were having to spend a few extra minutes doing a job that they did not have to do before but that these minutes were invested rather than wasted as they would collectively save the company easily ten times more than the extra time it took them to update the system and therefore the task was worth doing. A similar line of argument was followed with the Stores personnel but with the additional suggestion that *they* would save more time as a result of the micro SCM system due to the fact that there would be far fewer enquiries at the Stores counter asking about metalwork deliveries since the information would be accessible from any PC.

The author was very pleasantly surprised that the piloting of the system was as well received as it was and considers it fortuitous that the CAD team and Stores personnel were open-minded and able to put the agenda of Saftronics ahead of their own.

There are some common themes in the recommendations offered by the various authors to boost the likelihood of system implementation success, these 'critical success factors' and the steps that were taken to ensure they were in place are given in Table 1.

Table 1 Steps taken to ensure the Project's Critical Success Factors were in place

Critical Success Factor	Measures Taken To Facilitate Success
Education and Training	• Stakeholders / key users of the system were 'educated' i.e. informed of the theories behind having one central source of information, getting info from the PC rather than from pieces of paper, performance measurement and so on. • Latterly the same groups were 'trained' i.e. showed what the system could do and how to use it in order to be able to fulfil their duties efficiently.
Top management commitment	• Senior management involved from both companies.
Change management	• The author did 'tours' of all the main players, shop floor up to directors, and spent time explaining what was going to happen and why paying fair attention to the problems with the current way of doing things and the problems being experienced. • Significant time and effort was invested getting people to feed their ideas into the system specification process. • Communication in the form of meetings, presentations, training, email, newsletter and intranet articles ensured that the people involved in the change informed, up to speed and on-side.

Critical Success Factor	Measures Taken To Facilitate Success
Involvement and commitment	• Stakeholders were frequently involved throughout the whole process ranging from the education / training sessions outlined above, requests for input / feedback, announcements in the company newsletter and on the intranet and even a competition to name 'the system' where the winner received a bottle of wine – this helped to foster feelings of ownership of the micro SCM system or 'SAFTS' as it was christened (Saftronics Application For Tracking Steelwork).
Software suitability	• Detailed functional requirements analysis was conducted. • Users were invited to provide feedback and suggestions.
Data accuracy and integrity	• Relational database used. • Error checking built into the system. • Extensive testing of the system was undertaken prior to and during pilot.
Other factors	• Close liaised with Saftronics IT department. • Goals of the system clearly defined.

As can be seen from the micro SCM system usage profile shown in Fig. 3. testing and piloting of the system was complete by the end of June 2003. The redevelopment / further development of the system was undertaken for a short period then July saw the start of the full roll-out of the micro SCM system. With the exception of the lull around the Christmas period usage of the website has remained at a high level i.e. 150 page views per day.

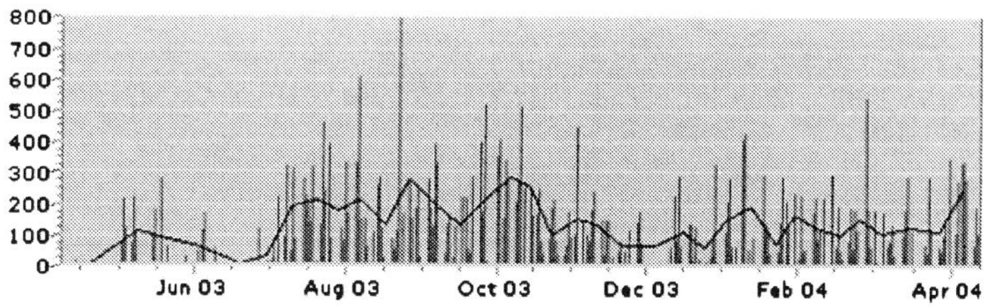

Fig. 3 Web Pages Viewed on the Micro SCM System per Day

The fact that the number of people using the system each day has recently been on the increase indicates that the system is becoming more used and that it is useful.

Companies attempting to install an IT system rather than implement a new way of doing things will be destined for trouble. Realising the incompleteness and myopia of the 'installation' approach, the focus should widen to incorporate change management, organisational and 'softer' issues as well as the more obvious technical issues such that systems can be properly and successfully implemented and accepted.

8 EVALUATION OF THE MICRO SCM SYSTEM

The micro SCM system has proven itself to be reliable, robust and capable of outputting recommendations that stand up to close scrutiny. The on time delivery performance of the fabricator has not improved beyond doubt since the implementation of the system due in part to the fabricator relocating their operations and the only very recent implementation of the late delivery penalty charges that Saftronics is now imposing on the fabricator. However, thc quality of the information that is provided to Saftronics from the fabricator for the purposes of production planning has improved markedly as can be seen in Fig. 4.

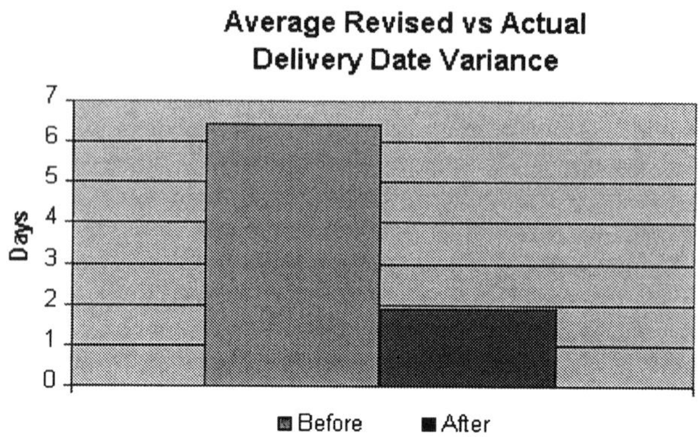

Fig. 4 People Using the Micro SCM System

The bar chart shows that the date that metalwork parts would previously have been delivered at Saftronics was over six days different from the date that Saftronics were advised the parts would arrive by the metalwork supplier. Now this date variance has been improved by more than a factor of three and currently stands at less than two days. While this improvement is very marked and positive in itself there is still room for improvement and the target is to have completely accurate delivery schedule information from the fabricator, i.e. 0 days variance.

The benefits that the micro SCM system has yielded / will yield are numerous, significant and heavily intertwined, they include:

- reduced visits by Engineers to the Stores counter (as previously Stores were responsible for tracking the delivery status of metalwork parts using a cumbersome paper-based system)
- improved working efficiency of the Engineers (fewer disruptions to work / disruptions of a shorter length)
- improved working efficiency of the Stores function (fewer internal queries being dealt with, time spent kitting the correct jobs, less time spent progress chasing / expediting)
- improved working efficiency of the fabricator (as fewer interruptions received from Saftronics chasing deliveries and progress chasing)
- improved information to Saftronics production planning resulting in more realistic plans
- reduced need for replanning
- improved morale due to more stability and visibility of upcoming work

- improved efficiency of shop floor working (reduced double handling due to increased stability of work priorities)
- leading to reduced shop floor overtime working
- reduced WIP leading to more shop floor space freed up (working on jobs until complete)
- improved profitability per job
- better service provided to the customer both in terms of swifter responses to questions about progress and also in terms of greater adherence to scheduled delivery dates

As can be seen the positive effects of the micro SCM system are quite far-reaching and considerable. Most of the above benefits could not be easily or reliably translated into financial figures that could credibly be used to help justify the costs of implementing such a system. However it is often necessary for companies to do an (expected) cost – benefit analysis before undertaking the expense and upheaval associated with system implementations hence it was important to quantify at least some of the benefits that the system had brought about.

In order to quantify the benefits of the micro SCM system, in financial terms, we considered which groups of personnel were affected most by its introduction and use and tried to quantify the time saving that system usage would bring. To calculate time saving the interruption time and time taken to visit Stores by an Engineer and then the time taken by the Stores person to handle the enquiry were judged / measured. This information was compared to the time that the same enquiry would take if done via the micro SCM system so as to arrive at a value for the average time saving per metalwork delivery status enquiry.

Phone data logs and internet access logs combined to show how often metalwork delivery status enquiries were being made. The product of this enquiry frequency figure and the average time saved per enquiry gives the total time saving per unit time that usage of the micro SCM yields. This time can be multiplied by the cost associated with the personnel to derive a financial value of the benefit of using the micro SCM system. Undertaking such an analysis gave rise to an annual saving of over £44,417 for using the micro SCM system and this is a conservative figure since it quantifies only one or two of the listed benefits of system usage.

9 CONCLUSIONS

- Paying due attention to the system implementation 'critical success factors' helps to ensure that the likelihood of system implementation success is maximised and that the common pitfalls associated with implementations are navigated round.
- The benefits of implementing and using an effective (micro) SCM system are significant and far-reaching, in this particular instance the benefits included:
    - reduced visits to Stores
    - improved working efficiencies (of Engineers, Stores personnel and fabricator)
    - more realistic production plans
    - morale raised on the shop floor
    - improved efficiency of shop floor
    - reduced shop floor overtime

- improved profitability per job
- better service provided to the customer

- Usage of the micro SCM system has led to a saving of over £44,417 per year.
- Effective and mutually beneficial communications necessitate that information flow in both directions.
- More details about the micro SCM system that has been developed can be found online at: www.leanscm.co.uk/saftronics

New technology enables SMEs to do things that were not feasible for them a few years ago. Low risk high return projects having payback times of months rather than years are achievable when applying the potential of new technology to solving old problems in innovative and novel ways. Reluctance to seek out and exploit these opportunities for increased efficiencies will lead to stagnation and the continued decline of the manufacturing industry in the UK. SMEs – this is your call to action!

REFERENCES

1. Barker, R., (1996) "Value chain development: An account of some implementation problems", *International Journal of Operations & Production Management*, Volume 16 Number 10, pp23-36.
2. Dobbeck, D. & Woods, W., (1994) "Mapping industrial activity", *OECD Observer*, June-July Number 188, pp19-24.
3. EC Commission & WEFA, (1994) "UK has the competitive edge", *Management Today*, February, pp27.
4. Jones, A., O'Toole, B., Webb, D. and Robinson, J., (2003) "Impact and Effectiveness of a Change Agent in the Development of the Operations Systems in an SME", *Advances in Manufacturing Technology XVII*, pp423-428. Professional Engineering Publishing.
5. McClenahen, J., (1999) "Hot Tip: Manufacturing", *Industry Week*, Volume 248, Issue 1, pp35.
6. Poirier, C. and Bauer, M., (2000) "E-Supply Chain", Berrett-Koehler Publishers Inc.

The electronic-enabling of product specifications for the digital environment of the automotive supply chain

C KERR, R ROY, and **P SACKETT**
Enterprise Integration, Cranfield University, UK

ABSTRACT

Within the e-Design environment of an automotive supply chain, there is a lack of support for upstream design activities. One of these critical activities is product specification as it is a necessary input for collaborative product development. However, product specification between the vehicle assembler and suppliers involves numerous iterations before a shared understanding is reached. The reason for this is various omissions, ambiguities and assumptions have to be resolved. This paper will present a prototype tool for the electronic-enabling of product specifications founded on ontology-based constructs to provide a shared conceptualisation of the knowledge needed for the specification of a given automotive system.

1 INTRODUCTION

The move to global sourcing is prevalent in the automotive industry (1). For example Ford, General Motors and DaimlerChrysler are all engaged in this activity (2). Currently, 70-85% of the cost for a vehicle is attributed to contracted out or outsourced products in the supply chain (3). Although, this trend was initially justified on the basis of cost reduction, Collins *et al.* (4) report that "the need for increased speed to market and for new business development" are the driving force. One avenue taken by the industry to address these drivers has been the implementation of digital product development across the supply chain. There have been great efforts made in the harmonisation of CAD systems, the utilisation of simulation tools and the realisation of the digital mock-up for a complete vehicle. PSA Peugeot-Citroen, for example, use their "Co-Conception" application for digital modelling with their suppliers (5). Additionally, design engineers in the vehicle assemblers and the suppliers of the respective supply chains have support from an extensive number of computer-based tools for modelling, simulation and analysis at the detailed design stages when the product attributes are well-defined. For example:

- CAD (2D and 3D)
- Product data management
- Finite Element Analysis
- Dynamic model simulation

However, there are relatively few applications for the pre-conceptual and early conceptual stages. According to Wood *et al.* (6) "the earliest stages of the design process have steadfastly resisted systematic formalisation." Thus, to have an 'all inclusive' digital product development environment, tools are needed for the upstream activities and the area of requirements has been overlooked with the advances made in providing support for downstream detailed design. Considering a requirements management perspective, the actual agreement and interpretation of what has to be developed and designed is a troublesome area for an automotive supply chain since the actual evolution of the product requirements, embodied in specifications, involves numerous time consuming interactions between the vehicle assembler and their suppliers. Omissions, ambiguities and incorrect assumptions often have to be resolved resulting in increased work effort and cost. To address this problem, research into the use of ontology for achieving a shared understanding of specifications is being conducted. The application of ontology to allow a shared conceptualisation of the knowledge needed for the specification of automotive products will be presented through a case study on car seats. A prototype tool will demonstrate how these ontology-constructs can be used for the electronic-enabling of the specification activities for future integration into the e-Design environment of an automotive supply chain.

2 PRODUCT SPECIFICATION

Product specification is becoming evermore important as vehicle assemblers increase the degree of 'black box' and 'grey box' responsibilities to their suppliers. The specification in itself is a means of communication (7). However, the business processes of detailing and negotiating specifications involves numerous iterations resulting in a negative impact on lead-time. The principal reason for this is that the details contained in the specifications need to be clarified between the vehicle assembler and respective supplier. They have to engage in frequent formal and informal discussions in order to resolve the semantic mismatches and thus come to a shared understanding of what exactly needs to be developed.

Fagerstrom and Johannesson (8) report that "many delays in product development projects are related to inadequate specifications". A specification is traditionally represented in the form of text documents (9) and Effendi *et al.* (10) recognized that the principal means of representation is through natural language. However, the use of natural language leads to ambiguity in interpretation (11). When a specification has to be prepared by a vehicle assembler for a supplier then the ambiguities are even more apparent since the companies do not use a common language.

Most automotive projects involve the collaboration of engineers from different organisations and different areas of expertise. However, they will all be working to a single specification. Thus they all need to have the same shared understanding of that specification. However each of these different groups of people have their own way of conceptually understanding the product requirements since they have different ways to express their needs, their own representations and differences in interpretation. One of the great difficulties is the use of different language, which is readily apparent in the form of different terminologies. This often results in the imprecise, approximate and incomplete expression of design requirements in specifications. It then takes valuable time to resolve the contextual mismatching in the specification vocabulary and the associated definitions. Numerous informal discussions through telephone, e-mail and meetings are then needed in order to receive clarification and confirmation. Thus, long periods in the product development process can be attributed to the need for many iterations in the specification phase as a direct result for the necessity to resolve ambiguities.

3 APPLICATION OF ONTOLOGY

According to Toye *et al.* (12) design occurs as a result of reaching a "shared understanding" of the design problem, the requirements and the process. Thus, the specification must be understood in the same manner by all participants. Failure to achieve this leads to wasted effort (13). For the specification to be shared, the product requirements knowledge has to be externalised in order to communicate the information to the organisations in the supply chain. This can be described as making the requirements knowledge accessible, conceptually, through shared 'concrete' representations that facilitate common understanding.

A shared conceptualisation of the product requirements will be realised through the application of ontology. "An ontology is an explicit specification of a conceptualization" (14). A conceptualization is a set of definitions that allows one to construct expressions about some physical domain (15). The explicit specification means that the concepts and relationships of the abstract model are given explicit terms and definitions. This is effectively "a formal and declarative representation" (16,17). Ontologies are viewed as specifications of domain knowledge conceptualisations (18). An ontology enumerates the concepts that are relevant in a domain (19) and are equated with taxonomic hierarchies of classes, objects and relationships. Thus, an ontology is therefore a "content theory" (20) that encompasses knowledge of the types of entities, functions, attributes, constraints and relations (21). An ontology can therefore act as a strict work of reference when communicating specifications through the supply chain.

3.1 e-Enabling Framework

To apply ontology to the domain of product specifications, an electronic framework has been developed. Figure 1 provides an overview of the web-based integration platform for e-enabling specifications in the automotive supply chain. The centre of the platform is the vehicle assembler who houses the 'global' system. This system is the one single source of information for a vehicle project and is based on an ontological constructed repository to enable the structuring and documenting of product requirements in a common standardized electronic format. A vehicle can be decomposed into a number of systems and the product requirements can be assigned to each respective system. For each of the systems, a 'product' ontology can be developed between the vehicle assembler and the respective suppliers. The vehicle assembler then allows their suppliers authorized access to the systems for which they are responsible for developing and manufacturing. Thus, the suppliers have a portion of this requirements data housed in their own 'local' clients for internal dissemination in their respective organizations. The operational data that is then transferred between the vehicle assembler and the suppliers conforms to the ontology (Figure 2). A secure web-based front-end, which is being developed using Visual Basic .NET, is used by both the vehicle assembler and suppliers for the uploading, browsing and downloading of the requirements for specific products.

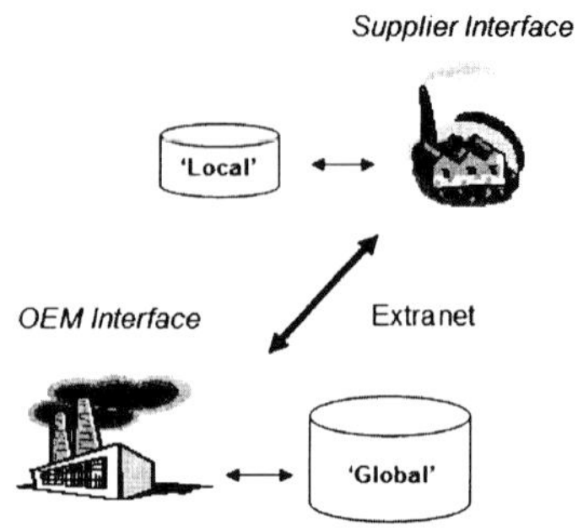

Fig. 1 Electronic framework

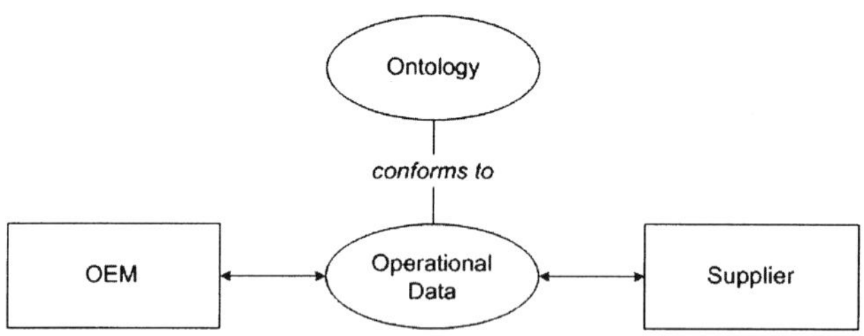

Fig. 2 Operational data conforming to the ontology

4 PROTOTYPE AND CASE STUDY

To illustrate the principle of ontology application to the domain of product specifications, a case study has been created for the specification of the seating system for a car. In order to elicit the information to create an ontology for seating systems, a series of one-to-one semi-structured interviews were conducted with personnel within a vehicle assembler, a Tier 1 seating system supplier and two Tier 2 seating component suppliers. The personnel interviewed were from the three principal stakeholder groups within each organization, namely: Design, Manufacturing and Purchasing. In total 28 participants were interviewed. In addition to the interviews, documentation was an important and rich source of information for the development of the ontology. The types of documentation material that was analyzed is shown in Figure 3.

The ontology was developed according to the guidelines proposed by Noy & McGuiness (22). The first step was to determine the domain and scope of the ontology. Then the important terms in the ontology were enumerated. This allowed the classes and the class hierarchy to be determined. For each class, the properties were defined in the form of slots and finally, the facets of the slots were defined. The ontology editor used was Protégé-2000 developed by the SMI group of Stanford University. Following the series of interviews and the content analysis of the documentation, in-depth workshops were conducted with senior designers in the vehicle assemblers and suppliers to refine and then validate the ontology.

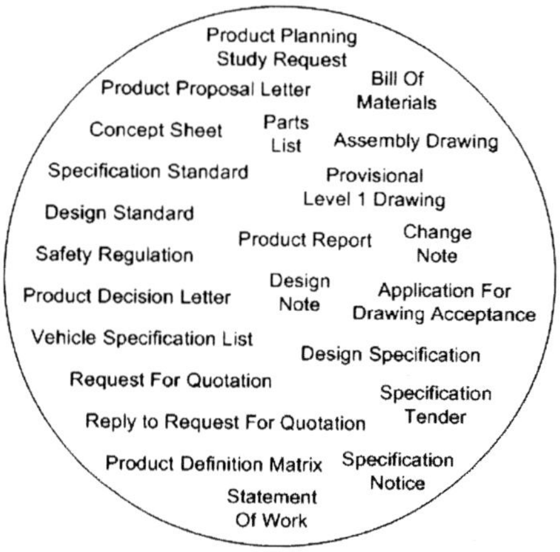

Fig. 3 Documentation sources

Figure 4 depicts a part of the ontology that was developed to provide a common and shared definition of the product requirements for the specification of the seats. The seating assembly has been decomposed into a number of high-level conceptualizations such as vehicle configuration (i.e. how the seats are orientated to the car); seated occupant (i.e. how the seats are orientated to a seated occupant); functionality of the seats; safety features and regulations; comfort features and levels.

Consider the basic functionality of the driver's seat which is to seat an occupant in the correct driving position. There is the shared understanding between the vehicle assembler and suppliers, as encapsulated in the ontology, that the basic occupant positioning functionalities are recline, slide and lift (Figure 4). For a design engineer to specify the requirements for the slider mechanism, the ontology defines that they need to supply values for the slider inclination angle, travel length and travel pitch. Additionally, the principal specification parameters of the recline function are the forward-most and rear-most reclining angles and the reclining pitch. Thus, applying the ontology for the specification of a product allows the vehicle assembler and suppliers to work to a standard requirements vocabulary and the associated definitions for each of the individual requirements.

The ontology is accessed and used for the specification of the product through a front-end that is being developed using Visual Basic .NET. A series of dialog boxes takes the engineer through the process of specifying the necessary elements of the seating system. For example, Figure 5 illustrates the recliner and slider specification. For these functionalities the ontology has encapsulated the key parameters that allow the vehicle assemblers and suppliers to have a shared understanding of the specification elements.

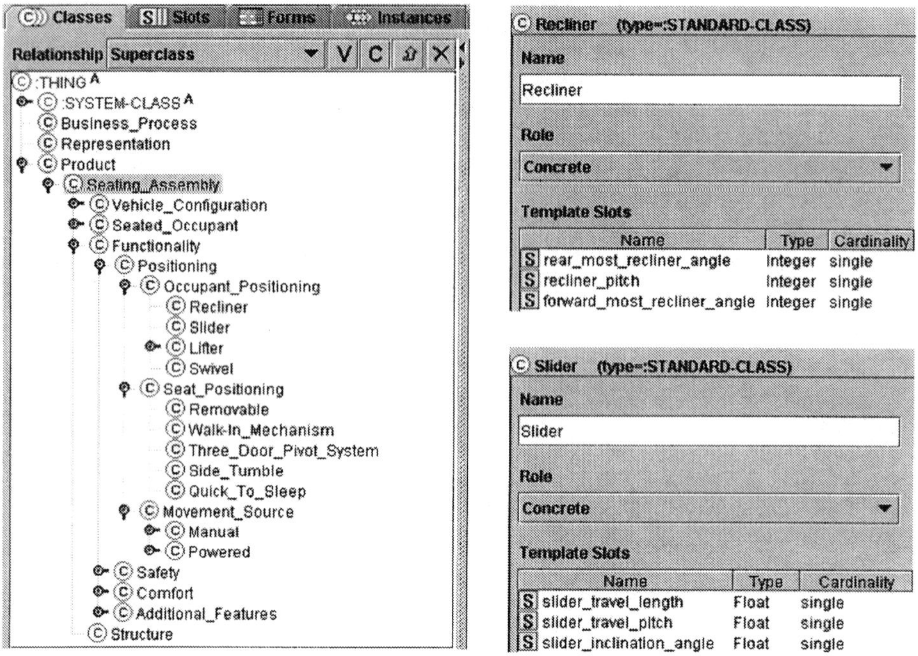

Fig. 4 Ontology for the seating system

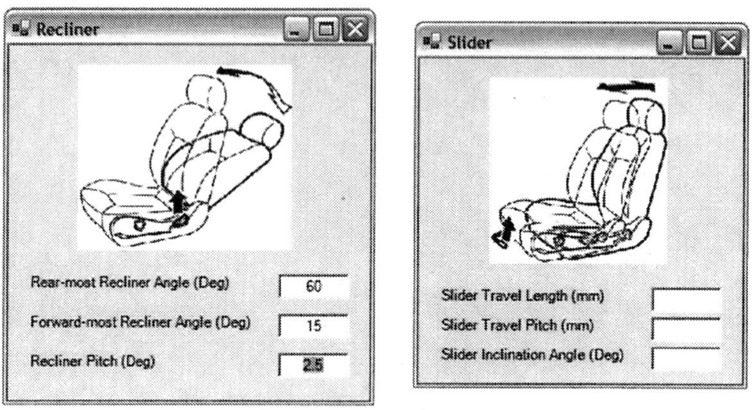

Fig. 5 Recliner and slider specification

5 CONCLUSIONS

When working collaboratively through the supply chain, an important enabler is a clear and shared understanding of the product requirements. However, the area of requirements management has been overlooked with the advances made in the digital product development environments of the automotive industry. Product requirements knowledge has to be externalised in order for communication through the supply chain. The different groups that are involved in the specification activities each have different ways to express their needs, their own representations and differences in interpretation. Such differences result in the imprecise, approximate and incomplete expression of design requirements in specifications. This means that numerous formal and informal discussions are then needed for clarification and confirmation. Making the requirements knowledge accessible, conceptually, through shared 'concrete' representations that facilitate common understanding can be achieved through the application of ontology. An ontology for a product allows the different stakeholders a shared conceptualisation of the product requirements in terms of mutually accepted and understood abstractions and associated language. The knowledge encapsulated within an ontology includes product attributes, functions, relations & constraints. Thus, specifications of high quality can be obtained since an ontology will ensure a coherent and consistent structuring of the semantic content, logical content and physical presentations.

ACKNOWLEDGEMENTS

The authors would like to acknowledge the UK Engineering and Physical Sciences Research Council (EPSRC); Nissan Technical Centre Europe, Johnson Controls Automotive, EDS and the Society of Motor Manufacturers and Traders (SMMT) as the e-RM project sponsors; and the Decision Engineering team at the Enterprise Integration department of Cranfield University.

REFERENCES

1 Esterman, M. and Ishii, K. (2001) "Concurrent product development across the supply chain: Development of integrator/supplier risk and coupling indicies", *ASME 2001 Design Engineering Technical Conferences and Computer and Information in Engineering Conference*, Pittsburgh, USA, 9-12 September 2001.

2 Nellore, R., Chanaron, J .J. and Soderquist, K. E. (2001) "Lean supply and price-based global sourcing", *The interconnection. European Journal of Purchasing & Supply Management*. Vol. 7, pp. 101-110.

3 Schweitzer, L. (2003) Keynote speech, Odette International Conference 2003, Paris, France, 16-17 September 2003.

4 Collins, R., Bechler, K. and Pires, S. (1997) "Outsourcing in the automotive industry: From JIT to modular consortia", *European Management Journal*. Vol. 15, pp. 498-508.

5 Arozamena, C. M. (2003) "A strategic gamble with ENX", *Odette International Conference*, Paris, France, 16-17 September 2003.

6 Wood, W. H., Yang, M. C., Cutkosky, M. R. and Agogino, A. M. (1998) "Design information retrieval: Improving access to the informal side of design", *ASME 1998 Design Engineering Technical Conferences and Computer and Information in Engineering Conference*, Atlanta, USA, 13-16 September 1998.

7 Shefelbine, S. J. (1998) "Requirements capture for medical device design", M.Phil. thesis, Cambridge University, UK.

8 Fagerstrom, B. and Johannesson, H. (2001) "A product and process model supporting main and sub-supplier collaboration", *13th International Conference on Engineering Design*, Glasgow, UK, 21-23 August 2001.

9 Svensson, D. and Malmqvist, J. (2001) "Integration of requirements management and product data management systems", *ASME 2001 Design Engineering Technical Conferences and Computer and Information in Engineering Conference*, Pittsburgh, USA, 9-12 September 2001.

10 Effendi, I., Henson, B., Agouridas, V. and de Pennington, A. (2002) "Methods and tools for requirements engineering of made-to-order mechanical products", *ASME 2002 Design Engineering Technical Conferences and Computer and Information in Engineering Conference*, Montreal, Canada, 29 September - 2 October 2002.

11 Greer, J. L., Stock, M. E., Stone, R. B. and Wood, K. L. (2003) "Enumerating the component space: First steps toward a design naming convention for mechanical parts", *ASME 2003 Design Engineering Technical Conferences and Computer and Information in Engineering Conference*, Chicago, USA, 2-6 September 2003.

12 Toye, G., Cutkosky, M. R., Leifer, L. J., Tenenbaum, J. M. and Glicksman, J. (1993) "SHARE: A methodology and environment for collaborative product development", *Proceedings of the IEEE Infra-structure for Collaborative Enterprises*, CDR-TR #19930507.

13 Thomson, G. A. (2001) "Requirements engineering - Laying the foundations for successful design", *13th International Conference on Engineering Design*, Glasgow, UK, 21-23 August 2001.

14 Gruber, T. (1993) "A translation approach to portable ontology specifications", *Knowledge Acquisition*. Vol. 5, pp. 199-220.

15 Schreiber, G., Wielinga, B. and Jansweijer, W. (1995) "The KACTUS view on the 'O' word," *JCAI Workshop on Basic Ontological Issues in Knowledge Sharing. International Joint Conference on Artificial Intelligence*, Montreal, Canada, 19-20 August 1995.

16 Fikes, R. (1997) "Reusable ontologies: A key enabler for electronic commerce", Knowledge Systems Laboratory, Stanford University, USA.

17 Gruber, T. and Olsen, G. (1994) "An ontology for engineering mathematics", *Proceedings of The Fourth International Conference on Principles of Knowledge Representation and Reasoning*, Bonn, Germany, Edited by Doyle, J.; Torasson, P. & Sandewall, E., Morgan Kaufmann Publishers, pp. 137–144.

18 Van Heijst, G., Schreiber, A. T. and Wielinga, B. S. (1997) "Using explicit ontologies in KBS development", *International Journal of Human-Computer Studies*. Vol. 45, pp. 183-192.

19 Musen, M. A. (1998) "Domain ontologies in software engineering: Use of Protégé with the EON architecture", *Methods of Information in Medicine*. Vol. 37, pp. 540-550.

20 Chandrasekaran, B., Josephson, J. R. and Benjamins, V. R. (1999) "What are ontologies, and why do we need them?" *IEEE Intelligent Systems*. Jan/Feb, pp. 20-26.

21 Devedzic, V. (1999) "A survey of modern knowledge modeling techniques", *Expert Systems with Applications*. Vol. 17, pp. 275-294.

22 Noy, N. F. and McGuiness, D. L. (2001) "Ontology development 101: A guide to creating your first ontology", Stanford Medical Informatics, Stanford University, Report No. SMI-2001-0880.

Dynamic planning with a wireless product identification technology in food supply chains

D LI, D KEHOE, and **P DRAKE**
Management School, University of Liverpool, UK

ABSTRACT

An innovative planning model has been developed for perishable food supply chain planning. The model maximises product values for consumers instead of merely reducing costs. The information which can be derived from RFID enabled control systems for product units (at palette or crate level) has been used to identify dynamic product quality status. Then, the information is used to optimise product allocations from manufacturers to distributors and retailers. The objective is to minimise lost value of products and maximise profits for supply chain partners. The real-time control towards value maximisation becomes possible with the real-time product quality information form RFID systems. An illustrative case has been described for simulating the dynamic planning process.

1 INTRODUCTION

Food supply chain deliver products under more strict legislation and safety requirements today. Product quality is significantly affected by the environment of supply chain processes. Real time tracking and traceability of product information are important to improve the supply chain efficiency and product quality. Tracking locations by GPS approach can provide location identification of facilities. It does not automatically provide dynamic information about loadings unless all details of variations are collected manually into integrated systems. Barcodes have been the primary technology for product-level identification. It needs visual alignment of each product with a scanner. The technology only provides limited manufacturer and brand data which are static and do not reflect impacts of supply chain processes on the product properties. The difficulty of catching barcode information in fast moving and bulky product flows has further prevented tracking real-time information of individual food products for quality assurance and high efficiency.

Due to decreased hardware costs and increased demands on product tracking, the wireless product identification technology, named radio frequency identification (RFID) or AutoID has attracted increasing attentions in supply chain management. RFID consists of electronic devices called tags that are attached to products to be identified, and tag readers that communicate with the tags via electromagnetic waves and connected to control systems. Many trials (WalMart, Sainsbury, Finnair, Ford Motor Co. and many others) have been implemented recently. Recognised benefits include improved traccability, information accuracy, and operation efficiency. The improvements lead to reduced labour costs, increased speed, greater responsiveness, and better product quality control. Existing research has concluded that overall supply chain cost may be reduced by five to ten percent [1].

The research in this paper focuses on innovation of food supply chain management by developing dynamic supply chain planning approaches based on real-time information received from RFID enabled control systems. The objective is to minimise lost value of products and maximise profits for supply chain partners. The real-time control towards value maximisation becomes possible with the product quality information form RFID systems. In the research, an illustrative case has been described for simulating the dynamic planning process.

2 LITERATURE REVIEW

Food supply chain management has been an important issue. In particular, the research on perishable food products in supply chains has been reported in the literature. To maximise profits or minimise the loss of product values in supply chains, models optimising order quantities, stock allocations and other parameters have been developed. Bhattacharjee and Ramesh [2] developed an optimisation model which maximises profits by optimised quantities of products allocated in different periods with different prices and relevant operational costs (holding and ordering costs). The single product in question has a fixed perishing life. The perish ability of the product was considered for a constraint of time period over which total demands are allocated, and is not directly related to the variation of product prices in supply chain stages. Blackburn and Scudder [3] developed a model for optimising order quantity with extended traditional EOQ model which considers the loss of perishable product values at some supply chain stages. The research investigated the value deterioration characteristics of some fresh produce at a harvest stage and a cooling stage in a supply chain. As the characteristics of lost values are different at the supply chain's harvest stage and cooling stage, the proposed model separates the two stages with different solutions, i.e. different order quantities and transportation modes. The research concluded that perishable products can be managed partly as innovative products as proposed in Fisher's model, and their values can be maximised by appropriate management of production and delivery processes. Lin and Chen [4] proposed a dynamic logistics control model for a perishable product. The research investigates an issue of dynamic allocating fresh food supplies to retailers with given remaining shelf life of products, expected sales and relevant costs (holding costs, disposal costs, penalty costs due to re-allocate orders to retailers). The model focuses on maximising profits according to the limited shelf life of perishable products, uncertainty of supplies and demands, and costs incurred during a shelf life period. The logistics control model is the most similar research to our model proposed in this paper. Some other inventory control oriented research relating perishable food products can be found in Chakrabarty, *et al.* [5] and Fujiwara and Perera [6], etc.

Most of the above research on supply chain or inventory control about perishable food has been based on expected product shelf life, sales and relevant costs. The automatic product identification technology introduced in section one provides opportunity that real-time information can be automatically collected about the dynamic product quality status based on key environment parameters, e.g. temperature and the time period during which products have been exposed to the temperature in a supply chain process. Such a technology enables more delicate approaches to control product flows with maximised product values to consumers and minimised operational costs, rather than planning merely based on estimated product shelf life. Some research and case studies on applying RFID technology to industries have been reported in the literature. Smart Manufacturing Forum [1] introduced detailed concept, technical specifications and possible solutions in applying RFID to food industry for improving food security. The white paper identified possible benefits to supply chains from the technology such as information accuracy, operation efficiency and cost reduction, etc. McFarlanea, *et al.* [7] proposed a concept of intelligent products which have their "information content permanently bound to its material content", and are able to "participate in or making decisions relevant to its own destiny". The research developed a manufacturing control model with Auto-ID enhanced product and resource holon interactions. The intelligent products can be identified dynamically in manufacturing processes in the individual bases, and are controlled intelligently according to information embedded with themselves. The research especially explores the impact of such technology on manufacturing shop floor control.

In this research, the proposed food supply chain planning model is built based on a situation that product identification information can be automatically collected from RFID systems and can be fed straight into supply chain planning systems. With RFID technology, the proposed model is able to identify up to date product perishing status at all supply chain control points. Therefore, it is capable of exploring more delicate analyses on product value maximisation through stage-wise dynamic planning. The next section describes the model details.

3 DYNAMIC PLANNING IN PERISHABLE FOOD SUPPLY CHAINS

In retail stores, prices of food products are affected by the closeness to expiration date which is the only information indicating possible spoilage of food. In fact, some food spoils before or around the expiration dates, but some others may be displayed on shelf much longer. This is the result of impacts of supply chain processes on the perishable product quality. To identify the actual impacts of the supply chain environment on food quality, parameters such as temperature and related time period should be recorded at each important supply chain points. The RFID technology provides this possibility that the required information can be recorded automatically and attached to each product unit, e.g. pallet or crate. Then, as the products pass through a supply chain, the product property information can be collected by control systems at each stage and is used to make decisions by control systems on current product values and the best place where the products should be delivered. This information is crucial to avoid wastage and facilitate dynamic planning for maximising product values to consumers.

3.1 Supply chain process control with RFID technology

The proposed model in this research dynamically allocates perishable products from food processing packaging companies to distributors, and then from distributors to retailers. The decisions on product allocation are made based on order quantities, shipping costs, losses of product values, and costs of re-allocating the products. The model is similar to the research of Lin and Chen [4] in allowing penalty costs for re-allocating product supplies against original orders of retailers. However, with RFID technology, the optimisation approach in this paper is capable of identifying product values at real time, and estimating product prices at various supply chain stage. The main factor affecting value deterioration of fresh products is temperature. The higher the temperature, the quicker the product value deteriorates. To provide product quality information, data dynamically stored in a RFID chip should at least include identification, manufacturer, packaging, codes of the processes (manufacturing, packaging, delivery, storing, sale on shelf, etc.) including the codes for preceding and succeeding processes, dates and times of entering and leaving the processes, temperature (and other necessary quality parameters) at these times, and so on. The data should be designed in a standard schema so that all systems in the supply chain can understand the information in RFID systems. Figure 1 describes the function of RFID systems in a supply chain.

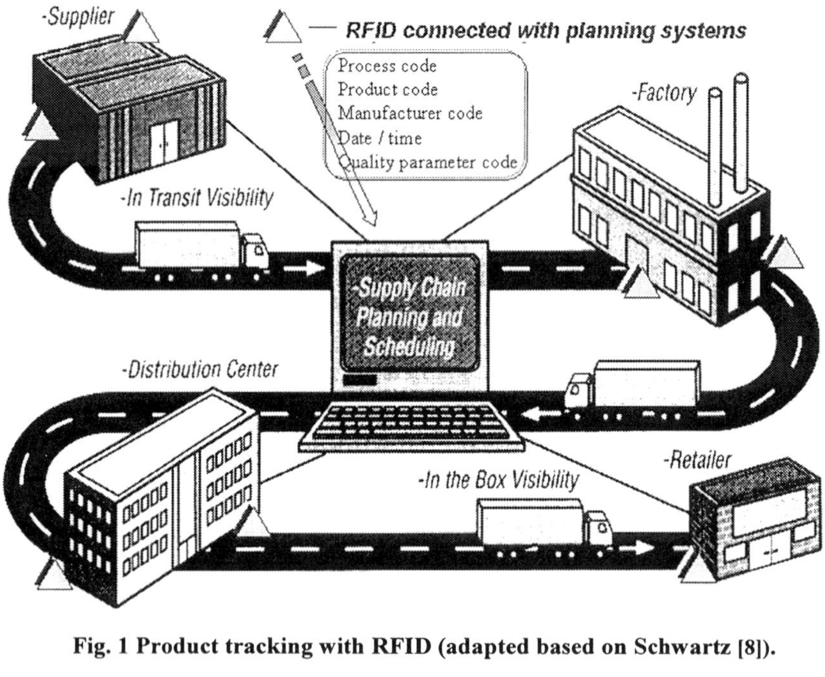

Fig. 1 Product tracking with RFID (adapted based on Schwartz [8]).

As seen in figure1, a RFID system is installed at key control points of each supply chain process to record both product and process details. This information is then transferred with products to the next process. Control systems automatically read the information with RFID chips on product (units). The systems dynamically plan for processing routes based on the received messages from current products and products in the next supply chain process. The information in current product RFID systems provides product quality situations after passing through previous supply chain processes. The information transferred from the next supply chain process provides current situations which will affect product quality in the future. The collection of RFID information is used for optimisation of product flows in following supply chain processes. Without this wireless identification technology, such information cannot be collected automatically for bulky product flows.

3.2 Measuring the lost values of products with RFID

As RFID systems can provide real-time product quality information, the proposed model optimises product flows in a dynamic forwarding process which always look at succeeding processes based on current status. The key factors in the decision model are demands, orders, product value deterioration rates, time spent at each supply chain stage, and shipping costs. The actual order placed by each supply chain partner is the decision variable. To simplify the problem and focus on the product value assessment issue, in this research, demands are assumed as known and deterministic. The total product quantity supplied from food growers are all delivered through the supply chain, and is therefore equal to the total order quantity of supply chain members. This assumption is used as a constraint in the model. Assuming that unit stock holding costs are the same at all firms and have no impact on the product allocation decision, stock costs are not considered in the model.

The quality and shelf life of perishable products are mainly affected by temperature at harvest fields, processing workshops, warehouses and on the shelf in stores. To measure the loss of product values, deterioration rates and time periods are used in functions representing product value changes. In the research of Blackburn and Scudder [3], Value variations were represented by exponential functions indicating quick decreasing of product values over time under high-temperature environment. A value deterioration parameter is used with a time variable to assess the value reduction. The research also concluded that the difference between the exponential presentation and a linear approximation is slight. In our research, this presentation of product value variances is adopted as seen in equation 1 and equation 2. α is the value deterioration parameter which is dependent mainly on environment temperature. T is the time period during which the product is exposed to the environment.

$$\text{Lost value} = \text{Original Value} \cdot [1 - \text{EXP}(- \alpha \cdot T)] \qquad (1)$$

(Derived from [3]).

Approximation in linear deterioration:

$$\text{Lost value} = \text{Original Value} \cdot [1 - (1 - \alpha \cdot T)] = \text{Original Value} \cdot (\alpha \cdot T) \qquad (2)$$

3.3 Minimising loss of product values

To minimise loss of values and costs, an optimisation model is developed to allocate products to different transit routes according to their current quality situation (i.e. loss of values in previous supply chain processes) and expected quality assurance conditions in succeeding supply chain processes. Considering a supply chain case shown in figure2, before a planning period starts, orders are placed by retailers. Distributors, packers and growers consequently place orders with their suppliers to satisfy the expected demands. The orders can be determined either based on long-term contracts with the supply chain members, or by the proposed optimisation approach based on estimated process and product data. In the process of delivering products form growers down through a supply chain, actual deliveries can be changed according to product quality data stored in FRID chips with each product unit. The dynamic routing activities lead to uncertainties in supply chain operations. The costs have to be considered and partners have to be compensated for the changes. A penalty cost component is used in this research to reflect this consideration as the return on collaboration. The dynamic planning model is described in equation. As the dynamic planning can be made at any supply chain stage based on real-time RFID data, it is not always the case that all supply chain stages need to be included for optimising product allocations. For instance, to assess dynamic product quality at a distributor warehouse, a retailer does not need to consider product allocation alternatives at packaging companies and distribution centres as the processes have been already completed. However, the products need to be traced back for assessing their previous value losses. The model is described in equation 3 for planning allocations of products transferred from distribution warehouses to retail stores.

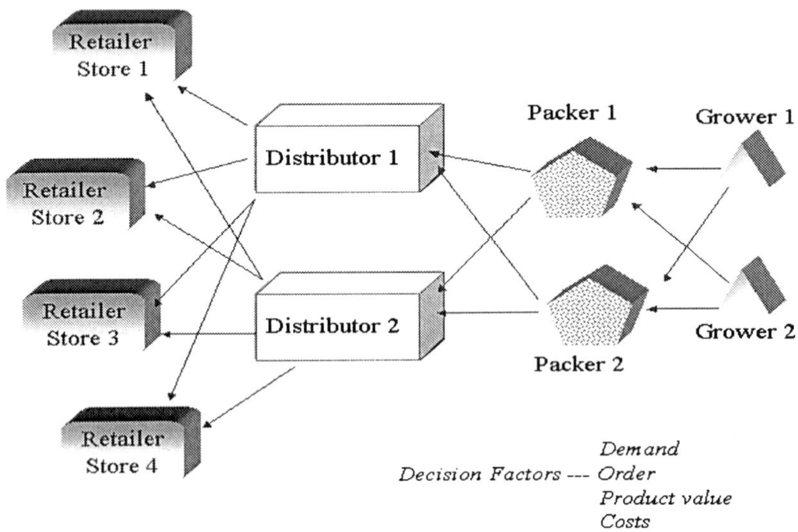

Fig. 2 A case of food supply chain.

Objective:

Min Costs = Loss of product values in processing + Costs in transit + Penalty

$$= \{\Sigma_i D_{or,i} V_{r,i} \cdot [1 - EXP(-\alpha_{r,i} \cdot T_{r,i})] + \Sigma_j D_{od,j} V_{d,j} \cdot [1 - EXP(-\alpha_{d,j} \cdot T_{d,j})]$$
$$+ \Sigma_k D_{om,k} V_{m,k} \cdot [1 - EXP(-\alpha_{m,k} \cdot T_{m,k})] \}$$
$$+ \{ \Sigma_{i*j} CS_{d-r,\, i*j} \cdot D_{or,i}\, b_{i*j} + \Sigma_{i*j} D_{or,\, i}\, V_{d-r,\, j} \cdot [1 - EXP(-\alpha_{d-r,\, i*j} \cdot T_{d-r,\, i*j})]\, b_{i*j}$$
$$+ \Sigma_{j*k} CS_{m-d,\, j*k} \cdot D_{od,\, j}\, b_{j*k} + \Sigma_{j*k} D_{od,\, j}\, V_{m-d,\, k} \cdot [1 - EXP(-\alpha_{m-d,\, j*k} \cdot T_{m-d,\, j*k})]\, b_{j*k}$$
$$+ \Sigma_{k*g} CS_{f-m,\, k*g} \cdot D_{om,\, k}\, b_{k*g} + \Sigma_{k*g} D_{om,\, k}\, V_{f-m,\, g} \cdot [1 - EXP(-\alpha_{f-m,\, k*g} \cdot T_{f-m,\, k*g})]\}\, b_{k*g}$$
$$+ \{\Sigma_i C_{p,i} (OD_{r,i} - D_{or,i}) + \Sigma_j C_{p,j} (OD_{d,j} - D_{od,j}) + \Sigma_k C_{p,k} (OD_{m,k} - D_{om,k})\}$$

Subject to: (3)

$\Sigma_k D_{od,k} = \Sigma_j D_{od,j} = \Sigma_i D_{or,i} = \Sigma_i OD_{r,i}$;

$V_{m,k} = \Sigma_g V_{f\text{-}m, g} \cdot EXP(-\alpha_{f\text{-}m, k*g} \cdot T_{f\text{-}m, k*g})$; $V_{m\text{-}d, k} = \Sigma_k V_{m,k} \cdot EXP(-\alpha_{m,k} \cdot T_{m,k})$;

$V_{d,j} = \Sigma_k V_{m\text{-}d, k} \cdot EXP(-\alpha_{m\text{-}d, j*k} \cdot T_{m\text{-}d, j*k})$; $V_{d\text{-}r, j} = \Sigma_j V_{d,j} \cdot EXP(-\alpha_{d,j} \cdot T_{d,j})$;

$V_{r,i} = \Sigma_j V_{d\text{-}r, j} \cdot EXP(-\alpha_{d\text{-}r,i*j} \cdot T_{d\text{-}r,i*j})$;

$D_{om, k} >= MinD_{om, k}$; $D_{od,j} >= MinD_{od, j}$; $D_{or,i} >= MinD_{or, i}$.

$i = 1, 2, \ldots, nr; j = 1, 2, \ldots, nd; k = 1, 2, \ldots, nm; g = 1, 2, \ldots, nf.$

(In this case, nr = 4, nd = 2, nm = 2, nf = 2)

In equation 3, i, j, k and g denote a particular retailer, distributor, packer and grower. i*j, j*k and k*g denote transit routes between adjacent supply stages. b_{i*j}, b_{j*k}, b_{k*g} = 0,1 which denote selection of a route and destination. $MinD_{om}$, $MinD_{od}$, $MinD_{or}$ are minimum order quantity placed by each company. V_r, V_d, V_m (the grower or farmer's internal processes are not analysed in this research) are initial product values when a product enters a supply chain stage, i.e. a retailer, distributor and a packaging company (packer). $V_{f\text{-}m}$, $V_{m\text{-}d}$, $V_{d\text{-}r}$ are initial product values at a transit through growers (f), packers (m), and distributors (d) to retailers (r). α_r, α_d, α_m are value deterioration parameters at a retailer, distributor and packer. T_r, T_d, T_m are time at the stages. OD_r, OD_d, OD_m are initial orders or predicted demands of a retailer, distributor and packer. D_{or}, D_{od}, D_{om} are planned or actual orders to be optimally placed as decision variables. $\alpha_{d\text{-}r}$, $\alpha_{m\text{-}d}$, $\alpha_{f\text{-}m}$ are the value deterioration parameters for the time periods $T_{d\text{-}r}$, $T_{m\text{-}d}$, $T_{f\text{-}m}$ which are time periods of product transit between a farmer, packer, distributor and retailer. $CS_{d\text{-}r}$, $CS_{m\text{-}d}$, $CS_{f\text{-}m}$ are transportation costs between the partners. C_p is a penalty cost for re-allocating orders against the original orders placed at the supply chain partners which involves costs for re-schedule deliveries and operations, cost of extra stocks or shortages, etc. As seen in the constraint section of equation 3, product value variance can be traced back through retailers up to produce growers. The value tracing process only presents the situation that any node in a supply chain only receives products from a single supplier. When products are mixed in the process, the tracing becomes more complicated. A model for dynamic planning of allocating product flows form the distribution centres is described in equation 4. The objective function is the same as in equation 3. As seen in the model constraints, the tracing process with mixed products from different suppliers has to be analysed with the associated products and their order quantities. The value and demand with subscript notations $V_{r, i*j, j*k, k*g}$, $D_{or, i*j, j*k, k*g}$ denote a tracing path through a retailer (*r*) up to a grower (*f*). With RFID information, the model assesses values of products by tracing back through the chain, and assesses costs forward of delivery, demands, value deterioation and penalty to retailers. Then, the best routes can be identified. The next section illustrates this dynamic planning case with a numeric simulation.

Objective: (see equation 3)

Subject to: (4)

$V_{m, k*g} = V_{f\text{-}m, g} \cdot EXP(-\alpha_{f\text{-}m, k*g} \cdot T_{f\text{-}m, k*g})$; $V_{m\text{-}d, k*g} = V_{m, k*g} \cdot EXP(-\alpha_{m,k} \cdot T_{m,k})$;

$V_{d,j*k, k*g} = V_{m\text{-}d, k*g} \cdot EXP(-\alpha_{m\text{-}d, j*k} \cdot T_{m\text{-}d, j*k})$; $V_{d\text{-}r, j*k, k*g} = V_{d,j*k, k*g} \cdot EXP(-\alpha_{d,j} \cdot T_{d,j})$;

$V_{r, i*j, j*k, k*g} = V_{d\text{-}r, j*k, k*g} \cdot EXP(-\alpha_{d\text{-}r,i*j} \cdot T_{d\text{-}r,i*j})$; $D_{or,i} V_{r,i} = \Sigma_{i*j, j*k, k*g} (V_{r, i*j, j*k, k*g} \cdot D_{or, i*j, j*k, k*g})$; $D_{or, i*j} V_{d\text{-}r, i*j} = \Sigma_{j*k, k*g} (V_{d\text{-}r, j*k, k*g} \cdot D_{od, j*k, k*g})$; $D_{or,i} >= MinD_{or, i}$; $\Sigma_i OD_{r,i} = \Sigma_i D_{or,i}$;

$i = 1, 2, 3, 4;\quad j = 1, 2;\quad k = 1, 2;\quad g = 1, 2.$

3.4 Simulation of dynamic supply chain planning: an illustrative case

Based on the food supply chain case shown in figure 2, simulation is implemented with sample data in table 1. To control the produce flows from distribution centres, the upstream partners do not need to be included for the order allocation analysis. OD is the initial order. SC is the transit costs. P is penalty. T is the periods when products stay at a stage. α is the deterioration rate.

Table 1. Sample data of supply chain parameters in the simulation

	α	T	OD	P		α	T	SC
	£/hour	Hour	SKU	£/SKU		£/hour	Hour	£/SKU
Retailer store1	0.003	48	216	2	**Distribut1-Retailer1**	0.002	4	7
Retailer store2	0.005	40	250	4	**Distribut1-Retailer2**	0.001	7	9
Retailer store3	0.004	46	381	1	**Distribut1-Retailer3**	0.003	6	6
Retailer stroe4	0.006	42	215	3	**Distribut1-Retailer4**	0.002	5	10
Distributor1	0.002	24			**Distribut2-Retailer1**	0.001	3	6
Distributor2	0.001	28			**Distribut2-Retailer2**	0.003	2	7
Packer1	0.007	6			**Distribut2-Retailer3**	0.002	4	7
Packer2	0.008	8			**Distribut2-Retailer4**	0.002	2	5
Packer1-Distribut1	0.005	3			**Farmer1-Parker1**	0.009	0.5	
Packer1-Distribut2	0.004	2			**Farmer1-Parker2**	0.008	1.2	
Packer2-Distribut1	0.006	1			**Farmer2-Parker1**	0.007	0.6	
Packer2-Distribut2	0.007	2			**Farmer2-Parker2**	0.006	1	

The most distinctive feature of the RFID-based approach is its tracing capability. The significance and effectiveness of the approach is tested by comparing control of product flows with and without tracking product value statues. Without tracing previous impacts, the products are usually directed to places based on distance and demands. Figures 3 and 4 describe solutions (from spreadsheet solver) for the simulation case with three approaches: optimisation without tracing value losses backwards (A1), optimisation without assess any value losses (A2), and optimisation with tracking all value variations (A3). A comparison among costs including value losses simulated with the three approaches is described in table 2.

With the approach A1, costs and potential value losses at retail stores are assessed. Then, products are optimally allocated. The approach A2 analyses the issue similarly, but ignores value losses of perishable produce at all processes. The optimal product allocations through A1 and A2 are then re-assessed with ignored value losses at relevant stages. The approach A3 considers all value losses based on the model in equation 4. The actual costs are recorded in table 1. The solution of A3 attained 2% and 7% savings against A1 and A2.

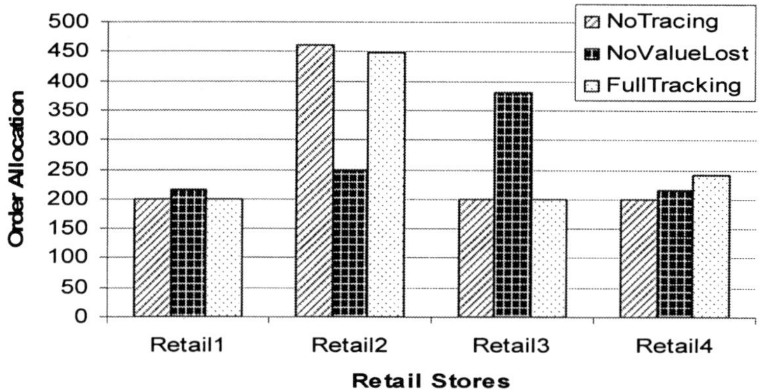

Fig. 3 Product allocations at four retail stores.

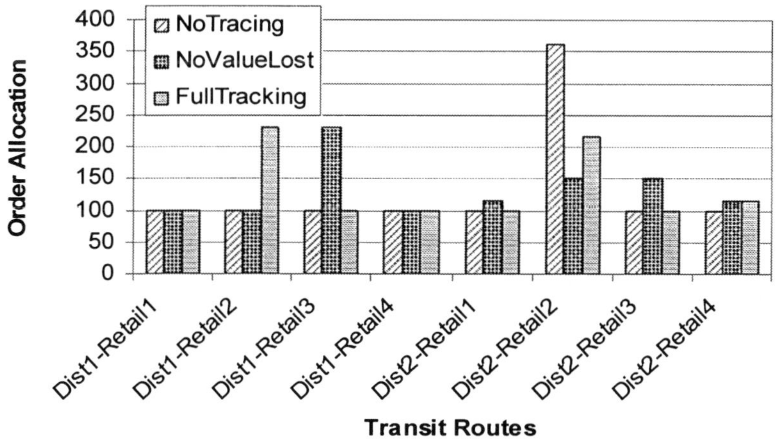

Fig. 4 Planning product flows in transit.

Table 2 Result of the simulation case

	Actual Costs	Savings %
No Value Loss Traced Back (A1)	54434.21688	2.071246443
No Value Loss Assessed (A2)	57375.77523	7.091886967
Full Value Tracking	**53306.7501**	

The above result shows that appropriately assessment of product values in supply chain planning for perishable produce can create savings and increase product values to customers. Assessment of product values is particularly useful for dynamic product pricing. The model facilitates integration of supply chain planning with product pricing and marketing strategies.

Analyses are in progress for investigation on sensitivities of value deterioration to control decisions.

4 CONCLUSION

The research developed an innovative planning model on a perishable food case. RFID technology underlies implementation of such dynamic planning approach, as real-time and traceable product information is a prerequisite for tracking and tracing products in fast moving, bulky product flows. The future research should be given to case studies to identify more detailed constraints and sensitivities to product value losses, so that the model quality and applicability can be improved. Investigation in accurate relationships between product values and store shelf life is also required.

REFERENCE

1 The Smart Manufacturing Forum (2003) "Optimising Food Safety Using RFID", September, www.smartmanufacturingforum.com.

2 Bhattacharjee, S. and Ramesh, R. (2000) "A multi-period profit maximizing model for retail supply chain management: An integration of demand and supply-side mechanisms", *European Journal of Operational Research*. Vol. 122, pp. 584-601.

3 Blackburn, J. and Scudder, G. (2003) "Supply Chain Strategies for Perishable Products: The Case of Fresh Produce", OGSM Working Paper #03-14, Vanderbilt University, Nashville, TN.

4 Lin, C. R. and Chen H. S. (2003) "Dynamic allocation of uncertain supply for the perishable commodity supply chain", *International. Journal of Production Research*. Vol. 41, No. 13, pp. 3119–3138.

5 Chakrabarty, T., Giri, B. C. and Chaudhuri, K. S. (1998) "An EOQ Model for Items with Weibull Distribution Deterioration, Shortages, and Trended Demand: An Extension of Philip's Model", *Computers and Operations Research*. Vol. 25, No.7/8, pp. 649-657.

6 Fujiwara, O. and Perera, U.L.J.S.R. (1993) "EOQ Models for Continuously Deteriorating Products Using Linear and Exponential Penalty Costs", *European Journal of Operational Research*. Vol. 70, No. 1, pp. 104-114.

7 McFarlanea, D., Sarmab, S., Chirna, J. L., Wonga, C. Y. and Ashton, K. (2003) "Auto ID systems and intelligent manufacturing control", *Engineering Applications of Artificial Intelligence*. Vol. 16, pp. 365–376.

8 Schwartz, B. M. (1997) "Tracking down need", *Transportation & Distribution*. Vol.38, No.10, pp.50-52.

Customer-oriented supply chain management practices in Pearl River Delta – a preliminary study

V H Y LO, D SCULLI, and **A H W YEUNG**
Department of Industrial and Manufacturing Systems Engineering, University of Hong Kong, Hong Kong

ABSTRACT

The Pearl River Delta (PRD) in Southern China is a major manufacturing area for the global clothing market. The traditional mode of original equipment manufacturing (OEM) is still used extensively, but major overseas customers are now constantly making additional demands that require enhanced capabilities in handling value-added activities such as R&D, product development, fashion design, and prototyping. This paper examines how the PRD clothing industry is starting to integrate customer expectations into their business models. A customer-oriented supply chain management (CoSCM) model is presented and used to uncover emerging industry practices. The findings are based on feedback from sixty well established clothing manufacturers in the PRD.

1 BACKGROUND

The global recession of the late 1990s has reinforced the need for companies to re-consider their business strategies by placing more emphasis on adding value and reducing costs throughout all supply operations. In recent years, many of the larger enterprises have adopted the concepts of global manufacturing as their business mode, thereby enhancing their competitiveness in terms of cutting costs, opening markets and improvement effectiveness.

In a general sense, global manufacturing refers to the series of activities involved in managing the internal and external components of the supply system, including material, cash and information flow along a global supply chain. More specifically, global manufacturing represents a managerial paradigm that links up suppliers (including both internal and external suppliers) and customers (including both industrial and consumers) through conducting various management activities throughout a global supply chain in order to ensure the effectiveness of material, cash and information flows around the world (1). This view indicates that the global strategies of individual companies and their corresponding supply chain are critical to the performance and success of the entire supply chain.

There is a current trend towards greater reliance on customer-oriented strategies, i.e. strategies that respond quickly to the needs of customers. Morash and Lynch show that global supply chain trends substantially increase reliance on delivery flexibility, responsiveness to customers, and customer service flexibility (2). Such capabilities support the increasing popularity of customer-oriented supply chain strategies.

Hines et al. (3). state that integrating customer expectations into the supply chain design can enhance organizational effectiveness across a global supply chain (3). This is because traditional Supply Chain Management (SCM) only ensures reliable delivery with minimum operating cost and a lower inventory buffer by effectively managing the supply function on the upper stream across the chain (4), but does not consider the real needs of customers. Hoover et al. also point out that having competitive products and the right supply chain for the average customer is not enough, suggesting that it is better to design supply chains for the individual customers as well (4). Customer-oriented SCM practices should thus help companies to better integrate their customer expectations into the supply chain design.

Customer-oriented Supply Chain Management (CoSCM) refers to a series of management practices that place emphases on the marketplace and then construct business strategies across the various business units on a global supply chain with the aim of satisfying customer expectation (4). It is suggested that by redesigning business strategies across the chain of business units, all chain partners can improve asset performance, enrich cost efficiency, and become more customer-oriented.

China's accession to the WTO will require China to open its domestic market to other countries. During the last two decades, many Hong Kong manufacturers have gradually relocated their productions facilities in Southern Mainland China, especially in the PRD region. The main reasons behind the relocation were, and still are, to take advantage of lower operating costs and the plentiful supply of labour. Over the past 20 years, the PRD region has rapidly developed better transportation networks that connect to Hong Kong and has gradually and firmly established itself as one of the major manufacturing hubs in the world. This growth in economic significance of the PRD region has encouraged many more companies to locate/relocate their manufacturing facilities in region, and especially so in the last few years with the added promise of quick access the local China market.

Hong Kong has the same language and culture as that of Southern China and manufacturers are well able to manage operations in the PRD. Familiarity with the China market and the recent Closer Economic Partnership Arrangement (CEPA) agreement between Hong Kong and China can offer considerable opportunities to many Hong Kong-based companies and to subsidiaries of international companies. All this has helped the PRD to develop into a strong production base in the global supply chain (5). It is thus timely that a study should be conducted on how manufacturers in the PRD can enhance their effectiveness across the global supply chain by integrating customer expectation into their strategy development.

Literature on manufacturing practices in the PRD is very scarce, especially on customer-oriented supply chain management practices. The primary purpose of this paper is to explore how the manufacturers in the PPD are attempting to integrate customer expectations into their business strategies. This paper will focuses on the clothing industry, one of most significant industries in the PRD region in terms of labour employed and value of output. The paper firstly describes the steps in developing a model to investigate the CoSCM practices in the PRD clothing industry. This is followed by a discussion on how to ensure the validity and reliability of the model developed in relation to the CoSCM practices of clothing manufacturers in the PRD.

2 RESEARCH METHODOLOGY

This study used a four-phase approach to develop the research instrument, see Figure 1. A previously developed framework (6) was used as the stating point, while results from the published literature and industrial interviews formed the foundations.

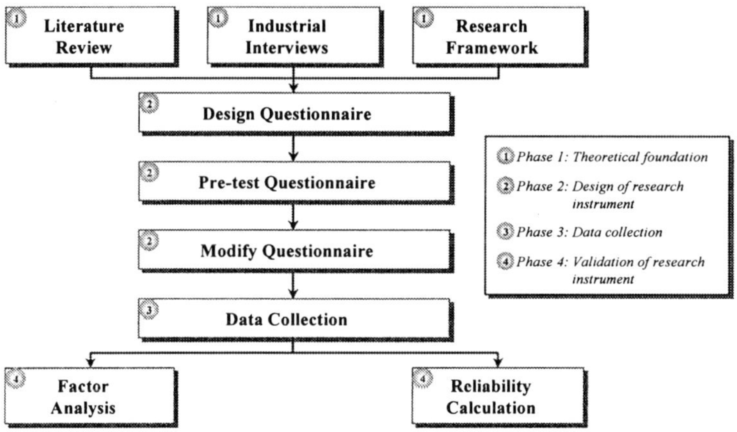

Fig. 1 Flowchart for portraying the research methodology

Phase 2 is in fact a draft of the instrument and consists of five parts. Parts A and B were designed to assess the Industrial Customer Perceived Values knowledge and Consumer Perceived Values knowledge of potential respondents respectively. Part C evaluates the performance of internal operations. Part D portrays the management philosophies adopted by the potential respondents. Finally, Part E records the profile of potential respondents.

To strengthen the content validity of the research instrument, ten local clothing manufacturers were involved in pre-testing, mainly by asking them to comment on the appropriateness and the ease of comprehension of scale items. After reclassifying certain items, the scale items were subjected to a preliminary study. The preliminary study used a sample of clothing manufacturers in the PRD region. Eighty potential respondents were randomly selected from the Hong Kong Directory and sixty responses were received, corresponding to a 75% rate. The collected data were subjected to statistical validity and reliability analyses.

2.1 Validity Analysis

Previous studies interpret validity of a measure by how well the measure represents the concepts under study (7, 8). Black and Porter suggest considering the content validity and the construct validity in establishing the validity of a measure, though there may be other types of validity (9).

Content validity refers to the extent of the correspondence between the scale items selected to constitute a construct and its conceptual definition (8). As mentioned earlier, the research instrument developed was initially based on a framework developed by Lo et al. (6), results published in the literature and industrial interviews. The content validity was scrutinized by the 60 clothing manufacturers used in the preliminary study, thus ensuring that instrument developed was valid to an acceptable degree.

Construct validity measures how well scale items represent individual constructs. Construct validity can be ensured by evaluating the unidimensionality of an individual construct (9). With reference to the conceptual framework developed by Lo et al. (6), individual constructs were subjected to a Principal Axis Factoring (PAF) analysis with varimax rotation separately. The empirical findings of PAF analyses are summarized in Table 1.

Table 1 Findings for the PAF analysis

Construct	Description	KMO	Bartlett's Test (Significant)	Variance Explained (%)
Industrial Customer Perceived Values Knowledge (ICPV)				
IC_PQ	Performance Related to Product Quality	0.726	0.0005	54.542
IC_OH	Performance in Order Handling	0.745	0.0005	78.289
IC_MS	Performance of Sales Force	0.805	0.0005	59.589
IC_OS	Interaction with Internal Staff	0.777	0.0005	72.502
IC_SC	Social Compliance	0.835	0.0005	80.682
Consumer Perceived Values Knowledge (DCPV)				
DC_CS	Customer Satisfaction	0.738	0.0005	85.450
DC_PP	Product Price	0.757	0.0005	86.342
DC_CI	Corporation Image	0.750	0.0005	71.369
DC_CB	Compliant Behaviour	0.738	0.0005	88.290
DC_RC	Relationship Commitment	0.838	0.0005	68.843
DC_CL	Customer Loyalty	0.614	0.0005	72.921
Management Philosophies (MP)				
MP_LP	Long-term Perspective	0.741	0.0005	65.719
MP_CF	Customer Focus	0.739	0.0005	79.033
MP_CI	Continuous Improvement	0.793	0.0005	66.573
MP_TI	Total Involvement	0.878	0.0005	74.065
Internal Operations Performance (IO)				
IO_SPP	Strategic Planning for Supply Chain	0.739	0.0005	63.110
IO_PSP	Plan for Sourcing Function	0.795	0.0005	74.804
IO_PMP	Plan for Manufacturing Processes	0.752	0.0005	66.894
IO_PDF	Plan for Delivering Function	0.794	0.0005	76.256
IO_SF	Sourcing Function	0.840	0.0005	74.196
IO_MP	Manufacturing Processes	0.737	0.0005	72.261
IO_DF	Delivering Function	0.794	0.0005	73.545

This study measured the adequacy of the sample with KMO measures which are used to assess the suitable of the sample for each unifactorial determination. In this study, all KMO measures are better than 0.5, the acceptable level (8, 9). Construct validity is further demonstrated with Bartlett's Test, used for testing the presence of inter-correlations among scale items for individual construct; all Bartlett's tests are significant at 0.0005 levels in this study. In addition, all constructs accounted for at least 54% of the variance of their corresponding variable sets, reflecting a good representation of scale items to individual constructs. The findings therefore reinforce the construct validity of the research instrument.

2.2 Reliability Analysis

Reliability is a measure for evaluating the homogeneity of scale items of the same construct (10), commonly measured by Cronbach's Alpha. Table 2 summarizes Cronbach's Alpha for individual constructs in this study. All reliability measures range from 0.7644 to 0.9385; all better than the acceptable level of 0.7 (7, 11).

Table 2 Cronbach's Alpha of individual construct in this study

Construct Variable	Critical Constructs of SQM Practices	Initial No. of Items	Deleted Items (by variable)	Cronbach's Alpha	
				Initial	Finalized
Industrial Customer Perceived Values Knowledge (ICPV)					
IC_PQ	Performance Related to Product Quality	5	IC_PQ1 IC_PQ4	0.7644	0.8464
IC_OH	Performance in Order Handling	4	Nil	0.9050	0.9050
IC_MS	Performance of Sales Force	5	Nil	0.8238	0.8238
IC_OS	Interaction with Internal Staff	5	Nil	0.9022	0.9022
IC_SC	Social Compliance	5	Nil	0.9385	0.9385
Consumer Perceived Values Knowledge (DCPV)					
DC_CS	Customer Satisfaction	3	Nil	0.9143	0.9143
DC_PP	Product Price	3	Nil	0.9201	0.9201
DC_CI	Corporation Image	4	Nil	0.8782	0.8782
DC_CB	Compliant Behaviour	3	Nil	0.9337	0.9337
DC_RC	Relationship Commitment	5	Nil	0.8886	0.8886
DC_CL	Customer Loyalty	3	Nil	0.8140	0.8140
Management Philosophy (MP)					
MP_LP	Long-term Perspective	4	Nil	0.8187	0.8187
MP_CF	Customer Focus	3	Nil	0.8627	0.8627
MP_CI	Continuous Improvement	4	Nil	0.8221	0.8221
MP_TI	Total Involvement	6	Nil	0.9295	0.9295
Internal Operations Performance (IO)					
IO_SSP	Strategic Planning for Supply Chain	5	Nil	0.8531	0.8531
IO_PSF	Plan for Sourcing Function	5	Nil	0.9146	0.9146
IO_PMP	Plan for Manufacturing Processes	5	Nil	0.8746	0.8746
IO_PDF	Plan for Delivering Function	5	Nil	0.9219	0.9219
IO_SP	Sourcing Function	5	Nil	0.9099	0.9099
IO_MP	Manufacturing Processes	5	Nil	0.9035	0.9035
IO_DF	Delivering Function	5	Nil	0.9089	0.9089

The reliability of the construct "performance related to product quality" (IC_PQ) can be improved to 0.8464 by deleting two scale items (IC_PQ1 and IC_PQ4). In addition, the item-to-total correlations of these two scale items are relative low, 0.4163 for IC_PQ1 and 0.3862 for IC_PQ4. Churchill (11) suggests improving the reliability of a construct by deleting scale items with a lower item-to-total correlation, and so it was decided to eliminate these scale items from the research instrument, and, as a result, other scale items designed for evaluating individual constructs became more significant.

3 CUSTOMER-ORIENTED SUPPLY CHAIN MANAGEMENT IN PRD

This study has clustered CoSCM practices into four major groups, namely industrial customer perceived values knowledge (ICPV), consumer perceived values knowledge (DCPV), management philosophies (MP) and internal operations performance (IO). Their definitions are given in Appendix I. This section first presents a summary of the respondents' background information, including job position and company size, and then explores current CoSCM practices.

3.1 Characteristics of respondents

Over 70% respondents were senior executives and close 20% were functional managers. More than 90% respondents were directly responsible for business strategy development, while the others were indirectly involvement. The respondents were thus well qualified to participate in this study.

Table 3 shows that over 30% of companies employed more than 1,000 workers. This is in line with the fact that the manufacture of clothing in the PRD is labour intensive. Approximately 27% of companies employed less than hundred people. This indicates that small and medium sized companies (SMEs) still play a significant role, even in such a labour intensive industry. The clothing industry one of the few mature industries that does not require very advanced technologies, enabling SMEs with a low level of investment to take part.

Table 3 Respondents by company size

Company Size (in number of employee)	Number of Company	Percentage	Cumulative Percentage
Less than 100	16	26.7%	26.7%
101 to 500	17	28.4%	55.1%
501 to 1,000	8	13.3%	68.4%
1,001 to 2,000	6	10.0%	78.4%
2,001 to 5,000	5	8.3%	86.7%
More than 5,000	8	13.3%	100.0%

3.2 CoSCM practices

The sample standard deviation is a measure to of how well the mean has been estimated from the sample. As shown in Table 4, all standard deviations range from 0.486 to 0.816, demonstrating that most of the mean scores have close confidence limits. All this lends more confidence to the results. This study has compared the mean scores of all CoSCM practices with the scoring guideline suggested by the Malcolm Baldrige National Quality Award Program. Using this guideline, any management practice with at least 90% of total score is "excellent" and with half of total score is "average", corresponding to a score of 4.5 as "excellent" and a score of 2.5 as "average" on our survey instrument.

Table 4 Mean score and standard deviation of individual CoSCM practice

Construct	CoSCM Practice	Mean Score	Standard Deviation
Industrial Customer Perceived Values Knowledge (ICPV)			
IC_PQ	Performance Related to Product Quality	4.056	0.573
IC_OH	Performance in Order Handling	3.942	0.816
IC_MS	Performance of Sales Force	4.167	0.486
IC_OS	Interaction with Internal Staff	3.797	0.687
IC_SC	Social Compliance	4.220	0.734
Consumer Perceived Values Knowledge (DCPV)			
DC_CS	Customer Satisfaction	3.994	0.594
DC_PP	Product Price	3.770	0.710
DC_CI	Corporation Image	4.170	0.596
DC_CB	Compliant Behaviour	4.103	0.573
DC_RC	Relationship Commitment	3.947	0.532
DC_CL	Customer Loyalty	4.246	0.599
Management Philosophies (MP)			

MP_LP	Long-term Perspective	4.229	0.638
MP_CF	Customer Focus	4.228	0.787
MP_CI	Continuous Improvement	4.225	0.637
MP_TI	Total Involvement	4.150	0.710
Internal Operations Performance (IO)			
IO_SPP	Strategic Planning for Supply Chain	3.650	0.612
IO_PSP	Plan for Sourcing Function	3.727	0.673
IO_PMP	Plan for Manufacturing Processes	3.671	0.593
IO_PDF	Plan for Delivering Function	3.720	0.674
IO_SF	Sourcing Function	3.613	0.617
IO_MP	Manufacturing Processes	3.657	0.646
IO_DF	Delivering Function	3.763	0.606

*: *"1" as the worst, and "5" as the best.*

3.2.1 Industrial customer perceived values knowledge

All practices in the section Industrial Customer Perceived Values Knowledge (ICPV) were designed to evaluate how well respondents recognise the needs of their industrial customers. As shown in Table 4, no practice under ICPV reaches the "excellent" level, but some practices under ICPV such as social compliance with a score of 4.220, indicate a good level of awareness of social and environmental issues. Since the major industrial markets of respondents are the United States and Europe, their supplier selection criteria tends to emphasize social responsibilities, working conditions, and environmental concerns. In order to gain access into these markets, the PRD clothing manufacturers must constantly put more efforts in the improvement of measures relating to social compliance.

Respondents, see Table 4, disagreed on the importance of setting up communication channels between their operational staff and industrial customers. However, some interviews with the industrial customers indicate that they would like to have direct contact with their suppliers' operational staff. They feel that better communication can speed up the efficiency of overall production and minimize the risk of making serious mistake, see also Lo and Yeung (12).

3.2.2 Consumer perceived values knowledge

Under the Consumer Perceived Values Knowledge (DCPV) section, all practices are designed to evaluate how well the respondents understand the domestic market. It was found that respondents were very conscious of "corporation image" (DC_CI), "compliant behaviour" (DC_CB) and "customer loyalty" (DC_CL), scoring 4.170, 4.103 and 4.246 respectively. This suggests that the respondents would like to secure their domestic market through building up their own company image. This also indicates that respondents would like the opportunity to develop their own brand name. Compared with other practices in the DCPV section, respondents appeared to be less concerned with "product price". This suggests that the PRD clothing manufacturers are gradually shifting towards emphasizing "quality" instead of "price" in their marketing strategy recently.

3.2.3 Management philosophies

All practices in Management Philosophies (MP) section were designed to reveal the management philosophies adopted by the respondents. This study sees management philosophies in terms of taking a long-term perspective on three total quality management (TQM) fundamental principles. In general, most practices in this section recorded a better score than practices in other sections, ranging from 4.150 to 4.229, and reflecting the existence well developed management philosophies in respondents' companies. Lo and Yeung (12) demonstrate that the long-term perspective is one of the core values for the

PRD clothing manufacturers. This value encourages manufacturers to improve their business continuously for better result. The good TQM-related scores suggest that most respondents have adopted the concepts of TQM as their core management philosophies. It is most likely that the TQM promotion efforts of the Hong Kong Government in the late 1980s would have had an impact on this.

3.2.4 Internal operations performance

The section Internal Operations Performance (IO) was designed to evaluate the effectiveness in maintaining internal operations in terms of reliability, responsiveness, flexibility, cost controlling and resource managing. The seven practices in this section scored from 3.613 to 3.763, suggesting considerable room for improvement of internal operations.

The delivery function (IO_DF) is the highest scoring operation, 3.763, indicating that manufacturers are well able to deliver the right products at the right places punctually. This is possibly because the logistic infrastructure has improved steadily in the PRD during the past 10 or more years, resulting in a highly efficient and integrated transport network that connects various destinations in the PRD region with airports and container ports. This has greatly enhanced the efficiency and effectiveness of delivering from the PRD to customers in the major industrialized countries of the West. The planning process of delivery function (IO_PDF) has a relative high score of 3.720. This suggests that respondents have and are putting more effort into planning the delivering processes, and also possibly explains the importance of the planning process in improving operational efficiency.

A relatively high score, 3.727, was obtained for "plan for sourcing function", while a lower score, 3.613, for "sourcing function". This seems to contradict the previous argument. However, these findings may indicate that respondents were aware of the significance of their sourcing functions, but that they may have had difficulties in managing them. An obvious recommendation here is that the PRD clothing manufacturers should manage their sourcing functions using supply quality management (SQM) principles. Previous studies have shown that SQM practices can improve overall organizational performance, see Lo and Yeung (12).

4 CONCLUSIONS

Using materials collected from the published literature, industrial interviews and the results of previous research, this study has developed a research instrument for investigating CoSCM practices. The instrument was refined through validity and reliability analysis and we are quite confident of its validity. This research instrument was used to collect data from 60 PRD clothing manufacturers for the purpose of investigating their current CoSCM practices.

The mean score of all CoSCM practices obtained in this study suggests that the PRD clothing manufacturers are not fully aware of the need to establish effective communication channels between their industrial customers and their operational staff. The PRD clothing manufacturers have already adopted TQM principles as their management philosophies and are also starting to take a long-term perspective of their business. This study has explored their weakness in managing their internal operations, especially their sourcing function, and the results can be used to make improvements.

In conclusion, this study recommends that the PRD clothing manufacturers adopt supply quality management (SQM) practices in managing their sourcing functions and to gradually build a long-term cooperative relationship with their suppliers. SQM practices can also improve the overall organisational performance through forming strong alliances with crucial suppliers. In addition, this study recommends the setting up of more communication channels for collecting customer data, including both industrial customers and consumers. This should help the PRD clothing manufacturers to recognise and better understand customer needs.

Future research in this field should be conducted using the research instrument developed and with a larger-sample size. A larger sample size will allow the cause-and-effect relationships among constructs to be better investigated through path analysis. The influential factors can then be identified and a path model constructed. Such a path model can act as a reference model for clothing manufacturers to improve their operations.

REFERENCES

1. Spekman, R.E., Kamauff, J.W. and Myhr, N. (1998) "An empirical investigation into supply chain management: a perspective on partnerships" *International Journal of Physical Distribution and Logistics Management*. Vol. 28, pp. 630–650.
2. Morash, E.A. and Lybch, D.F. (2002) "Public policy and global supply chain capabilities and performance: a resource-based view", *Journal of International Marketing*. Vol. 10, pp. 25–51.
3. Hines, P., Silvi, R. and Bartolini, M. (2002) "Demand chain management: an integrative approach in automotive retailing", *Journal of Operation Management*. Vol. 20, pp. 707–728.
4. Hoover, W.E., Eloranta, E., Holmström, J. and Huttunen, K. (2001) "Managing the Demand-Supply Chain: Value Innovations for Customer Satisfaction", Wiley.
5. Hong Kong Trade Development Council (2003) "Introduction — a new momentum of growth", *Pearl River Delta*, http://www.tdctrade.com/prd/intro.htm (22 January 2003).
6. Lo, V.H.Y., Sculli, D. Yeung, A.H.W. and Yeung A.C.L. (2004) "Integrating customer expectations into the development of business strategies in a supply chain environment", *International Journal of Logistics: Research & Applications (unpulished)*.
7. Ahire, S.L., Golhar, D.Y. and Waller, M.A. (1996) "Development and validation of TQM implementation constructs", *Decision Sciences*. Vol. 27, pp.23–56.
8. Hair, J.F., Anderson, R.E., Tatham, R.L. and Black, W.C. (1998) "Multivariate Data Analysis", Prentice Hall
9. Black, S.A. and Porter, L.J. (1996) "Identification of the critical factors of TQM", *Decision Sciences*. Vol. 27, pp.1–21.
10. Oppenheim, A.N. (1966) "Questionnaire Design and Attitude Measurement", Heinemann.
11. Churchill, G.A. (1979) "A paradigm for developing better measures of marketing constructs", *Journal of Marketing Research*. Vol. 16, pp.64–73.
12. Lo, V.H.Y. and Yeung, A.H.W. (2004) "Practical framework for strategic alliance in Pearl River Delta manufacturing supply chain: a total quality approach", *International Journal of Production Economics*, Vol. 87, pp.231–240.

APPENDIX I: DEFINITIONS FOR COSCM PRACTICES

CoSCM Practice	Definition
Industrial Customer Perceived Values Knowledge (ICPV)	
Performance Related to Product Quality	Characteristics of the degree to which the customer's requirements have been fulfilled with regard to the physical performance of the deliverable.
Performance in Order Handling	Efficiency and effectiveness in handling a customer order.
Performance of Sales Force	Level of satisfaction with regard to the knowledge of the sales force of the product and its conditions of use, as well as their ability to solve problems.
Interaction with Internal Staff	Communication with staff in operations through different channels.
Social Compliance	Corporate social accountability and responsibility with regard to human rights, working conditions, and environmental concerns.
Consumer Perceived Values Knowledge (DCPV)	
Customer Satisfaction	Ability to recognize and satisfy customer needs and expectations.
Product Price	Value of the products for which customers are willing to pay.
Corporate Image	Overall perception of the providers of the products.
Complaint Behaviour	Satisfaction on the attitude of the providers of the product in handling customer complaints.
Relationship Commitment	Commitment to keep customers loyal to the company that provides the products, even though with low customer satisfaction.
Customer Loyalty	Willingness to develop long-term relationships with the providers of the product.
Management Philosophies (MP)	
Long-term Perspective	The thinking of senior management with regard to developing long-term strategic plans for the organization.
Customer Focus	Recognize and satisfy customer needs and expectations.
Continuous Improvement	The philosophy of continuously looking for ways to improve people processes in order to meet the expected demands of customers.
Total Involvement	The philosophy of involving people from all levels, including suppliers, in all activities related to improving performance.
Internal Operations (IO)	
Strategic Planning for Supply Chain	The development of an action plan over specified periods of time, which represents a projected appropriation of the infrastructure in a demand chain in order to satisfy the end-customer.
Plan for Sourcing Function	The development of an action plan over specified periods of time, which represents a projected appropriation of sourcing in order to satisfy the end-customer.
Plan for Manufacturing Processes	The development of an action plan over specified periods of time, which represents a projected appropriation of resources in production in order to meet the requirements of production.
Plan for Delivering Function	The development of an action plan over specified periods of time, which represents a projected appropriation of resources in delivery in order to meet the requirements of delivery.
Sourcing Function	Operational processes are designed for the purchase of raw materials, subassemblies, products, and services to meet the expected demands of customers.
Manufacturing Processes	Operational processes are designed to add value to purchased parts through mixing, assembling, separating, forming, machining, and chemical processes to meet the expected demands of customers.
Delivering Function	Operational processes are designed to deliver finished products or services to meet the expected demands of customers.

Research on the method of multi-level fuzzy evaluation for performance of logistic centre processing

X-L WANG
Mechanical Engineering Department, Zhejiang University, Hangzhou and Jiliang University of China, China
Y ZHANG
Zhejiang University of Science and Technology, China
Z-C CHEN
Mechanical Engineering Department, Zhejiang University, Hangzhou, China

ABSTRACT

The logistics center processing consists of extensive contents. The index system of logistics center processing is analyzed and modeled by fuzzy evaluation method. The performance of the logistics center processing is evaluated synthetically by assigning weight to a set of influence coefficients. By means of derivation and demonstration, the definite evaluation result is given.

1 INTRODUCTION

Recently, analysis and control of logistics chain have been playing an important role in industrial and distribution systems. Logistics can be defined as a process that deals with the planning, handling, and control of the storage of goods between the manufacturing point and the consumption point (1). Almost every operation in logistics center processing involves different manpower and facilities, accordingly emphases for their evaluation vary. Some of operations behave in manpower paying out, another operations lie on decision method and level, and others are closely relative to facility number, equipment magnitude. The effective method is dividing the logistics processing system according to operations' functions, and then evaluating different parts by functions or performance.

After determining functions of every part, function evaluation evaluates the system project's performance and presents ways to improve the object, which are based on the comparison between function coefficient and cost coefficient, or target cost and actual cost of the whole system and every part. The goal of function evaluation is discussing object value, finding out function zone of appropriate value, determining improved concrete range and processing corresponding improvement work.. Through function evaluation, the improvement path and method is presented for those objects, such as service structure , technology, material, facility, organization and so on, which value coefficient is low or deviate people's expectation value. And function system structure can be perfected more (2).

For every operation of the logistics center, detailed indices should be deduced firstly. Only through analyzing and evaluating these data, it can be observed in which factor the key of problems may be resulted. If some of operation performances are dissatisfactory, the firm can set about improving them exactly, consequently to increase its productivity availably. Therefore, the task for the firm to complete is performance evaluation based on the principle of function evaluation.

Performance measurement(evaluation) was identified as one of the top three area of logistic research needs (3). Generally, logistics performance measurement, contains internal performance measurement, external performance measurement and comprehensive supply chain measurement. Internal performance measurement is analyzed by the firm. External performance measurement is illustrated through customer's perceptions. While comprehensive supply chain measurement focuses on supply chain performance(4). In this paper, a model is established according to the requirements of internal performance measurement, whose performance is evaluated based on fuzzy theory.

2 METHOD OF FUZZY COMPREHENSIVE EVALUATION FOR LOGISTICS CENTER PROCESSING PERFORMENCE

Logistics center mainly accomplishes purchase, inventory, order treating, sorting, distribution, and procurement, et al. Further more, Each processing involves person, facility, space, goods order, decision plan, time, cost, quality and the rest factors.

According to operation function, the architecture of evaluation index is composed of hierarchy that starting at bottom and deducing level by level. In Evaluation course, the second level operation indices are firstly deduced by means of bottom indices, then deducing the third level indices by the second level indices, analogically, reducing the final integrated indices.

One of the approaches resolving the hierarchy problem is selecting and using a multi-object, multi-level or multi-factor comprehensive evaluation method, which can synthetically consider various factors and ideas. The complex multilevel problem was preferably resolved in literatures (6, 7, 8).

2.1 Fuzzy comprehensive evaluation of single factor

2.1.1 Establishing evaluation element factor set U

$U = \{u_1, u_2, \cdots, u_i, \cdots, u_M\}$

M in formula is the evaluated objects' number. Object i, u_i (i=1, 2, ⋯ , M), can be divided further. $u_i = \{u_1, u_2, \cdots, u_i, \cdots, u_N\}_i$

N in formula is the evaluated factors' number on objector i. Factor j, u_j(j=1, 2, ⋯ , n_N), can be divided further. $u_{ij} = \{u_1, u_2, \cdots, u_k, \cdots, u_P\}_{ij}$

P in formula is the evaluated element factors' number of factor j on object i.

The rest may be deduced analogically, and continued division can get sub- element factor.

2.1.2 Establishing evaluation grade set V

$V = \{v_1, v_2, \cdots, v_n\}$

2.1.3 Establishing fuzzy matrix R_{IJ}

R_{IJ} is denoted the fuzzy matrix of element J on object I.

$$R_{IJ} = \begin{vmatrix} r_{11} & r_{12} & \cdots & r_{1n} \\ r_{21} & r_{22} & \cdots & r_{2n} \\ \vdots & \vdots & \vdots & \vdots \\ r_{p1} & r_{p2} & \cdots & r_{pn} \end{vmatrix}_{IJ}$$

Where each row $(R_i)_{IJ}$ of R_{IJ}, $(R_i)_{IJ} = (r_{i1}, r_{i2}, \cdots, r_{in})_{IJ}$, is a evaluated result of element i, r_{ij} denotes the degree of membership on evaluation element i to evaluates grade j. It shows fuzzy relationship in terms of degree of membership between evaluation element and evaluation grade.

2.1.4 Establishing weight coefficient matrix A

In fuzzy evaluation, the important step is to determine the weight coefficient of object, factor, and element factor. Weight coefficient matrix of evaluation object

$A = \{a_1, a_2, \cdots, a_i, \cdots, a_M\}$

Where a_i is the weight coefficient of evaluated object i, satisfying

$$0 < a_i \leqslant 1, \quad \sum_{i=1}^{M} a_i = 1$$

The weight coefficient matrix of evaluated factor is represented as follows.

$A_I = \{a_1 \quad a_2 \quad \cdots \quad a_j \quad \cdots \quad a_N\}_I$

Where a_{jI} is the weight coefficient of evaluated factor j on object I, and satisfying

$$0 < a_{jI} \leqslant 1, \quad \sum_{i=1}^{N} a_{jI} = 1$$

Weight coefficient matrix of evaluated element factor is represented as follows.

$A_{IJ} = \{a_1 \quad a_2 \quad \cdots \quad a_k \quad \cdots \quad a_P\}_{IJ}$

where a_{kIJ} is the weight coefficient of evaluated element factor k of factor j on object I, and satisfying following condition.

$$0 < a_{kIJ} \leqslant 1, \quad \sum_{i=1}^{P} a_{kIJ} = 1$$

2.1.5 Comprehensive evaluation matrix of single element factor

Comprehensive evaluation of factor J on object I can be denoted as follow.

$$B_{IJ} = A_{IJ} \times R_{IJ} = (a_1 \quad a_2 \quad \cdots \quad a_P)_{IJ} \begin{vmatrix} r_{11} & r_{12} & \cdots & r_{1n} \\ r_{21} & r_{22} & \cdots & r_{2n} \\ \vdots & \vdots & \vdots & \vdots \\ r_{p1} & r_{p2} & \cdots & r_{pn} \end{vmatrix}_{IJ} = (b_1 \quad b_2 \cdots b_j \cdots b_n)_{IJ}$$

$i = 1, 2, \cdots, n$. Where $b_{jIJ} = \sum_{i=1}^{P} a_{iIJ} \bullet r_{ijIJ}$

2.2 Fuzzy comprehensive evaluation

The higher level evaluation matrix R_I can be constructed by Colligating B_{IJ} ($J = 1, 2, \cdots, N$), the fuzzy evaluation result of every single factor. Adopting the same method, colligating R_I with weight coefficient A_I to resolve the comprehensive evaluation result R_I ($I=1, 2, \cdots, M$) of factor I., the higher level evaluation matrix R is formed. And then the comprehensive evaluation matrix B can be resolve. Matrix B is the comprehensive evaluation result. Namely

$$B = A \times R = (a_1\ a_2 \cdots a_M) \begin{vmatrix} (a_1 \quad a_2 \quad \dots \quad a_N)_1 & \bullet & \begin{vmatrix} A_{11} \bullet & R_{11} \\ A_{12} \bullet & R_{12} \\ \dots & \dots \\ A_{1N} \bullet & R_{1N} \end{vmatrix}_1 \\ \dots & \dots & \dots \\ (a_1 \quad a_2 \quad \dots \quad a_N)_I & \bullet & \begin{vmatrix} A_{I1} \bullet & R_{I1} \\ \dots & \dots \\ A_{IJ} \bullet & R_{IJ} \\ \dots & \dots \\ A_{IN} \bullet & R_{IN} \end{vmatrix}_I \\ \dots & \dots & \dots \\ (a_1 \quad a_2 \quad \dots \quad a_N)_M & \bullet & \begin{vmatrix} A_{M1} \bullet & R_{M1} \\ A_{M2} \bullet & R_{M2} \\ \dots & \dots \\ A_{MN} \bullet & R_{MN} \end{vmatrix}_M \end{vmatrix} = (b_1\ b_2 \cdots b_k \cdots b_n)$$

Evaluation result B not only considers all factors' influence, but also reserves all evaluation information of every level. Furthermore, percentage score can show the comprehensive evaluation result also. Generally, by defining evaluation criterion degree of membership set to be $\beta = \{\beta_1 \quad \beta_2 \cdots \beta_k \cdots \beta_n\}$, the score grade for comprehensive evaluation result can be calculated. So, evaluation sequencing is accomplished for variant processing scheme. Namely

$$\text{Score} = 100B \times \beta = (b_1 \quad b_2 \cdots b_k \cdots b_n)(\beta_1 \quad \beta_2 \cdots \beta_k \cdots \beta_n)^T \times 100 = (\sum_{k=1}^{n} b_k \bullet \beta_k) \times 100$$

Tab. 1 **Performance evaluation table for logistics center processing**

Evaluation. object			Evaluation factor			Evaluation element factor			Evaluation degree				
No. i	u_i	W a_i	No. j	u_{ji}	W a_{ji}	No. k	u_{kjj}	W a_{kji}	v_1	v_2	v_3	v_4	v_5
									Fuzzy matrix				
									big	bigger	mid.	smaller	small
1	Inventory operation	0.25	1	Space utility ratio	0.25	1	Store zone area-ratio	0.25	0.2	0.2	0.3	0.2	0.1
						2	Capacity util. ratio	0.25	0.2	0.2	0.4	0.2	0
						3	Custodial area	0.25	0.2	0.2	0.4	0.2	0
						4	Storage in unit area	0.25	0.1	0.2	0.4	0.3	0
			2	Stock efficiency	0.55	1	Inventory turnover	0.65	0.1	0.25	0.4	0.15	0.1
						2	Inventory control level	0.35	0.2	0.4	0.2	0.2	0
									low	lower	mid.	higher	high
			3	Checking goods	0.2	1	Quantity error rate	0.35	0.2	0.3	0.4	0.1	0
						2	Sort error rate	0.35	0.25	0.25	0.4	0.1	0
						3	Fund error rate	0.3	0.25	0.4	0.25	0.1	0
									big	bigger	mid.	smaller	small
2	Distr. operation	0.25	1	Person efficiency	0.15	1	Per-capita distrib.quant.	0.25	0.25	0.3	0.25	0.2	0
						2	Per-capita distrib.space	0.25	0.2	0.4	0.2	0.2	0
						3	Per-capita distrib.heavy	0.25	0.2	0.3	0.25	0.25	0
						4	Distrib.car-freq./person	0.25	0.15	0.25	0.4	0.2	0
			2	Car efficiency	0.4	1	Per-car distrib. space	0.15	0.3	0.4	0.2	0.1	0
						2	Per-car distrib. heavy	0.15	0.2	0.3	0.4	0.1	0
						3	Per-car distrib. ton/km	0.3	0.2	0.3	0.4	0.1	0
						4	Non empty car ratio	0.4	0.2	0.35	0.3	0.15	0
			3	Distribution Plan	0.2	1	Stowage ratio	0.2	0.2	0.4	0.2	0.2	0
						2	Per distribution. quant.	0.25	0.1	0.2	04	0.2	0.1
						3	Per distribution. space	0.25	0.2	0.3	0.3	0.2	0
						4	Per distribution. heavy	0.2	0.2	0.25	0.3	0.25	0
						5	Distribution. speed	0.1	0.2	0.3	0.4	0.1	0
			4	Distribution benefit	0.25	1	Distribution interest/hou	0.6	0	0.3	0.4	0.3	0
						2	Distribution quant./hour	0.4	0.3	0.4	0.2	0.1	0
3	Order treating operation	0.15	1	Order efficiency	0.5	1	Order quantity/day	0.3	0.2	0.25	0.3	0.15	0.1
						2	Customer quant./day	0.3	0.3	0.4	0.2	0.1	0
						3	Goods quantity/order	0.4	0.2	0.2	0.3	0.2	0.1
									low	lower	mid.	higher	high
			2	Client service quality	0.5	1	Order lag ratio	0.15	0.2	0.4	0.2	0.2	0
						2	Untimly delivery ratio	0.2	0.2	0.4	0.2	0.2	0
						3	Sales return ratio	0.15	0.4	0.3	0.2	0.1	0
						4	Deficit ratio	0.3	0.4	0.3	0.2	0.1	0
						5	Scarcity ratio	0.2	0.4	0.3	0.2	0.1	0
4	Proc. operation	0.20	1	Procuremer cost	0.55	1	Proc&Manag. fee rate	0.6	0.2	0.4	0.3	0.1	0
						2	Fund retention ratio	0.4	0.4	0.3	0.2	0.1	0
			2	Purchasing quality	0.45	1	Purch. quant. error ratio	0.3	0.3	0.4	0.2	0.1	0
						2	Fraction defective	0.3	0.3	0.4	0.2	0.1	0
						3	Delay ratio	0.4	0.2	0.3	0.4	0.1	0
									big	bigger	mid.	smalle	small
5	Sorting operation	0.15	1	Person efficiency	0.4	1	Per-capita sorting ability	0.25	0.2	0.4	0.2	0.1	0.1
						2	Treating bill quant./shift	0.25	0.1	0.4	0.3	0.2	0
						3	Sorting quant. in unit time	0.25	0	0.3	0.5	0.2	0
						4	Sorting freq. in unit time	0.25	0.2	0.4	0.2	0.2	0
									low	lower	mid	higher	high
			2	Facility cost	0.6	1	Sorting facility cost	0.3	0.2	0.4	0.3	0.1	0
						2	Sorting cost/bill	0.3	0.1	0.3	0.4	0.2	0
						3	Input-output cost	0.4	0.2	0.3	0.3	0.2	0

3 REALITY EVALUATION EXAMPLE

According to logistics center' working condition, it is divided 5 evaluation objects, 3 level structures. Detailed layout is shown in table 1(5). Every evaluation object consists of variant evaluation factors, which are composed of a variety of element factors. Different weight coefficients are assigned for every evaluation object, factor and element factor according to each importance. The importance grade is divided into 5 degrees $V = \{ v_1 \quad v_2 \quad v_3 \quad v_4 \quad v_5 \}$.

Evaluation process of logistics center processing performance evaluation begins from the element factor level in bottom, then evaluating level by level upwards. Firstly, evaluating begins from the factor of vehicle efficiency on distribution processing object.

3.1 Establishing car efficiency evaluation factor set U_{21}

U_{22} = { Per-car distributed space (u_1) Per-car distributed heavy (u_2) Per-car distributed ton/km (u_3) Non empty car ratio (u_4) $\}_{22}$

3.2 Establishing fuzzy evaluation matrix R_{22} of car efficiency evaluation factor

Fuzzy evaluation matrix is given in table 1. All of the values in table are obtained by statistical method and expert survey, and denoted by degree of membership. R_{22}, the fuzzy evaluation matrix of car efficiency evaluation factor, denotes

$$R_{22} = \begin{vmatrix} 0.3 & 0.4 & 0.2 & 0.1 & 0 \\ 0.2 & 0.3 & 0.4 & 0.1 & 0 \\ 0.2 & 0.3 & 0.4 & 0.1 & 0 \\ 0.2 & 0.35 & 0.3 & 0.15 & 0 \end{vmatrix}_{22}$$

3.3 Establishing Weight coefficient matrix A_{22} of car efficiency evaluation factor

$$A_{22} = (0.15 \quad 0.15 \quad 0.3 \quad 0.4)_{22}$$

3.4 Calculating comprehensive evaluation matrix B_{22} of car efficiency evaluation factor

$$B_{22} = A_{22} \times R_{22} = (0.215 \quad 0.335 \quad 0.33 \quad 0.12 \quad 0)_{22}$$

B_{22} is the comprehensive evaluation result of car efficiency evaluation factor. According to the same method, the person efficiency B_{21}, the distribution plan B_{23} and the distribution benefit B_{24} can be resolved individually. Results are as follows.

$$B_{21} = (0.20 \quad 0.3125 \quad 0.275 \quad 0.2125 \quad 0)_{21};$$

$$B_{23} = (0.175 \quad 0.285 \quad 0.315 \quad 0.20 \quad 0.025)_{23};$$

$$B_{24} = (0.12 \quad 0.34 \quad 0.32 \quad 0.20 \quad 0)_{24}$$

3.5 Fuzzy comprehensive evaluating for each evaluation object

Utilizing evaluation results B_{21}, B_{22}, B_{23} and B_{24}, higher-level fuzzy matrix with distribution operation as its object can be established.

$$R_2 = \begin{vmatrix} 0.20 & 0.3125 & 0.275 & 0.2125 & 0 \\ 0.215 & 0.335 & 0.33 & 0.12 & 0 \\ 0.175 & 0.285 & 0.315 & 0.20 & 0.025 \\ 0.12 & 0.34 & 0.32 & 0.22 & 0 \end{vmatrix}_2$$

Weight coefficient matrix, which is composed of 4 evaluation factors belong distribution operation, is

$A_2 = (\ 0.15 \quad 0.40 \quad 0.20 \quad 0.25\ )_2$

So, the comprehensive evaluation result for distribution operation object can be obtained.

$B_2 = A_2 \times R_2 = (\ 0.181 \quad 0.323 \quad 0.316 \quad 0.3175 \quad 0.005\ )_2$

By the same method, the evaluation results for other objects can be resolved as follows.

Inventory operation $B_1 = (0.165 \quad 0.279 \quad 0.346 \quad 0.168 \quad 0.042\ )_1$

Order treating operation $B_3 = (0.28 \quad 0.305 \quad 0.235 \quad 0.145 \quad 0.035\ )_3$

Procurement operation $B_4 = (0.271 \quad 0.36 \quad 0.269 \quad 0.1 \quad 0\ )_4$

Sorting operation $B_5 = (0.152 \quad 0.348 \quad 0.318 \quad 0.172 \quad 0.01\ )_5$

The higher-level fuzzy matrix can be formed using these results. Then combining with corresponding weight coefficient matrix A, the final fuzzy evaluation result is obtained.

$B = A \times R = (0.176 \quad 0.324 \quad 0.308 \quad 0.168 \quad 0.024\ )$

If the result is expressed with percentage score, with the degree of membership set for evaluation criterion β = {good(1) better(0.8) middle(0.6) worse(0.4) bad(0.2)}, the score result of performance evaluation for logistics center processing is calculated as follows.

$S = B \times \beta \times 100 = (\ 0.176 \quad 0.324 \quad 0.308 \quad 0.168 \quad 0.024)(1 \quad 0.8 \quad 0.6 \quad 0.4 \quad 0.2)^T \times 100 = 69.71$

4 CONCLUSION

The system of logistics center processing is a complex hierarchy model. According to different attribute, it can be divided into three levels as evaluation object, evaluation factor, evaluation element factor, even more. Divided processing system of evaluation example, it can be gotten 45 element factors in bottom level, 13 factors in second level, 5 objects in top level. All of them include many logistics processing items.

In the selection of two primary parameters, the weight coefficient and fuzzy matrix values of element factor in 5 grades, the key values for evaluation are gotten with experts' idea and statistical method. So, the final evaluation result is able to reveal real situation objectively.

No matter how many factors and element factors involved, the system's evaluation can be perfectly achieved with multi-level fuzzy comprehensive method by dividing it according to operation attributes and establishing the hierarchy model.

REFERENCES

1 **Silva, C. A., Sousa, J. M., Runkler, T. and Sa da Costa, J. M. G.** (2003) Evolved genetic algorithms with fuzzy aggregation applied to priorities in logistics systems. *2003 IEEE*. pp. 775-781.

2 **Qian-li, D.** (1999) Advanced logistics. *Peoples' traffic press*. Peking.

3 **Bowersox, D. J. and Closs, D. J.** (1996) Logistical Management: The Integrated Supply Chain Process. McGraw-Hill.

4 **Chen, Y. C.** (2002) An application of fuzzy set theory to the external performance evaluation of distribution center in logistics. *Soft Computing 6*. pp. 64-70.

5 http://www.56source.com, 100 Evaluation indices of logistics center productivity.

6 **Wang, Y. J. and Meng, X. Y.** (2000) Research on the method of multilevel fuzzy evaluation of the green products. *China Mechanical Engineering*. Vol. 11, No. 9, pp. 1016-1019.

7 **Korpela, J. and Tuominen, M.** (1996) Benchmarking Logistics Performance with an Application of Analytic Hierarchy Process. *Transaction on Eng. Management*. Vol. 43, No. 3.

8 **Chan, F. T. S., Chan, H. K. and Kazerooni, A.** (2002) Fuzzy Multi-Criteria Decision-Making Technique for Evaluation of Scheduling Rules. *Int. J. Adv. Manuf. Technol*. Vol. 20, pp. 103-113.

Information systems capabilities influencing partnership integration within supply chain management – an empirical study

E A WILLIAMSON and **M JORDAN**
BIM Division, Glasgow Caledonian University, UK
D K HARRISON
ESD, Glasgow Caledonian University, UK

ABSTRACT

Information systems development has resulted in enhanced information sharing and communications with partners. This has resulted in an increased focus on partner relations and organisational factors affecting partner collaboration. However, while close strategic partnerships can create new value, they are costly to develop and maintain. The number of close partnerships a company can build and maintain is also limited. Therefore it is important that factors affecting partner integration are analysed, in order to maximize supply chain effectiveness. This paper evaluates organisational factors influencing partnership integration at varying levels of information systems capabilities within Supply Chain Management (SCM). Studies carried out within three companies identify these factors.

Keywords: Supply chain management, inter-organisational information systems, information sharing, trust, partner relations

1 INTRODUCTION

An inter-organisational information system (IOS) is a collection of IT resources, including communications networks, hardware IT applications, standards for data transmission, and human skills and experiences. It provides a framework for electronic cooperation between businesses by allowing the processing, sharing and communication of information [1]. The development of IOS has increased rapidly in recent years and has meant a shift in the role of IT, from a competition weapon to a cooperation enabler [2]. This has also led to businesses in the supply chain forming partnerships and alliances. The Internet has improved inter-organisational information systems capability and therefore is being adopted as a routine platform for information systems development. However, although the Internet improves information flows and allows better communication between parties, organizational factors are also important as they affect the levels of integration and communication between businesses. These factors include trust, power and organisational culture. Also, while close strategic partnerships can create new value, they are costly to develop, nurture and maintain. The number of close partnerships a company can build and maintain is also limited.

Therefore it is important that the right type of relationship with a partner is built, in order to maximize SCM effectiveness and to allow high levels of information sharing and participation [3].

This paper evaluates organisational factors influencing partnership integration at varying levels of information systems capabilities within Supply Chain Management (SCM). Studies carried out within three companies illustrate the above factors.

2 REVIEW OF PREVIOUS WORK

Parke [4] defines partner relations as

> *"relatively enduring inter-firm cooperative agreements, involving flows and linkages that use resources and/or governance structures from autonomous organisations, for the joint accomplishment of individual goals linked to the corporate mission of each sponsoring firm"*

Supply Chain Management (SCM) creates a virtual organisation composed of a number of independent businesses with the common goal of efficiently and effectively managing and integrating the functions of purchasing, demand management, new product design and development, and manufacturing planning and control [5]. This virtual organisation can be a compromise between vertical integration and a fully open market situation. However, a number of variables affect the functioning of this virtual organization. Two major factors which have considerable influence on any partnership relation are trust and power. Trust may be defined as

> *"a general expectancy held by a channel member that the word of the other can be relied upon"*
>
> (Ballou, Gilbert & Mukherjee) [6]

Trust can improve co-operation and confidence between companies who are working together within the supply chain and, by working together, companies can achieve much more than the sum of their individual efforts [7,8].

In order to build trust, some conditions are necessary. One of the most important is sharing information or communication [6,9]. This may involve passing formal or informal information between companies to assist in decision making, resolving disputes and confirming the benefits of coordination to all parties. Thereby, there exists a "virtuous" circle, with information sharing building the level of trust and thereby improving supply chain co-operation.

Another variable identified in research as a requirement for trust is shared values [6]. As beliefs influence actions, partners with similar values and policies will act according to these values, which will be in line with the philosophies of the second company.

Trust and cooperation between companies can also be increased by adapting their behaviour. Research carried out [10,11] showed that high levels of adaptation feed back into a relationship, increasing levels of trust and co-operation and enhancing commitment to the relationship. Also, as trust and commitment grow, there is a greater likelihood of mutually advantageous adaptive behaviour. Given these findings, reciprocal adaptation would be expected from the parties to high trust/high commitment relations.

IOS enables information sharing [12] and even increases information flow in distribution channels, thereby allowing relations to develop with partners [13]. Therefore IOS can help establish trust between partners in the supply chain. However, it was found [14] that the loss of independence and lack of trust are the most cited reasons why organisations avoid or end arrangements.

Power is an informal device which may be used to influence co-operation in a supply chain. Alternative forms of power include reward power, expert power and referent power. Reward power can be used to support acceptable behaviour and give additional benefits, such as preferential delivery schedules or prices. A manufacturer, when trying to tie customers to the products, may use expert power by providing free product training. Referent power is used when a company is capitalizing on a strong brand or image to influence another whether it is on persuading them to increase their order amounts or expand products ordered. Some writers [15] consider the presence of power as the main influencing factor in achieving co-operation. For example, a large customer may require a number of small manufacturers to join an IOS if they wish to supply them. Also, large customers may exert power over smaller customers of a single supplier. Therefore the "less powerful" partners have subordinate roles, which may affect their trust in the more powerful partners. Power and misuse may be seen as a "vicious circle", with decreasing levels of trust being experienced.

Implementation of IOS can also lead to changes in bargaining power of participants which can affect the success of the IOS [13]. The implementation of an IOS can also affect previous stable relations and in fact, those businesses that have developed good long term working relations are more likely to have trouble in adopting IOS than those businesses which have weak or transaction based relations [16].

A business may require its suppliers or partners to adopt IOS to maximize its effectiveness and therefore initiate the move to IOS systems implementation. This business is therefore exerting power over its smaller partners. This 'push from one side' approach results in relations disruption as suppliers are less aware of benefits which may accrue to them, they feel that the power-base seems to lie with the other party and because of the resource commitment which they must make. Over time, however, the relations can improve if the suppliers realize the long term financial and relations benefits and can use the IOS with several partners [17].

Some researchers [18, 19] argue that the most important variable in managing the supply chain to the benefits of all businesses and partners involved is the fit of their respective organizational cultures. An organizational culture that emphasizes short-term, company-focused performance is at odds with effective SCM.

This paper evaluates the relationship between partner integration and electronic co-operation in three particular companies and identifies significant organisational factors.

3 RESUME OF CASE STUDIES

Three companies, Cisco, Smurfit and Company C were studied to investigate organisational factors influencing partnership integration at varying levels of IS capabilities. Information was gathered by carrying out lengthy interviews with at least one key personnel member from each company, as well as completion of a questionnaire. In particular, the questionnaire gave information on

- The effectiveness of IS in supply chain functions
- The impact of IS on partner relations
- IS implementation and partner integration issues
- Barriers to the use of IS

The following analysis focuses on the impact of IS on partner relations and IS implementation and partner integration issues.

3.1 Company 1: Cisco

Cisco commenced Internet development in 1990 and committed more than $100million to its development. This has resulted in a very high percentage of Cisco's employee, customer, supplier and partner interaction being network based; the initial contact being made through Cisco's home page.

Cisco has differentiated their suppliers into major suppliers who routinely supply most of their products and minor suppliers who supply fewer products less frequently. They have developed very close collaboration and longer term partnerships with their core manufacturing partners who supply high volumes of critical products (products which have a high impact on their core business). This has involved, not only ensuring that their 'partner' can supply suitable, satisfactory products which meet their product specification, but also that the 'partner' fits with Cisco's culture and mindset, that they hold similar values, for example, with regard to the importance of customer service and satisfaction. These close collaborators include Selectron, Motorola, Jabil and Celestica. They have 15 second tier partners and around 500 third tier partners. The third tier partners supply low volume, 'critically low' products, such as spot logistic companies. They therefore use a 'pro-rata investment of time and effort' with respect to partner importance.

Cisco has used advanced information systems extensively within supply chain management (SCM) since the mid 1990's to give them a distinct competitive advantage. These include advanced supplier portals and an integrated supply-base to allow collaboration with partners. There are 3 levels of connectivity relating to the levels of suppliers given above. The impact of IS on SCM, both its functioning and effectiveness have been substantial. For example, with regard to its Engineering and Manufacturing Connection Online portal (myMCO), there are:

- 125000 hits per day
- 150 web-based tools
- 19000 users
- and has led to a 25% increase in productivity

Current developments include their eHub Initiative to which 40 suppliers have currently connected. It will give:

- Central point for *all* supply chain information including forecasts, committed resources, inventory, supply, demand, purchase orders, sales orders, product master

- Leverages internet "exchange" technology and open standards
- Supplements and enhances existing execution systems
- Manages supply/demand mismatches
- Open to all Cisco supply chain partners

Questionnaire Analysis

Cisco's overall business strategy has included the implementation of advanced IS and this strategy has impacted on the number of partners, rather than the use of advanced IS as such. For example, Cisco has restructured/reorganised virtually every department to improve effectiveness, efficiency and customer service and this has been enabled by web-based approaches where-ever this adds value. The number of external suppliers has not been affected because of IS. The establishment of ubiquitous web-based access removes constraints on supplier number at one level but Cisco's preferred approach is to partner with fewer suppliers and work closely with mutual investment to maximise the 'win-win' approach.

The use of advanced IS has moved Cisco towards partner integration in all functions with both suppliers and customers. This involves partners carrying out many core processes instead of Cisco and is supported by frequent two-way flow of information. IS has had a significant positive effect on partner communication confidence in partners and the level of trust between partners has increased.

In forming partnerships, the culture, core processes, information systems, employee skills and trust in the partner have a significant positive effect.

3.2 Company 2: Smurfit

Jefferson Smurfit Group (JSG) began as a small Irish box-maker in 1934, and became a public company in 1964. JSG is now one of the largest European-based manufacturers of containerboard, corrugated containers and other paper based packaging products, with operations in Europe, Latin America and North America in 20 countries. It employs some 30,000 people and is a significant player in Europe. JSG had annual sales of £4.7 billion in 2002, and produces a total of 4 million tonnes of paper and paperboard and 3.5 million tonnes of paper-based packaging products annually, in 285 plants throughout the world. JSG also produces décor paper, with around 25% of the Global market. Décor paper is used in the manufacture of wall coverings and other interior furnishings. It also has interests in several associated companies (20). Their IS suite is similar to many companies, in that it is made up of a number of bespoke and off-the-shelf systems of different ages which are not fully integrated. For example, a SAP ERP system has not been fully implemented and only some modules, such as Accounting and Logistics are used. Sales and Distribution functions are included in the SAP system but are not yet implemented. The SAP system is organised as a wide area network with the server in Holland and is used by all 18 Smurfit plants. Access to information is controlled by tiered access and allows sharing of information to the benefit of each plant within the company. The accounting module allows the Chief Accountant and accounting staff to access income and expenditure information in all account budgets. The Logistics module includes Materials Management, which is further sub-divided into Purchasing and Inventory Management functions.

The Sales Order Processing System is a legacy, bespoke system. It has an over-night batch input to the SAP system. Details of the customer's order are usually received by fax or email. Some customers allow Smurfit to calculate their requirements for cardboard cartons by allowing Smurfit to access, via the Internet, their stock management system. This shows the customer's current production plan and current stock levels of cartons. The Smurfit employee then calculates the customer's carton requirements and raises an order. The order details are then inputted into the customer's stock management system and stock delivery is arranged. An email is sent to confirm order details. This integration of information includes tangible benefits for both partners, such as:

- continuity of supply, as responsibility is taken by Smurfit, rather than the customer,
- weekly delivery schedules as required by the customer
- less face to face contact although better information
- automatic email to the customer every time their system is entered by Smurfit
- trust built up between partners

Smurfit is also able to grasp new business opportunities. They hold stock for the customer as the customer-specific product is purchased when produced and the customers do not have room to hold stock. They are also following a Just-In-Time process and the stock is being pushed down the supply line and held by the producer. This business has expanded in recent years, to such an extent that a new warehousing facility was developed and other stock items, including plastics, held for customers. Current stock at this plant is now 3 times the previous value.

Questionnaire Analysis
The questionnaire showed that the company had not reorganised/restructured in any way and that there was no change in the number of permanent suppliers, although there were more temporary suppliers in manufacturing

There is a move towards contract relationships in purchasing and towards partnerships (investment of resources by both partners) in warehousing and logistics.

There is a move towards partner integration, with partners having access to the same information system, in warehousing and customer service.

The use of information systems had a positive effect on the trust levels between partners in warehousing, logistics and customer service.

3.3 Company 3

This company was founded in 1890 and developed as manufacturers of its main products in the 1950's. It now sells more than 300 million products, to all parts of the world, including Japan, Middle East, Caribbean as well as Europe and Canada. Its turnover is in excess of £26 million and employs about 600 employees in its one Scottish production plant. It also has a sales office in London and 80% of its turnover is taken up by the major supermarket chains.

The main use of information systems is EDI in the receiving and processing of customers' orders. 80% of orders are received via EDI (20% fax and email). This standardised system is simple to operate as the same required information is given in a pre-formatted layout. Therefore it is a very efficient system. A semi-automated EDI system was installed in the early 1990's, due to pressure from a large supermarket chain and this required some manual input. An updated, fully automatic EDI system was implemented in 1999. Now, the orders are received via EDI and flow straight into the Sales Order Processing System (SOPS). In fact, 3 different EDI systems are used to support three different customers systems. The SOPS is a server-site system with a number of dumb terminals in the factory, which are used for entering production details. Order information is sent from the SOPS to radio terminals held by Production Planners who use this data to allocate orders from stock and plan production schedules, taking into account stock being held in the warehouse. (Only 12/14 hours stock is held – due to the type of product and also, the customers require 95% of shelf life of the product. The computer terminals in the main office are used to post invoices (not fully automated), into the sales ledger system.

Large customers exert power by insisting that they operate their systems in a particular way, for example:

Supermarket A require a despatch note to be sent before the order is received. This is exacting as the order is picked and then sent to the warehouse in the same day, only 15 miles away , therefore it is difficult to send the despatch note before the order.

Supermarket B – Company 3 must subscribe to their Information Exchange System. This is carried out by with Sales Manager in England but the Information Exchange System is not used by the production plant. Orders are received through the older EDI and SOPS systems.

It is acknowledged that information is important and that more advanced systems give more timeous and accurate/useful information to plan and produce products, for example information from suppliers websites but separate systems and organisational factors mean that traditional less effective systems are used. For example, with regard to the purchasing system, Company 3 uses information from suppliers' web-sites, although purchase orders are sent using mostly 'phone and fax and then tied into the accounts system

Use of information systems has resulted in less human contact with suppliers and customers, with the telephone being used for problems.

The development of the current system, to include a web-based system and opening up systems to customers, was envisaged in the near future as it was felt that the powerful customers would abuse the access given, for example, they would take orders destined for other customers. This was also unlikely due to the strong and direct influence of the owner and his close control on all operational and strategic matters.

Questionnaire Analysis

IS has had no impact on the number of external suppliers and no internal restructuring has been done to improve relationships. There is a move towards contract relationships with customers in Customer Service function. This company has little partner integration and IS has had no change on the level of partner integration.

IS has had a significant positive effect on information flows within Customer Service and a positive effect in Logistics and also in the integration of processes within these two functions. These two functions have also seen some increase in the trust levels between partners and confidence in partners.

4 REVIEW OF FUTURE WORK

Ongoing work includes developing the above case studies by gathering information from their partners up and down the supply chain, to confirm information gathered as well as expanding the number and type of companies studied. Opportunities to collaborate with a number of companies exist. For example, current findings seem to suggest the following:

Table 1: The Importance of Organisational Factors at varying levels of IS capabilities

Factor	Low Level (eg EDI)	Medium Level	High Level	Impact
Technology				As technical sophistication and integration increases, trust in the technology increases Technology is also integrated into the business strategy
Influence of Power				Power has a stronger influence at a high level of IS implementation, although some partners can influence others, even at the lowest level of partner integration (eg EDI)
Partner integration and trust				Experience in IS can enhance trust in partners.
Management/ Leadership style				Top management beliefs can have a strong influence at all levels of integration, whether it is to act as a barrier or driver towards IS integration

It is hoped to develop a three-stage model of IS development and to highlight a number of factors at each of these stages. Figure 1 shows an embryonic three-stage model of IS development, categorising this development into Elementary, Intermediate and Advanced stages.

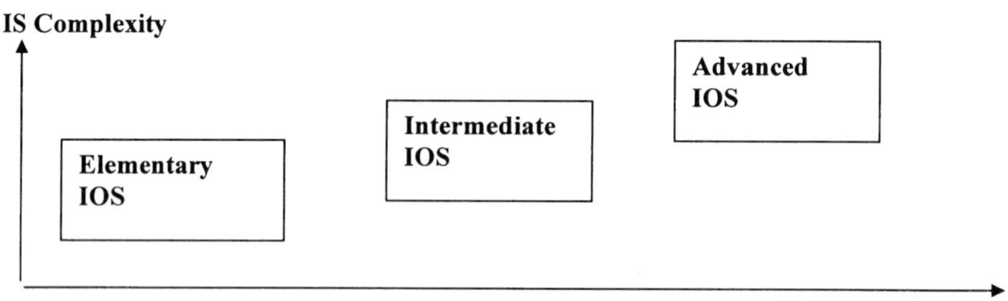

Figure 1: Information Systems Development

Time Months/Years

The following tables define each of these three stages in terms of four variables – Technology, Use, Partner Factors and Organisational Factors.

Table 2: Characteristics of Elementary IOS Development

Characteristics			
Technology	**Use**	**Partner Factors**	**Organisational Factors**
Many different IS used, including EDI and third party networks. Technology seen as an operational tool, rather than as a key strategic component. Automation of processes gives efficiency gains.	Transactions only. Partners enter information into their own systems. Limited information communication and co-ordination. Limited benefits from use of technology. Collaboration at operational level.	Companies may be working towards their own agenda and for their own benefit. Limited trust may exist between partners. Partner collaboration is weak, which does not allow much power to be exerted.	Organisational culture fit of partners is not important. Technological, financial or organisational barriers may exist to hinder further development.

Table 2 shows that within Elementary IOS development, there is a low level of systems integration, information co-ordination, partner collaboration and of benefits gained. Significant barriers include technological development, lack of financial resources and management/leadership style. Insignificant factors to the use of IOS include partner trust, power and organisational culture.

Table 3: Characteristics of Intermediate IOS Development

Characteristics			
Technology	**Use**	**Partner Factors**	**Organisational Factors**
Many different IS used, including EDI and third party networks. Technology seen as an operational tool, rather than as a key strategic component. Automation of processes gives efficiency gains.	Transactions only. Partners enter information into their own systems. Limited information communication and co-ordination. Limited benefits from use of technology. Collaboration at operational level.	Companies may be working towards their own agenda and for their own benefit. Limited trust may exist between partners. Partner collaboration is weak, which does not allow power to be exerted.	Organisational culture fit of partners is not important. Technological, financial or organisational barriers may exist to hinder further development.

Table 3 shows that there is some systems integration, information co-ordination, partner collaboration and benefits gained within Intermediate IOS development. Significant barriers and insignificant factors seem to be similar to those in the Elementary stage.

Table 4 shows that there is a high level of systems integration, information co-ordination, partner collaboration and benefits gained within Advanced IOS development. The insignificant factors of the previous two stages become very significant and influential at this stage, namely, partner trust, power and organisational culture.

Table 4: Characteristics of Advanced IOS Development

Characteristics			
Technology	**Use**	**Partner Factors**	**Organisational Factors**
Technology viewed as a key strategic component. Advanced IT systems allow controlled access to extensive information	At all levels, strategic to operational – use, collaboration and benefits	Tiered partners according to 'value' of partnership. Power may be exerted over partners in the chain	Organisational culture fit of partners is important. Well developed management systems and knowledge. Trust developed through use of IS and organisational factors

5 CONCLUSION

The development of IS has increased rapidly in recent years and has meant a shift in the role of IT, from a competition weapon to a cooperation enabler [2]. This has also led to businesses in the supply chain forming partnerships and alliances. However, although IS improves information flows and allows better communication between parties, organizational factors such as power, trust, leadership and organisational culture are also important as they affect the levels of integration and communication between businesses. This paper analysed organisational factors, such as trust, power and culture influencing partnership integration within three companies who have different levels of information systems capabilities. This study showed that the development of IS has encouraged partner integration and the development of trust between partners. Ongoing work includes developing the above case studies by gathering information from their partners up and down the supply chain, to confirm information gathered as well as expanding the number and type of companies studied.

ACKNOWLEDGEMENTS

The authors would like to thank the following for their assistance in completing the case studies:
Cisco: Mr K Johnston, Country Sales Manager; Mr Gordon Thomson, Business Support Manager
Smurfit: Mr D Hamilton, Financial Controller; Mr Alec Greenshields, Sales Manager
Company 3: Mr B Henderson, Computing Manager

REFERENCES

1 Haiwook (2001) "The Effects of Interorganisational Information Systems Infrastructure on Electronic Cooperation: An Investigation of the 'move to the middle'", PhD Abstract. Proquest Digital Dissertations. www.lib.umi.com/dissertations, accessed 13.3.2003
2 Hong I.B. (2002) "A new framework for interorganisational systems based on the linkage of participants' roles", *Information & Management*. Elsevier Science Ltd. B.V. No 39, pp 261-270
3 Heikkila J. (2002) "From supply to demand chain management: efficiency and customer satisfaction", *Journal of Operation Management*. Elsevier Science Ltd. Vol 20, Issue 6, pp 747-767
4 Parkhe A. (1993) "Strategic alliance structuring: A game theoretic and transaction cost examination of interfirm cooperation", *Academy of Management Journal*. Briarcliff Manor. Vol 36, Issue 4, pp 794-829
5 Tan K.C. (2001) "A framework of supply chain management", *Journal of Purchasing and Supply Management*. Elsevier Science Ltd. New York. Vol 7, Issue 1, pp 39-48
6 Ballou R.H., Gilbert S.M. & Mukherjee A. (2000) "New Managerial Challenges from Supply Chain Opportunities", *Industrial Marketing Management*, Elsevier Science Ltd. New York. USA. Vol 29, Issue 1, pp 7-18

7 Mariotti J.L. (1999) "The Trust Factor in Supply Chain Management. Supply Chain", *Management Review* 3 (1, Spring), pp 70-77

8 Anderson J.C. & Narus J.A. (1990) "A Model of Distributor Firm and Manufacturing Firm Working Partnerships", *Journal of Marketing.* 54 January. 99 42-58

9 Lalonde B.J. & Cooper M.C. (1989) "Partnerships in Providing Customer Service: A Third Party Perspective", Oak Brook, IL: Council of Logistics Management in Angeles R. & Nath R. (2001) Partner Congruence in Electronic Data Interchange-enabled Relationships. *Journal of Business Logistics.* Oak Brook. Vol 22, Issue 2, pp109-128

10 Hallen L., Johanson J. & Seyed-Mohamed N. (1991) "Interfirm adaptation in business relationships", *Journal of Marketing.* Vol 55, April, pp 29-37 in Wilson D.D. & Vlosky, R.P. (1998) Inteorganisational information system technology and buyer-seller relationships. *The Journal of Business & Industrial Marketing.* MCB University Press. Vol 13, Issue 3, pp 215-234

11 Brennan R. & Turnbull P.W. (1999) "Adaptive Behaviour in Buyer-Supplier Relationships", *Industrial Marketing Management.* Elsevier Science Ltd. New York. USA. Vol 28, Issue 5, pp 481- 495

12 Choi C. (2001) "The Effects of Interorganisational Information systems Infrastructure on Electronic Cooperation: An Investigation of the 'move to the Middle'". PhD Abstract. Proquest Digital Dissertations. www.lib.umi.com/dissertations, accessed 13.3.2003

13 Clemons E.K. & Row M.C. (1993) "Limits to interfirm coordination through information technology: Results of a field study in consumer packaged goods distribution", *Journal of Management Information Systems.* Armonk. Vol 10, Issue 1, pp73-89

14 Williams T. (1997) "Interorganisational information systems: issues affecting interorganisational cooperation", *Journal of Strategic Information Systems.* Elsevier. No 6, pp 231-250

15 Thorelli (1986) "Networks: Between Markets And Hierarchies", *Strategic Management Journal* Chichester: Vol. 7, Issue 1; Pg 37 - 52

16 Wilson D.D. & Vlosky, R.P. (1998) "Inteorganisational information system technology and buyer-seller relationships", *The Journal of Business & Industrial Marketing.* MCB University Press. Vol 13, Issue 3, pp 215-234

17 Vlosky R.P., Wilson D.T. & Vlosky R.B.(1997) "Closing the interorganisational information systems relationship satisfaction gap", *Journal of Marketing Practice: Applied Marketing Science.* MCB University Press. Vol 3, Issue 2, pp 75-86

18 Hahn C.K., Kim K.H. & Kim J.S. (1986) "Costs of competition: implications for purchasing strategy", *Journal of Purchasing and Materials Management.* Vol 14, Issue 6, pp 40-51

19 Fairley, G.A. (1997) "Discovering supply chain management: a roundtable discussion", *APICS – The Performance Advantage.* Vol 7, Issue 1, pp 38-39

20. www.smurfit.ie, accessed 26.2.04

Research on mass customization of MIS software based on common business application templates warehouse

X-Y ZHU
Department of Information Management and Information Systems, Zhejiang Institute of Science and Technology, China

ABSTRACT

Most techniques and methodologies that are currently used by the mass customization of mechnical products can be also applied to MIS production. One practical approach for MIS mass customization is using templates, which are nearly complete applications where the completed parts include applications architecture and reusable components. In this paper, I first present a brief survey of the common business application models derived from various e-business systems, then I conceptualize templates-based approach by presenting the templates warehouse and the architecture of an integrated software mass customization toolkit. I conclude by an example showing the practicability of this approach.

Keywords: MIS; software mass customisation; templates warehouse; toolkit

1 INTRODUCTION

Computer technologies have not only established new production lines for mass customization of many mechanical products, but have also provided new means of production for management information systems (MIS). So far, mass production of MIS software has always been in form of commercialized software packages, which claim to fit for nearly all business processes in almost all organizations. Undoubtedly, commercialized software packages can meet certain requirements for management and decision-making in some organizations, but as pre-defined, pre-built and highly standardized software systems, they are mainly targeted to those business processes which can be easily normalized or standardized, such as accounting and inventory. Moreover, these software packages seem to be rather inflexible as to user's specific requirements and demands, because any minor change to the functions of these packages may result in great modifications to their source programs. With urgent need to ease the maintaining of their products and to meet the increasing demand for diversities in MIS systems from users, many MIS vendors has begun turning to software mass customization techniques.

Software mass customization focuses on the means of efficiently producing and maintaining multiple similar software products, exploiting what they have in common and managing what varies among them [1]. With more and more successful implementation of business systems in various kinds of organizations, developers have found that many functionally similar application models exist among different kinds of systems in varied organizations. These application models are always functional modules or subsystems of most typical or common business processes in e-business systems. They have clearly-defined function boundaries and certain accepted programming schema with little variation in their internal architectures. By using software mass customization techniques to exploit what these application models have in common and to effectively manage the variation, it is possible to implement mass customization to these general-purpose application models and automatically generate their programs.

2 TECHNIQUES FOR MIS MASS CUSTOMIZATION

2.1 What to be customized in an MIS

MIS mass customization involves managing the internal and external varieties of an e-business application. Fundamentally, these varieties can be divided into four domains: (i)Functionality of an e-business application including business logics, business processes, function boundaries, and other user's requirements at the system analysis stage; (ii)Presentation or the appearance of an e-business application such as menu styles, input and output layouts, graphics, colors, fonts, reports, etc.; (iii)Software development platforms or running environment including operating systems supporting the software, programming languages, network architecture, DBMS, and other specifications at the system design stage; (iv)Other software features such as data integrity, system reliability, system security, etc. Software mass customization tools usually provide users effective and simple enough means, such as domain modeling language or visual programming environment, to help users define the varieties.

2.2 How to accomplish MIS mass customization

As to the techniques used in mass customization, software is analogous to what is practiced in the manufacturing industry, where the focus is on creating a singular product line, out of which many customized but similar variations of a product model are produced. The key aspect of MIS mass customization is to identify and use what is common in data and procedures among different e-business applications. Another aspect is to reduce the internal variety of MIS by using standardization, and to create external variety by modularization and the reusability of program modules. The fundamental techniques include standardization, modularization, modeling, and automatic code generation[2].

Standardization is to (i)construct business logic, establish common business objects through abstraction and normalization, and make the e-business systems more customizable; (ii)create general-purpose and standardized parts of e-business applications and maintain a collection of their prototypes (or templates); (iii)define what can be varied in each application template, and effectively manage the variations to meet user's specific requirements. Object-oriented technology is desirable for standardization and rationalization of e-business logic and other implementation issues.

Modularization is used to create variety, to utilize similarities and reduce complexities of e-business systems. Modularization helps a programmer to break a complex program into smaller subprograms called procedures. Furthermore, modularization increases clarity, reduces complexity and provides flexibility. Application module is a set of instructions that can be tested and verified independently of its use in a larger program. An expansive variety of e-business software products can be produced by combining a limited number of modules[3]. Component-based technology and reusable components are supposed to be helpful to modularization.

Domain modeling language (DML) is suitable for defining user's specific requirements for the software to be produced. The DML language is usually a list of attributes. Each attribute is a key-value pair. The attribute key is basically an identifier. DML is suitable for building object-oriented software models, and more importantly, for building original types of models. It usually includes a graphical front-end, and a powerful back-end language for generating code, analysis and documentation.

Automatic Code Generation is essential to software mass customization. Usually, automatic code generation is realized by a compiler called code generator, by which user's specifications of an e-business application are translated into source code programs. The code generator acts as a mapping between business object model and implementation languages.

2.3 Common business application templates

Components technology seems to be a promising approach towards software mass customization by enabling application construction through "plug and play". However, the middleware supporting this approach is still complicated and difficult to use. So far, the concept of components is still more limited to the basic objects or units within an application, such as GUI forms, buttons, list boxes, check boxes, grid lists, etc. A software developer has to use and reuse these controls to construct an application. Therefore, the efficiency of component-based software development still needs to be improved.

Template is a nearly complete application where the completed parts include an application's architecture and reusable components. The architecture is generic to fit any application in the class, and the reusable components are designed to work with each other within the architecture[4]. Common business application templates represent those business components that are common to a very large set of business domains. An application template serves at least one specific or even very comprehensive business function, it is general-purpose, flexible, and can be easily combined to build e-business applications.

After years of experience in the analysis, design and implementation of e-business systems such as finance and accounting systems, sales and marketing systems, customer relationship management(CRM) systems, and supply chain management(SCM) systems, the author of this paper has abstracted and derived a series of general-purpose and typical programming models from those e-business systems, and has established an application templates warehouse that contains numbers of preprogrammed application prototypes. A toolkit oriented to the MIS mass customization has also been created upon the construction of this warehouse

3 TEMPLATE-BASED APPROACH FOR MIS MASS CUSTOMIZATION

3.1 The application templates warehouse

Over the time, the meaning of the term template has changed from being defined by the physical presence into being defined by structure and functionality. A common business application template here is defined to be a technically typical and functionally general-purpose module or subsystem that is abstracted or derived from various e-business systems. Each application template has static and dynamic parts of attributes. The static part consists of data and procedures that can be pre-defined, designed, programmed, documented and stored, such as function boundaries, programming flowcharts, input and output, related database or tables, and so forth; The dynamic part consists of parameters and procedures that need to be specified later because they are not general-purpose and cannot be pre-determined. Fundamentally, an application template can be seen as a 'big' class or component where data, procedures, and system architectures are embedded into it.

The application warehouse here is defined as a set of common business application templates. So far, the warehouse contains numbers of various application templates derived from Web-based finance and accounting systems, sales and marketing systems, CRM systems, SCM systems, and government office automation systems. Typical examples of these templates are: (i) highly generalized and highly cohesive subsystems in e-business applications, such as database-based advertisement rotators, chat rooms, electronic discussion groups, Web-linked call centers, spreadsheets, and online training centers, etc., (ii) highly generalized but less cohesive functional modules, such as code classification designing modules, bill input, validation and checking modules, account books looking-up modules, reporting, database backing-up and restoring modules, user authentication modules, user rights management modules, system log file maintenance modules, etc., (iii) less generalized but highly cohesive subsystems, such as customer resources management systems, pricing systems, contracts management systems, invoices generation systems, and network management systems, etc.

3.2 Mechanism for MIS mass customization

The MIS mass customization toolkit prsented here is a set of comprehensive and integrated software development tools, it consists of five key parts: the Application Templates

Warehouse, the Warehouse Management System, the Domain Modeling Language, the Code Generator, and the Data Dactionary. Figure 1 illustrates the overview structure of these tools.

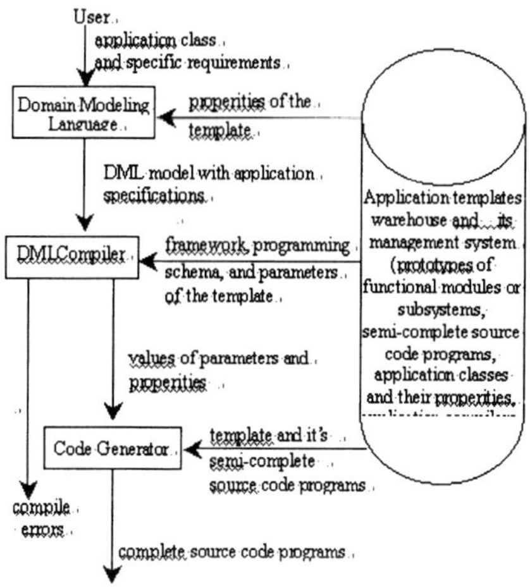

Fig. 1 An Overview of the Development Tools

The process of MIS mass customization can be divided into two main stages: the application definition stage, and the application generation stage. In the application definition stage, users need only to select the class of an application to be produced and to specify the requirements through domain modeling language. In the end of this stage, a DML model file is created to store the specifications of the application. In the application generation stage, an analyzer first checks out the DML model file for any possible errors, then extracts the assigned values of parameters previously set in the DML file. The Code Generator retrieves semi-complete programs from the template warehouse and tailors these programs to customize them for user's requirements and the specifications. In the end of this stage, appropriate, complete, and executable source code programs are produced in Delphi or Java language.

3.3 Example for MIS mass customization

The Window shown in Figure 2 is an application template basically designed for the catalog management of customers with comprehensive functions of adding, updating, deleting, locating, and printing records into or from customer tables. The template can be widely used in CRM or SCM systems to create and maintain various kinds of catalogs, such as products, suppliers, raw materials, department, etc. By defining the DML file listed in Table 1, the toolkit can automatically and quickly produce another application for catalog management of products shown in Figure 3. These two applications created from the same template are similar in functionality but vary in contents.

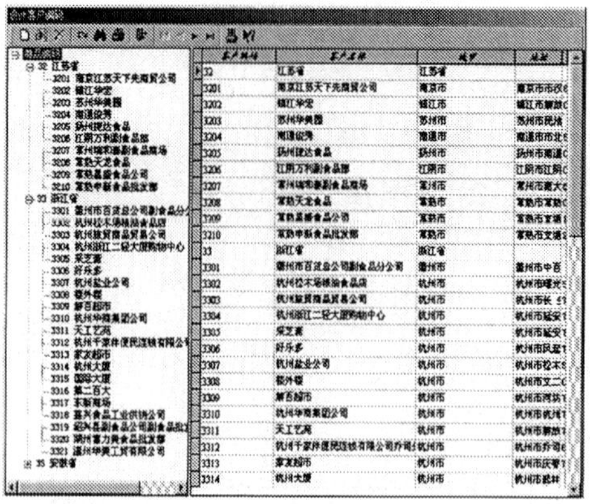

Fig. 2 An Template for Catalog Management

Table 1 DML Model for Products Catalog Management

Template Class "DBBrowse"
TargetLanguage "DELPHI"
DBMS “MSSQL"
DBFName "KECRM"
TableName "x_products"
Caption "设计产品分类"
Columns "产品编码,产品名称,规格型号,计量单位"
Fields="fcpdm, fcpmc, fgg, fdw "
ToolBar "TOP"
TreeView "LEFT"
Keyfields for TreeView "fcpdm, fcpmc"

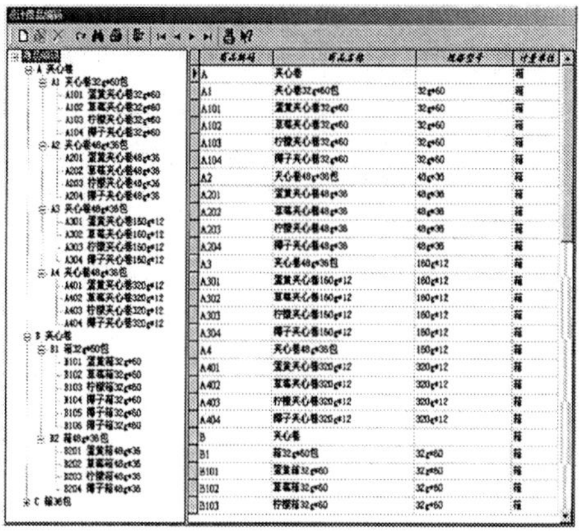

Fig 3 An application produced from the template for catalog management of products

4 CONCLUSIONS

With respect to software engineering, object-oriented methodology has become one of the principal techniques that are widely used throughout the entire software development lifecycle. However, traditional concepts of classes or objects are more limited in small components or units within an application, therefore the efficiency of object-oriented programming (OOP) still needs to be improved. This paper combines the concepts and techniques of OOP and UML into an MIS mass customization toolkit with focus on modeling the key and typical business processes in an organization and the generation of programs. Compared with other UML tools, the toolkit introduced in this paper is more concise, business-oriented, and practical.

ACKNOWLEDGEMENT

*This work is funded by the Science and Technology Department of Zhejiang Province of China under grant 2004C31115, and was partially funded by the Chinese National Funds for Innovation under grant 00C26213300787.

REFERENCES

1. Clements,P.,Northrop,L. *Software Product Lines.* Practice and Patterns, Addison-Wesley, Reading, MA..2001.
2. Michael Blaha and William Premerlani, *Object-Oriented Modeling and Design for Database Applications.* Prentice Hall, Upper Saddle River, N.J., (1998).
3. Brian W. Embley, Barry D. Kurtz and Scott N. Woodfield, *Object-Oriented systems Analysis: A Model-Driven Approach.* Yourdon Press, Englewood cliffs, N.j., (1992).
4. Hemeixun, Huangfenggang, and Songxiaoyang, *A Third Generation Object-Oriented Development Methodology Based Upon UML*, Applied Science. No 9(2002) p90

Web-based CAD/CAM/CAPP/PDM

The development of a web-accessible databank of material behaviour characteristics for use in virtual design environments

P BROOMHEAD, J KULON, R J GRIEVE, and **D J MYNORS**
Department of Design and Systems Engineering, Brunel University, UK

ABSTRACT

Continual pressures, such as reducing time to market, have resulted in the increased use of virtual design and manufacturing tools such as the simulation of manufacturing process using FEA packages. The simulation of processes such as forging allows an assessment of the manufacturability of components to be made. The reliability and accuracy of results from process simulation depend on the quality, appropriateness and accuracy of the representation of material behaviour characteristics.

This paper describes the development of a web accessible databank of mechanical and thermal material properties for direct linkage into virtual design environments for metalforming applications.

1 INTRODUCTION

A web-enabled databank of material properties is being produced in collaboration with the Confederation of British Metalforming (CBM), the Department of Trade and Industry (DTI) and various collaborating companies. Its purpose is to provide the British metal working industry with an easily accessible source of material property data appropriate for use with metalforming simulation packages. It follows that the data should reflect the wide range of strain, strain rate and temperatures found in metalforming for the metals used by that industry. For example, mechanical properties will be required up to and beyond yield through to fracture.

This paper extends previously published work [1] on developing a web-enabled databank and describes the final structure using XML.

Recent developments in FEA packages have resulted in the availability of computer-based systems specifically for use with metalforming applications. The user interfaces have been designed for metalforming applications and are relatively easy to use. No detailed knowledge of FEA techniques is required but inputs from the user are needed in response to prompts from the system. The workpiece and die geometries, with the operating characteristics of the forming machine are needed and should be known by the user. Workpiece and die material properties (mechanical and thermal) are also needed but may be unknown by the user. Typical of the mechanical properties is the flow stress vs strain curve, whilst thermal properties include specific heat capacity and thermal conductivity. Metalforming operations may be conducted at high temperatures and high rates of strain, which means that the mechanical and thermal properties should also be appropriate for those conditions.

Typical mechanical and thermal properties required as input to simulation packages, with their dependency on temperature (t) strain (ε) and strain rate ($\dot{\varepsilon}$), are:

- Flow stress $f(\varepsilon, \dot{\varepsilon}, t)$
- Young's modulus f(t)
- Poisson's ratio f(t)
- Yield stress f(t)
- Density f(t)
- Fracture criteria $f(\varepsilon, \dot{\varepsilon}, t)$
- Thermal conductivity f(t)
- Specific heat capacity f(t)
- Heat transfer coefficient f(t)

The description of computer based FEA simulation packages and the material properties required has been described elsewhere [1].

2 WEB ENABLED DATABANK

The previous work [1, 2] argued that the Extensible Markup Language (XML) should be used to overcome some of the problems associated with variable primary data formats, interoperability between software platforms and applications, and legacy problems. This was seen to be an advantage over the traditional method of structuring a web page using the HyperText Markup Language HTML as the primary data structure for the databank.

The argument presented [1, 2] was as follows:

"XML is a meta-language, a means by which users can define their own data description languages consisting of markup tags that convey a meaning appropriate to their use. A schema (XSD) or Data Description Type (DTD) can be used to ensure that the use of the markup follows the rules defined by that language. Tag ordering and duplicity is tightly controlled and leads to consistent documents that can be easily parsed and used in other applications. In the present case an XML document (essentially plain text) containing material property data could be parsed and the information therein combined with HTML to dynamically create a web page. Any browser could then display this page. Although the final result will still be a HTML page, a page that presents the same readability problems when viewed outside of the browser, however in this case HTML is not the primary data format. XML is the primary data format, a format that provides a well-formed structure that is self-documenting and can be verified against a schema or DTD. Domain specific XML languages have been developed in recent years; femML for finite element modelling, MathML for mathematics and CML for chemical descriptions are examples. Of relevance to the work described in this paper is MatML [2], a recent development that aims to provide an XML markup language for use by the materials community. Version 3 of MatML has now been completed and registered with www.XML.org. MatML allows material data to be presented in a structured manner using relevant tags describing the content. MatML 3.0 uses a schema to ensure the application conforms to the syntax and semantics defined by the markup."

The original intention was to use the MatML structure for the databank and present the appropriate tables of data [1]. Whilst this would be possible it has a number of drawbacks.

Although MatML is a comprehensive language for the description of material properties, it suffers from the limitation of only supporting the markup of basic numerical data. It is also a verbose implementation that extends far beyond the current requirements of a databank to support metalforming simulation, which typically will only contain the data described in Section 1. The tag and attribute names used in MatML to describe data properties do not conveniently match those in common use by the metalforming industry and there is much to be said for supporting a common language that is understood by all users of the databank.

Additionally, MatML does not allow the interrelationships between metalforming properties to be presented in an equation form. A simple example is the well-known Holloman's equation relating stress 'σ' to strain 'ε'; $\sigma = A\varepsilon^n$ (where A and n are constants). Equations of this form (and for that matter more complex forms including other terms such as temperature and strain rate) are used in metalforming simulation packages. If the form of the appropriate equation for a particular metal was known, together with the constants, then presenting the equation along with the constants and the range over which they apply, would be sufficient. Such a facility would eliminate the need to represent data in a tabular format as required by the equivalent form in MatML. Presenting equations in such a way would therefore be a useful capability of any databank addressing the requirements of the metalforming industry.

A typical XML document for illustrative purposes (showing only the main elements and a reduced data set with no temperature or strain rate influences) using an equation to provide interrelationships between properties is shown below.

```
<FormingMLDoc>
  <Material>
    <BulkDetails>
      <Name>Demonstration</Name>
      <Class>steel</Class>
      <Subclass>stainless steel</Subclass>
      <PropertyData description="de6" representation="re6" source="ds1">
        <EquationData parameter="pa3"/>
        <EquationData parameter="pa4"/>
```

```
                <DataSet>
                    <Data parameter="pa27">
                        <Value>2.0</Value>
                    </Data>
                    <Data parameter="pa25">
                        <Value>0.20</Value>
                    <Data>
                </DataSet>
            </PropertyData>
        </BulkDetails>
        <Metadata>
            <DataSourceDetails id="ds1">
                <Name>Source of data</Name>
            </DataSourceDetails>
            <PropertyDetails id="de6">
                <Name>True stress v.s. true strain</Name>
            </PropertyDetails>
            <RepresentationDetails id="re6" type="equation">
                <Name>flow stress equation</Name>
                <Display>hollomans.gif</Display>
            </RepresentationDetails>
            <DataParameterDetails id="pa3">
                <Name>true flow stress</Name>
                <Symbol>σ</Symbol>
                <Units name="MPa"/>
            </DataParameterDetails>
            <DataParameterDetails id="pa4">
                <Name>true strain</Name>
                <Symbol>ε</Symbol>
                <Units name="Unitless"/>
            </DataParameterDetails>
            <DataParameterDetails id="pa25">
                <Name>work hardening exponent</Name>
                <Symbol>n</Symbol>
                <Units name="Unitless"/>
            </DataParameterDetails>
            <DataParameterDetails id="pa27">
                <Name>constant</Name>
                <Symbol>A</Symbol>
                <Units name="MPa"/>
            </DataParameterDetails>
        </Metadata>
    </Material>
</FormingMLDoc>
```

The structure of the markup follows the usual hierarchical tree format found in all XML documents. The hierarchical form defines the materials name etc at the higher level while allowing the detail of the properties to be defined in the lower levels. The root element is

Material, which contains the elements *BulkDetails* and *Metadata. BulkData* contains elements that provide the material's name, class, subclass, and property data. *Metadata* contains descriptions of the data sources, data parameter properties (name, symbol, units) and representation details (tabular or equation).

The markup conforms to the following DTD:

```
<?xml version="1.0" encoding="UTF-8"?>
<!ELEMENT FormingMLDoc (Material)>
<!ELEMENT Material (BulkDetails, Metadata)>
<!--
   BulkDetails
-->
<!ELEMENT BulkDetails (Name, Class?, Subclass?, PropertyData+)>
<!ELEMENT Name (#PCDATA)>
<!ELEMENT Class (#PCDATA)>
<!ELEMENT Subclass (#PCDATA)>
<!ELEMENT PropertyData (EquationData*, EquationNotes?,DataSet*, Notes?)>
<!ATTLIST PropertyData description CDATA #REQUIRED representation CDATA
#REQUIRED source CDATA #REQUIRED>
<!ELEMENT DataSet (ConstantParameterValue*, Data+, ParameterValue?, Notes?)>
<!ELEMENT ConstantParameterValue (Value)>
<!ATTLIST ConstantParameterValue parameter CDATA #REQUIRED>
<!ELEMENT Data (Value+)>
<!ATTLIST Data parameter CDATA #REQUIRED>
<!ELEMENT EquationData EMPTY>
<!ATTLIST EquationData parameter CDATA #REQUIRED>
<!ELEMENT ParameterValue (Value+)>
<!ATTLIST ParameterValue parameter CDATA #REQUIRED>
<!ELEMENT Value (#PCDATA)>
<!ELEMENT Notes (#PCDATA)>
<!ELEMENT EquationNotes (#PCDATA)>
<!--
   Metadata
-->
<!ELEMENT Metadata (DataSourceDetails+, PropertyDetails+, RepresentationDetails+,
DataParameterDetails+)>
<!ELEMENT DataSourceDetails (Name)>
<!ATTLIST DataSourceDetails id CDATA #REQUIRED>
<!ELEMENT PropertyDetails (Name)>
<!ATTLIST PropertyDetails id CDATA #REQUIRED>
<!ELEMENT RepresentationDetails (Name, Display?)>
<!ATTLIST RepresentationDetails id CDATA #REQUIRED type (tabular | equation)
#REQUIRED>
<!ELEMENT Display (#PCDATA)>
<!ELEMENT DataParameterDetails (Name, Symbol, Units)>
<!ATTLIST DataParameterDetails id CDATA #REQUIRED>
<!ELEMENT Units EMPTY>
```

```
<!ATTLIST Units name CDATA #REQUIRED>
<!ELEMENT Symbol (#PCDATA)>
```

It is proposed that this XML structure will form the basis for structuring the material property data required for metalforming simulation. Each XML document will contain all the relevant thermal and mechanical data; in effect a data sheet for that material described in XML.

3 IMPLEMENTATION

The solution is based on Sun's Java Servlet technology and is implemented as a server-side variant of the classic Model View Controller (MVC) architecture, the so-called Model 2 web-architecture [3]. Thin clients, web browsers request material property data via the Internet, the HTTP request is serviced by a Java servlet. The servlet, a server-side component then accesses the corresponding XML file on the local storage and unmarshals the XML into an object, a serializable Java bean. The bean is passed to a JSP page for view processing by adding it to the original HTTP request received by the servlet and then forwarding that request on to the JSP page as part of an HTTP workflow.

The JSP page extracts the bean from the request scope, accesses the bean properties and dynamically generates a HTML page 'on-the-fly' in response to the remote clients request. The HTML 'view' reflects the property data contained in the XML databank, but formatted for visual presentation. It should be noted that the beans containing the unmarshalled XML content are cached in an object pool for subsequent reuse. Beans are automatically removed from the pool after a configurable period of inactivity suggests they are now 'stale'. The key elements of the system are shown in figure 1.

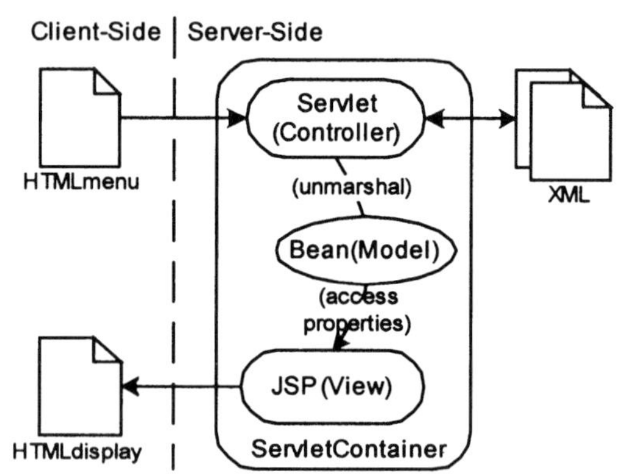

Figure 1: MVC Client–Server relationship

Figures 2 and 3 show typical views of the rendered XML data displayed on a web browser. Figure 2 illustrates an equation as the property data form, this view is corresponds to the XML shown in Section 2.

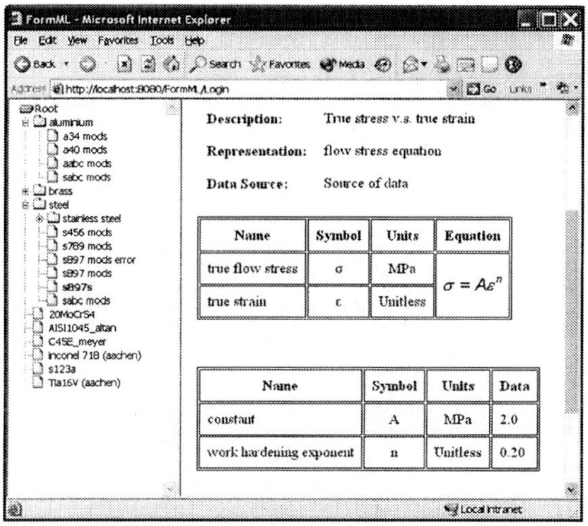

Figure 2: Equation representation

Whereas figure 3, illustrates the alternative XML rendering in the tabular property data form.

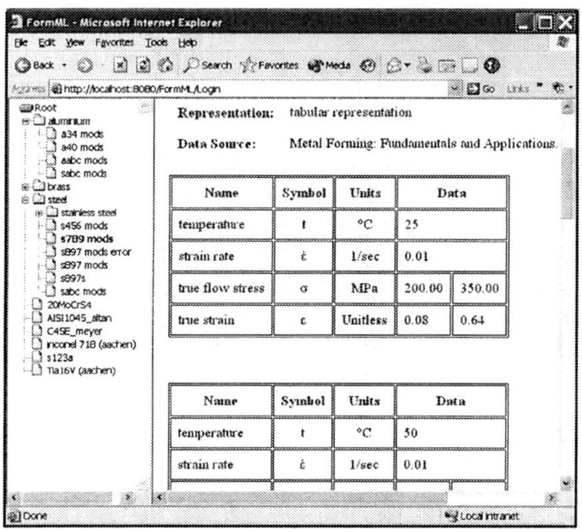

Figure 3: Tabular representation

4 CONCLUDING REMARKS

The XML structure described in this paper is similar in many respects to MatML but has been simplified and structured to the specific requirements of the material properties required by FEA simulation packages used for analysing metalforming applications. XML is a popular markup method for web based applications and is seen as an appropriate vehicle in the present work for structuring material data for display on web browsers.

If material data exists over the range of temperature, strain and strain rate required it would be useful if it were available for manufacturers to use with metalforming simulation packages. Activities using the web are now commonplace and offer a relatively simple way of collating data for dissemination. It can be easily accessed and downloaded and if structured in an appropriate format could be used directly with the simulation.

There is the problem of where the primary data will come from and its format. Testing under quasi-static conditions for tension, compression or plane strain compression is well developed and available from a number of sources. Testing methods at high strain rates and temperatures are not so well developed. They require a high degree of control and are not generally available. Unfortunately it is materials properties at high temperatures and strain rates that are, in many cases, required for the simulation of metalforming processes.

The fact that the databank is web based also provides an opportunity to allow direct acquisition of the data from materials testing. It is a relatively simple task to take the output from a testing machine (say force against extension in a tensile test), convert it to a useful form (stress and strain), format it immediately into the XML structure and download into the databank.

Using the approach described above has allowed the creation of a system to fulfil the needs of the web based environment and the simulation package users.

ACKNOWLEDGEMENTS

The authors would like to thank their industrial collaborators Confederation of British Metalforming, W H Tildesley, Clevedon Fasteners, Mills Forgings, Bloxwich as well as the United Kingdom's Department of Trade and Industry for funding ICT Carrier Project 0017.

REFERENCES

1. Mynors, D.J., Broomhead, P. and Grieve, R.J., "Web-based material properties for use with metalforming simulation packages", *Journal of Materials Processing Technology*, in print, (2004)
2. Begley, E.F., "Matml version 3.0 schema, NISTIR 6939", *National Institute of Standards and Technology*, (2002)
3. Seshadri, G., "Understanding JavaServer Pages Model 2 Architecture", *Java World*, http://www.javaworld.com/javaworld/jw-12-1999/jw-12-ssj-jspmvc.html (1999)

A study and development of PDM system based on visual studio .NET

Y HOU, S LIN, and **Y LV**
Mechatronics College, North China University of Technology, Beijing, China
H OUYANG
Department of Precision Mechanical Engineering, Shanghai University, China

ABSTRACT

The paper puts forward the solution of PDM technology. A new type of PDM system in a three-level structure has been developed based on Visual studio .NET and SQL Server technologies. It can produce BOM file in the form of XML by combining the dataset object of ADO.NET model, which can be browsed by Internet Explorer. Some CAD systems such as Pro/E, AutoCAD can be integrated by this platform of system with interface to realize collaboration design and real-time management of products and datum. A case study is given.

KEYWORDS: PDM Visual Studio.NET ADO.NET XML

1 INTRODUCTION

With the coming of economics globalization and the Internet era, enterprises are being faced with a fact of survival and development, so as to keep the lead. So, it is important for enterprises to pursue to shorten the circle of product R&D, reduce cost, improve quality and service, ensure clean environment and seize hold of market as soon as possible. It becomes a terrible problem for enterprises how to manage effectively these datum (data file, design data, project drawing, NC program, process design) and share them, they are mainly from mature, isolated and intellectual technologies [1] such as CAD, CAE, CAPP, CAM. In addition, every automatism cell is independent, they hardly have effective information communication and coordination, so all of these result in "information island". Now enterprises have realized that they must keep the lead in the future by realizing management of information. The PDM system provides a better solution for enterprises.

Present PDM systems usually have low data integrity, heavy data dependence, inadequate openness and expansivity, besides which, they are affected by all kinds of development tools (VB, VC, Dephi etc). These shortcomings bring about much trouble during the process of implementation and application in enterprises. In the future, the main development trend of PDM is Internet/Intranet, object-oriented (OO), distributed technology, STEP, XML, portability etc, so we try to research and develop a PDM system with .NET, ADO.NET and SQL Server.

2 THE ARCHITECTURE OF BEIFANG PDM SYSTEM BASED ON .NET

2.1 The Solution Based On PDM Technology

In order to well develop a PDM integrated framework, two questions must be understood, the first is the importance of PDM system in enterprise informatization, the second is the data flow, every isolated system can exchange and share datum by PDM platform. PDM system can exchange datum with CAD/CAM system, and manage datum in CAD/CAM system via SQL Server and TCP/IP. It also can integrate downstream software such as CAPP, ERP via BOM interface. A product is mainly make up of document and product graphics. For the network design and manufacture, PDM system outputs documents in XML, it adopts international standard STEP to save the related datum in product graphics and outputs them by the exchange STEP and XML (for example ST-Developer 10™[2]). It also outputs product graphics in VRML. Figure 1 illuminates the system whole framework as follows.

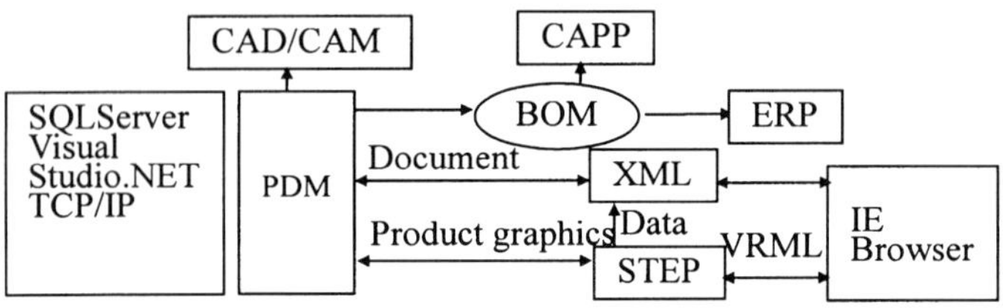

Fig.1. The Enterprise Informatization Solution Based On PDM Technology

2.2 Beifang PDM System Based On .NET Technology

Former developing tools of the PDM system cannot accomplish the seamless links to the technologies such as XML, WEB etc very well. Therefore, we adopt Visual Studio.NET to construct PDM integrated framework and ADO.NET to operate database. The new ADO (ActiveX Data Object) data object model strengthens class safety, accomplishes the reusing of code, code standardization, multi-language development and memory management (reclaiming garbage), it also can solve the sharing and exchange of heterogenous data [3].

The development of Beifang PDM system (Fig.2) is based on Visual Studio.NET, and adopts SQL Server database to construct bottom database of Beifang PDM system, which consists of product database, foundation database, version database etc. Based on those databases, PDM system framework is constructed, and then realizes a series of functions of orienting technology center, including project management, right management, graphics file management, document management, material management, user management, product browser, version management etc. Finally this system integrates all kinds of softwares such as CAD, CAM, ERP, CAPP.

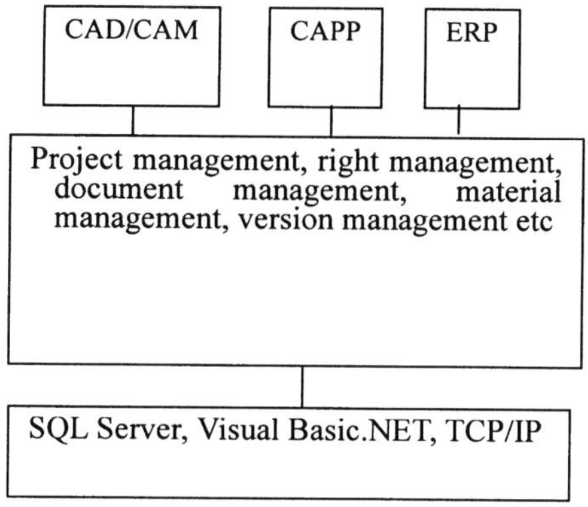

Fig.2. Beifang PDM System Based On Visual.NET

3 THE MAIN TECHNOLOGIES IN BEIFANG PDM SYSTEM

3.1 .NET

As a distributed computing framework, based on the current infrastructure of network, specification and new techniques such as database, .NET supplies an intelligent information platform for customers. In order to dispel "information island", solve software-crisis and realize the intelligent sharing of the computing resource, data resource, information resource and application resource, .NET sets up a kind of new generation internet information platform and software infrastructure on the OS (Operating System) and Web. .NET technology completely realizes the transformation of compute pattern from single computer, C/S, Web site to Distributed Computing. NET has two key parts: Web Service and XML, thereinto XML can resolve data translation in heterogenous system.

3.2 ADO.NET

.NET data provider has four core objects, which lie in system structure of ADO.NET management provider. There are Connection Command DataReader and DataAdapter object[3]. Connection object is very easy, because its role is to connect database; Command object provides Parameters aggregate, including parameters information of executing order; DataReader object provides quickly access to read-only unilateral data; DataAdapter object comprising Command object, can realize the mapping from specific operation to data resource, DataAdapter is a mechanism connecting management provider and cutting DataSets. ADO.NET can be looked upon as two parts: .NET data provider and data access, they belong to the connection of data access and specification and cutting model respectively. They can naturally alternate between data provider or management provider and database. Management provider is similar to the former OLE DB provider or ODBC driver. Fig.3 illustrates working mechanism of ADO.NET.

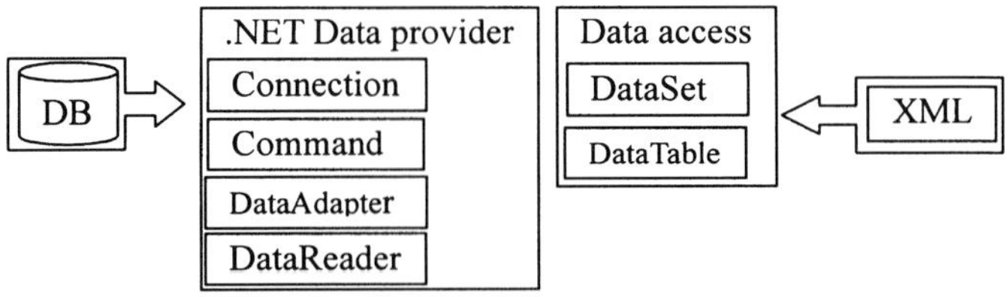

Fig. 3. ADO.NET Framework

4 DETAILED DESIGN OF BEIFANG PDM SYSTEM

From the view of OO, the system mainly adopts DLL and designs in a three-level structure [4], every part is an independent mechanism, but change of any part will not affect others. Take creating a new project as an example, bottom level adopts ADO.dll, mainly include database connection, operation methods such as 'create', 'delete', 'update', 'insert' [5]; physical level mainly realizes 'delete', 'add', 'edit' function by CreatePj.dll; interface level is user interface, and realizes parameter enactment and transfers to physical level, as shown Fig.4. By Ado.dll, we can realize any operation on database.

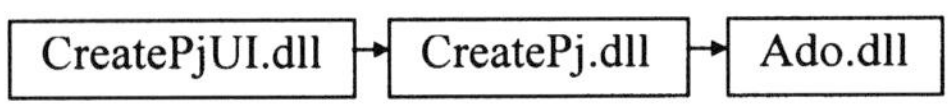

Fig.4. The Design Schema Of Creating New Project

4.1 Data Service Level

Data service level adopts SQL Server because working by network should be accomplished, outputting document in XML and enormous datum. Considering the importance of bottom database design, so before design, author does detailed function analysis and partition, the main study aspect has four parts: dividing this system into five modules based on the whole module partition figure; analyzing self-function and related function of every module; analyzing property and method of every module and designing database. Now design and creation of database is explained.

Data source is connected by Connection method, Command object provides the implementation order access to data source, DataReader provides read-only unilateralism data flow; Dataset provides denotation of data source memory; DataAdapter is an agency between DataSet and data element, provides their mapping order.

According to requirement of present enterprise and the main function of PDM system, when product data is managed, database should become more and more large along with increasing of data information quantity, moreover, all kinds of data tables of product information can produced by itself, so we create new table by CREATE TABLE method, for example creating project definition table.

```
Public Sub CRT_PJDEF()
        Dim cmmstr As String
          cmmstr = "create table dbo.PJDEF"
          cmmstr += "( PJID char(10) not null   primary key"
```

```
        cmmstr += ",PJCODE char(50) not null"
        cmmstr += ", PJNAME char(50)"
        cmmstr += ",MNGNAME CHAR(50)"
        cmmstr += ",ISDELETE CHAR(5)"
        cmmstr += ",SAVERPATH CHAR(254))"
        CRTtable(cmmstr)
    End Sub
```

Beifang PDM system has eighteen tables, eight tables are original, others simultaneously come into being along with every new project creation, as following table 1.

Table.1. Data Structure Of Beifang PDM Sytem

Public tables		New tables	
PJDEF	Project base property	PAX	Patulous name of All files
PJLCK	Project lock	CMPATT	Component physical attribute
PJDOC	Project document	CMPDOC	Component document
PJBOM	Project related BOM	CMPTIP	Component related BOM
MATANG	Material information	CMPLCK	Component lock
USERREG	User definition	DOCDEF	Document type
USERTYPE	User type	VER	Version information
SEG	Physical property Self-description	ASM	Assemble relation
		CMPDEF	Component flag, working environment, state etc

4.2 Logical Service Level

Logical service level uses interface encapsulated at data service level and realizes commercial function via reencapsulating.

At first, user management module and project management are regarded as a design team, component management module and material management as a design team, document and report forms management as a design team. So we can simultaneously concurrently develop. Depending on module encapsulation and entity realization, the system has eight modules, there are project management module, user type management module, user management module, component management module, material management module, creation database management module, data management module, common logic module. For example Project class mainly includes project ID, project code, project name, creator, project compose (component information), project member, project state, project document, project save path etc

```
ClassProject{
PrjID;
PrjCode;
PrjName;
PrjCreater;
PrjAsm;     project assemble relation
```

```
PrjUers;
PrjState;     lock
PrjDoc;
PrjSavePath;
……}
```

4.3 User Service Level

That is user interface level. It realizes parameter enactment and transfers to physical level, that is using the interface of logic service level.

5 CASE STUDY

In order to verify feasibility of Beifang PDM system, a product (hand-action rocker) is managed by it, which has been designed by Pro-E, the system basically realize the browser and query function, realize user right enactment and also can edit material of components. It has realized to output BOM of component. Fig.5 is the interface of it.

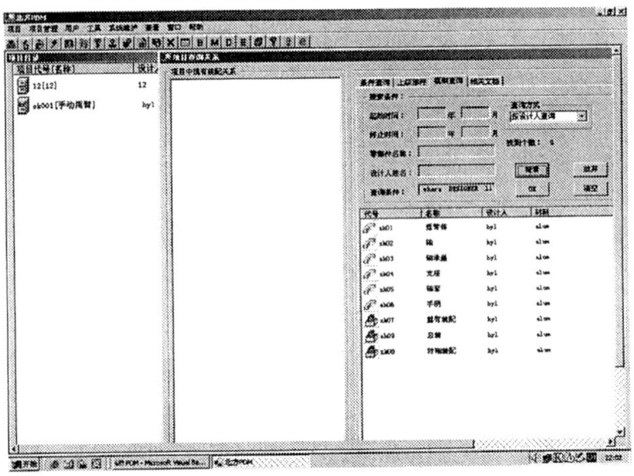

Fig.5. Beifang PDM System Interface

```xml
  <?xml version="1.0" encoding="utf-8" ?>
- <productdef>
    <code>shoudongyaobl</code>
    <name>手动摇杆</name>
    <creater>hyl</creater>
  - <compnentAttName>
      <attnumber>8</attnumber>
      <att0>代号</att0>
      <att1>名称</att1>
      <att2>设计人</att2>
      <att3>材料</att3>
      <att4>规格</att4>
      <att5>质量</att5>
      <att6>来源</att6>
      <att7>备注</att7>
    - <Compattributes>
        <code>char(254)</code>
        <name>char(254)</name>
        <creater>char(254)</creater>
        <material>char(254)</material>
        <size>char(254)</size>
        <mass>double</mass>
        <flag>int</flag>
        <memo>char(254)</memo>
      - <flagdef>
          <data0>0</data0>
          <memo0>自制件</memo0>
          <data1>1</data1>
          <memo1>外购件</memo1>
        </flagdef>
      - <toplevelComp>
          <TopAssmNum>1</TopAssmNum>
          <TopAssm0>A0A00008</TopAssm0>
        </toplevelComp>
      </Compattributes>
```


Fig. 6. A Part Of Outputting XML File

Fig.6 is a BOM file in XML about a hand-action rocker, which has been managed by Beifang PDM system.

6 CONCLUSIONS

By studying present PDM system and the network development trend of PDM technology, the paper puts forward the enterprise informatization solution based on PDM technology. We try to develop a new type of PDM system adopting .NET and ADO.NET technologies. It primary realizes data transmission and collaboration on Internet. The feasibility of Beifang PDM system is verified via a case study. Of course, there are some research to continue, such as providing well BOM interface for downstream software such as ERP, CAPP, providing Integrated development interface for CAD/CAM software and so on.

REFERENCES

1 Huangqing Tang et al (2002) "CAD/CAPP/CAM integrated method study based on PDM platform, combined tool and automatization machining technology", pp.22-25, (in Chinese).

2 ST-Developer 10™, http://www.steptool.com: products/stdev/index.html

3 Michael Halvorson (2002) "VB.NET Program Design", Beijing: Tsinghua University Press, in Chinese).

4 Jing Su et al (2002) "The study of CAD/CAM information integrated environment based on Internet", R&D, 40(452), pp.26-30 (in Chinese).

5 Tony Bain Denise Gosnell, (2002) "VB.NET and SQL Server 2000 Advanced Program", Beijing: Tsinghua University Press (in Chinese).

Applying knowledge-based engineering to traditional manufacturing design

J KULON, P BROOMHEAD, and **D J MYNORS**
Design and Systems Engineering Department, Brunel University, UK

ABSTRACT

The paper outlines the development of a KBE system for hot forging design using state-of-the-art technology and the Internet. The benefits of a KBE approach over a traditional design process are emphasized. The aim of the proposed KBE system is to integrate the hot forging design process into a single framework for capturing knowledge and experience of design engineers. The KBE application guides the design engineer through different stages of the design process enabling the generation of forgeable geometry from a component profile taking into account machine, material and forging company specific data, and design considerations.

1 INTRODUCTION

Today's economy presents a new set of challenges for the manufacturing industry. In the UK and other developed countries the manufacturing sector is under siege because developing countries offer quality products at lower prices. Global competition has empowered customers who now constantly demand new and better quality products. To meet the challenge of foreign competition and survive in the market, manufacturers in general and the metalforming industry in particular are now under tremendous pressure to improve their efficiency in terms of product development and resource utilisation. This means a dramatic reduction in the overhead cost of redundant engineering work, eliminating repetitive design tasks, reallocating technical resources to new product development. To stay competitive in today's global economy a company must build on its intellectual capital and create time to innovate and assess more advanced, complex types of tools to eliminate uncertainty and repetition from the design process. With an ageing workforce it is vital that experience is not forgotten or allowed to drift away through retirements etc, but is captured and built upon within a company. More manufacturers are now exploring new ways to harness the advantages of new technologies and communications tools such as Internet. KBE systems, which complement traditional CAD/CAM systems by adding the engineering knowledge that drives the product design process, are becoming an essential part of strategy for improving effectiveness (1-2).

This paper outlines the development of KBE system for hot forging design using state-of-the-art KBE technology and the Internet. Currently most of the activities and costs related to planning forging processes depend strongly on human expertise, creativity and intuition as well as time consuming experimental, trial and error, work. The numerous variables required to describe forging processes make the knowledge capture and the encapsulation of the decision-making and process planning rules difficult using conventional methods (3-7). The aim of the proposed KBE system is to integrate the whole design process into a single computer model and to create a framework for efficient capturing of a company's most important asset - the knowledge and experience of its engineers. In the course of the design the KBE application guides the design engineer through different stages of the hot forging design process enabling the generation of forgeable geometry from a component profile. It also supports computation of the forging load, complexity factor, component volume, flashland ratios etc. A relational database is used to encapsulate the design rules as well as their complex interdependencies on material, production unit, and manufacturer capabilities. As part of the system a cost-estimating procedure for hot-forged parts is being developed, taking into account the size of production volume, the production unit, the production time and the dimensions of the billet and the component to determine the total forging cost.

2 TRADITIONAL ORGANISATION OF HOT FORGING DESIGN PROCESS

As illustrated in Figure 1 the traditional organisation of hot forging design and process planning contains several stages starting with the component design and concluding with the cost estimation and the process parameter selection and optimisation. The component design stage involves the transformation of the component geometry to obtain forgeable shape. The design tasks at this stage involve the determination of a parting line, specification of appropriate radii, fillets and corners, internal and external drafts, design of holes, ribs, and webs. The empirical design guidelines, which are used, often depend on the material, production unit and manufacture's experience. The creation of a forgeable shape is followed by the determination of the process sequence and designing all the tools needed during the production process, including those used for die profiling and trimming. The most critical and expensive tools are the forging dies, of which, for a particular forging, there may be several preform dies. The main objectives of the design of preform dies are to insure a good distribution of material in order to fill the finishing dies without any defects, minimise metal losses and minimise die wear, i.e. minimise metal flow in the last die so as to extend its life. The perform design stage is generally considered to be the most creative, requiring the highest level of expertise. According to some estimates approximately 80% of the final product cost is decided at the design stage. Finally after all the necessary tools are created, the design engineer generates drawings and reports that document the forging design. Currently in the forging industry the design process is still mainly based on very inefficient iterative 'trial and error' procedures. After taking successive measurements of the forgings, gradual corrections to the die geometries are made, which eventually lead to the production of so called 'good parts'. Since many repetitive engineering tasks are not automated, they need to be done by hand, which creates a bottleneck that slows down the design cycle. As shown in Figure 1 this design process is really a loop, which is repeated until a satisfactory design is found, therefore the effect of the bottleneck is multiplied by the number of times the loop is repeated.

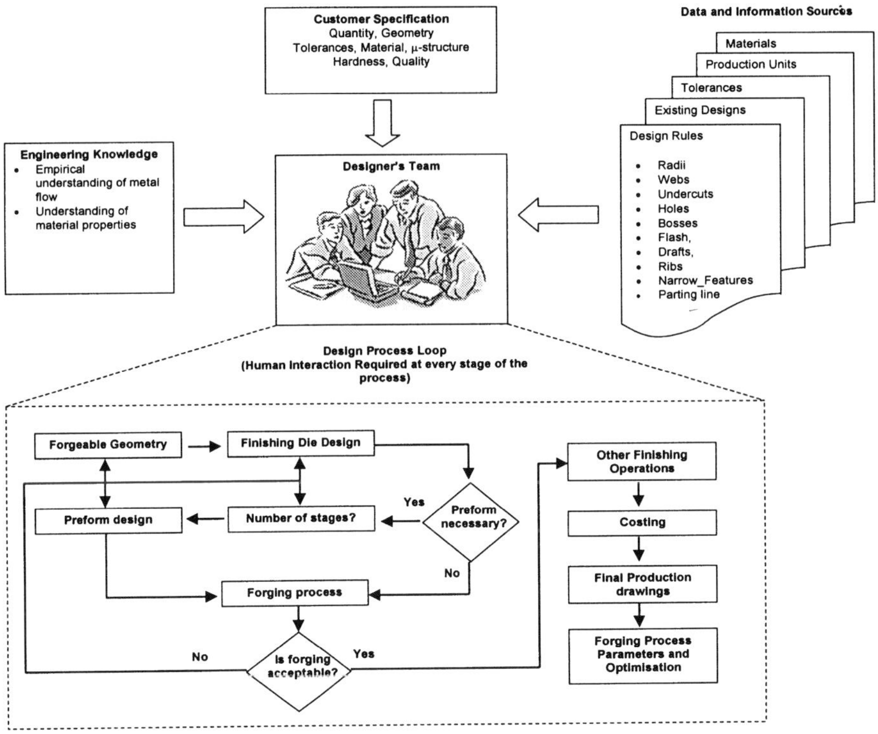

Figure 1. A traditional, 'manual' organisation of forging design process

In order to speed up the design and optimise the process sequence more often software tools such as CAD packages (8) and numerical methods like the finite element method (FEM) are used. Many studies have demonstrated that numerical analysis can help the forging industry to define the best process sequence, reducing time and costs involved (9-11). Typically, CAD software is used to display, manipulate, and transform the component geometry, which subsequently is exported to FEA packages to simulate the forging process. Although, traditional CAD systems allow engineers to process designs faster than when designs were done completely by hand, they go only part of the way toward providing a completely flexible tool for adapting to changes in a product design and provide no means of capturing the engineering expertise behind the design. While the geometric information can be handled using CAD systems other design knowledge, including catalogues, databases, and engineering standards, must be managed separately without integrated tools. In addition, if the input specifications change at any time, the entire process must be repeated. Even with all of engineering team's experience, the process of developing a single design is still laborious, repetitive, and subject to error. The following sections will show how this situation can be significantly improved using KBE environment.

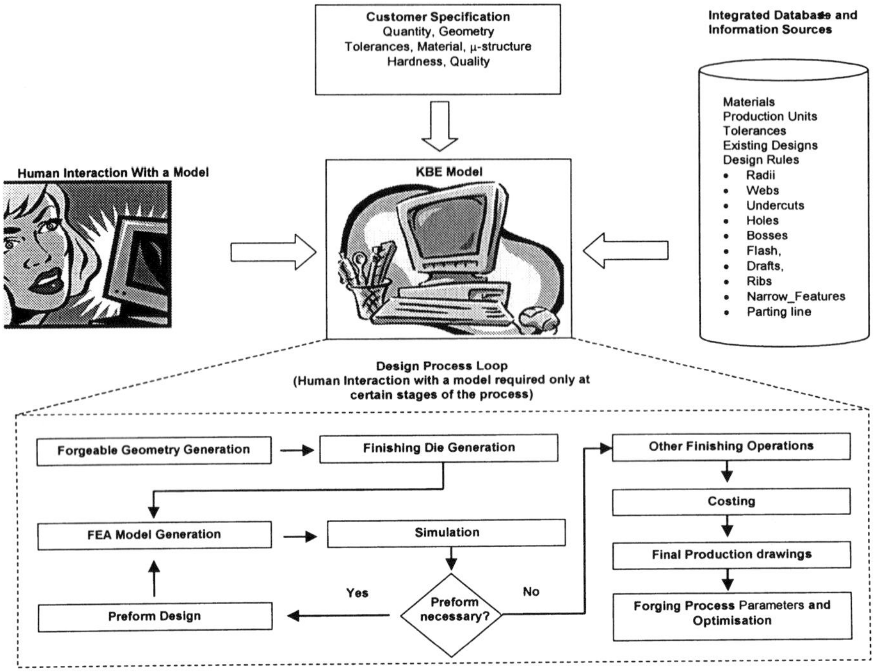

Figure 2. KBE model of hot forging process

3 KBE APPROACH

3.1 KBE basic principles and benefits

The conventional definition of KBE is that it is "the process of combining engineering knowledge, methodologies, rules and best practices with process knowledge and best practice to create product models that describe how product designs are created or engineering analyses are undertaken" (12). Unlike the traditional CAD systems, which mostly contain geometric information for a single design, KBE does not just create a single design; it models the whole process of generating the design. A KBE system stores knowledge about a product in a comprehensive product model composed of engineering rules that describe how products are designed, analysed and manufactured. The rules can be design rules, standard engineering rules, or experiential 'rules of thumb', which reflect years of design experience. They can be expressed in the form of physical equations, graphical and tabular relationships describing the attributes of the physical product such as geometry, functional constraints material type etc (13). Once engineering knowledge about the product is collected and stored as a product model, design engineers can generate and evaluate new designs quickly and easily by changing the input specifications for the product model, or modify designs by extending or changing the product model. This frees the engineer from repetitive time-intensive and often tedious design tasks such as repetitive calculations and allows more time for creative design work. Another strategic benefit of KBE is that the product model also includes various outputs such as reports of engineering results, data of engineering analysis, 3D geometric models as well as non-geometric output such as bill-of-materials, costs and mass estimations and manufacturing instructions (14). When the design changes, so do the outputs. Information is always up-to-date and reflects the current state of the design, automatically. A KBE system is also able to integrate other engineering programs such as FEA software in a wider organised system working as a kind of pre-processor for the analysis applications.

3.2 Proposed KBE model of hot forging

The diagram shown in Figure 2 illustrates a proposed KBE model of the hot forging design process. The centre of the design process is the product model, which manages all the routine engineering tasks. The role of the engineer who interacts with the model is to provide the input specifications such as component geometry and make the important design decisions. Databases of materials, standards books, production units' specifications, design rules encoded in the system by the forging engineer etc., are all integrated into the product model through the relational database. First the KBE application automatically constructs forging geometry from component geometry using design specifications. Once a design has been generated from the product model, the shape can be exported to a FEA for a forging simulation allowing the forging designer to test several process parameters in order to improve the forging process. According to the simulation results, intermediate forgings (performs) may appear to be necessary. At any stage the geometric information can be transferred back to a CAD tool for further detailing, modifications. A repository of the previous designs can be used to retrieve the relevant cases and exploit analogies and similarities with previously solved cases. The designer of preform shapes can use not only the information from the simulation results but also the solutions of a family of previously solved cases. The redesigned preform may be again validated by FEA simulation. Finally, the KBE generates automatically the document set for the forged part, e.g., the drawings, reports etc. The design engineer can easily modify the design either by changing the answers to some of the key design questions or by modifying the input specifications. In this scenario, the bottleneck of the traditional design process the repetitive tasks that had to be done by an engineer is eliminated. All such tasks are managed by the product model. The design loop can be repeated efficiently as many times as necessary without incurring the expensive trial-and-error procedure and experimental work. With the product model, generating a new design can be done much faster in a matter of hours by a single design engineer, as opposed to several days or weeks with traditional methods.

4 KBE TECHNOLOGY

4.1 KBE server architecture

Traditionally, all computer aided design tools were designed as interactive applications for a single user. However, with the fast development of the Internet and database technologies, it is advantageous to build an integrated system based on the web in order to better implement knowledge share and to distribute designs (15). The web-based environment offers several advantages in comparison to more traditional software platforms, such as: an open architecture, uniform information model, and object-oriented structure. All of these can help to eliminate the time and space restriction of engineering design and realise maximum knowledge share. The web environment also provides the user interface, file-transfer and can display graphics representing the geometry through any standard web browser.

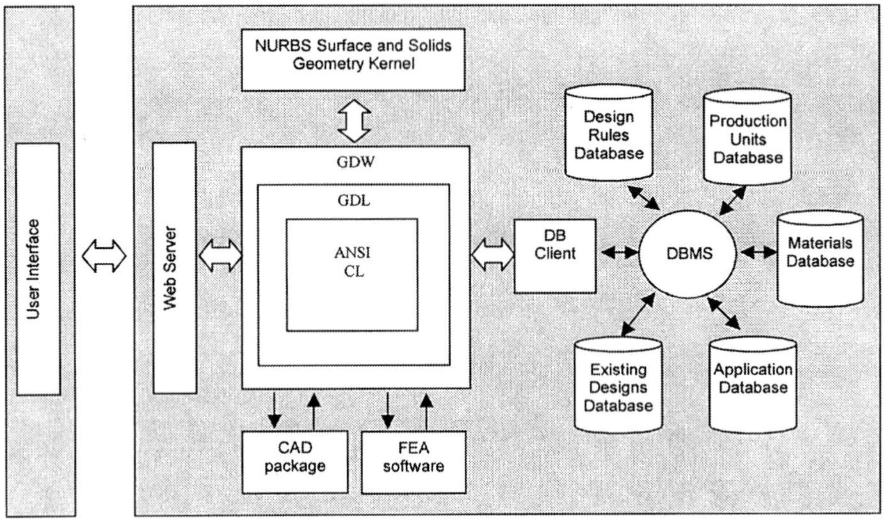

Figure 3. KBE server architecture

The proposed web-based KBE system architecture is shown in Figure 3. The left side of the diagram in Figure 3 indicates user interface, the right side corresponds to the processes run and the steps carried out on the KBE 'server' machine. The core of this KBE application is the GDL/GWL (General-purpose Declarative Language/ Generative Web Language) developed by Genworks International (16). GDL/GWL is a knowledge base programming environment, which implements the features of caching and dependency tracking as well as demand-driven evaluation. Further, because GDL is based on a full object-oriented language model, a superset of ANSI Common Lisp, it follows a non-procedural paradigm for application development and execution, with multiple threads of execution and the ability to load new code and redefine objects dynamically. The server side framework is implemented using GWL as an integrated web server-based user interface that manages data and controls other software tools on the server. The main application tasks carried out on the server are: accessing the database of rules, design features and materials, facilitating file transfer between the user and CAD package or a FEA software and performing the KBE computations. GDL/GWL integrates the application logic and databases through a direct MySQL interface and handles geometry, through its library of 3D geometric objects. It is able to export and import component surfaces directly embedded in IGES format. The upcoming integration to NURBS surface and solid modelling libraries will add new powerful surface and solids capabilities to GDL's built-in 3D wireframe facilities (16).

4.2 Integrated database of materials and design rules

The schematic diagram of the integrated database system is shown in Figure 4. The database of the KBE application is designed as an independent separate part of the whole system containing the design rules, production units' specifications and material properties.

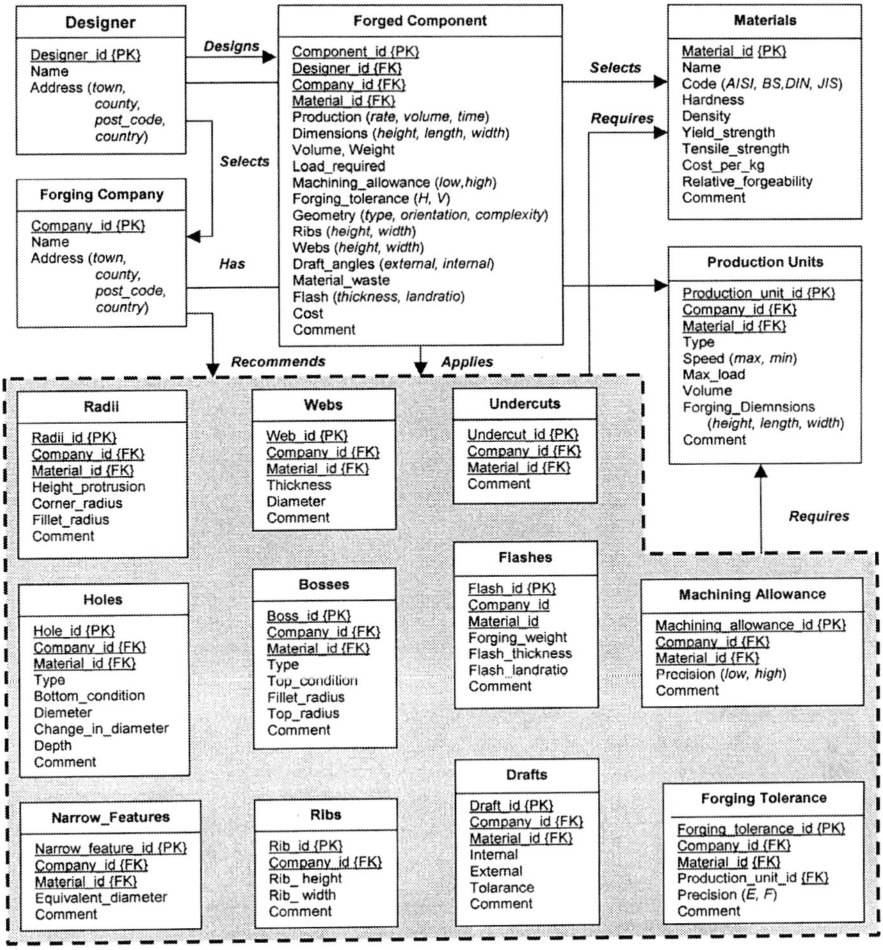

Figure 4. Integrated database of materials and design rules

Knowledge (design rules) has been obtained from handbooks, plasticity theories, relevant references and empirical know-how of field experts in hot forging companies. The database also takes into account the dependency of the design rules on materials, production units (such as a mechanical press, screw press, hammer etc) and component geometry (such as axisymmetric, plain strain, 3D etc). Rules are arranged in the form of tables or equations.

4.3 User interface and visualisation

The browser-based user interface for accessing the design knowledge is shown in Figure 5. The tree menu on the left allows the selection of several expandable submenus, which in turn update the main viewing window to guide the user interactively through the design process.

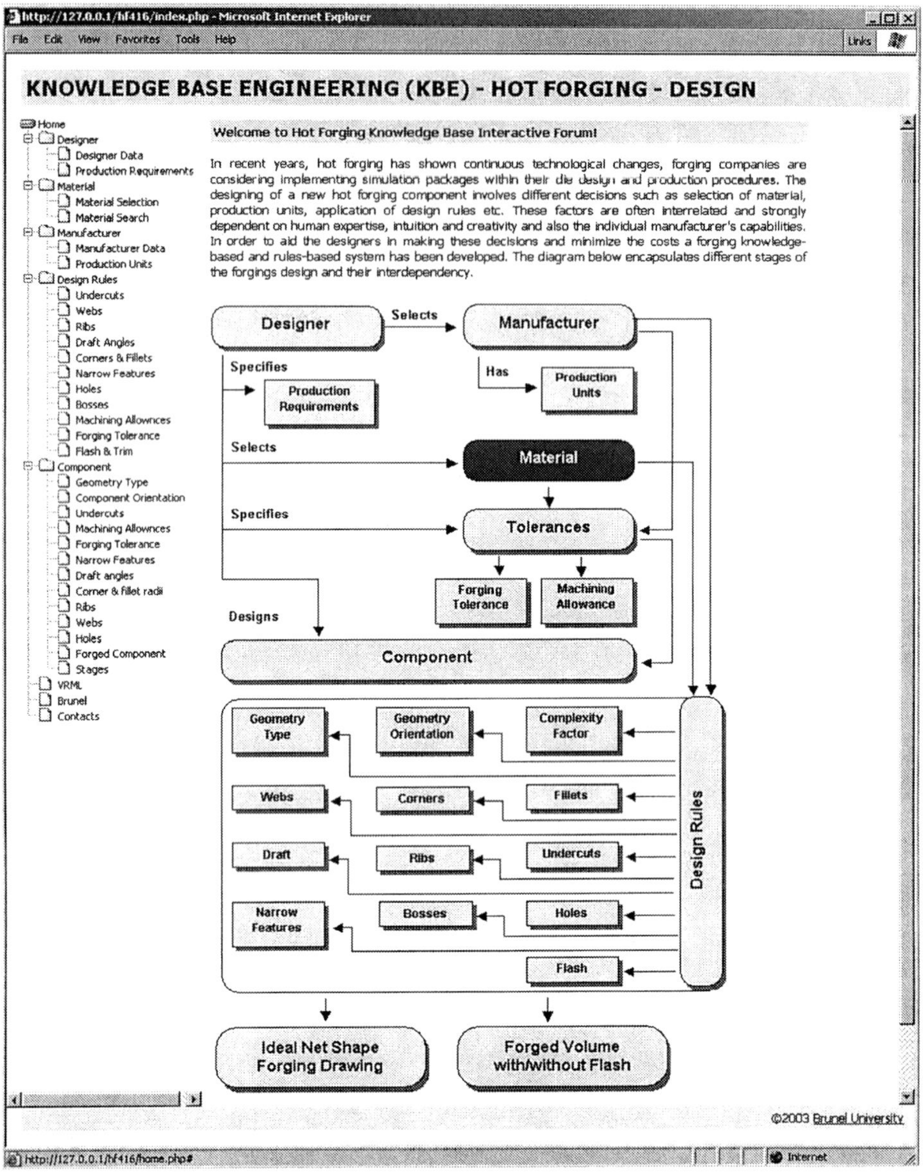

Figure 5. User interface

The same can be achieved by using the block diagram, shown in the main window in Figure 5. Various boxes represent different design decisions and guidelines, which are being implemented in the system, such as material selection, determination of component geometry, parting line, specification of appropriate radii, internal and external drafts, design of holes, ribs, webs etc. Most of the web pages consist of information entered directly using standard

HTML form objects. Visualisation of the component geometry on the client's machine is achieved using the Virtual Reality Markup Language (VRML) (17). The VRML is currently the best way to create interactive, 3D environments on the WWW. VRML is generally viewed within a Web browser like Netscape Navigator or Microsoft Internet Explorer. These browsers don't support VRML natively therefore a plug-in such as Cosmo Player is required to view VRML content.

5 SUMMARY AND CONCLUSIONS

This paper presented the preliminary stage of the development of a KBE system for hot forging design using state-of-the-art KBE technology and the Internet. Instead of trying to fully automate the forging design the KBE leads designers through the design process and enables appropriate manufacturing rules as well as company capabilities to be complied with. The key feature of KBE application is that it integrates the whole design process within one computer model. The relational database is used to encapsulate the design rules as well as their complex interdependencies on material, production unit, and manufacturer capabilities. Design rules have been obtained from handbooks, plasticity theories, relevant references and empirical know-how of field experts in hot forging companies. The KBE also takes into account the dependency of the design rules on materials, production units (such as mechanical presses, screw presses, hammers etc) and component geometry (axisymmetric, plain strain and 3D). The work is on going aiming at further implementation of design rules, transfer of simulation results as well as the storage and retrieval of a large number of case studies.

ACKNOWLEDGMENTS

The authors would like to thank their industrial collaborators on this project and especially Mike Twelves of Corus Automotive, Neville Turner of George Dyke Ltd. and John Tildesley of W.H.Tildesley Ltd. as well as the United Kingdom's Engineering and Physical Sciences Research Council: Grant: GR/N21611.

REFERENCES

1 Mynors, D. J., Griffiths, B. J., Twelves, M., Etheridge, D., Moore, J., Griffiths, T. and Piddington, C. (2004) "Final report of the Tooling and Experiential Design Initiative (TEDI)", (ICT Scoping Study 0116), March 2004, Obtainable from either the UK Department of Trade and Industry or the Department of Design and Systems Engineering, Brunel University, Uxbridge, Middlesex, UB8 3PH, UK.

2 Cooper, S., Fan, I. S., and Li, G. (1998) "Achieving Competitive Advantage Through Knowledge Based Engineering - A Best Practice Guide", *Department of Trade and Industry*.

3 Tisza, M. (1995) "Expert systems for metal forming", *Journal of Materials Processing Technology*, 53: pp 423-432.

4 Kim, D.Y. and Park, J.J. (2000) "Development of an expert system for the process design of axisymmetric hot steel forging", *Journal of Materials Processing Technology*, 101: pp 223-230.

5 Zhang, X., Peng, Y. and Ruan, X. (2004) "A web-based cold forging process generation system", *Journal of Materials Processing Technology,* 145: pp 1–6.

6 Cheok, B. T. and Foong, K. Y. (1994) "Some aspects of a knowledge-based approach for automating progressive die design", *Computers in Industry*, v. 24. N. 1. pp 81-96.

7 Caporalli, A., Gileno, L.A. and Button, S.T. (1998) "Expert system for hot forging design", *Journal of Materials Processing Technology*, 80–81: pp 131–135.

8 Shin, Y., Han, S. and Bae, D. (2000) "Integration of heterogeneous CAD databases using STEP and the Internet", *Decision Support Systems,* 28: pp 365–379.

9 Mynors, D.J. and Bramley, A.N. (1997) "The consequences of incorporating computer simulation within the die design procedure for closed die forging", *Proc. Int. Conf. Des. Prod. of Dies and Molds*, Istanbul, Turkey, pp 277-282.

10 Mynors, D. J., Snape, R. G. and Bramley, A. N. (1996) "A proposal for implementing complex simulation techniques into the forging industry", *Proceedings of the 12 NCMR Conference*, University of Bath September 1996.

11 Bramley, A. N. (1987) "Computer aided forging design", Annals of the CIRP Vol. 36: pp 135–138.

12 Javed. Y. "Knowledge Based Engineering", (2004) MSc Project Dissertation, Systems Eng. Dept, Brunel University, to be submitted.

13 Chapman, C. B. and Pinfold, M. (1998) "Design Engineering - A Need to Rethink the Solution Using Knowledge Based Engineering. Knowledge Based Systems" Volume 11 Numbers 5-6 Pages 257-267.

14 Knowledge Technologies International (2000) "ICAD" (WWW Document) http://www.ktiworld.com.

15 Zeng, J., Chen, W. and Ding, Q. (2003) "A Web-based CAD system" *Journal of Materials Processing Technology*, 139: pp 229–232.

16 Genworks International: http://www.franz.com/success/customer_apps/knowledge_mgmt/genworks.lhtml.

17 Wagner, R., Castanotto, G. and Goldberg, K. (1997) "FixtureNet: interactive computer aided design via the WWW", *Int. J. Hum. Comput. Stud.* 46: pp 773-788.

An integrated software framework for synchronizing internal business functions of a ship-repair yard

G CHRYSSOLOURIS, S MAKRIS, N PAPAKOSTAS, and **D MOURTZIS**
Laboratory for Manufacturing Systems and Automation, University of Patras, Greece

ABSTRACT

The ship-repair sector is characterized by uncertainty and unpredictability, since the work to be performed is specified roughly before the arrival of the ship at the shipyard and a lot of new work is identified during the execution of its repair. Therefore, the specification that is related to the repair of the ship evolves during the lifecycle of the ship repair contract, from the initial enquiry to the final invoicing of the work.

This paper demonstrates an integrated software system, capable of modelling the evolution of the work list during the ship-repair, while supporting decision making, in a set of critical business functions, such as cost estimation and production planning of a ship-repair contract. The proposed software environment allows for following up on the modifications, according to the newly identified work.

The proposed software system is demonstrated in a real life scenario, and proves its capability of dealing with the particular requirements of the ship-repair environment. It is shown how the ship-repair work can be modelled so as to facilitate the capturing information that is necessary to perform critical functions, such as those of estimating the cost of a ship-repair contract and of planning its execution.

1 INTRODUCTION

The operations that take place in the shiprepair industry, involve the interaction of individual business units of the same company, with each unit performing its own activities, required for the accomplishment of the shiprepair contract. The activities that take place in a shiprepair yard can be modelled in the following functional areas that need to be addressed, in order for an integrated management of the shiprepair operations to be achieved:

- Data for support of marketing prior to enquiries.
- The estimating and tendering process when enquiries are received.
- The purchasing of materials and equipment from external suppliers.
- Booking the services of sub contractors for specialistic jobs, or when insufficient labour is available in-house.
- Maintaining records of all expenses for suppliers and sub contractors for each job, in each contract.
- The planning of jobs on a vessel when a contract is won.
- The scheduling of jobs on a vessel when a contract is won.
- Recording hours spent on each job, of each contract, by each employee.
- Maintaining appropriate levels of human resources, both permanent and casual (temporary).
- Recording basic employees' personal records and details.
- Recording all invoices from suppliers and sub contractors, and logging to correct jobs and contracts.
- Providing the owner with an invoice, at the completion of the job, and just prior to the vessel leaving the yard

The areas listed above have been identified as key functions for the integrated management of the shiprepair process, which are required to support the ship repair industry. This provides management information, allows for progress to be monitored, as well as to make known the expenditure, both financial and in man hours, thus, allowing for the invoice of the ship owner to be produced as soon as the contract has been completed.

2 THE SHIPREPAIR YARD BUSINESS PROCESS ANALYSIS

2.1 Description of shiprepair activities

In general, the activities in the shiprepair sector are evolved inside and in the premises of the shiprepair yard. The ship is accommodated in the shipyard's facilities and all manufacturing activities take place thereby, from the beginning of the repair work up to the ship's delivery to the customer. The activity is initiated by the submission of the shiprepair enquiry from the shipowner to the shipyard, which initiates the contract negotiation process. The overall procedure for contract negotiation is as follows:

The specification is received from the Owner, usually by fax or e-mail. The estimator tenders the specification, which is then typed, item per item. After confirmation of the repair contract, the working orders corresponding to each item of the specification and the general services are issued to the production. After the contract is initiated, the project begins (Figure 2). When the work is scheduled empirically, the ship manager takes into account three rules:

- The jobs that last longer (e.g. painting the ship, major steel works, etc) are scheduled to start first, because the deadline for completion of the overall work is strict.

- The jobs are scheduled according to the availability of resources (human and equipment). The ship manager has to schedule the repair jobs to take place in periods of time when the resources (labour, machines, material, etc.) will be available; if these are not available, Subcontracts are considered and scheduled as well.
- The interdependencies of the jobs have to be considered, because some jobs have to finish before others can start, and there are also physical constraints that do not allow some jobs to be carried out simultaneously (e.g. no work can be done on the deck while the hull is being painted).

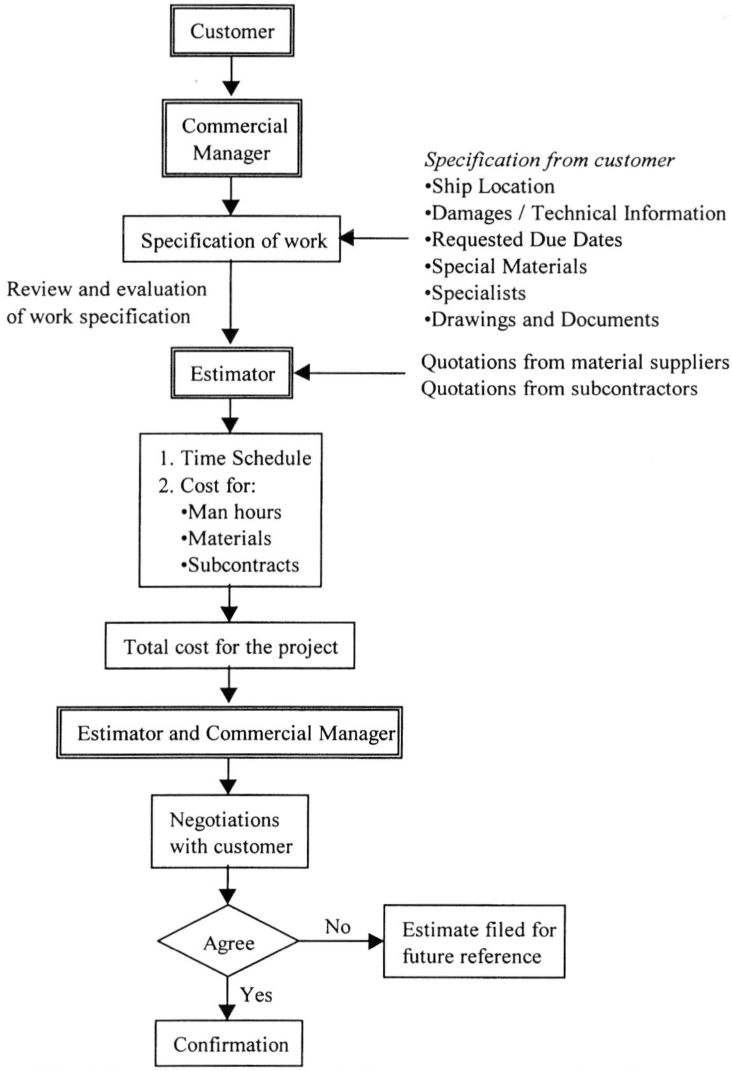

Fig. 1. Sequence of activities in the contract negotiation phase

The Commercial Manager and the ship-manager, after checking with the production, decide which jobs will be subcontracted. The decision is based mainly on factors others than cost, because the cost of subcontracted labour is usually low and therefore rarely can it be taken into consideration for deriving a decision.

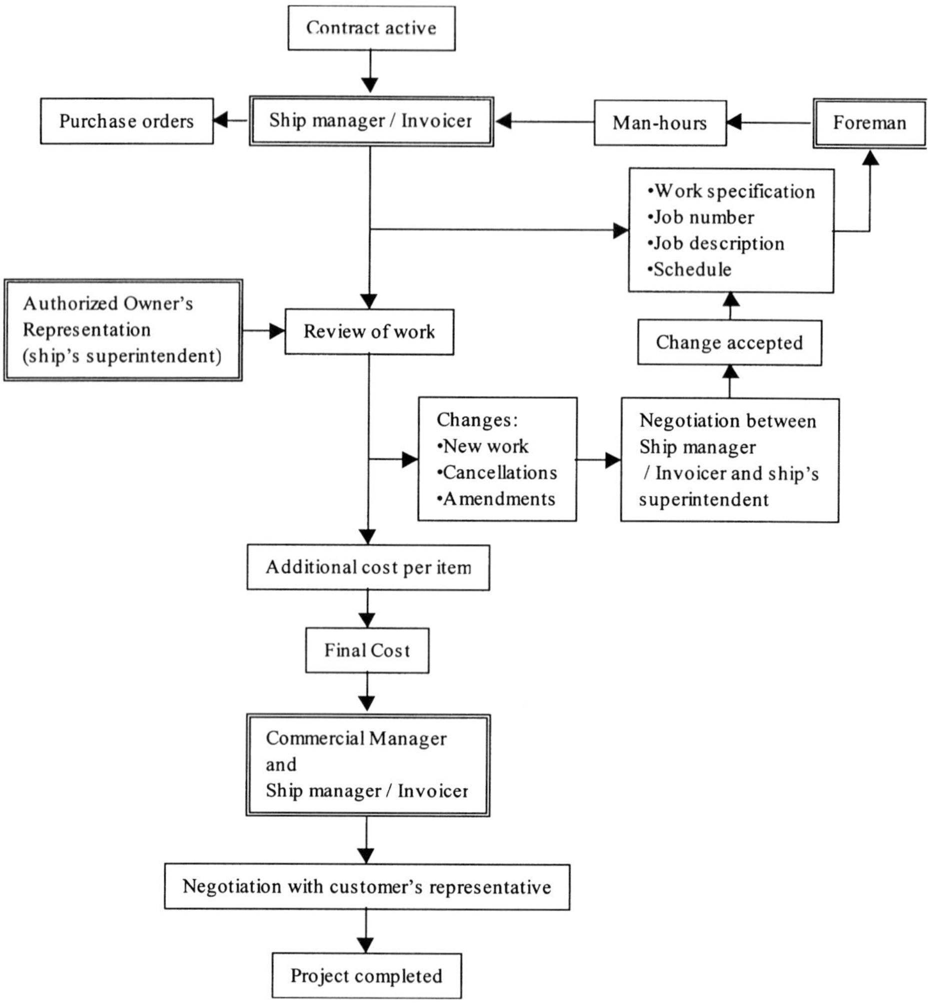

Fig. 2. Work planning and control

The ship manager assigns the job orders to the resources of the shipyard or the subcontractors. Changes are very common in the specification of the work. Some of the specified jobs are cancelled, others are only partly carried out and a lot of new ones, the requirement for which is identified, while others are processed, emerge quite often. In these cases, the ship manager negotiates with the superintendent of the ship and decides which work will be carried out and how it will influence work planning, progress and scheduling.

At the end of the day, the foreman of each workshop files a report, which contains the information related to labour consumption per job. Materials are requested to the warehouse by the means of material requisitions. A few days prior to the completion of the repair, a description of the works, per item is included in a report for the Owner's approval and afterwards, the corresponding invoice is issued.

The business process corresponding to material supply or external service is as follows:
For Materials, internal material requisitions are issued for the shiprepair contracts. If the intended material is on stock, it is used for the repair; otherwise if the intended material has to be purchased outside, a purchase order is issued by the Purchase Department to the Suppliers.
For Subcontracts, a purchase order is issued to the Subcontractor for the respective shiprepair contract. The person-power of the subcontractor is considered during the allocation of the shiprepair work and it is the manager who decides which work will be assigned to the subcontractors, according to the following rules:

- Some types of work are always subcontracted.
- Taking into consideration the availability of the shipyard's own resources, some jobs are subcontracted before the start of the repair job. This decision is usually made approximately one week before the start of the repair.
- In some cases the need to subcontract some jobs emerges while the work is being done. In such cases the subcontract is arranged within six to ten hours. There are even cases when the work can start almost immediately, because the subcontractor's crew may be present in the yard, working on another subcontract.
- There are cases when the shipyard makes an "open contract" with a subcontractor for all the jobs of a certain type (e.g. a contract for steel works).

It becomes apparent, that the worklist evolves during the contract execution, from an initial form upon the submission of the enquiry, it is enhanced during the visit of the vessel at the shipyard, when new shiprepair items are identified and finally it gets to the final list when the repair is finished. During the contract execution, the worklist is complemented with information related to the shiprepair, such as information on the purchase of materials, labour consumption and subcontracts issued. Therefore, the problem of synchronisation of business functions, inherits the management of the worklist during its evolution and is tightly coupled with the issue of planning the manufacturing operations within the shipyard.

A major difficulty in planning the operation of shipyards is that between the start and the completion time of the work there are typically many work changes which include added and cancelled work as well as priority changes. Ship repair jobs, in particular, are often done at a fixed budget and are considered therefore, as "Fixed Cost" exercises (1). Moreover, the planning of the work in a shiprepair yard, was considered as a form of dispatching available resources to production tasks (1). Dispatching is concerned with the detailed assignment of operations to production resources (2). Modern software systems often fail to grasp the whole picture of complicated and interrelated scheduling activities as well as the dynamic spatial layout, i.e. spatial resources with material handling equipment such as cranes (3). Therefore, the major requirement for planning a shipbuilding/shiprepair yard is to have the ability of producing a good schedule quickly with a flexible and adaptive system to changes of the production data. The most commonly adopted approaches to the planning of shipyards, is the approach of modelling the problem as a job-shop scheduling one (4), or a resource constraint project scheduling problem-RCPSP (5), which is a generalisation of the job-shop problem (6).

This paper, demonstrates an integrated software system that models the execution of the activities, taking place during a shiprepair contract, by means of modelling the worklist evolution and by offering an integrated approach to follow up the changes in the worklist. It demonstrates a way to implement synchronisation of individual shiprepair business functions and a concrete way to plan the shiprepair manufacturing activities.

2.2 The shiprepair yard business model

According to the business model, a customer may own one or more ships. The ship submits the shiprepair *enquiry* to the shiprepair yard. The enquiry includes specification for the repair of the ship that is a list of jobs (Figure 3).

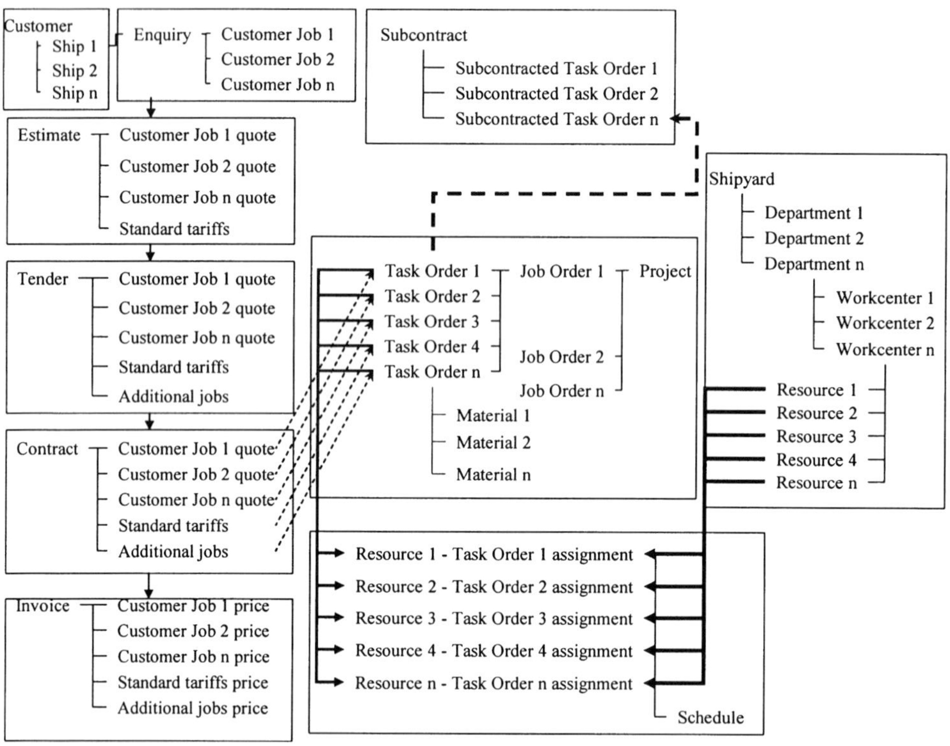

Fig. 3. Business process model

Based on the enquiry, the shipyard *estimates* the cost and duration of the repair of the ship that is a list of quotes per customer job. Then, the shipyard submits the so called *tender*, which is a quote per customer job and includes also pricing of standard jobs that are not included in the original specification of the shipowner. Upon the acceptance of the tender by the shipowner, the shiprepair *contract* is signed, which includes the specification of the repair of the ship, with some additional jobs which occur during the stay of the ship in the shipyard and the inspection, carried out by the shipyard specialists.

Having the contract specification, the shipyard produces the related production orders, which are in the form of a shiprepair *project* that includes a number of *jobs orders,* which in turn, include the *task orders* that are to be performed by the shipyard resources. Each task that has to be performed, either it is assigned to the appropriate shipyard resource or it is assigned to a subcontractor, in which case it is a *subcontracted task order*. Furthermore, for every task, the required materials are ordered and purchased. Based on the existence of other projects in the shipyard, a total workload has to be carried out at any time of the operation of the shipyard. Taking into account the available resources as well as the existing workload, a *schedule* is produced for the work to be performed, which is represented in the form of Task Order *assignments* to Resources of the shipyard (8).

The materials, the subcontracts and the labour that are related to each task order, are recorded, thus making it possible to compare the tender with the actual cost of the repair. Based on this information, the final invoice is produced and in case that it declines from the initial tender, this decline may be justified by the information collected for each task order.

3 THE SOFTWARE IMPLEMENTATION

The business model described in the previous sections was considered and implemented in the form of a software framework, which is based on the client-server paradigm. The approach discussed above, has been implemented entirely in Java (Java 2) and it operates over any system for which there exists a Java virtual machine. Object-oriented techniques have been used in order to develop the software in a modular fashion (7).

Specifically, the software design has been carried out with the use of the Unified Modeling Language (9). Internally, the system is divided into two layers (client-server approach); The User Interface “front-end” and the Data Source “back-end”. The communication between these two layers is accomplished through Java DataBase Connectivity (JDBC). Externally, the User Interface is structured around three areas of functionality (modules). Moreover, the Database of the software tool stores its data in a central database, implemented in Oracle 8i or Microsoft SQL Server 7.0. Multiple instances of the application can access at the same time the Database of the software system.

The system enables the user to construct a hierarchical model of the shipyard’s facilities and their workload. The facilities’ model includes the definition of the Factory, the Job Shops, the Work Centers and the Resources. The user "fits" the workload model to the factory model, by specifying the Orders, the Jobs and the Tasks. Furthermore, the system allows the user to specify which Resources are suitable for performing each Task, the precedence relationships, the processing times and the set-up times. Additionally, the system allows the user to enter information about shipyard customers, their ships and the enquiries submitted to the shipyard. It offers rich functionality to prepare an estimate by offering libraries of jobs and tariff pricing per job. It permits making a cost estimation of each job, for a wide range of categories, such as subcontract costs, labour costs, materials costs and other costs. It implements a mechanism that will be handling the evolution of the worklist from its reception, at the enquiry stage through the estimation and tendering, during the contract and planning stage, up to the final invoicing of the work.

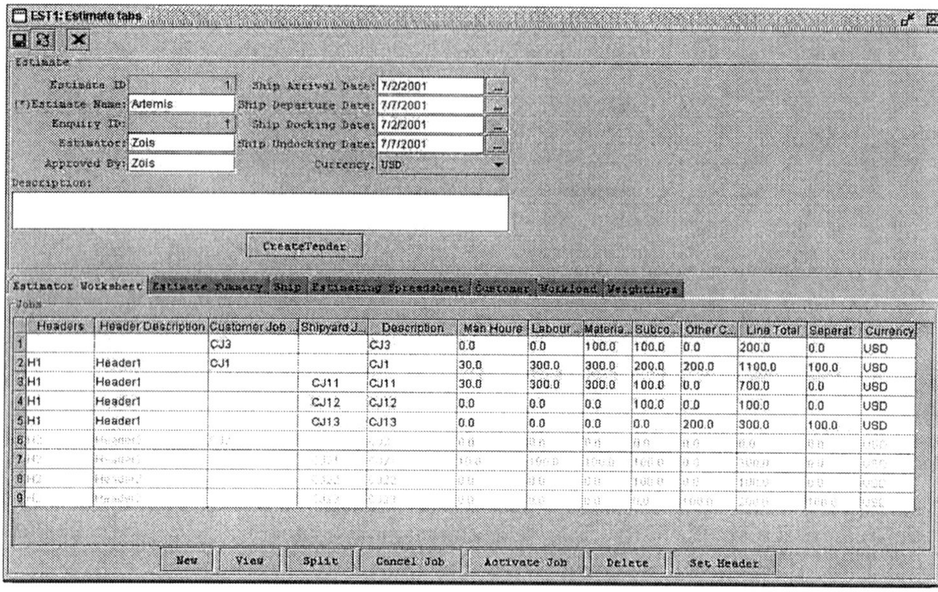

Fig. 4. Estimate preparation dialog

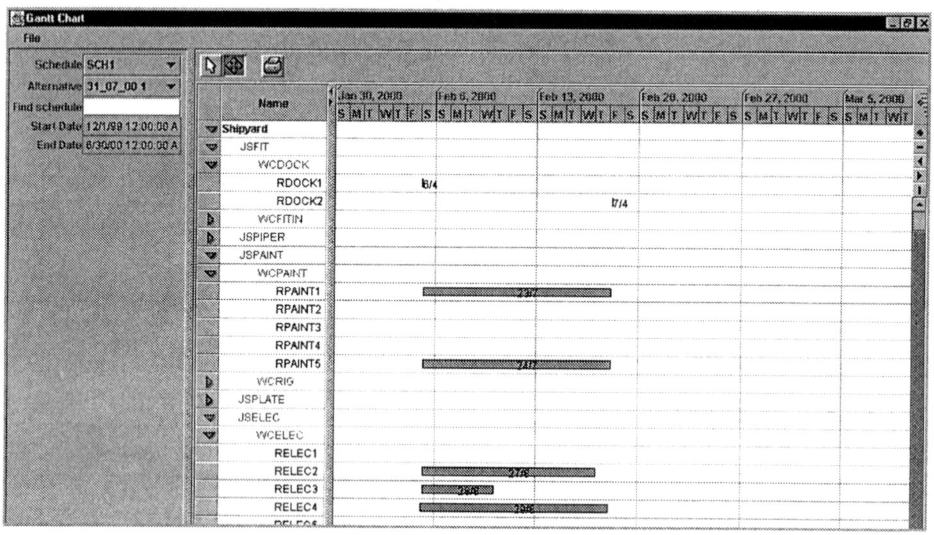

Fig. 5. A simple shiprepair plan produced by the system

The graphical user interface is menu-driven, providing Java Swing dialogues for guiding the user through the modelling process (Figures 4 and 5).

4 CONCLUSIONS

This paper demonstrates an integrated framework, implemented in the form of a software system, to manage the evolution of a shiprepair enquiry, from its initial submission, to the final invoicing of the work. It demonstrates the means by which the work can be planned in an integrated way, utilising modern software technology. Advanced scheduling methods have been implemented in the software framework, providing software generated schedules for the execution of the shiprepair. The information regarding the shiprepair is provided to the system at the beginning of the shiprepair business process, and it evolves during the implementation of the shiprepair contract, thus, offering the functionality to follow up the progress of the repair while minimising the effort to manage the information flow.

The software framework has been implemented in a shiprepair yard, to manage a shiprepair case in an integrated way. A set of estimates had been produced for a specific repair, before an agreement was reached with the shipowner. The shiprepair work execution was monitored and orders for materials and extra personnel were monitored through the system. Schedules regarding the use of the available resources were regenerated by the software system to aid the planning of the work. At the end, an invoice was produced summarising the work done and was being compared with the initial tender. The above scenario demonstrated the feasibility of the suggested approach to manage the shiprepair in an integrated way.

Further work includes planning of the activities of cooperating subcontractors, while taking into account the availability of materials and the suppliers' delivery conditions thus, managing the extended shiprepair enterprise.

REFERENCES

1 Chryssolouris, G., Papakostas, N., Makris, S. and Mourtzis, D. (1999) "Planning and Scheduling of Shipyard Processes", in *Application of Information Technologies to the Maritime Industries*, C. Guedes Soares, J. Brodda (Eds.), pp. 255-274.

2 Chryssolouris, G. (1992) "*Manufacturing Systems – Theory and Practice*", Springer-Verlag.

3 Lee, J. K., Park, J. K., Hong, J. S. and Lee, J. S. (1997) "Developing scheduling systems for DAEWOO Shipbuilding: DAS project", *European Journal of Operational Research*, pp. 380-395.

4 Stidsen, T., Kragelund, L. B. and Mateescu, O. (1996) "Job shop scheduling in a shipyard", *Proceedings of the 12th European Conference on Artificial Intelligence*, Published by John Wiley & Sons, Ltd.

5 De Boer, R., Schutten, J. M. J. and Zijm, W. H. M. (1997) "A Decision support System for Ship Maintenance Capacity Planning", *Annals of CIRP*, Vol. 46, pp. 391–396.

6 Herroelen., W., Reyck, B. and Demeulemeester, E. (1998) "Resource-Constrained Project Scheduling: A Survey of Recent Development", *Computers Ops Res.*, Vol. 25, No. 4, pp. 279-302.

7 URL sun.com (2004). http://java.sun.com.
8 Chryssolouris, G., Makris, S., Xanthakis, V. and Mourtzis, D. (2004) "Towards the Internet based Supply Chain Management", *International Journal of Computer Integrated Manufacturing.*
9 Quatrani T. (1998) "*Visual Modelling with Rational Rose and UML*", Addison-Wesley.

Business barriers to the e-sourcing and re-use of mechanical components

H J REA and **J R CORNEY**
School of Engineering and Physical Sciences, Heriot-Watt University, Edinburgh, UK
D E R CLARK and **N K TAYLOR**
School of Mathematical and Computing Sciences, Heriot-Watt University, Edinburgh, UK

ABSTRACT

Manufacturing enterprises around the world have archives of tens of thousands of 3D CAD models that define the components they manufacture. Technologies are emerging from a number of research projects that suggest it will be feasible in the near future to conduct global 'shape' searches on these collections. In principle such "3D-search" engines could revolutionise the sourcing of mechanical components and enable a dynamic market for reselling existing production tooling.

This paper presents an investigation of the commercial and business aspects of this emerging technology (i.e. 3D e-Sourcing). State-of-the-art part-sourcing is outlined and estimates of the potential benefits of reuse are discussed. The paper then outlines ownership issues of CAD/CAM data and tooling and identify barriers, both technical and business, to forms of 3D e-Sourcing which would enable secondary use. The paper identifies the ability to prohibit the physical production of publicly circulated 3D models as key to the viability of this trading system. Consequently the paper concludes with details of a proposed "3D Model Public Licence" which is specifically designed to protect the IP of manufactures while allowing 3D search engines to index components.

1 INTRODUCTION

The collaborative re-use of design and manufacturing data is one way that e-commerce can significantly reduce development costs of new products. However the proliferation of web-based catalogues for standard components (e.g. brochure-ware for nuts and washers etc.) only hints at the productivity gains possible if effective forms of 3D e-Sourcing become available (see Table 1). One study (1) reported that the potential cost savings from reuse of "plastic clips", for example, in a single automotive supply business ranged between \$1 to \$5 million per year. However, obtaining and surveying catalogues of existing parts, even within a single enterprise, is a costly and time-consuming task.

Table 1. On -line sources

	Company	Model Type	Web address
3D Models	Trace Parts	Standard parts libraries	http://www.traceparts.com/
	Cadalog	Standardised parts and manufactures catalogues	http://www.cadalog.com/
	3D Model Sharing	Custom built mechanical parts, architecture etc.	http://www.3dmodelsharing.com/
3D Search Engines	Princeton	General objects for graphics	http://shape.cs.princeton.edu/search.html
	Heriot-Watt	Mechanical Engineering components	http://www.shapesearch.net/index.html
	LMU	3D Proteins	http://www.dbs.informatik.uni-muenchen.de/Forschung/Similarity/Demos/protein/

Given that recent estimates suggest that over 20 billion CAD models already exist (2) it is highly probable that almost every engineering component has a close geometric relative somewhere in the world. The Internet already offers the necessary technology for every jobbing shop, foundry and injection moulding company throughout the world to display 3D models of their components (3). However, as yet there is no efficient means of searching a vast number of parts (4) (i.e. content based retrieval).

Currently 3D models (like engineering drawings) are indexed by alpha numeric 'part-numbers' with a format unique to each individual organisation. Although this indexing system works well in the context of on-going maintenance and development of individual parts, it offers little scope for 'data-mining' (i.e. exploration) of an organisation's inventory of designs. This paper has been motivated by a research project concerned with an on-line system that allows collections of 3D models to be searched for components geometrically similar to a user defined target-model (1). The system aimed to enable much wider searches to be made for suppliers and increase re-use of existing designs within enterprises and promote collaboration between manufacturers. This approach is referred to as "3D e-Sourcing".

In addition to the sourcing of parts the application of a 3D similarity matching algorithm to large collections of parts would allow many other efficiency gains, such as the following (see Ng and Chan (4)):

- **Cost Estimation for Machined Parts**. For some manufacturing domains such as rapid prototyping, reasonably accurate estimates of cost can be achieved by estimating the volume or weight of the part. But for other manufacturing domains, such as machining, cost estimation depends on the geometric details of the object, and automated procedures are not available for particular accurate cost estimation. However, the cost of manufacturing a new part can be estimated by finding previously manufactured parts that are similar in shape to the new part. If a sufficiently similar part can be found in the database of previously manufactured objects, then the cost of the new part can be estimated by suitably modifying the actual cost of the currently manufactured similar part.

- **Part Family Formation**. In many manufacturing domains such as sheet metal bending, machine tools can be setup to produce more then one type of part without requiring a setup or tool change (5, 6). However, parts need to be shape compatible in order for them to share common tools and setups. Therefore, in order to find common tools and setups, geometrically similar parts need to be grouped into families. Shared tools and setups can be used to manufacture objects in the same family, resulting in significant cost savings.

- **Reduction in Part Proliferations by Reusing Previously Designed Parts**. Reusing design/manufacturing information stored would result in a faster and more efficient design process. While designing a new part the designer can refer to existing designs and utilize previously used components.

1.1 Objectives and Scope

The objectives of this paper are to identify commercial, cultural and technological barriers to 3D model-based sourcing associated with the computational and algorithmic problems of automated 3D similarity assessment. The discussion specifically addresses the issues involved in sourcing individual components (rather than assemblies) which have functions characterized largely by their shape (e.g. housings, covers, frames, brackets, and fixtures). Often the role of such components is simply to occupy a space in order to protect, support or guide, other more critical parts. 3D shapes are independent of particular company part numbers and human language. Consequently prototype 3D search engines let users query them by either sketching or uploading a target model against which a similarity match is performed (see Figure 1).

The next section outlines the size and complexity of modern manufacturing sourcing activities. Then the recent impact of web-based sourcing systems is discussed before the issues surrounding engineering re-use are presented. Having established the context, sections 3 and 4 describe technical and business barriers to 3D model-based sourcing. Finally, in section 5, the issues raised are discussed and some conclusions are drawn.

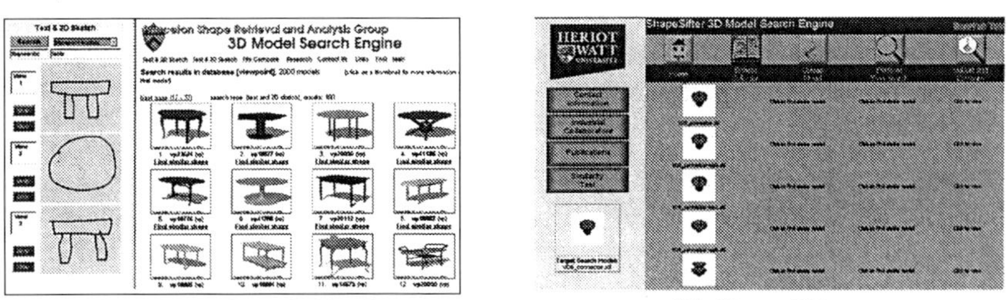

(a) 3D Model Search Engine **(b) ShapeSifter**

Figure 1. Online 3D Search Engines

2. REUSE AND THE MANUFACTURING SUPPLY CHAIN

Almost every manufactured product is created from a mixture of bespoke and standard components. Standard components are typically functional elements like bolts, bearings, gears, seals, electric motors, and pumps, which are produced in a range of sizes by specialist manufacturers. Essentially these are catalogue products. The manufacturers of such products are eager to see their components incorporated in the designs of new products and so make 2D drawings and 3D models freely available in many formats. Such components are identified by part numbers and websites providing online access to the component usually allow searches by manufactures, part number or by generic category/function (e.g. washers).

In contrast, bespoke components are, in general, unique to a particular product or product range. They are designed by OEMs (Original Equipment Makers) for a particular task. Within a company these bespoke parts will be referenced by part numbers. Often their production will be outsourced to subcontractors, who specialise in a particular method of manufacturing (e.g. casting, pressing, etc.). The economies of scale, so evident in the price of standard parts, do not apply to bespoke parts, and consequently the bespoke parts represent a significant proportion of a product's cost.

Whereas standard parts are purchased from catalogues with well-defined prices and performances, the production of bespoke parts is, typically, put out for tender to several manufacturers who compete on price, quality and delivery time. This process is called "sourcing" and since large companies (e.g. car manufacturers) frequently source from medium sized companies (e.g. radiator manufacturers), who in turn buy components from small companies (e.g. pipe manufacturers), the suppliers are referred to as forming a "supply chain". The complexity and size of supply chains in major industries is known to be impressive:

> *"Further complicating an analysis of the automotive supply chain is the complexity of the relationships between customer and suppliers. OEMs design and produce only some of the 15,000 parts and accessories that make up an automobile; they procure others from first tier suppliers. A company's position in the supply chain may differ depending on the part and the customer. Thus, a company that is a first-tier supplier of transmissions to one OEM may be a subtier supplier of other parts to the same or other OEMs. Furthermore, these companies, especially the subtier suppliers often supply parts to customers outside the auto industry."* Brunnermeier and Martin (7)

The productivity gains of employing an e-commerce approach to component sourcing and supply chain management have been recognised and been given the generic "web-based sourcing". The process found at FirstIndex (8) typifies the on-line subcontracting of bespoke components. Here, buyers supply FirstIndex information about the required item in the form of a standardised RFQ (request for quotation) including drawings (in pdf format). FirstIndex then identify a suitable supplier in their database and pass on the necessary information to them either by email or post. The buyer can view the status of each suppliers progress with the RFQ, make amendments and communicate with the suppliers online (9).

2.1 Design Reuse

Clearly sourcing is much easier if existing components can be re-used. Academic attempts to quantify the benefits of design reuse have produced inspiring numbers:

"In engineering, it is conservatively estimated that more than 75% of design activity comprises case-based design – reuse of previous design knowledge to address a new design problem" Regli (10)

Consequently no one doubts the value of reuse, and the few attempts to quantify the benefits of reusing existing components all suggest there are significant productivity gains to be had. For example, a study (11) reports:

"Considering design reuse as a whole we can see that time, quality and performance are all currently benefiting from reuse to a similar degree at 10-11%, with cost nearly half as much at 6%. The greatest overall potential benefit is seen to be cost reduction with 28%, followed closely by improved quality with 27% and reduced time at 25%, while performance is at 20%. Thus cost reduction has by far the greatest scope for improvement of 367%. Improved quality on the other hand has a potential for improvement of 170%, with time at 1127% and performance at 82%."

However, despite the productivity gains to be had, there is currently little commercial support technology to enable increased re-use. It is apparent that most reuse that takes places today is on an ad hoc basis enabled largely by the memory of individuals directly involved in the production process. The failings of CAD systems to support re-use are well described by Fowler (12):

"Yet current CAD systems exhibit shortcoming for a variant design environment. The first shortcoming is design retrieval. The only mechanism which CAD systems provide to designers to retrieve existing designs is by part or assembly name. These names are textual, they may have been automatically generated by the CAD system itself or specified by the designer, and typically correspond to a filename convention supported by the host computer platform. While design rules in force at a particular company may provide naming guidelines to designers, there is no guarantee that a given textual name has any relationship to the functionality, intended use, or context for the design.

Finding an existing design product using a current CAD system without any mechanism other then a part or filename becomes a chore not significantly different than locating a drawing in a storage room of archived (shelved) drawings. Like its paper-based predecessor, a CAD system and the host computer operating systems provide features for organising design file. Designs may be indexed by ownership to a designer, as member components in an assembly hierarchy, as the root of an assembly as a particular design project, by revision status, and so forth depending on the features offered by the particular CAD system. But such indices may offer only vague guideposts to a designer who is wondering whether any existing compressor housing designs provide adequate clearance for a redesigned compressor. Finding a selection of existing compressor housing designs may take longer than designing a satisfactory housing from scratch ('Where is that housing we used on the T2 project? Is it in Beth's directory ...Is this the release part? I thought it had a flat mounting plate...')."

It is possible that there has been little pressure to provide better search facilities because of the strong commercial motives for concealing re-use from customers. For example, the creation of production tooling is a cost explicitly charged to a customer. If existing tooling can be adapted (or reused) without the customer being aware of the fact, a charge can still be made for the tooling!

3. TECHNICAL BARRIERS TO 3D E-SOURCING

Our work has identified only two technical barriers to the part sourcing system envisaged, namely data translation and data protection.

1) Data translation problems arise with the difficulties of reliably importing and exporting 3D data from CAD systems produced by different vendors. For example, a 3D model created on AutoCad would have to be translated via a neutral format before it could be imported into ProEngineer. This translation process is well-known to be difficult and unreliable. Even using existing approaches to part sourcing the problems of interoperability are significant. It has been estimated by Brunnermeier and Martin (7) that the US automotive supply chain incurs costs of $1billion a year in the labour required to repair incomplete or inaccurate models.

2) Data protection problems arise from the issues of ensuring the protection of copyrighted CAD data. A problem frequently discussed in the context of digital watermarking techniques 3D CAD models:

> *"Now imagine the following scenarios. Is the CAD data you just received from you colleagues in a branch office (or vendor) via the Internet genuine, or is it subtly tampered with so that your project fails to meet the deadline? If you are a vendor of a part, how do you prevent a thief disguised as a customer copying your design provided for pre-production 'evaluation'? How do you detect, prove, and put forth a legal claim regarding such IP-violations?"* Ohbuchi and Masuda (13)

Neither of these issues are insoluble problems. Data translation in particular has improved markedly in the last few years. There is also hope that the new ASME standard for solid modelling (14) will improve the quality of the 3D data generally available. The issues of data protection are more complex and are considered in the discussion.

4. BUSINESS BARRIERS TO E-SOURCING

A workshop held by the authors (15) identified three principle business barriers to part-sourcing systems based on 3D shape retrieval:

- Ownership of Tooling;
- Ownership of 3D CAD/CAM Data; and
- Determination of Pricing Models.

Each of these is considered in turn in the following sections.

4.1 Ownership of production tooling

The issue of ownership arises from the great variety of practices in subcontracting. Although a customer will supply a design, it is often the task of the subcontractor to develop the tooling (e.g. mould and fixtures) needed for its production. The costs of creating moulds and other production tooling are significant (often in excess of £10000) hence, ownership is an important issue, current practice breaks down into three possibilities:

- the customer owns tooling, and can insist on its destruction or relocation;
- the subcontractor owns tooling, with restrictions on use (i.e. can only be used for original contract);

- the ownership is not specified.

What must be appreciated is that the tooling for the production of a given design is a costly and complex task in its own right. Not surprisingly, some of the most stringent controls on tooling ownership and reuse are found in the aerospace and defence industries. A typical example of such controls can be seen in Vought Aircraft Industries on-line Supplier Tooling Manual (16). Similarly Boeing subcontracting contracts explicitly prohibit the reuse of special tooling. This is evident in online supplier-documentation (17) where special tooling is defined as:

"All jig, fixtures, patterns, taps, gauges, other equipment and manufacturing aids, and replacements thereof, which are of such a specialized nature that without substantial modifications and alteration their use is limited to the development or production of particular supplies or parts thereof, or the performance of particular services."

The ownership of such tooling is defined as:

"Special Tooling acquired on this or prior Purchase orders used to perform this Purchase Order/Subcontract is Government Property and shall be used only for this Purchase Order/Subcontract. Tooling will not be modified (except as specified herein), altered, reworked, destroyed, disposed of, or used on any other contract/program or transferred/shipped to a subtier supplier without the express authority of the Buyer. Invoices for special tools fabricated to Buyers source control drawing will be payable until such time as Seller has completed and submitted a Boeing-Mesa Certified Property List (CPL), to Buyer."

4.2 Ownership of 3D Models

The issues of the IPR associated with 3D CAD data is well described by Dean (18):

"Now consider the modern supply chain environment, whereby the OEM carries out the majority of design for a new product. Once the component's form has been finalised, these will then be sent to the preferred suppliers in a common 3D geometry format, such as mould makers. The mould maker then takes the 3D part form and develops the mould required to create the part and historically, the mould would then be manufactured and sent to the OEM for use in production.

Whereas in days gone by, the common communication method would have been drawings, (which hold only the 2D drawings and annotation required to manufacture the part) the standard has now become the 3D model, which can potentially hold a great deal more information, perhaps of most relevance, the design history of that part. In many supply chain environments, it is a requirement that any data is sent back to the OEM for approval and once complete, for archiving. It's at this point that data ownership becomes somewhat fuzzy as communication of vital engineering data potentially poses massive risks to any organisation's intellectual property."

4.3 Pricing Models

If production tooling is to be re-used both customers (i.e.OEM's) and suppliers (i.e. subcontractors) must benefit economically. Such motivation could arise in the following way: Potential customers could get cheaper parts either because:

- They are reusing existing tooling; or
- The cost of original tooling created for the job is offset, overtime, by secondary usage.

Potentially suppliers could benefit by being able to exploit the value of the tooling currently stored. However, to unlock this mutual benefit both customer and suppliers must be willing to agree contracts that enable this virtuous circle. Currently there are several different contractual models which could enable this, these include:

- Fixed price for supply of physical parts, where the originator has no rights to tooling, or say in its reuse;
- Subcontractor buys rights to tooling (or has option); and
- Vendor specifies reuse right at time of initial contract. These could be graded:
    1. No reuse allowed;
    2. Limited reuse, possibly after consultation; and
    3. No restriction on reuse.

Perhaps the easiest and most pragmatic model is one where the originator simply agrees to allow their tooling to be entered into a 3D sourcing database, on the condition that a royalty agreement will be negotiated before any physical components are produced.

5. DISCUSSION AND CONCLUSIONS

It is apparent from our study that because subcontractors are often contractually prevented from re-using either tooling or CAD/CAM data they cannot exploit their archives. However, we have also seen that, if use of less critical components was available, at a price, there is a great opportunity to increase manufacturing productivity.

What is required is a method allows potential re-use of existing designs to search through collections of 3D models, while assuring the owner's commercial interests. An important question is, how could such security be assured? It is arguable that existing IPR laws provide much of the protection required. For example, although it is clear that most industrial parts will not be sufficiently novel, or inventive, to be patented (or qualify for copyright as artistic works), they could qualify for copyright as original "computer aided" works, in contrast to "computer generated" works. These latter works require no significant human contribution and do not qualify for copyright in themselves. Computer aided works, on the other hand may qualify for copyright.

Furthermore, a computer file holding the data for a 3D Model might be regarded as a recipe for producing that model in a CAD package. The computer file could thus be treated as an original literary work in the same way that a computer program does. Such a computer file might also be regarded as providing the content for a database system in the form of a geometric modeller. Modellers permit the storage, retrieval and manipulation of data just as do database systems. Thus, the computer file can be treated as the content of a database, which may also qualify as an original literary work.

Protection for design histories could be obtained through copyright of the histories themselves or design right of the intermediate designs produced during the historical development. Both copyright and design right attach automatically so all that is required is a scheme to record when the works covered came into existence. Legal deposit libraries offer an example of how this might be achieved. An honest broker or trusted third party could maintain a library of design histories date-stamped upon receipt.

So, the current legal framework underpinning traditional IPR could be used to enable 3D e-Sourcing. Indeed, it has been shown that the ownership of design IPR need not be a barrier to e-trade in its functionality. For example, the VCX (19) brokers a market in chip designs between various semiconductor manufacturers.

There are, however, alternatives to simply applying existing IPR law. Perhaps 3D e-Sourcing will not emerge from large established networks of manufacturers, but rather from a combination of the less industrialised nations, the increasing flexibility of modern manufacturing technology and the established benefits of the open source model

Ultimately industry might achieve most economic benefit by allowing the "sharing" and "collaborative reuse" of design data (i.e. 3D CAD models), while restricting the physical manufacture of parts to revenue generating activities. To enable this, a version of a GNU type public license has been drafted by the authors (20). The license is analogous to the MOZILLA Public License (MPL) where the following transformations have been applied:

- Code to Model (i.e. 3D electronic representation of a components);
- Source Code to Source Model (i.e. proprietary CAD format); and
- Executable Code to Facetted Model (i.e. approximate model).

The licence makes the distinction between physical and electronic copies of a 3D CAD model. This is significant because almost uniquely amongst digital contents (i.e. images, text, sound etc) CAD data has two roles:

1. Generation of 3D Images for visualisation; and
2. Generation of manufacturing instructions.

In other words a 3D model can be used to visualize its placement within an engineering assembly but it can also be used to create moulds, tool paths or STL files which drive various manufacturing processes. We propose that to enable 3D e-Sourcing, components should be licensed in such a way that generation of physical models or manufacturing data free electronic use is granted. The aim is to define a contractual framework analageous to that used in the entertainment industry where ownership of the text of a play is distinct from performance rights.

The authors believe that significant 3D e-Sourcing could be enabled by making the distinction between "virtual" and "physical" use of CAD data. This approach has been explicitly defined in the author's open licence agreement. Further work is required to investigate how such an approach could be enforced with-in data exchange formats.

REFERENCES

1 Berchtold, S. and Kreigel, H.-P., "S3: Similarity Search in CAD Database Systems", in *SIGMOD 97*. 1997: ACM Press.

2 Regli, W.C. and Circirello, V.A., "Managing digital libraries for computer-aided design", *Computer Aided Design,* 2000. 32: p. 119-132.

3 Ng, V.T.V. and Chan, S.C.F., "A Solid Modelling Library for the World Wide Web", *Computer Networks and ISDN Systems,* 1998. 30(20-21): p. 1835-1863.
4 Cardone, A., Gupta, S.K., and Karnik, M., "A Survey of Shape Similarity Assessement Algorithms for Product Design and Manufacturing Applications", *Journal of Computing and Information Science in Enginering,* 2003. 3(2): p. 109-118.
5 Gupta, S.K. and Bourne, D.A., "Sheet Metal Bending: Generating Shared Setups", *ASME Journal of Manufacuring Science and Engineering,* 1999. 121(November): p. 689-694.
6 Alva, U. and Gupta, S.K., "Automated Design of Sheet Metal Punches for Bending Multiple Parts in a Single Setup", in *Robotics Computer Integrated Manufacturing.* 2001. p. 33-47. 2.
7 Brunnermeier, S.B. and Martin, S.A., "Interoperability Cost Analysis of the U.S. Automotive Supply Chain". Research Triangle Institute. North Carolina.1999 . 7007-03.
8 FirstIndex, : http://www.firstindex.com. .
9 Percival-Straunik, L., *E-Commerce.* 2001, London: Profile Books Ltd.
10 Regli, W.C., "Digital Library Suppoer for Engineering Design and Manufacturing", in *ASME Design Engineering Technical Conferences.* 1999. Las Vegas, Navada.
11 Duffy, A.H.B. and Ferns, A.F., "An analysis of design reuse benefits", in *International Conference on Engineering Design.* 1999. Munich.
12 Fowler, J.E., "Varient Design for Mechanical Artifacts - A State of the Art Survey". US Department of Commerce, National Instute of Standards and Technology. Gaithersburg, MD.1994 . NISTIR 5356.
13 Ohbuchi, R. and Masuda, H., "Managing CAD Data as Multimedia Data Type Using Digital Watermarking", in *FourthWorkshop on Knowledge Intesive CAD.* 2000. Parma, Italy.
14 "ASME Y14.41-2003:Digital Product Definition Data Practices". ASME International. .2003 .
15 Corney, J.R. and Rea, H.J., "Business Issues Workshop Report". Heriot-Watt University. Edinburgh, UK.2000 . 001.
16 "Supplier Tooling Manual". Vought Aircraft Industries Inc. .2003 . http://www.vought.com/suppliers/contract/ suppliertooling.pdf.
17 "SP Section 6000 - Tooling and Test Equipment". McDonnell Douglas Helicopter Systems. .1999 . http://www.boeing.com/companyoffices/doingbiz/tcmdhs/sp60.pdf.
18 Dean, A., "CAD SYSTEM IP RIGHTS - ENTERING A NEW DIMENSION", in *Thomson Derwent.* 2002.
19 VCX, "Virtual Component Exchange", : http://www.albacentre.com/about/vcx.asp. .
20 Corney, J.C. and Rea, H.J., "3D Model public License". Heriot-Watt University. Edinburgh.2003 . http://www.shapesearch.net/frames/reports/3DModelLicense.doc.

Product life-cycle management – an essential component for agile manufacturing

K K PUSTHAY and **Z WANG**
Nevada Manufacturing Research Centre, University of Nevada Las Vegas, USA

ABSTRACT

This paper describes the need for an effective product life cycle management (PLM) system coupled to a shop floor control system. PLM is enterprise wide web centric application software that can manage all product related data throughout its life cycle. Continuous change in customers demand makes it difficult for modern manufacturing companies to rapidly adapt to these changes and sustain competition from their counterparts over the globe. Even though computer integrated manufacturing tries to address this issue to some extent, the integration of all units in manufacturing system in order to adjust to the rapid changes in the products design specifications with respect to the customers needs is still a problem faced by the modern manufacturing units. Use of proper networking system and good shop floor control software can definitely address these issues to a major extent. Apart from good shop floor control system there is a need of coupling product life cycle management component for achieving continuous improvement and profitability.

Keywords:
Product life cycle, Shop floor control, Agile-manufacturing

1 INTRODUCTION

Agile manufacturing refers to the ability to rapidly respond to unpredictable changes occurring in the market environment without going for any major overhaul of the existing production setup. The agile manufacturing should constantly develop innovative ways to interact with the customers and suppliers and should provide them with a chance to assess the capabilities & competencies of the enterprise. This will in turn help the enterprise to better envision the customer needs.[1] The present market demands a systematic approach right from early product conception stage to the end of the product's life i.e., until its disposal. Recent market trends shows that majority of the companies are out sourcing their operations to other parts of the world. To remain competitive they must have reliable product life cycle management tools and techniques that can track their products from the shop floor to the end customer. The networking system must be able to connect each of the following units design, production, scheduling, sales, supply chain etc. With all these units in an organization well coupled it is possible to continuously monitor the production and there is enough room for making online decisions. The effective feedback system, which couples the entire manufacturing system, not only enables a designer in California to modify the design and pass the information to shop floor manager somewhere in Asia but also update the existing product design and implements the new product process plan for that part. The online information exchange will help to run the production continuously .The scrap and rework rate is also reduced to major extent and the companies can gain competitive advantage of early to market.

CIMdata defines PLM as:" A strategic business tool that applies a reliable set of business solutions in support of the collaborative creation, management, dissemination, and use of product data across the extended enterprise from concept to end of life-integrating people, processes, business systems, and information."[2] The term product data management (PDM) was used to link any information related to the product and this term is often referred as the foundation for the product life cycle management. This evolution is not limited to the terminology only but it is also evident in terms of the technology too. The PDM which was initially about how to maintain the engineering design data and custom implementations focused on product related data & technology. The custom implementations developed into tool kits that automated certain function and these generic applications resulted in standard data models and predefined workflow templates that are essential to solve certain key business problems. Now the focus is shifted towards complete business solution and focus directed towards bottom line issues.

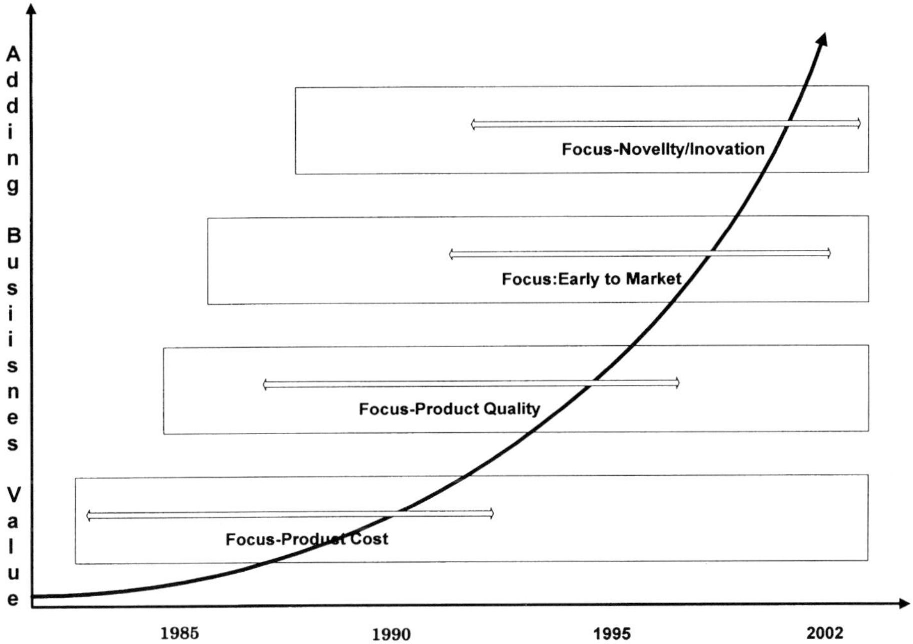

Figure 1: Evolution of PLM

The above figure shows the evolution of product life cycle management in last two decades. It is evident from the above graph that the focus has shifted from minimising the product cost to achieving novelty/innovation. It is clear that the current day focus is on achieving complete business success by incorporating best practices that will help the organizations to migrate their business process toward de facto industry standards. [2]

Design of shop floor management in present customer driven market aims at not only joint optimization of technology, human resources but also effective management of product through out its life cycle, a concept popularly known of Product life cycle management (PLM). The PLM keeps track of all the changes associated with the product through out its life cycle, right from conception to retirement/disposal and makes this information available to everyone in the organization. A typical product life cycle is shown in the figure below. In a traditional organization the shop floor control system tries to address only few components of the product life cycle. In other words the traditional shop floor control system does not directly account for any changes to the product once it leaves the shop. As we can see that the product has just finished half of its lifecycle when it is deployed, there is still another half part in its lifecycle before it is retired or disposed, and this half can decide the profitability and survival of the product from the competitors.

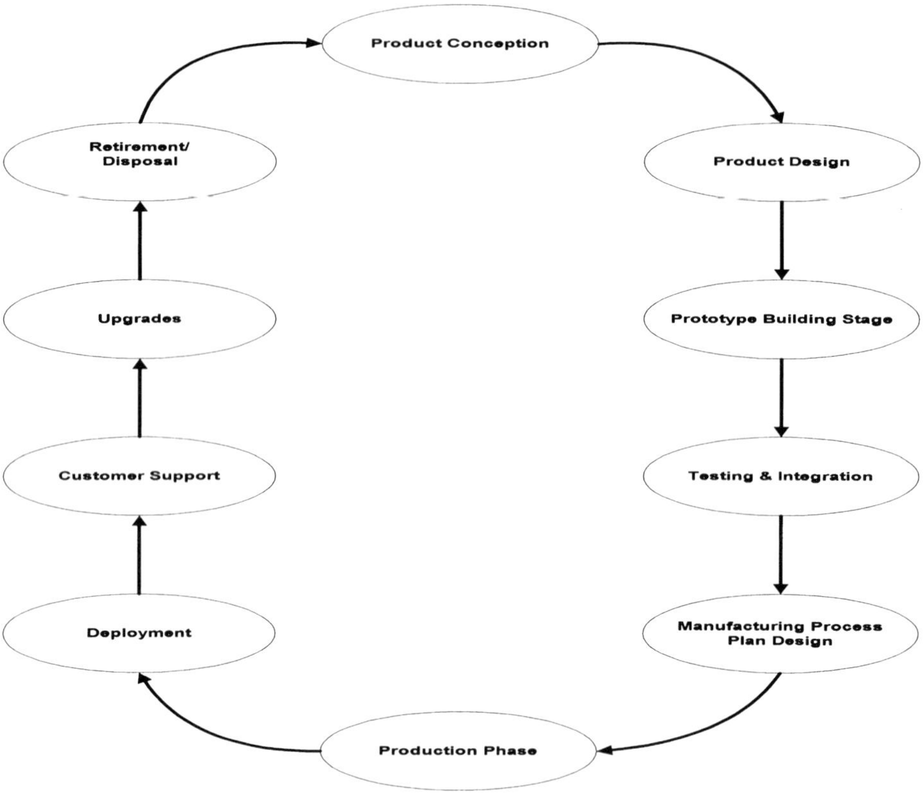

Figure 2: Typical Product Life Cycle

Figure 2 shows the various stages in product's life cycle. Waller (1999) defines the following PLC stages [3]:

Development: This is a period prior to commercialization, in which the product is being designed and tested, which includes conception, design, prototype building and testing& integration phases as shown in the figure 2.

Introduction: This phase is when the consumer starts to see the product as it is commercialized and launched to the market. This corresponds to the deployment stage in figure 2.

Growth: Period when the product is well accepted in the market, revenues are growing rapidly and support costs decline somewhat as the product starts to become self-marketing. This phase covers the upgrades and customer support phase of the figure 2.

Maturity: In this phase the product is at a point where its existence is never given a second thought.

Decline and death: Phase when the product reaches the end of its life and is eventually retired. Final stage is retirement/disposal phase that marks the last phase of the product life cycle.

2 PLM ARCHITECTURE AND PROPOSED SYSTEM

Product life cycle management tools require highly sophisticated, secure and intelligent network infrastructure that is available at any time thus enabling effective collaboration among the internal resources and external partners of the organization. The PLM software's are built on a three-tier architecture, in which a central server can be accessed from a remote corner of the globe. In three-tier architecture, the front end which is a browser can be easily loaded into individual client machines. The back end consists of application softwares and database. The main advantage of this type of architecture is any change in version of front end softwares does not require upgrading all client machines as in case of traditional client-server architecture wherein all client machines will be loaded with the .exe files of the front end in case of any version upgrades.[5]

Thus there is need to for coupling this information repository to a centralized database, which connects all other units of an organization. The web based PLM applications offers the manufacturer to communicate effectively not only among various units within the organization but also with the outside world and also better manage product development. The use of Internet enables information sharing in a collaborative environment where a product is designed in one part of the world and manufactured in another part of the world. The product knowledge is shared between engineering, design, manufacturing, supply chain, sales and customer support. For example a manufacturer has eliminated large number of inefficiencies and reduced the cost by employing lean manufacturing techniques but if the customer service detects a problem which may need immediate change in product, may remain undetected for many days or weeks in a traditional organization but by employing internet based PLM this information is immediately supplied to all units and necessary change is incorporated without creating any additional scraps/inefficient parts. This concept can be practically applied to the automobile industry where the amount of recall may be as large as several billion dollars annually.

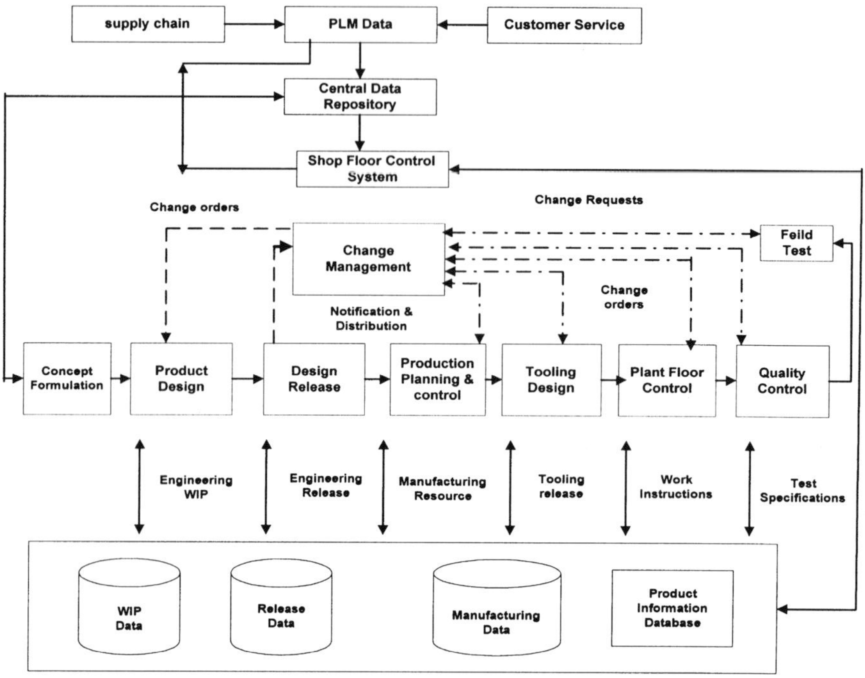

Figure 3: Flow diagram for the proposed PLM system

The flow diagram in figure 3 represents PLM system coupled to the existing shop floor system. [4] A brief explanation of various components is given below. The proposed model not only couples PLM data with the traditional shop floor system but also other units of the organization like supply chain management, sales, marketing and Customer support. The advantage of coupling PLM to shop floor control system allows the production line to be updated with customer needs and rapid market changes. Modern companies which initially started enterprise resource planning (ERP) and supply chain management (SCM) systems to optimize their internal operations are now focusing on front-end solutions to manage relationships with customers, suppliers, partners, and employees.

It is well know fact that traditionally integration of shop floor with other functional departments of the organizations has been neglected. In the early 90's with the advances in ERP technologies and Internet boom in the later part of the century the shop floor has received some attention. Many agent based paradigms which help realize SCM, MES & ERP solutions have been developed to tackle the emerging market trends; popular among them are Holonic manufacturing systems, Random manufacturing systems, and Bionic or Genetic manufacturing systems. [6] E manufacturing has further revolutionized the traditional shop floor by integrating it with the supply chain systems ERP & CRM. But in today's market were technologies and customer demands change overnight; it is imperative to have an effective PLM system, which can integrate the shop floor with the central data repository. This central data repository contains all the product related data through out its life and gets it input from various functional units of the organization. This data is made available to each individual who has to deal with the product at some stage in its life cycle. Any changes in customer demands can be effectively communicated to the shop floor. Similarly if there are any quality issues that needs to be corrected immediately can be answered in short period and thereby reducing the rework.

2.1. Advantages of this model:

The advantages of PLM are it reduces time to market, reduces product development time by up to 40% ultimately reducing cost, since about 85% of products life cycle cost is incurred during product design stage. PLM also minimizes resources used in the enterprise by reducing the waste or inefficient parts. One of the biggest benefits of PLM is that it can help resolve as many problems in design phase as possible. According to Erica Rugullies, (Giga information group) [7] Successful implementation can cut down as much as 90% of document production and shipping costs and about 50 to 80% of travel costs. The specific advantages of this model are

1. Reduced development times: With the central data repository maintaining continuous and updated information about the product at each stage of its life cycle, it is easy to identify quality issues. If these quality problems require immediate changes then the design team can issue the change notices to the plant floor without any delay, thereby reducing the development times.

2. Reduced rework/scrap*:* The efficient integration of shop floor with the central data repository will help the production managers to monitor the customer demands. Any change in product specification can be immediately identifies and the production line can either halt the production or make necessary changes and resume the production without any scrap.

3. Control over production: The production team can gain control over the entire production operations by using the updated PLM data. By knowing to produce right quantity at right time they can reduce the inventory and WIP.

3 CASE STUDY

The following case studies demonstrate the importance of PLM in modern manufacturing environment. It is seen that product life cycle Management will be a cutting edge technology for the manufacturing industries in general and automobile industries in particular. The recession in developed countries, collapse of south east economies and poor or under performance in emerging markets resulted in over production i.e.70 million units produced where as demand was only for 55 million units, this resulted in emergence of a global consortium of major automobile giants like Ford, GM, and Toyota. The companies are under continuous pressure of producing cars that cater to needs of global market. This is possible if companies can develop and manufacture cars at different locations. For example GM, which produces approximately 8.5 million vehicles annually, equal to 15% of worlds production, has to affiliate itself with Fiat of Italy and Suzuki of Japan for developing new cars that cater to variety of markets around the globe. GM will develop the platform in Detroit, USA, simultaneously Fiat will develop Engine at Turin, Italy and Suzuki will develop the transmission in Gurgaon, India [5]. Now PLM applies to this situation where different components are being developed at different geographic locations around globe. For example, let us look at the latest global cooperation in 3D product lifecycle management in the automobile industry, which aims to build world-class collaboration around PLM solutions. Toyota Motor Corporation, IBM & Dassault systems recently announced that they would use the PLM solutions to support Best-In-Class processes across design, development and production engineering. Based on the new Version 5 Product Suite -- CATIA for design, ENOVIA for collaboration, and DELMIA for digital manufacturing -- the transformation will cover end-to-end vehicle development processes worldwide. It will enable Toyota to extend its legendary ability to deliver the best products on time to global customers. [8]

The second case study is also related to automobile industry. In this case we look at the Bridgestone/Firestone tyres, which had to be recalled, as they were considered responsible for number of accidents, which resulted in many deaths in the year 2000.Firestone has to recall nearly 6.5 million tyres and it cost them around $350 million, as it had to provide free replacements. Most of these tyres were used on Ford Explorers, and some models of GM and Toyota. The company had already produced 14.4 million tyres. The cause of these accidents was due to tyre-thread separation (tread peel off). Ford was already replacing these tyres on its vehicles in west Asia and Asia in response to customer complaints. If the PLM system was in place the management could have identified the potential problem in earlier stages through customer service or market research. Thereby providing a scope for immediate action to either halt the production or introduce necessary changes and provide improved and safe tyres. This could have resulted in savings of millions of dollars and customer satisfaction.

4 CONCLUSIONS

With the help of a web centric architecture described above, which woks in a similar fashion as Internet the information can be stored in a centralized information storage/retrieval system and this centralized system will also enable information to be accessed from different parts of the world. This system can be equally applied to other units of an enterprise, for example if a customer service detects any problem in the product already in the market, then this information is sent to central data storage/retrieval system which in turn alerts the design, engineering and production units. This problem in a traditional system will remain unnoticed for several weeks and months eventually leading to rework or scrap rates of high volume. But with a PLM system in the enterprise this problem can be addressed in earlier stage and the rework or scrap can be avoided by taking necessary action in time. Thus product development can be continuously monitored with the status information available at anytime. This allows good communication and coordination among designer who are at physically distant location but has access to a common database, hence time required to develop new product is considerably reduced and also allows the organization respond quickly to customer needs.

However there is certain limitation that applies to the implementation of PLM in medium and small-scale organizations. First, the cost involved in implementing PLM is very high. The PLM software and associated hardware are expensive. Even though there are many PLM commercial softwares that are compatible with Personal computers, the cost of integration of various departments within and outside the organization is very high; cost of converting the information into electronic form is also considerably high. Secondly to gain full advantage of PLM all units of an enterprise must agree to be a part of it. Thus effective implementation of PLM system will result in organizations achieving competitive advantage through earlier sales, short product development and product change processes, implement new features faster, cut scrap and rework costs, improve customer relations and profitability. Hence it can be concluded that product life cycle management system is an essential component for agile manufacturing.

REFERENCES

1 Kidd, T, P, 2000,"Agile Manufacturing: Key Issues" [online] available from internet http://www.cheshirehenbury.com/agility/agilitypapers/paper0794.html (Accessed May. 7.2004)

2 Product Lifecycle Management A CIMdata Report: Empowering the future of business [online] available from internet http://www.cImdata.com/publIcatIons/PLMDefinitIon_0210 pdf (Accessed May.5.2004)

3 Garcia Luna, E, F & Aguilar-Savén, R, S, 2004, "Manufacturing Strategy Linked to Product Life Cycle", in second world conference on POM, Mexico.

4 Tsao, S," An Overview of Product Information Management" [online] http://www.pdmic.com/articles/pimintro.html (Accessed May.1. 2004)

5 Product Life Cycle Management, 2003, [online], from Internet, http://www.indiainfoline.com/bisc/ sys4.html (Accessed December.5.2003)

6 Ritter, H. Richter, E. Westkämper, 2003,Agent Life Cycle Management for Manufacturing Systems, in the 36th CIRP-International Seminar on Manufacturing Systems, Germany.

7 Mello, A, 2003 "Better Products, Faster through Product Life-Cycle Management", iQ Magazine, [online] from Internet, http://business.cisco. com/ prod/ tree.taf %3Fasset_id=8879 6&ID=48297&ListID=44753&public_view=true&kbns=1.html (Accessed March.12-2004)

8 Griffith. J, 2002, e-Newsletter, 03-29-2002, Aventec Inc, [online] http:// www.aventec.com/newsletter/03-29-02.html. (Accessed April.24.2004)

A pragmatic system to support interactive assembly planning and training in immersive virtual environment (I-VAPTS)

Y X YAO, P J XIA, J S LIU, and **J G LI**
School of Mechanical and Electrical Engineering, Harbin Institute of Technology, China

ABSTRACT

Assembly planning for complex product is a difficult task with which both intensive knowledge and experience are needed. Computer aided assembly planning (CAAP) systems have been the subject of considerable research in recent years without achieving a wide application in manufacture industry. An alternative approach based on the adoption of immersive virtual reality is presented to generate optimal assembly planning scheme in this paper. A product is assembled from CAD models by providing a CAD interface to transfer assembly constraint information from CAD system to virtual environment; In virtual environment an efficient geometry constraint dynamic recognition and management method based on geometry surface is present, and a process-oriented assembly task model is established to support interactive assembly planning and evaluation. The system is implemented using object oriented methodology, and has been successfully applied to train and guide the assembly workers in a pump assembly process.
Keywords: Virtual reality; Assembly planning and training; CAD

1 INTRODUCTION

Assembly planning is a difficult task for complex product. The traditional assembly design is finished by experienced workers by means of fabricating a physical model, which leads to increase product lifespan and cost, especially in the case of lacking necessary advanced method for checking and evaluating the assembly process. Recently even if computer aided assembly planning (CAAP) with CAD systems has been the subject of considerable research, it doesn't achieve significant up-take in manufacturing industry. The main reason is that the 3D visualization is limited in CAD system, and automatic generation of assembly sequence for complex product is too expensive, meanwhile, the human assembly experience and knowledge is difficult to exert in traditional CAD environment, and some factors such as quality testing, shop floor layout and human ergonomics can't be taken into account for assembly evaluation. Now the successful applications of Virtual Reality (VR) in engineering provide a new low-cost method to solve these problems. In immersive virtual environment, the designer can not only visualize the product to appreciate the inner structure and spatial relationships of components, but also plan assembly sequence and path for product interactively, and then obtain an optimal assembly scheme for assembly workers.

A number of research groups in different parts of the world have proposed the use of virtual reality systems for engineering assembly tasks. For example, Fraunhofer-Institute for Industrial Engineering (IAO) in Germany proposed the first virtual assembly planning prototyping system by applying a virtual model of a person VirtualANTHROPOS [1] to carry out assembly operation, and based on the user interactions with the virtual objects, a precedence graph is generated and the time of assembly and cost is determined. Jayaram et al[2] developed a virtual assembly design environment (VADE) to allow engineers to plan, evaluate, and verify the assembly of mechanical systems. Rayan et al [3] designed a virtual assembly planning and jig design system (JIGPRO) to evaluate the alternate assembly sequence and the jig design, and Chryssolouris et al [4] developed a virtual assembly work cell. Bound et al [5] focused their research on assembly task training in traditional environment (2D drawings), immersive virtual environment (VE) and augmented reality environment (AR), their results indicated that the assembly quality and assembly efficiency in immersive virtual environment and AR environment are much better than in traditional environment.

Although these studies have made evident progress in this area, there is still a long way to go before VR-based virtual assembly technology gets a wide application in industry production. This paper proposes a CAD-based virtual assembly planning and training system, and introduces its application in a pump assembly process.

2 SYSTEM ARCHITECTURE OF I-VAPTS

An immersive virtual assembly planning and training system (I-VAPTS) proposed in this paper integrates the following three functions:

- Visualization: the system can provide the user with a 3D display for navigating and manipulating the models in virtual environment. With a view to visualize the entire system and its environment, the planner can exert assembly experience and knowledge to carry out proper assembly operations, and with the interactive hardware, the engineers can also navigate in the virtual environment freely.
- Operation: the user can not only view the generated computer simulation worlds, but also can interact with it actively using data glove, and work in the virtual world as he would act in reality. He can point a base part, perform virtual assembly and disassembly, and rotate the part or assembly to get a better sight. Based on interactive assembly operation, a process-oriented assembly task model is established to express and record assembly process information, and further to evaluate and select the optimal assembly scheme.
- Training: virtual reality can be an efficient tool to support training for complex tasks. The system can train the assembly staff before actual assembly production. On the other hand, it can guide and teach the assembly workers in assembly floor.

The system architecture of I-VAPTS, which is modularized based on the functional requirements, can be also summarized into three aspects (Fig. 1).

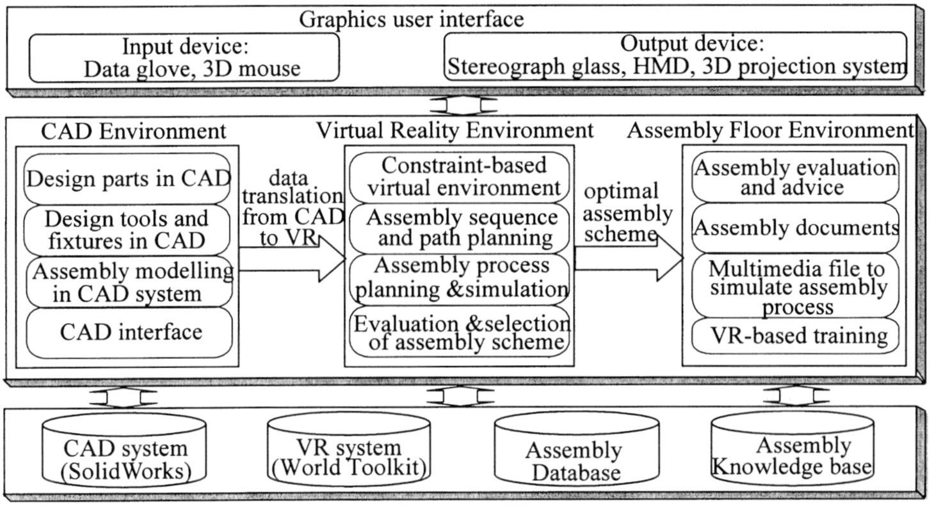

Fig. 1 System architecture

In CAD environment, the parts, tools and fixtures are designed and stored in assembly database. By means of defining a series of mate constraint relationships, which realize the relative positioning of parts to one another, the parts are assembled together and a CAD model of the product is obtained. Only the final position of the part in assembly model is important, the assembly sequence and assembly process are not taken into account in CAD system. These CAD models are loaded into virtual environment. Because virtual reality using polygon representation to visualize objects, some useful information must be extracted from CAD and exported into virtual environment to support interactive assembly planning and evaluation. All these work are called assembly modeling, and now is finished in SolidWorks, a commercial CAD software which provides powerful API functions to access its inner CAD database.

In virtual reality environment, an efficient geometry constraint recognition and management method is supplied to support interactive assembly planning and evaluation. By simulating the process of how the constraints are applied and realized, the system can provide a valuable way to analyze and evaluate the assembly design, and further optimize the assembly operation processes, which involves planning assembly sequence and path, selecting assembly tools and fixtures, determining convenient point for quality testing and surface preparing, and obtaining a time and cost estimation for assembly operation, then provides an optimal assembly planning scheme and reference documents to assembly workers.

In assembly floor environment, the workers carry out the actual assembly production after assembly task training in virtual environment. For a complex product, there are a number of assembly operations that require dexterous operation skills, and it is a time-consuming process for workers to master these skills, especially for a new staff or when a new product comes into production. A multimedia file is clipped and synthesized to simulate the dynamic assembly process derived from the optimal assembly scheme obtained in virtual environment, including the optimal assembly sequence, assembly path and assembly operation. This multimedia video file is output and played in assembly floor to teach and guide the workers for assembly production.

3 OBJECT ORIENTED SYSTEM IMPLEMENTATION OF I-VAPTS

The proposed virtual assembly planning and training system is designed using object oriented software design methodology and implemented with an object oriented programming language, Visual C++ and World Toolkit.

3.1 Data translation from CAD to VE

The virtual reality software (World Toolkit) uses polygonal representation to visualize objects, the polygon model used in most VR systems provides the illusion of being immersed, but it may not be able to precisely define the object geometry. On the other hand, it is difficult to carry out assembly modelling operations due to the lack of topological relationship and constraint information. So a method called IDM (Information Decomposition Method) is developed to translate CAD data into virtual environment. All the assembly information defined in CAD system is decomposed as geometry information, topology information and constraint information. These three different types of information are translated by a CAD interface from CAD to VR respectively, on the other hand they are related with each other to constitute an integrated assembly model to support interactive assembly planning and evaluation in virtual environment. The data flow of Information Decomposition Method is shown as Fig.2.

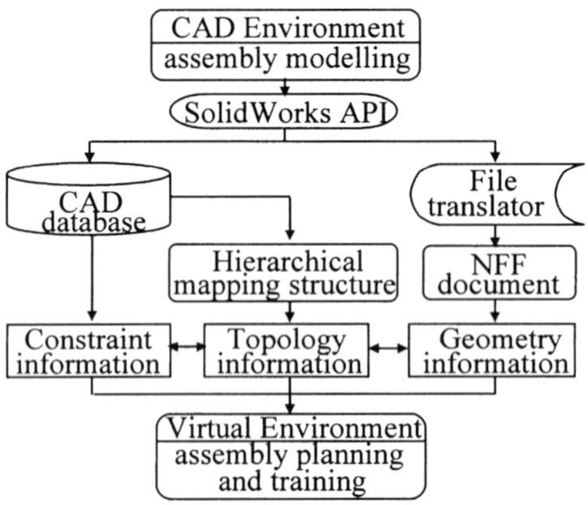

Fig. 2 Data flow of IDM

The parts are designed in a CAD package such as SolidWorks, the results can be saved as .wrl file or .sldprt file. Although WTK can load part from .wrl file, there are many limitations in .wrl file generated by SolidWorks, such as the material and color information are not included in this file. A neural file format (.nff) is created by reading triangle polygons from .wrl file, and a CAD database is connected to the .nff file through SolidWorks API.

A file translator is developed as Fig.3 to translate geometry information from CAD to VR. It reads geometrical data from .wrl file output by SolidWorks, and these geometry data is reordered to generate .nff document. At the same time, every geometry surface is given a name, every polygon is given an identity number, and the material, colour and texture information for every polygon is also written using SolidWorks API function. The generated .nff document (Fig.4) is a triangle-based polygon representation, it describes each geometry surface as a common array of triangles together with a connection list that defines each surface as a set of indices into the array of vertices.

Topology relationship of part must be rebuild in virtual environment. A hierarchical map structure is set up as shown in Fig.5. An assembly object is composed of many parts, a part object is the aggregation of geometry surfaces, and a surface object is the aggregation of a series of polygons. The polygons are the mesh unit to visualize in virtual environment, and the surface object is the foundation to define geometry constraint, such as parallel surfaces, perpendicular surfaces et al. The assembly object, each part object, each surface object and each polygon object are given an identity number as the same written into .nff file, for example, a polygon recognized by the identity number in virtual environment corresponds to a unique surface object, and a surface corresponds to a unique part, and a part object corresponds to a unique assembly object. The hierarchical map structure is an important representation for constraint recognition and assembly operation in virtual environment.

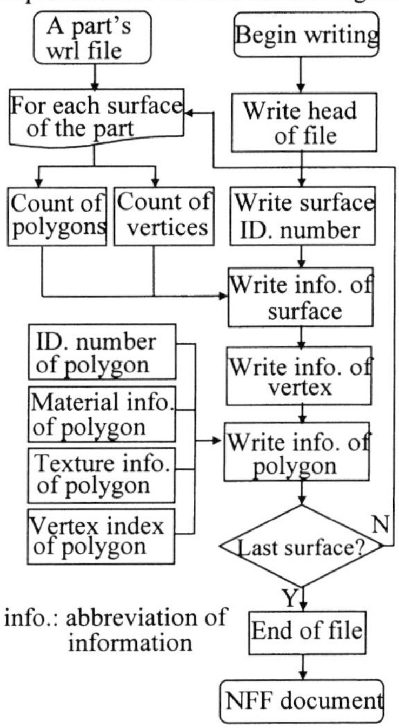

Fig.3 Geometry information translation

Surface Id: S_1
Total number of vertices in S_1
Vertex 1, V_{x1}, V_{y1}, V_{z1}, Normal_definition
Vertex 2, V_{x2}, V_{y2}, V_{z2}, Normal_definition
...
Total number of polygons in S_1
Type=3, V_1,V_2,V_3, material, colour, polygon id=P_{11}
Type=3, V_2,V_4,V_5, material, colour, polygon id=P_{12}
...

Fig.4 NFF file syntax

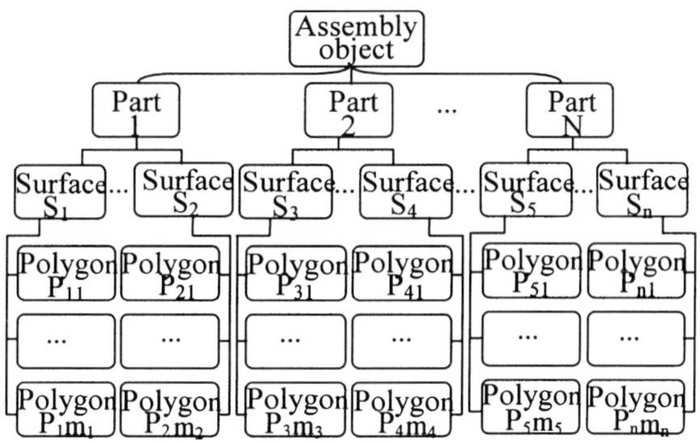

Fig.5 Hierarchical mapping structure

Geometry constraints are another important information defined in CAD system. Constraints play a critical role in assembly design, because the designers usually use constraints to realize the relative positioning of parts to one another. Most CAD systems represent the assembly using constraint relationships between parts. In SolidWorks, after the parts are assembled together, a mate feature tree is obtained. Using SolidWorks API functions the mate constraints information can be extracted by recursively traversing the mate feature tree object, then these geometry constraints can be represented in a constraint manager, which will be discussed in the next section.

In the constraint manager (Fig.6), mate surface is a geometry surface of a part that has mate relationship with other parts. A part may be composed of many surfaces, but only a few surfaces have mate relationships with other parts, these mate surfaces are included in the constraint manager, other surfaces are excluded in order to simplify the constraint manager structure so as to improve the efficiency during constraint recognition process. Each mate surface is represented by specific elements that define its parametric equation. For example, a point on the plane and its normal vector defines a planar surface while a point on the cylinder axis, its direction and the radius of the cylinder value define a cylinder surface. Besides surface specific elements, each surface has a bounding volume, this volume defines surface's borders because the equations define limitless surface. The constraint list includes all the geometry constraints defined in CAD model, each constraint node of the list includes the constraint type, such as concentric, against and collinear, and mate tolerance et al. All these information are extracted from CAD database by using SolidWorks API functions.

3.2 Dynamic constraint recognition and management in virtual environment

Virtual assembly applications now mainly serve as a visualization tool to examine the geometrical representation of the assembly design and provide a 3-D view of the assembly process. This technology is not yet at a stage to be implemented and used in industry for a number of reasons, one of which is the lack of physical constraints that the user depends upon in the real world. To compensate for this, the user needs to provide with constraints required for precisely manipulation virtual objects. On the other hand, to guide interactive assembly, the planner should analyze the mating constraints between the parts, their relative position and orientation et al.

The automatic recognition of geometric constraints can simulate realistic behavior interactively. While an object is being manipulated, the position of the moving object is sampled to identify new constraints between the manipulated object and the surrounding objects. This is performed in five steps:

Step1: When the user grabs a part using data glove, the system searches all the other objects that exist mate relationship with the manipulated object from the constraint manager, and the mate surfaces in these mate objects are highlighted.

Step2: When the user moves the object, detection of collision between the manipulated object and the surrounding objects is continued until a collision occurs.

Step3: Further collision testing is made to recognize possible geometry constraints between the surfaces of the collided objects. A constraint is recognized if the geometric surface elements of the collided objects satisfy conditions of a particular constraint type within a predefined tolerance.

Step4: When a constraint is recognized, feedback is provided to the user by highlighting the only two mating surfaces while the other highlighted faces in step1 are disappeared.

Step5: If the user decides to accept the new constraint, the precise position of the manipulated part is calculated and it is transformed to its proper position. If the user continues to move the object to invalidate the condition for that constraint, the newly identified constraint is ignored.

A constraint is recognized between two planes when the angle between their normals is less than the angular tolerance and that the distance between the planes is less than the linear tolerance. Two cylinders have a potential constraint when their axes make an angle within the angular tolerance and are less than the tolerance apart. The constraint between a plane and a cylinder is recognized if a plane's normal is perpendicular within the tolerance to the cylinder's axis and if the distance between the cylindrical and the planar surface is less than the linear tolerance.

A further surface collision detection is needed when two objects collide. Traditionally VR toolkits use polygonal representation to visualize objects; similarly, the collision detection for VR applications is based on triangles. When two objects are tested for collision, the algorithm returns all intersecting triangles first. These triangles are then processed to identify all the colliding surface pairs. This approach is time expensive because the potentially large amount of redundant computation is required to check collisions between all the possible polygonal pairs. A geometry surface based collision algorithm is implemented to associate polygons with their surfaces. According to the hierarchical mapping structure, each part object is considered as a collection of surfaces and each surface is maintained within the scène graph as a separate node. Each surface node in the scene graph maintains the surface information and its corresponding polygonal data. For example, for a planar surface it maintains a point, normal, bounding box and polygonal data for that surface. Then the algorithm is described as the following three steps.

Step1: collision detection between the bounding boxes of the two parts.

Step2: Does each surface in one part's surface list collide with the other part's bounding box? If collides, the surface is stored into a list called result surfaces.

Step3: surface collision detection between the two part's result surfaces, and return the colliding surface pairs for constraint recognition.

During the assembly process, the parts can be in various stages in virtual environment: only one constraint is applied, or two constraints are applied, or maybe just lies on the table. As the constraint status of different parts are different, it needs to know the status of the constraint information and the constrained degrees of freedom about parts to guide proper assembly operation in virtual environment. A constraint manager is set up as Fig.6. The constraints of a part have two different statuses: already applied or going to be applied, which are called applied constraints and unapplied constraints respectively. At first when a part is not constrained and can move in the global space freely, all the constraints will be unapplied constraints. When a new geometry constraint is recognized and applied, this constraint information moves from unapplied constraints to applied constraints. When a geometry constraint is released during disassembly, the corresponding constraint information moves from applied constraints back to unapplied constraints too. After the part is fully constrained and placed on the base part, unapplied constraints should be empty.

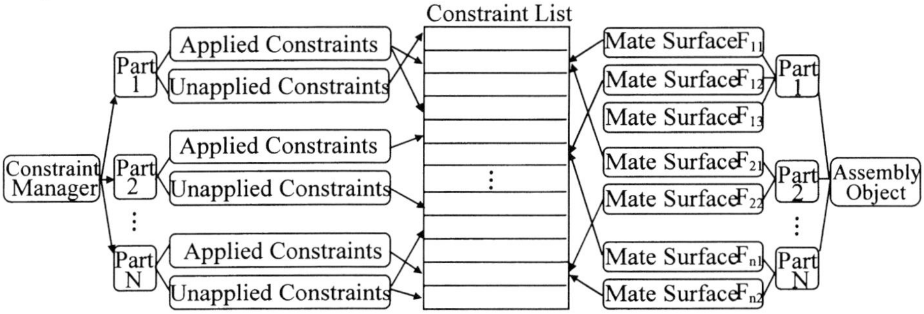

Fig.6 Constraint manager structure

The object hierarchy of I-VAPTS is shown as Fig.7. With the system module property identified, a set of objects can then be designed to implement it.

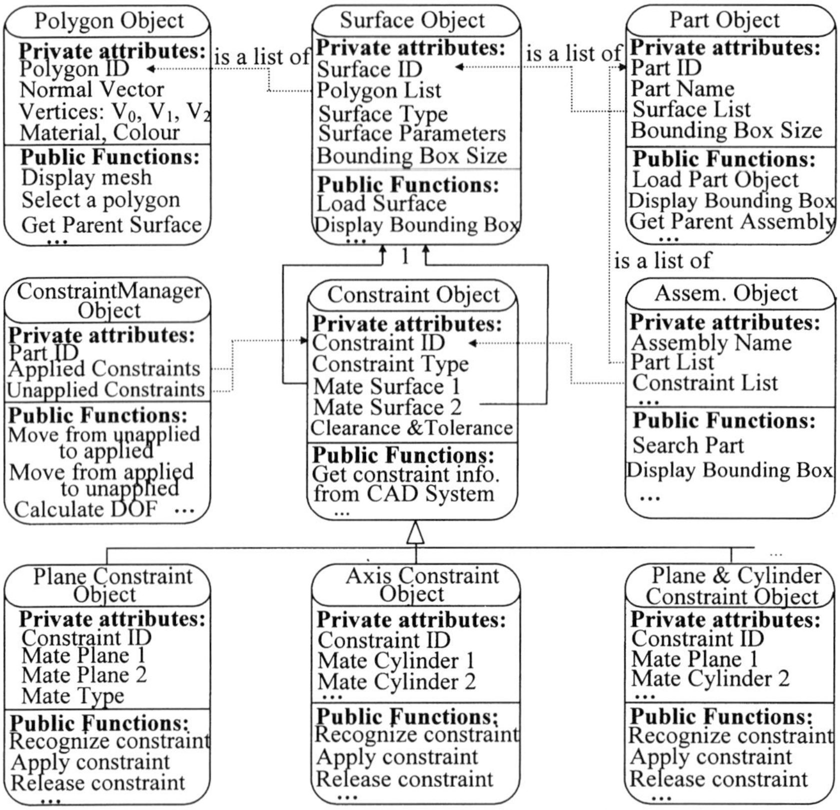

Fig.7 The object hierarchy of I-VAPTS

3.3 Process-oriented interactive assembly task planning and evaluation

In a constraint-based virtual environment, the planner should not only check the collision-free mate paths between the parts, the virtual tools and the environment during interactive assembly, but also ensure that the parts are designed for ease of assembly, assembly space is sufficient for tools, and ergonomics sound. The assembly knowledge base supplies on-line helps and advices to the planner, such as using the heavy and large part as the support base, working form heavy to light, working from inside to out, and installing identical parts simultaneously et al.

Traditionally virtual assembly simulations focus on 3D graphics display, and some useful assembly information and assembly knowledge during assembly process isn't taken into account. A process-oriented virtual assembly task planning and evaluation method is present in this paper. For each step in a product's assembly sequence, we define the assembly operations of installing a part in this step as an assembly task. An assembly task is described as Fig.8. Based on the user interaction with the virtual object, an assembly task model graph is generated interactively.

Process-oriented assembly task model supplies the following advantages to support interactive assembly planning and evaluation:

1) An assembly task includes a series of assembly actions. For example, an assembly task can be divided into six basic activities such as reach, select, grasp, move, position and assemble. The assembly operations, including their corresponding operation time and constraint status, are stored in an assembly action list.
2) The part's assembly path and sweep volume can be used to evaluate assembly space complexity.
3) Assembly hierarchical tree information, such as the division of subassemblies can be used to measure the parallelism of assembly task.
4) Assembly precedence constraint relationship between parts can be used to evaluate and select an efficient assembly sequence. For example, if part A must be assembled before part B, the precedence constraint relationship can be denoted as A>B; if part B must be assembled immediately following part A, then the precedence constraint relationship can be denoted as A→B et al.
5) Assembly tools and fixtures, assembly time et al can be used to evaluate and select an optimal assembly planning scheme.

<an assembly task> = <task NO.><task type><task object><action list><assembly path>
<precedence constraint> <tools and fixtures> <assembly time>
<task NO.>: the identity number of assembly sequence.
<task type>: point a support base, or assemble a part, or assemble a component.
<task object>: the name of the assembling part or component.
<action list>: the collection of assembly operations in an assembly task.
<precedence constraint>: the precedence constraint relationship between parts.
<tools and fixtures>: the tools and fixtures involved in an assembly task.
<assembly path>: the space path and sweep volume of part or subassembly.
<assembly time>: include start time and end time of assembly task.

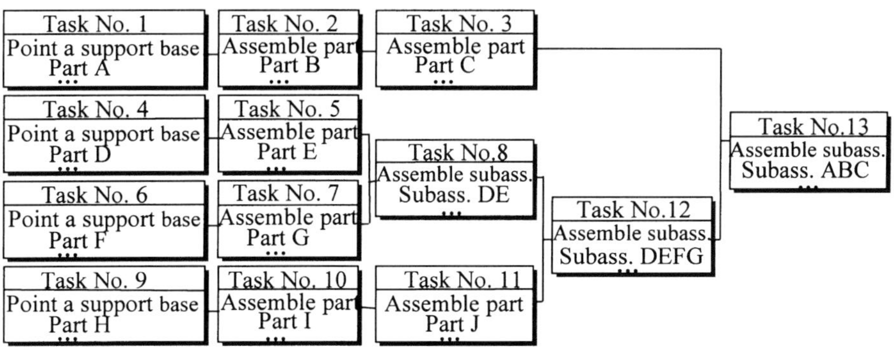

Fig.8 Assembly task description and assembly task model

4 APPLICATION

The system has been applied in industrial product successfully, and a case study for a pump assembly in aeronautic engine is used here to demonstrate the assembly planning and training functionality of the I-VAPTS system. The pump is a key subassembly in engine including many parts and components. The CAD model of the pump in SolidWorks is shown in Fig.9. After loaded the CAD models into virtual environment (Fig.10), the planner can interactively plan assembly sequence and path for the product, visualize assembly operation process, evaluate and select the optimal operation method, then an optimal assembly scheme and assembly documents are provided to assembly workers (Fig.11-12). An application structure of the system is given as Fig.13. An assembly process multimedia simulation movie file (including text, voice and graphics) is played to display the assembly operation process in workshop (Fig.14). The workers watch the assembly simulation movie for help on one hand, on the other hand to carry out the actual product assembly. The assembly performance and assembly efficiency can be improved evidently than traditional assembly that the workers can only reference the 2-D drawings and assembly handbooks.

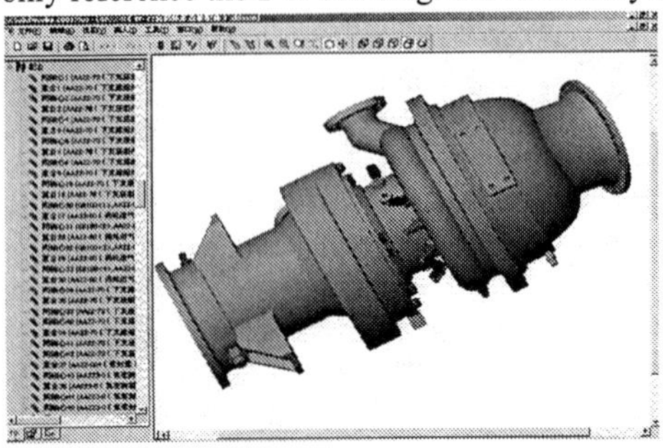

Fig.9 CAD model of pump in SolidWorks

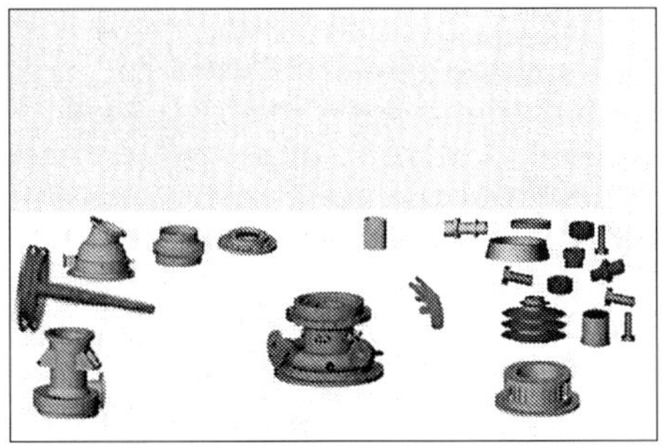

Fig.10 CAD models loaded in VR

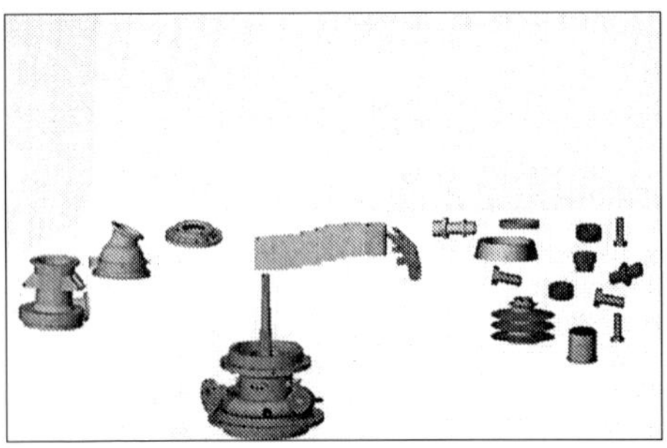

Fig.11 Interactive assembly planning in VR

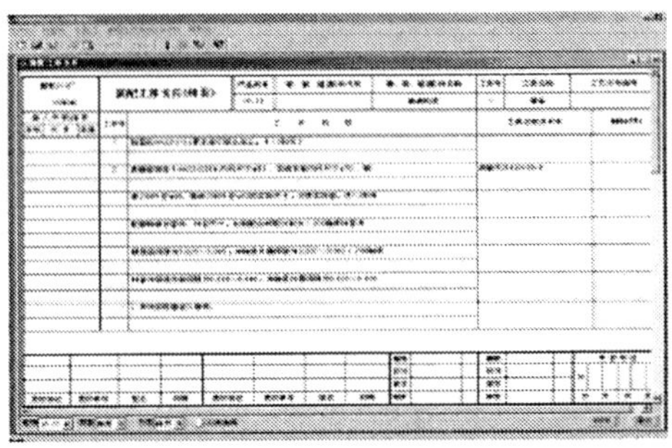

Fig.12 Assembly reference documents

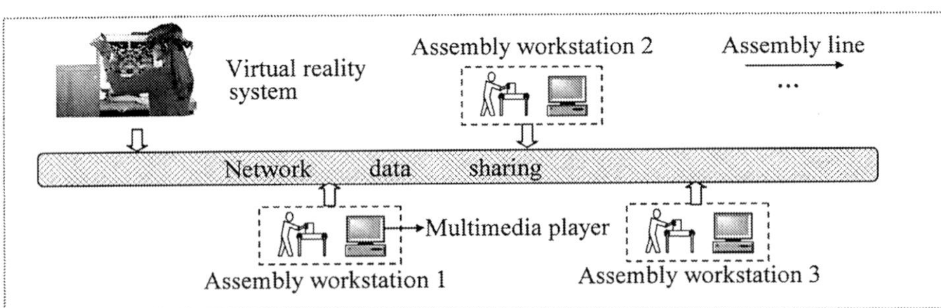

Fig.13 An application structure of I-VAPTS in assembly floor

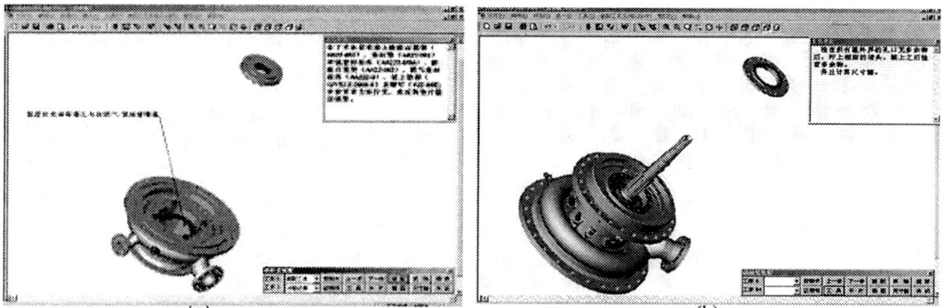

(a) (b)

Fig.14 Multimedia file simulating assembly process

5 CONCLUSION

Assembly process planning are both knowledge and experience intensive. Automated or semi-automated CAAPP systems do not appear to have made a significant impact within manufacturing industry. Immersive virtual environment developed in this research provides an alternative model for the generation of optimal assembly planning scheme. On one hand it can be used to train assembly staff, new staff in particular, to undergo rigorous training so as to improve their knowledge and skills for assembly tasks. On the other hand, it can be used to guide the workers to undertake practical assembly production in the workshop. The application in a pump assembly process indicates that the system is valuable and pragmatic.

REFERENCES

1 Bullinger, H. J., Richter, M., Seidel, K. A., "Virtual assembly planning. Human Factors and Ergonomics", in *Manufacturing* 10, 3, Summer 2000 John Wiley & Sons Inc. pp: 331-341.

2 Sankar Jayaram, et al., "VADE: A Virtual Assembly Design Environment", *Computer Graphics and Applications*, 1999, 19(6): 44-50.

3 Venkat N. Rajan, Kadiresan Sivasubrananian, Jeffrey E. Fernandez, "Accessibility and ergonomic analysis of assembly product and jig designs", *International Journal of Industrial Ergonomics,* 23, 1999, pp 473-487.

4 Chryssolouris G., Mavrikios D, Fragos D. et al., "A virtual reality-based experimentation environment for the verification of human-related factors in assembly processes", *Robotics and computer -integrated manufacturing,* 2000,16(4) pp 267-276.

5 Bound A C, Haniff D J, Baber C, Steiner S J., "Virtual reality and augmented reality as a training tool for assembly tasks", *Proceedings of IEEE international conference on information visualization*, London, 1999, pp 32-36.

6 Luis Marcelino, Norman Murray, Terrence Fernando, "A constraint manager to support virtual maintainability", *Computers & Graphics*, 2003, (27) pp 19-26.

Virtual and Collaborative Engineering

Fragmentary collaboration in a virtual world – the educational possibilities of multi-user, three-dimensional worlds

R ENGLAND and **I GIBSON**
School of Technology, Leeds Metropolitan University, UK

ABSTRACT

This paper discusses research into online, multi-user, interactive virtual worlds for use in education. A 3D virtual gallery with a shared whiteboard has been developed using several multimedia software packages. The results show that it is possible to create effective, intuitive virtual environments in which users may interact, communicate and collaborate with projects. Technical, social and pedagogical issues are discussed. Further work will be directed towards exploring applications for this environment.

1 INTRODUCTION

The three-dimensional virtual world is an emerging medium that offers the communication opportunities of a text-based environment with a visual representation of space for users to interact within. They also pose a number of questions. How technically feasible is it to create a multi-user 3D space? How is the user constructed in that space? Do these worlds allow for new methods of communication, learning and collaborative working?

As there is little existing research specifically in the area of multi-user 3D worlds, a large part of the research described in this paper is derived from data collected through observations and interviews. The research explores the design features that serve to represent the user in the world. Popular virtual world software packages have been used, namely Active Worlds and Adobe Atmosphere. The project's objective is, in the first instance, to develop a three-dimensional, multi-user collaborative world.

2 BACKGROUND

Shared 3D worlds are a relatively recent development. Therefore it is hard to find specific and readily available research into their educational use [1]. However, much can be gleaned from research based on two-dimensional text based environments [2], such as UseNET and other applications featuring visual chat interfaces with 2D avatars (The Habbo Hotel, for example).

Research into the roots and social issues associated with virtual worlds and communities is comprehensive. Issues arising include fragmentation [3], the process of creativity and 'collective' collaboration [4], reality and simulation [5] and the construction of the 'self' [6]. These subjects are referred to in this paper but are not discussed at length.

3 TOOLS USED IN THE CREATION OF A VIRTUAL COMMUNITY

The list of tools which were used to create and supply the virtual community is extensive. It includes software for a robust client-server architecture (Flash Communication Server), dynamic generation and streaming of 3D object-models, latency handling and video/audio streaming.

Macromedia Flash (available to a large number of users) provides dynamic content and an object oriented programming language. It was used to create the whiteboard as an 'intelligent' texture. 3D Studio Max and Bryce were used for 3D model and texture generation, the former used specifically to generate 'cubic' background textures and landscapes. Director/Shockwave was incorporated into the system since it is currently the only web-3D engine supporting dynamic textures via Flash.

4 AIMS AND OBJECTIVES

As this was an experiment to develop shared virtual spaces in an educational environment, the first step was to create a shared space. The objectives were:

1) To investigate user interaction using a virtual whiteboard.
2) To develop methods of communication either via text-based chat or via audio-visual communications.
3) To investigate ways of creating *immersion* in a three-dimensional space.
4) To identify innovative communication methods.
5) To consider the demand for virtual worlds in education.
6) To investigate the learning potential of a virtual world.
7) To look at the costs involved in the creation of a virtual space.
8) To investigate the technical and social limitations of a virtual 3D space.
9) To explore the potential of real world models in a virtual world.
10) To explore issues relating to communication bandwidth limitations.

5 SYSTEM DEVELOPMENT

A series of experiments was produced. Each experiment was designed to incorporate the ideas of social interaction and a sense of community via shared items and experiences. Multi-user environments were investigated utilizing object oriented programming languages to create shared objects and stream data. In addition, fast and efficient ways of generating 3D models in Lingo were developed.

Shared objects and data streams were stored and processed, server-side, by Flash Communication Server. Client applications connected to the server application via a network connection object. Any change to a shared object, either client or server side, initiated a synchronisation command to all connected clients. This provided a means of real-time collaboration and communication between many users.

A webcam video-stream rendered as a 3D texture provided a visual construction of the 'self' and an audio contact. The video texture concept was expanded using scripted (Flash) applications as textures (a whiteboard), thereby creating an interactive texture. The prototype system is shown in Figure 1.

Figure 1 Screenshot of the prototype system

3D visualisation was achieved through modelling techniques in 3D Studio Max and Bryce (for a QTVR style cubic skybox). In addition 'burnt' textures were generated, producing an atmospheric environment.

6 PROBLEMS AND RESTRICTIONS

Bandwidth capacity was a limiting factor of this system, and was eased by the streaming features of Flash. Other client-side hardware restrictions were overcome by creating efficient Lingo generated 3D images (which required less loading and processing time) and the use of a 2D portal. Figure 2 illustrates the connection of 2D and 3D portals through the server architecture. Maintaining 'intelligent' textures was also found to be CPU intensive. However, a synchronisation command was used to resolve performance issues by updating images only when necessary.

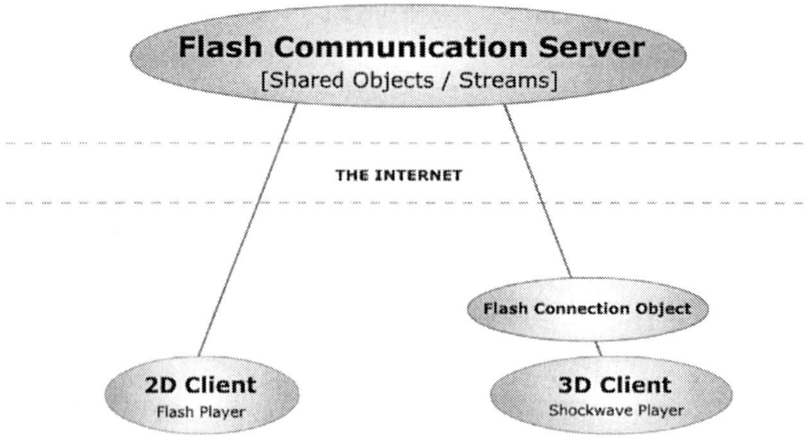

Figure 2 The system architecture permits communication between any combination of 3D and 2D portals

File size issues also needed to be considered. By streaming images after the initial download, avoiding over-complex models and utilising Lingo-generated 3D images, initial file download times were reduced. On the server side, multiprocessors acted as a solution to scalability concerns [7].

Educational institutions are subject to bureaucracy in the form of network and installation limitations. Common plug-ins (e.g. Flash and Shockwave) and HTTP 'tunnelling' features are a partial solution to these limitations because they are widely accessible to users.

Applications (such as 'Atmosphere') allow user-specific virtual worlds to be built quickly and cost effectively by using existing models as templates. It is important that such a feature is made available in any competing product. The experimental shared objects provide a potential for many users to develop a 3D world collaboratively, thereby reducing development time. However, the method of creation needs to be addressed.

7 ANALYSIS OF THE SYSTEM

The system incorporated the following features: a merging of 2D interactivity with 3D engagement, collision detection, first/third-person viewpoints and real-world models. It must be noted, however, that most 'real-world' settings did not offer enough potential for interaction or crowded areas. This is offset by the fact that many objects on-screen were found to reduce the frame rate.

Methods of control were based upon the well-known features of QTVR, such as a changeable focal length, 'hotspots' and mouse control. A graphical two-dimensional world was proven to work (albeit with limited immersion) as an alternative route to the same community features.

Real-time communication, produced via a microphone and web-cam, provided human interaction in 3D. A textual chat facility was also incorporated. Such real-time verbal and non-verbal communication allows for a greater degree of emotional engagement and *immersion*, which is not the result of sensorial input, but the result of humans interacting with one another in an environment [8].

Learning is viewed as a social activity. To that end the system provided communication and collaboration with an identity. Engagement, both among participants and with environments, is important for any educational medium.

In essence the system is an assimilation of QTVR engagement with 3D benefits. A content rich environment offers interactivity with other media such as web sites, while multi-user shared objects provide a constructive approach to interaction. As an instrument of public relations, a common problem of document-based websites is the very lack of such interaction. Virtual worlds can offer potential students information, combined with fun, chat and exploration. They can be a promotional tool combating the elitist metaphors usually associated with higher education. Additionally, a generation of students have been brought up with games (such as Doom). It is this generation that is most likely to relate to software environments with the metaphor of a place as opposed to those with a more conventional interface.

8 CONCLUSIONS

The following conclusions were reached:

1) Shared objects provide potential for online collaboration in a virtual space.
2) Audio, visual and text-based communications are all viable possibilities with this medium.
3) Worlds should be constantly changing and changeable, i.e. interaction is the key to immersion.
4) Real time video communication does not currently exist in the market for multi-user 3D worlds. Therefore this is an innovative feature of this system.
5) A generation of students brought up on games and environments that stress navigation through 3D spaces demands a virtual world that is built around the metaphor of a place.
6) In an educational establishment, a virtual world could be used as a way of promoting facilities.
7) It is possible to produce virtual worlds cost effectively.
8) Overbuilding creates a slow frame rate while large spaces offer less stimulation. A rich variety of objects close at hand (in a smaller world) are preferable.
9) Very few real world models are worth emulating in a virtual world, the most suitable being places designed for dense crowding and interaction.
10) Communication in a virtual world does not require a high bandwidth.

9 FUTURE RESEARCH

Issues to be investigated further include user authentication, ‘crowd control’, positional sound, management of objects/users via databases, and the dynamic building of user-made worlds. Importantly, further applications of 3D virtual worlds need to be explored.

REFERENCES

1 Dickey, M. (1999) "3D Virtual Worlds and Learning", [Internet] Available : http://michele.netlogix.net/side3/dissertation.htm

2 Bruckman, A. (1998) "Community Support for Constructionist Learning". *Computer Supported Cooperative Work*, pp. 47 - 48.

3 Jameson, F. (1991), *'Postmodernism'*, Durham.

4 Alison, F., et al. (1997) "*Process: A Tomato Project*", Thames & Hudson

5 Baudrillard, J. (1994) "*Simulacra and Simulation*", Ann Arbor (Ed.): The University of Michigan Press

6 Reid, E. (1991) "Electropolis: communication and community on Internet Relay Chat.", [Internet]. Intertek. Available from: http://citeseer.nj.nec.com/reid91electropolis.html

7 Manapragada, S. (2003) "Developing Scalable Macromedia Flash Communication Server MX applications", [Internet]. Macromedia. Available from :http://www.macromedia.com/devnet/mx/flashcom/articles/scalable.html

8 Heim, M. (1998) "*Virtual Realism*", Oxford University Press: New York.

A web-based system for collaborative electro-acoustic composition

I GIBSON
School of Technology, Leeds Metropolitan University, Leeds, UK
M DOVEY
e-Science Centre, University of Oxford, UK

ABSTRACT

This paper describes development of an online system for collaborative electro-acoustic music composition. Traditionally, software systems for music composition have been single user systems. As accessibility to the Internet has increased, some systems have been adapted so that musicians can produce collaborative work. However, many such systems are still based around the composition methods of the original single user systems. Various Service Oriented Architectures are emerging based on WebServices or GridServices and other Internet technologies for dynamically building systems out of distributed components. Many e-Science projects are using these architectures to build collaborative environments. The software described in this paper takes such concepts and tools from Service Oriented Architectures and e-Science, and applies them to develop software specifically for collaborative electro-acoustic composition on the Internet, allowing compositional tools and components to be published, discovered and used within a distributed environment. An objective is to explore and determine methods of composition appropriate for this environment.

1 BACKGROUND

In many cases, electronic sound synthesis techniques can produce a greater range of timbres than any one acoustic instrument. The composer-performer has access to a wide range of software and hardware tools for score writing, sound synthesis, sound editing and music arrangement.

Real-time control of synthesized sound is often achieved using a device compatible with the Musical Instrument Digital Interface (MIDI) protocol [1]. However MIDI devices (such as keyboards) tend to have limited degrees of freedom [2]. For this reason, many electro-acoustic composers choose to compose in non real-time using Music Computer Languages such as CSound [3] which allow the user to attain a higher resolution of control over parameters but at the expense of being more computationally intensive. Traditionally, these languages have taken the form of text-based interfaces whereby the computer is used to process a script. CSound stores 'instrument definitions' in an orchestra file and instructions for playing instruments in a score file.

2 SOUND SYNTHESIS TECHNIQUES

This section describes common techniques used by the composer-performer for synthesizing sounds.

2.1 Additive Synthesis

Additive or *Fourier synthesis* operates in the frequency domain. A waveform may be represented as a series of sinusoidal waveforms (*partials*). Fourier analysis may be used to analyse an audio signal for re-synthesis. Small sections of the original audio signal (*windows*) are taken and analysed in this way producing sets of partials with their amplitudes and phase. The size of the window must be larger than the longest pitch period to be analysed. To increase accuracy of the analysis stage, some fourier analysis programs allow interpolation between windows.

Re-synthesis becomes possible by generating sinusoidal oscillators which follow the amplitude envelopes generated in the analysis stage. When re-synthesising acoustic instruments in this way it is necessary to perform analysis over a range of frequency values. A note played on a lower register will typically have different partial envelopes to a note played on a higher register.

Fourier synthesis can also be used to generate new sounds and musical effects. A *morph* between two sounds may be achieved using interpolation between their respective envelopes, many examples of which can be found in the piece "Vox 5" by Trevor Wishart. Further interpolation may be applied within the time domain to resolve differences in attack or decay values between the two sounds. This method is also useful for creating a sound which has timbral qualities of both sounds.

The Digital Phase Vocoder [4] is an extension of fourier transform and has been used to control a synthesiser from a traditional instrument [5]. It is used to map spectral characteristics of the acoustic instrument directly onto a synthesized sound using additive synthesis.

2.2 Subtractive Synthesis

Subtractive synthesis is based around the use of one or more of the following filters on a source sound: low-pass, high-pass, band-stop, and band-pass. These filter out parts of the sound spectrum, shaping the amplitude and phase of each spectral component. A low-pass filter permits frequencies below a cutoff frequency (fc) to pass. Above the frequency fc, the spectral components are reduced significantly in amplitude. The cutoff frequency is defined at that point at which the power transmitted drops to one half (-3 dB) of the maximum power transmitted.

A high-pass filter allows only signals above the cutoff frequency to pass unaffected. A band-pass filter rejects frequencies either side of two cutoff frequencies. These are specified by a centre frequency and a bandwidth. The band-stop filter is the inverse of a band-pass, attenuating frequencies within the specified bandwidth.

Because no new frequencies are introduced subtractive synthesis is particularly effective with complex sound sources. Sawtooth and triangle waves produce sounds rich in harmonic spectra but due to their periodic nature they can sound predictable. Noise and pulse waves are also used due to their spectral richness. The pulse waveform has an amplitude which lasts just a brief period of time (determined by the pulse width). A narrow pulse produces a large amount of spectral energy in its higher frequencies relative to its low frequency components.

2.3 Frequency Modulation

Frequency modulation (FM) occurs when the output of one oscillator (the *modulator*) is used to modulate the frequency of another (the *carrier*). The output spectrum resulting from frequency modulation contains sidebands around the centre frequency. The faster the rate of modulation the more power is found over the sidebands. Time varying modulation produces complex and varying spectra. FM is ideal for bell-like sounds. Small changes in modulation frequency and amplitude can have a dramatic effect on the timbre of the resulting sound.

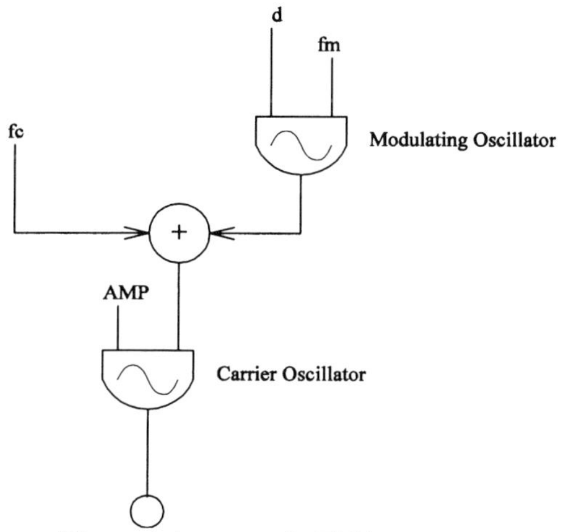

Figure 1 An example FM instrument

An example FM instrument is shown in figure 1. This instrument has one carrier oscillator and one modulating oscillator. f_c is the output of the carrier if the modulating oscillator outputs zero. When modulation occurs, the frequency f_c is modulated above and below the modulating oscillator output at a rate *d. AMP* is the amplitude of the output signal of the carrier oscillator. The maximum frequency produced is $f_c + d$, and the minimum is $f_c - d$. For example, with a value f_c of 400, d of 50, f_m of 100, and AMP=1000, then the carrier oscillator will be modulated between the values of 350 and 450 at a rate of 100 Hz, with a carrier oscillator output amplitude of 1000.

2.4 Amplitude And Ring Modulation

Amplitude modulation (AM) occurs when a sound's amplitude is modulated over time between 0 and a maximum value. The rate of modulation is determined by the frequency of the modulating oscillator. Ring modulation (RM) is similar to AM except amplitude (as well as the rate of oscillation) is determined directly by the modulating oscillator.

2.5 Waveshaping Synthesis

Waveshaping synthesis occurs when a waveform is taken and transformed. Usually a *transfer function* is used to produce an output waveform based on the input waveform. Like FM synthesis, the outcome of subtle changes in parameters can not be often predicted especially if the waveform is altered by a non-linear processor.

2.6 Granular Synthesis

Granular synthesis combines frequency and time domain synthesis techniques. Several grains of sound are produced each with fixed frequencies and amplitudes. Many hundreds may be output in a single second producing a rich and varying texture. This texture can be difficult to control due to the large number of events involved, however granular synthesis can produce a variety of sounds including those resembling shattering of glass or torrents of water.

3 THE SOFTWARE ENVIRONMENT

Work has begun on the implementation of a distributed environment to allow the sharing synthesis techniques and eventually to allow real-time collaborative composition. The current prototype system uses SOAP based Web Services to allow access to a database of audio samples, audio processors and compositions.

3.1 Service Oriented Architectures

The current methodology in developing distributed systems is Service Oriented Architecture (SOA), building upon methodologies such as Object Oriented programming, Components and Distributed Object Request Brokers. Within a SOA, systems are composed of multiple individual services located and maintained on different heterogeneous machines administered by different organizations. The key in SOA is that the component services should be loosely coupled i.e. be well-defined, self-contained, and should not depend on the context or state of other services [6]. To achieve this, a SOA should display the following properties:

- The services should implement a small set of simple, ubiquitous and well known interfaces which only encode generic semantics.
- The interfaces should deliver messages constrained by extensible schema for efficiency. This allows both services and consumers to work with well defined message structures, but allowing new versions of the services to be introduced without breaking existing systems
- The messages should be descriptive not instructive and the interfaces should not define system behaviour. This allows internals of a service can be viewed as a "black box".
- Service Oriented Architectures must have mechanisms for the discovery of services matching the consumers requirements

There are a number of emergent technologies which can underpin SOA, namely REST WebServices; SOAP WebServices and GRID Services:

- Representational State Transfer (REST) works on the basic of "resources" which can be references by URIs [7]. A REST web service is limited to using HTTP interfaces (GET to obtain a representation of the resource; DELETE to remove a representation of a resource; POST to update or create a representation of a resource; PUT to create a representation of a resource). REST messages are in XML, constrained by schema definitions in the XML Schema language (http://www.w3.org/XML/Schema) or Relax NG (http://www.oasis-open.org/committees/tc_home.php?w-_abbrev=relax-ng)

- SOAP Web Services use messages encapsulated in a structure defined by the SOAP specification (http://www.w3.org/2000/xp/Group/). This adds additional information in the form of headers for message routing scenarios and mechanisms for reporting errors using faults (a style similar to exceptions in various programming languages). SOAP Web Services use Web Service Description Language (WSDL) to define both the structures (again using schema languages such as XML Schema) but also messaging semantics such as whether the message is initiated by the client or the server, and what messages can be used as a response to a particular message.
- GRIDServices are based on WebServices but provide additional semantics. In particular they add some object-oriented and REST concepts. The object-oriented concepts are the ability to inherit service definitions (portTypes in WSDL terminology) and add new messages using a multiple inheritance model and the ability to add properties (or service data elements) to WebServices. The REST concept introduced is that of creating a new representation of resource. In the GRIDServices model this uses a factory model whereby a new instance of a GRIDService can be created by its corresponding factory GRIDService. GRIDServices also offer an extensibility model whereby part of the structure of the message can be left undefined, but the allowed structures can be determined dynamically by querying the appropriate service data elements.

The current prototype is based upon SOAP based WebServices, although future versions may need to take advantage of GRIDService based technology for some of the more advanced features such as remote execution of compositions.

3.2 Web Service Definitions

The Web Service definitions define how the client communicates with the server using XML based messages. The prototype definitions are defined within WSDL which would allow clients and servers to be implemented on different platforms using different programming languages. The definition defines three different "PortTypes" (which is the WSDL term for a collection of functions). The division into different functional groups allows for a system in which different servers implement different groups, although our current prototype server includes implementations of all three. The defined PortTypes are:

3.2.1 Processor Service PortType

This defined two functions: **processAudio** and **getProcessorsDescriptions**. The **processAudio** function allows the client to send audio data to be processed at the server. It takes as its parameters an identifier, a collection audio sample and a collection of parameters and returns a single audio sample. The identifier determines what code will be used to process the audio using the various mechanisms described above. The collection of audio samples consists of MIME based 64 encoded binary of the actual wave data plus a name to identify that particular input to the code – typical names might be "input1", "input2" etc. The collection of parameters also consists of a list of name, value pairs and allows fine control over the functioning of the server side sound synthesis code. The **getProcessorsDecriptions** returns a list of all the sound synthesis techniques available on the server. For each technique, it also returns the list of names for the input audio, the list of parameters including descriptions and type (e.g. integer, real or Boolean), and also the provenance of the sound synthesis technique using Dublin Core[] metadata such as title, creator and description.

3.2.2 Audio Service PortType

This defines functions for managing a server based database of audio samples. It implements the functions **getAudio** and **submitAudio** for adding new samples and retrieving existing samples. It also has the function **getAudioDescriptions** which returns a list of available samples and their provenance using Dublin Core metadata (title, description, creator, date etc.)

3.2.3 Composition Service PortType

This defines functions for managing a server based database of compositions. A composition is represented as an XML document detailing the identifiers of the audio samples and processors used (i.e. the identifiers to use as a parameter for the **getAudio** and **processAudio** functions respectively) and the workflow i.e. how the audio and synthesis are composed to form the final audio. It implements the functions **getComposition** and **submitComposition** for adding new compositions and retrieving existing compositions. It also has the function **getCompositionsDescriptions** which returns a list of available compositions and their provenance using Dublin Core metadata (title, description, creator, date etc.)

3.3 Server Implementation

A prototype server has been developed in Java implementing all three port types, i.e. performing the roles of a audio repository, composition repository and a repository of various audio synthesis processors. This has been developed using OpenSource components such as Apache Tomcat (http://jakarta.apache.org) to provide the base HTTP functionality and the Apache Axis (http://ws.apache.org) to provide SOAP functionality.

The Audio and Composition repositories have currently been implemented a simple file stores with additional metadata store in an XML index file. The processor web service dynamically loads new processors by compiling and loading java classes stored in the processor directory. A web interface has been developed to allow remote uploading of new processors.

Helper java classes have been written to hide the Web Service complexity. To write a new synthesis processor, you have to implement a java class which has the method **processAudio**. The functions **getAudioOutput ()** is available to return a ByteArrayOutputStream to which to write the audio output, and functions **getAudioInput(String name)** and **getGlobalParameter(String name)** are provided to access the audio input and parameters. Use is made of the the new Java 1.5 metadata feature to add descriptive metadata to the source code about the creator and title of the processor, the names of the audio inputs and the names and types of the parameters which are returned by the **getProcesssorsDescriptions** WebService function. A simple processor which just adds samples together is show below (the lines beginning @ are the descriptive metadata using the new Java 1.5 features):

```
public class SampleAudioProcessor extends AbstractAudioProcessor {

  @Provenance(creator="Dovey, Matthew", title="Simple Processor")
  @AudioStreams({@AudioStreamDescription(name="in1", description="first"),
                 @AudioStreamDescription(name="in2", description="second")})
  @Parameters({@ParameterDescription(name="parm1", description="", type="integer")})

  public void processAudio() {
    byte array1[] = new byte[in1.available()];
    byte array2[] = new byte[in2.available()];

    this.getAudioInput("in1").read(array1);
    this.getAudioInput("in2").read(array2);

    for (int i = 0; i < array1.length && i < array2.length; i++) {
      this.getAudioOutput().write(array1[i] + array2[i] + intParameter(getGlobalParameter("parm1")));
    }
  }
}
```

A new processor is added to the server by dropping the source code in the processor directory with the extension .jps (either directly or via a web form). The file is automatically and dynamically detected and compiled by the server and the processor is then available for use.

3.4 Client Implementation

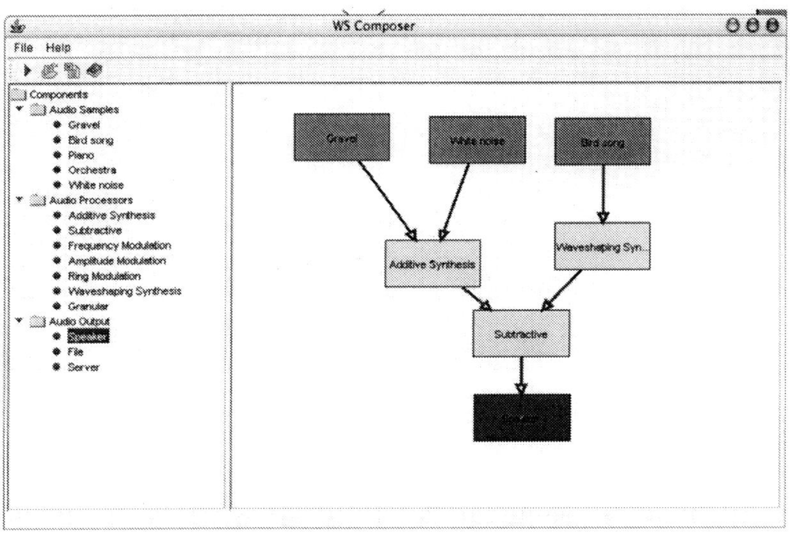

Figure 2 The prototype client user interface

A prototype client has been developed using Java Swing, and example of the user interface is show in figure 2. The client establishes a link to the server and uses the **getAudioDescriptions** and **getProcessorsDescriptions** WebServices to enumerate available resources. This are listed in the Java JTree control on the left hand side. Audio samples and audio synthesis processors can be dragged and dropped onto the flowchart control on the right hand side which is implemented using the open source JGraph Swing control[1]. The various components can be then linked up to form a composition. Right-clicking on a processor brings up a list of the parameters that can be changed for that processor. The file menu allows the saving and loading of compositions either to the server (via **submitComposition** and **getComposition** WebServices) or to the local machine. The output of the composition can be sent to the speakers of the local machine, a local wave file or to the server (via the **submitAudio** WebService).

The current system is in an early prototype and still needs testing and feedback by users. However there are a number of areas which have already been identified as requiring further development

4 FUTURE RESEARCH

4.1 Searching

At present, a client must request the data for all the resources (audio, processors or composition) on the server in order to discover what is available. This is clearly not scalable as the number of resources grows. For this it would be better for the server to support a search interface so that audio processors, audio clips or compositions can be found matching a user's query. The use of the generic SRW WebService protocol[2] will be investigated for this purpose.

4.2 Multiple Servers

The architecture should be expandable to support multiple servers, so that the client might use processors, audio, etc. from numerous servers. For this to work there needs to be a naming convention so that a resource (audio sample or synthesis processor) and the server on which it is located can be determined from an identifier stored in a composition. Various systems such as WS-Addressing will be investigated for this purpose. There are also various architectures as to how a client discovers the various servers and there resources. In one case a client might send a get...Descriptions web service (or a search) to multiple servers simultaneously. An alternative might be for the servers to replicate metadata descriptions between themselves so that a client need only send a get...Descriptions or a search to a single server which will then response with information of the resource of all servers. It is likely that a hybrid approach will be implemented for greater flexibility.

4.3 Server execution of compositions

The current prototype is inefficient in the use of network traffic. In a typical composition, the client will retrieve a number of audio samples from the server using the **getAudio** WebService, then send those audio samples back over the network to the server for a synthesis

---

[1] http://www.jgraph.com

[2] http://www.loc.gov/z3950/agency/zing/srw/

technique to be applied via the **processAudio** WebService. The resultant audio will be transferred back to the client. As this may be an input to another audio processor the resultant audio may be passed back over the network to the server, and so on. A more efficient solution would be for the composition to be sent to the server and executed on the server. With multiple servers it will be necessary to calculate the most efficient workflow. GRID technologies will be investigated to provide this functionality.

4.4 Peer to Peer technologies

The prototype architecture is a client-server approach. A more flexible approach might be to use a peer to peer approach whereby a user can publish audio samples and synthesis processors from their local machine. Combining the client with the server would enable this, once a multiple server architecture is in place, although this may have some security implications to be investigated.

4.5 Instant Messaging Technologies

The system described is meant to encourage collaborative composition. This can only be achieved if the collaborative composers not only have access to shared resources but also to real time communication tools such as text based chatting, whiteboards, video/audio conferencing etc. Various technologies will be investigated for integration into the client.

REFERENCES

1. IMA (1988) "MIDI 1.0 Detailed Specification, version 4.0", International MIDI Association.
2. Gibson, I.S. (1997) "Voice Analysis for Music Synthesis Systems", PhD Thesis (York, UK), 2.
3. Vercoe, B.L. (1986) *"The CSound Manual",* Cambridge Massachussets: Experimental Music Studio, Media Laboratory, MIT, Cambridge.
4. Flanagan, J.L. & Golden, R.M. (1966), "The Phase Vocoder", *Bell System Technical Journal*, 45, 1493-1509.
5. Bailey, N.J., Purvis, A., Bowler, I.W., Manning, P.D. (1993) "Applications of the Phase Vocoder in the Control of Real-time Electronic Musical Instruments", *Interface*, 22, 259 - 273.
6. "What is Service-Oriented Architecture", http://webservices.xml.com/pub/a/ws/2003/09/30/soa.html
7. Fielding, R. T. (2000) "Architectural Styles and the Design of Network-based Software Architectures", Ph.D. Dissertation (University of California, Irvine), Chapter 5.

Co-operative design based on multi-agent system

C HUA
Shaanxi University of Science and Technology and CAD/CAM Institute of Xi'an Jiaotong University, Shaanxi, China
L CHAO
Shaanxi University of Science and Technology, China
C YAN and **Z RUJIA**
CAD/CAM Institute of Xi'an Jiaotong University, Shaanxi, China
Q XU
School of Computing and Engineering, University of Huddersfield, UK

ABSTRACT

Competition demands better quality, faster delivery, and cost effective. Cooperative design based on multi-agent system is one of the recent technology developments benefited from the latest development of information technology and its application in conventional design process. In this paper, a review of the process of product design and its modelling and current state of art of cooperative design is given first; the concept of agent oriented system is deployed and a hierarchical architecture based on internet is proposed where the role of design manager as well as agents are clearly outlined; further consideration for system development is given to how to interactive with collaborators and customers. This approach relies on the experts to cooperate the works to be conducted by agents and is relative practical. The proposed system is to be implemented and tested. Future work may incorporate artificial intelligence to help experts in decision-making.

KEYWORDS: Multi-Agent System, Cooperative Design, System Development

1 COOPERATIVE DESIGN

With the development of global Internet and information technology, Computer Supported Cooperative Work (CSCW) is gaining more and more attention [1-2]. The trend is that product-developing teams are composed of people of different specialties. It is the key to coordinate the activities of the personnel involved [3-4] to harmonize the whole developing process.

The objective of cooperative design is to integrate the activities involved in the product design process by taking every aspect of product lifecycle into account at early stages of product development. Thus, unnecessary iteration is avoided. Traditional CAD/CAM systems are devoid of communication functions and cooperation mechanism that leads to ill concurrence and cooperation. There are two ways to solve the problem. One is to modify the existing software to support multiple analyses using common data structure. The emphasis is put on perfecting models to facilitate design tasks and eliminate unnecessary repetitious work. The functions and architecture of traditional CAD/CAM software need to be extended and consummated so that it can be applied to not only geometric and structural design, but also process design and engineering analysis. Thus, CAD/CAM systems are becoming more and more compositive and complex that calls for higher and higher skills to run them. The other is to exchange information between different models and data structures by using extended API, CORBA, DCOM, ActiveX, STEP, etc.

Because distributed cooperative design is carried out by teams located at different places together, the regional limit must be overcome to facilitate information exchange, knowledge processing, design result exchange, etc., among the teams. Multi-agent system can solve the communication problem in the distributed systems effectively. The agent is an abstract functional entity. Personnel, CAD tools, and CAM tools can all be regarded as agents. They have communication interface that distinguish them from general application programs, make them able to contact with other agents for information, and make them able to cooperate with others to achieve common objective. Every agent has certain attributes, and approaches to accomplish some functions. Their perceptive abilities are realized through message mechanism. In the multi-agent system, a message is generated when an event happens, and corresponding agents are activated to carry out some actions when the conditions of the message are met.

2 THE ARCHITECTURE OF COOPERATIVE DESIGN BASED ON MUTIL-AGENT SYSTEM

Multi-agent system is composed of many autonomous or semi-autonomous agents, and studies how the agents cooperate with each other to solve problems in the product developing teams.

2.1 Product Developing Process Management

Process management carries on top macroscopical control to cooperative product design and development. Based on information exchange and share, the key is to coordinate the activities of the agents when the multi-agent system is constructed. Along with the evolvement of modern design theories and Concurrent Engineering (CE), traditional serial design activities are replaced by concurrent ones to the greatest extent. In order to carry out the product development effectively, efficient and rapidly, product developing project has to be managed rationally. Tasks should be decomposed; organization structure should be adjusted aiming at the project in progress while various resources are reconstructed to make full use of limited resources.

2.2 Information Management

The exchange and share of product and process data in heterogeneous systems has been an active research field. To improve mutual operation that is the important characteristic of intelligent cooperative design, the information exchange at two levels should be realized. At product information level, exchange standards of product information, such as STEP, should be set down. Besides the exchange standards at information exchange level, a communication language at communication level should be defined that indicates where information is sent, how to sent information, and what the information receiver should do to receive the information. Thus, the agents can communicate with each other and process the information correctly through combining the two levels.

2.3 Conflict Resolution in Product Development

Though multi-agent system manifests great flexibility, conflicts can not be eliminated completely. The potential reasons to generate conflicts may be incorrect information or knowledge, different evaluation standards adopted by the agents, different reasoning mechanism, various objectives, and so on. Negotiation is an important approach to solve the conflicts in addition to approaches, such as restriction relaxation, arbitration, and so on.

2.4 The Architecture

The architecture is the important element in multi-agent system design that affects the performance of the communication pattern among the agents and workflow management system greatly. The familiar multi-agent system architectures are tree-like, netlike, and layered. The tree-like architecture is suitable for centralized managing organization where the hierarchical structure is formed according to layered task decomposition. The agents at lower levels accept commands and data from upper levels. The results computed are fed back to upper levels. In the architecture, the task structure is clear, and the control of information exchange among agents is comparatively easy. In the netlike architecture, the agents can communicate with each other directly. They are equals in the system and do not take orders from any superiors. Thus, the efficiency of information exchange is quickened greatly, and at the same time, the reliability of the system is improved because the dependence on upper levels is eliminated. The layered architecture has the advantages of both the other two where not only the control relations between lower and upper levels exist, but also the direct relation between interrelated agents can be established. Consequently, the flexibility of information communication is enhanced, which accords with product development practice, as shown in figure 1.

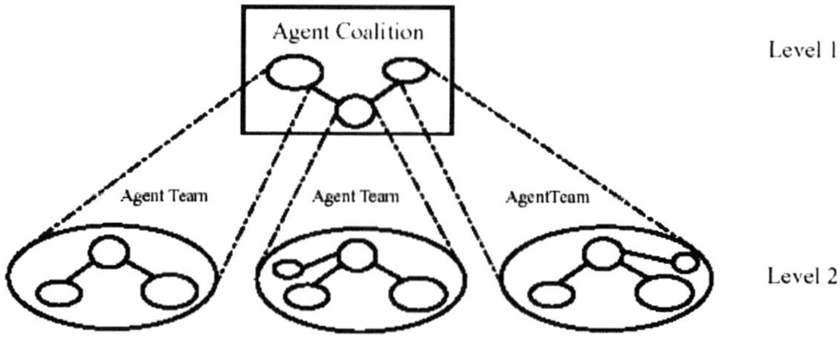

Figure1 Hierarchical Coalition

3 COOPERATIVE DESIGN BASED ON MULTI-AGENT SYSTEM ON INTERNET

The architecture of cooperative design based on multi-agent system on Internet is shown in figure 2. The core of the system is cooperative design manager that integrates many cooperative tools, such as electronic whiteboard, meeting manager, FTP, Email, etc. It coordinates the activities of the whole system through GUI interface. Its functions are mainly:

1) Manage the whole cooperative design process, and assign tasks to the agents.
2) Control the cooperation among the agents, and arbitrate the conflicts.
3) Communicate with other application systems.

The router in figure2 is a software router to accept and send messages that deals with agent registration, stores messages, etc. The application agents include mechanical CAD/CAM system agent, control system agent, engineering analysis agent, CAPP system agent, manufacturing agent, etc. Developing personnel can make use of familiar tools to work. They send requests to cooperative manager for cooperation and cooperate with registered agents. If there is no partner needed, cooperative design manager will release the demand to summon other agents to respond to the cooperation need. The communication language can encapsulate data and knowledge used by different applications. The architecture of cooperative design system can not only bring individual's knowledge and expertise into full play, but also make them work together effectively and efficiently.

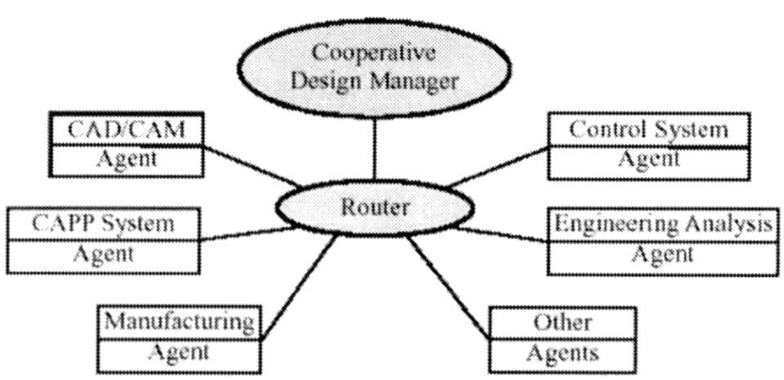

Figure 2 The architecture of cooperative design

4 SYSTEM DEVELOPMENT BASED ON WEB

Enterprises can not participate in the ever-drastic global competition effectively if they only utilize their own resources. Therefore, all resources from product developing partners, such as suppliers and distributors, have to be integrated. Considering users' participation in product development, a convenient and intuitionistic design approach is needed. Thereby, system development based on Web is put forward. At present, Web based developing tools, such as JAVA, JAVAScript, VBScript, etc, enhance the capability to manipulate database, transmit data, etc. For example, new functions of JAVA reduce the entry time of database, and what is transmitted is the data to be processed, not all the data on lists of each transaction processing. So the information transmission on network is also reduced. Three leveled client-server system can be used to achieve the data link between clients and servers that includes client browser, homepage application server, and database server. The client browser is used to visit user interface. The homepage application server is used to publicize the homepage to deal with users' requirements, and also carries out complex computation and sends SQL commands to database server. The database server carries on data operation and management. The structure to link homepage application server with database by using JDBC is shown in figure 3.

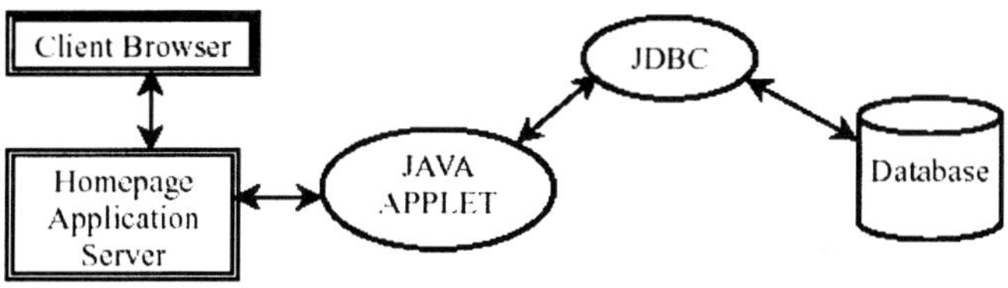

Figure 3 Link homepage application server with database by using JDBC

5 CONCLUSIONS

With the development of Internet technology and extensive, in-depth application of artificial intelligence and computer technology to engineering in 21st century, multi-agent technology provides an approach to the integration of cooperative activities of product development during all its lifecycle. In the paper, the architecture of cooperative design based on multi-agent system on Internet is proposed by analyzing cooperative design pattern, and its key technologies, such as product developing process management, information management, and conflict resolution, are discussed. Finally, system development mode based on Web is also put forward.

REFERENCES

1 Schrefl, M. and Kappelet, G., "Modeling collaborative behavior using cooperation contracts", *Data & Knowledge Engineering*, Vol. 26, pp 191-224, 1998

2 Bullinger, H-J. and Warschat, J., "Rapid product development – an overview", *Computers in Industry*, Vol.42, pp 99-108, 2000

3 Hideyuki Ando, "Study on the collaborative design process over the Internet - a case study on VRML2.0 specification design", *Design studies*, Vol.19, pp 289-308, 1998

4 Latch, D., Zimring, C., "Supporting collaborative design groups as design communities", *Design Studies*, Vol.21, pp 187-204, 2000

A co-operative software environment for planning and controlling ship-repair contracts

G CHRYSSOLOURIS, S MAKRIS, N PAPAKOSTAS, and **V XANTHAKIS**
Laboratory for Manufacturing Systems and Automation, University of Patras, Greece

ABSTRACT

The current trend of manufacturing companies is to maximize their communication and collaboration capacity by integrating their operations with those of their business partners. The Internet growth and the associated software technologies provide the means for the realization of this trend.

This paper presents the application of web-based software technologies for supporting a set of supply chain business processes, related to the maritime industry. In particular, the ship-repair business process is discussed, involving the transactions among the ship-owner, the shipyard, the material suppliers and the shipyard's subcontractors. The software implementation is based on the 3-tier paradigm, representing a cost effective, simple to configure and use, platform independent integrated environment, which allows small enterprises with moderate information technology infrastructure, to participate in the supply chain network.

The case study addressed, focuses on the business functions, taking place in the process of planning and controlling the preparation and execution of a ship-repair contract, in a number of critical ship-repair nodes. The web-based solution presented, demonstrates the way that members of the supply chain may work together, in the ship repair business processes, fostering the efficient communication and information sharing across the participating actors.

1 INTRODUCTION

A typical shiprepair job involves a large number of partners, with the shipyard being the main one. The activities required for the completion of a repair job, set the boundaries of a typical project shop (1), realised within the frame of a supply chain. The other participating partners, performing activities that are critical to the overall shiprepair operation, are comprised of:

- The material suppliers that supply material to the shipyard. These companies have stock available and bring it to the yard, whenever requested.
- The service suppliers that are called subcontractors, providing labour to the shipyard; these companies, offer specialised personnel, capable of carrying out part of the shiprepair activities in case the shipyard does not have such expertise or the capacity, required for the completion of a shiprepair contract.
- The ship owner is the customer that brings the ship for repair.

Additional partners, such as the classification society and the port authorities participate in quite rare incidents (e.g. accidents) or in cases, where their role does not affect the performance of the supply chain, since they do not contribute to the execution of the manufacturing activities. In this paper, these partners are not included in the analysis and modelling of the cooperative work.

The business process that takes place for the execution of a typical shiprepair contract will be analysed in the following sections. The business model is built, using the UML modelling approach, a standard way to model the activities that take place (2). The companies modelled as actors, communicate among themselves, exchange information related to the repair, play their own part and add their value to the entire chain, the Value Added Chain (3). This cooperation takes place through the exchange of ship repair related data, materializing the flow of information throughout the supply chain. This business process is complex for a number of reasons (4):

- The ship is on the sea and it is required to send the repair related data in a fast, easy and reliable way to a large number of actors.
- The repair work is not known from the beginning and a lot of repair items are identified during the inspection at the shipyard.
- A large number of actors are involved in the repair, performing manufacturing activities.
- The activities performed by the shipyard and by the suppliers are interrelated.

The shipyard is the main partner because it is directly involved in the work that is carried out by its own resources as well as in the work carried out by other partners, such as service and material suppliers. The shipyard is, therefore, aiming at coordinating production and levels in a way that work can be completed "in time", with minimal costs and inventory levels (1). Towards this direction, new business models are required and are currently emerging (3).

There is a trend for the implementation of computer-based environments that can support the process described above, thus allowing for automated communication among the actors. Value-chain integration uses Internet technology to improve communication and collaboration among all parties within a supply chain (3). The software environment has to be able to bring all related actors together. This paper discusses the application of web-based software technologies for supporting planning and monitoring activities of the ship repair supply chain. In particular, the ship repair business process is discussed, involving the transactions among the ship-owner, the shipyard, the material suppliers and the shipyard's subcontractors. The software implementation is based on the 3-tier paradigm, representing a cost effective, simple to configure and use, platform independent integrated environment, which allows small enterprises, with moderate information technology infrastructure, to participate in the supply chain network. This implementation is presented with the help of a "real world" application on a ship repair industrial application.

2 BUSINESS MODEL

The business process was analysed, with the help of a functional partitioning list. This list is used, to describe in detail, the functions performed by each partner. A sample of the functional partitioning list, is listed in table 1.

FUNCTION	ACTOR	FUNCTION Level 1	FUNCTION Level 2	FUNCTION Level 3
F2.1	Shipowner	Enquiry	Submit enquiry	
F2.2	Shipyard		Receive enquiry	
F2.3	Shipyard		Check data sufficiency	
F2.4	Shipyard		Request details	
F2.5	Shipowner		Submit details	
F3.1	Shipyard	Tender	Check need for sub-contract	
F3.1.1	Shipyard		Perform market survey	Identify sub-contractors
F3.1.2	Shipyard			Request for quotations
F3.1.3	Sub-contractor			Receive request for quotations
F3.1.4	Sub-contractor			Estimate production orders
F3.1.5	Sub-contractor			Estimate production plan
F3.1.6	Sub-contractor			Submit quotations
F3.1.7	Shipyard			Receive quotations
F3.2	Shipyard		Calculate material requirements	
F3.3.1	Shipyard		Perform market survey	Identify suppliers
F3.3.2	Shipyard			Request for quotations
F3.3.3	Supplier			Receive request for quotations
F3.3.4	Supplier			Check inventory
F3.3.5	Supplier			Estimate Stock Availability
F3.3.6	Supplier			Estimate Stock plan
F3.3.7	Supplier			Submit quotations
F3.3.8	Shipyard			Receive quotations
F3.4	Shipyard		Estimate cost/duration	
F3.5	Shipyard		Submit tender	
F3.6	Shipowner		Receive tender	
F3.7	Shipyard		Negotiation	
F3.8	Shipowner		Confirm/Decline	

Table 1. Functional partitioning list

There are a number of major business functions, considered as Level 1 functions. These functions are divided into sub-functions, the so called Level 2 functions, which in turn, are divided into Level 3 functions, capturing the level of detail that allows further modelling of the business process. This representation could be extended to include more levels of functions, depending on the desired level of detail. Table 1 is a part of the complete functional partitioning list, which models the whole shiprepair cooperation business process. Each function associated to a reference number and each function is performed by a specific partner (actor in UML).

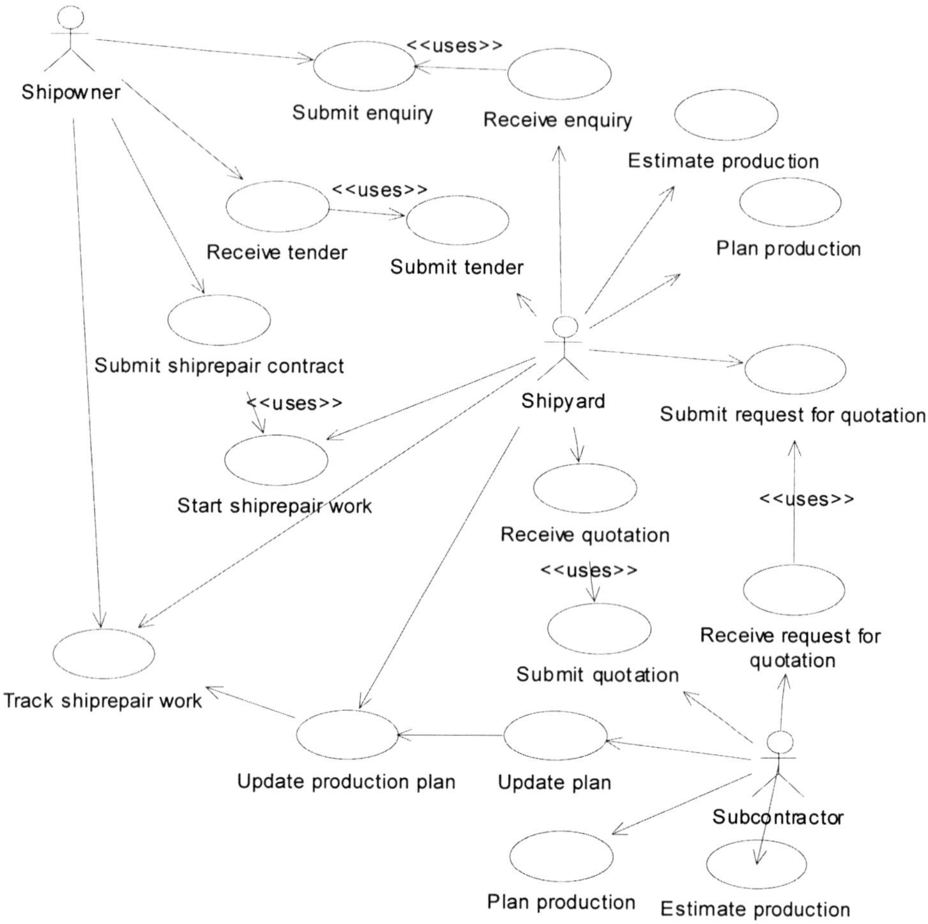

Fig. 1. Overall ship repair use case model

The ship owner initiates the business process, upon the submission of the expression of interest, called an *enquiry*. The shipyard receives this *enquiry*, and evaluates if the yard is in a position to perform the repair.

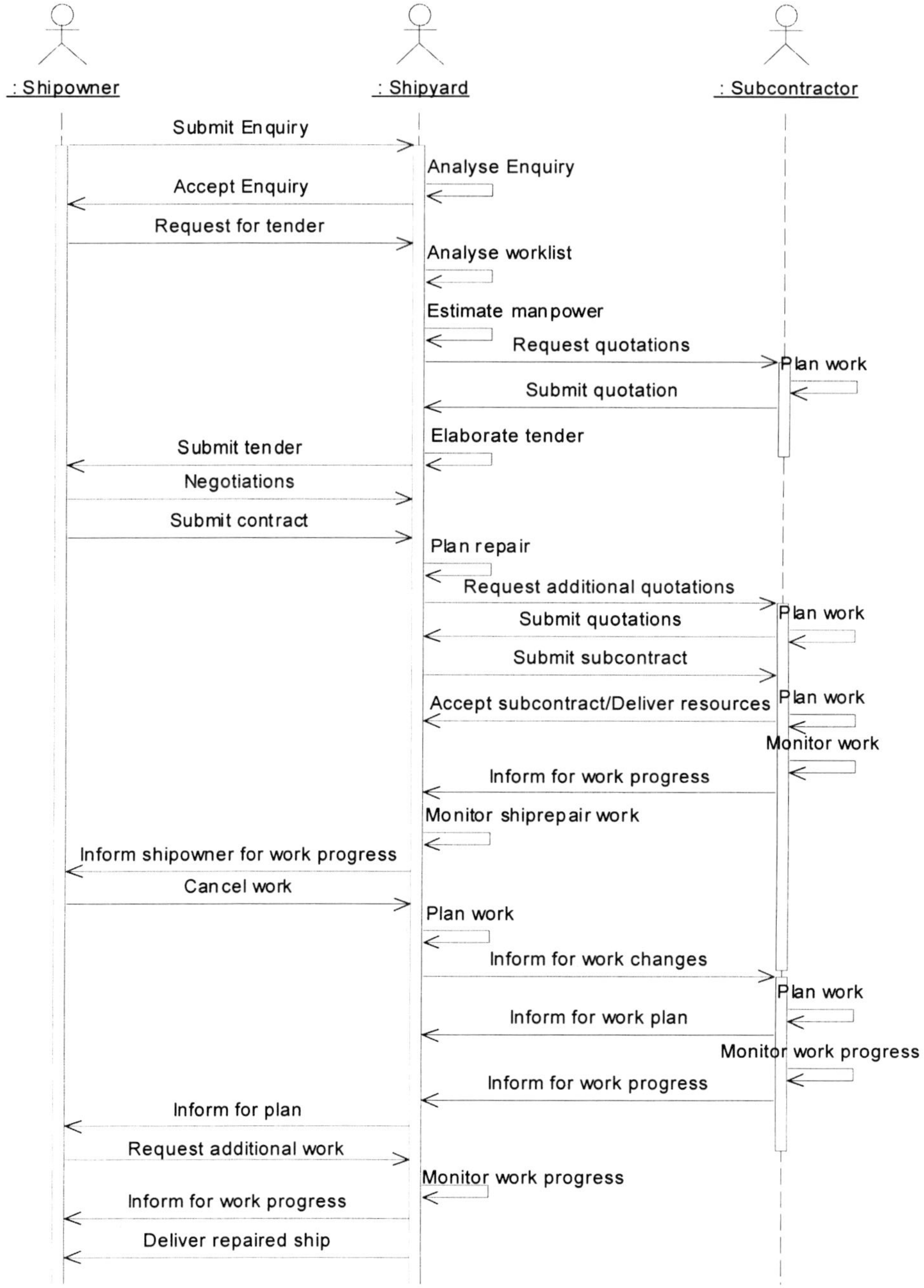

Fig. 2. Ship repair sequence of interactions

First of all, it examines if the required facilities, such as the size of docks, are available in the yard in order to accommodate the ship. Then the yard informs the ship owner about the feasibility of the repair in its facilities (figure 2). The ship owner submits a detailed list of work that must be performed and asks the yard to make an estimation of the cost and duration of the repair. The shipyard, based on the work list, carries out a market survey, if required, and estimates the cost and the duration of the repair. Based on these estimations, the shipyard submits to the shipowner the *tender*. The ship owner examines the tender and usually after several negotiations with the shipyard, they reach an agreement, namely the *contract* for the shiprepair. Based on the content of the contract, the shipyard plans both its internal resources and contacts the relevant suppliers that will supply materials, such as steel, pipes etc., needed for the repair. Furthermore, depending on the availability of the personnel required for the work, the shipyard usually plans a number of subcontracts. For all external companies required to contribute to the execution of the shiprepair, the shipyard places the necessary *requests for quotations*.

The external companies, namely the suppliers and subcontractors, provide their *quotations* that include the cost for the required material or service and the time to deliver the equipment or the time required for the service to be supplied. The shipyard selects both the most suitable supplier to bring the materials in and the most suitable subcontractor to perform the necessary work. In other words, the shipyard places a purchase order to the material supplier and a subcontract to the service supplier.

The ship repair work is executed and the partners inform each other about the progress and the relevant modifications made onto the original work list, because as the work evolves, new work items are identified, which were not recognized from the beginning. These changes, usually affect the initial plan regarding the execution of the work. Some of the new work items are carried out trivially, whereas some others are not, since the budget for the repair as well as the time for the ship to remain out of operation are usually fixed and of critical importance. The cost, for every single day, the ship is out of operation, can be too high, mainly in terms of the opportunity cost.

3 WEB-BASED IMPLEMENTATION

The software architecture deals with how, large software applications can be or should be organized for a better performance and ease of development. A commonly used option, in the last decade, is the so-called 3-tier architecture. Its main advantage is the separation of functionality from the presentation layer, thus providing a better understanding to the developers, while leading to well defined components and limiting the software changes, throughout its life cycle. This approach involves a "Presentation" layer, an "Application" layer and a "Data" layer. The "Presentation" layer implements the "look and feel" of an application. The "Application" layer implements the business logic of the application and the "Data" layer manages the persistence (storage) of application information.

The implementation of the 3-tier architecture nowadays, with the spread of Internet technologies has introduced a new kind of applications, the so called Web-Applications (6). In a Web-Application the "Presentation" layer is implemented by the web browsers (thin clients). The programming languages used, are a combination of HTML, CSS which stands for Cascading Style Sheets and Javascript. HTML is a Web authoring mark-up language for defining content structures and for rendering a web page. Javascript is commonly used for client-side validation and control over the look-and-feel of a page in dynamic HTML.

The “Application” layer is divided into sub-layers. The software technology, used in application design and development phase, is mainly implemented in the “Application” layer of the web application. The “Application” layer is responsible for the interface with the “Presentation” layer, the transactions between “Business Objects”, which are objects that represent business entities and finally the interface with the “Data” layer, which consists of Data Access Objects, implementing methods for "creating", "retrieving", "updating" and "removing" a business object from the database (Figure 3).

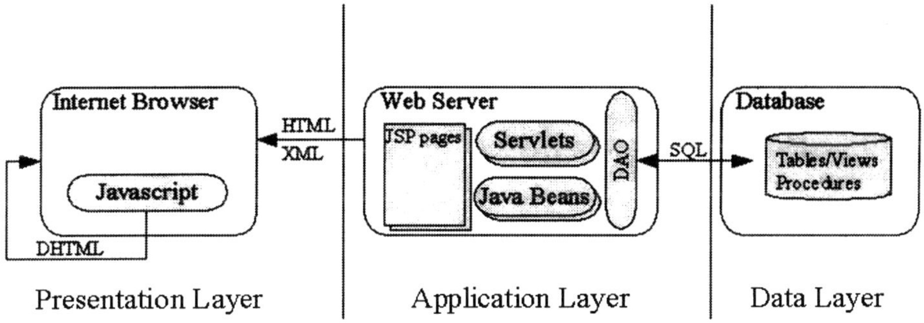

Fig. 3. Three tier application structure

During the last decade, the Internet has evolved from a network-based distributed information system, offering static information, to a marketplace for selling and buying goods and services. The increasingly sophisticated applications that enable this marketplace functionality, require a technology for presenting dynamic information.

Technologies that are quite popular in enterprise computing are the so called “Servlets”, which make it easier to implement server-side applications using Java technology. Combined with “Java Server Pages”, it is possible to generate a dynamic content, thus build web-based applications that work with a variety of other technologies: web servers, web browsers, application servers and other development tools (7).

3.1 Contract multi-site planning and control case study

In this paper, a web application is presented, which deals with contract planning and contract monitoring, in order to improve the partners’ collaboration and communication capabilities. Multi-site contract planning conceals the implementation of planning and monitoring of the operations of a number of companies, which cooperate for the execution of a shiprepair contract, taking into account the resources, the capacity and the availability of each company as well as the travel plan of the ship to be repaired.

Planning and monitoring is performed by each site independently, namely by the shipyard and its suppliers. The plan from each partner is facilitated through the allocation of tasks, to suitable resources, using various dispatching algorithms depending on the users’ demands and policies (FIFO, SPT, EDD) (1)(5)(8). Data from the shop floor are collected and are available for adjustments of the production plans. The allocation of the shiprepair tasks to the shipyard production departments is performed by utilising a state of the art modelling approach, where a generic hierarchical model is used to model the facilities of the shipyard, while the allocation of the tasks is performed through the so-called *release and assignment logic* (5)(8), resulting to a shiprepair production schedule. Following the same approach, the shipyard subcontractors produce a production plan for the tasks that they perform, and consequently, a production schedule is available for each company that participates in the shiprepair contract.

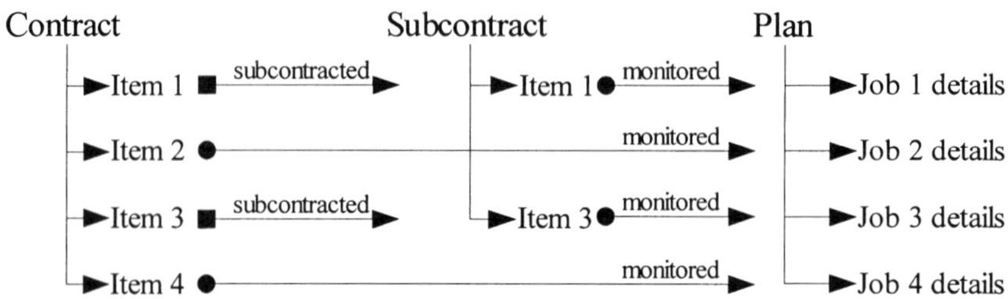

Fig. 4. The multi-site planning and control model

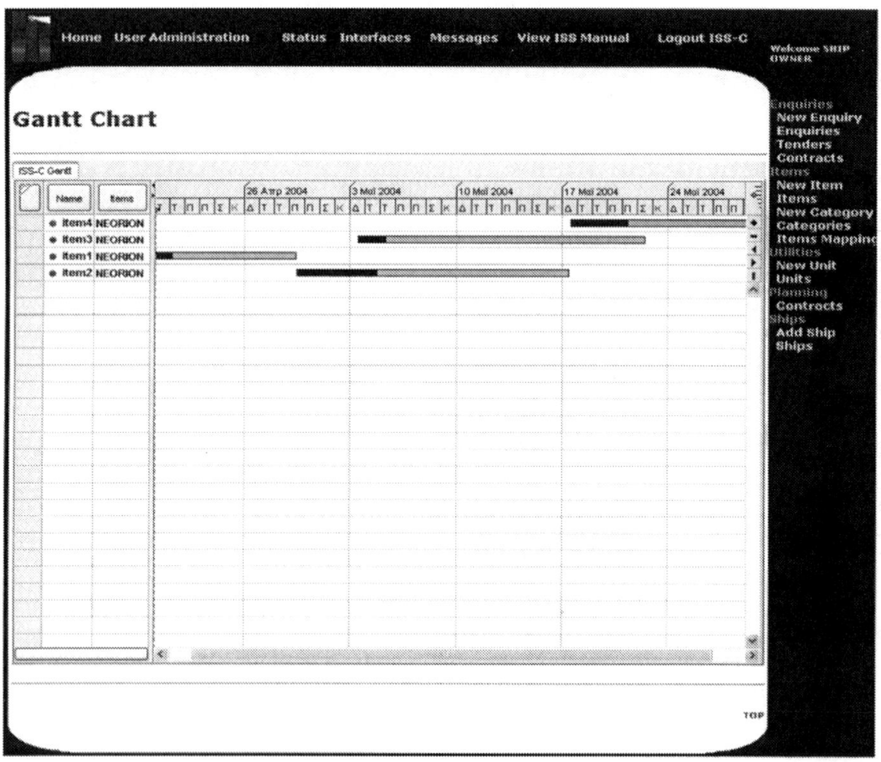

Fig. 5. The Gantt Chart

Planning and monitoring data, are then shared among the ship repair partners, through a collaboration module, that shares contract planning and monitoring information that is common to the participating partners. This information is used by each company, to adjust their production and to commit more resources or remove resources from the specific contract. This information is shared in the form of a graphical representation, i.e. a Gantt chart that shares the information of the activities, performed by each company (Figures 4 and 5).

The contract includes the ship repair items, which are actually the specification of the ship repair. Some of the items of the contract are subcontracted, therefore a number of subcontracts are created and each one includes one or more of the contract items. Also for each contract, a plan is created for controlling and monitoring the contract planning procedure. The plan consists of jobs that have a one-to-one relationship with the contract items. Each job is planned to be performed either by the shipyard or by the subcontractor executing the subcontract that includes this job (Figure 5).

4 CONCLUSIONS

Modern information technology was used in order to support the Value Added Chain in the maritime industry. A web based implementation was adopted to implement the contract planning and monitoring for a ship repair contract. This work shows that the adoption of such an approach is feasible and reduces communication efforts, while it improves the communication among the cooperating companies in a shiprepair contract.

A generic hierarchical model, implemented in the form of a web based software application, was used to implement planning and monitoring of manufacturing activities of the cooperating partners. A collaboration software module was implemented in the form of a web application, to share the planning and monitoring information. The overall scheme is implemented in a real shiprepair environment, thus proving its feasibility.

The implementation of web-based management of the supply chain, offers a competitive advantage, allowing for better and more effective collaborations, while opening up markets that were previously unapproachable.

REFERENCES

1 Chryssolouris G. (1992) "Manufacturing Systems Theory and Practice", Springer-Verlag, New York.

2 Quatrani T. (1998) "Visual Modeling with Rational Rose and UML", Addison-Wesley.

3 Papazoglou, M., Ribbers, P., Tsalgatidou, A. (2000) "Integrated value chains and their implications from a business and technology standpoint", *Decision Support Systems*, Vol. 29, pp. 323–342.

4 Chryssolouris, G., Makris, S., Xanthakis, V., Konstantinis, V. (2001) "An XML based implementation of the Value Added Chain in Manufacturing: A Ship repair Case Study", *CIRP 32th International Seminar on Manufacturing Systems* – Technology and Challenges for the 21st Century, Athens, 16-18 May.

5 Chryssolouris, G., Makris, S., Xanthakis, V., Mourtzis, D. (2004) "Towards the Internet based Supply Chain Management for ship repair industry", *International Journal of Computer Integrated Manufacturing*, Vol. 17, pp. 45-57.

6 Brown S. et al. (2001) "Professional JSP", 2nd Edition. WROX Press, ISBN: 1861004958

7 URL sun.com (2004) http://java.sun.com/developer/technicalArticles/xml/WebAppDev/

8 Chryssolouris, G., Papakostas, N., Makris, S. and Mourtzis, D. (1999) "Planning and Scheduling of Shipyard Processes", *Application of Information Technologies to the Maritime Industries*, C. Guedes Soares, J. Brodda (Eds.), pp. 255-274.

A methodology for agile partner selection

J REN
Department of Engineering, The University of Exeter, UK
Y Y YUSUF
Business School, The University of Hull, UK
N D BURNS
Department of Engineering, Loughborough University, UK

ABSTRACT

One of the prime responsibilities of an agility-based company is the evaluation and selection of partners. This paper proposes a decision-making methodology and the structure for the selection of Agile partners. The methodology allows for the evaluation of partners to help organisations become more agile. The use of the Analytic Hierarchy Process (AHP) as a decision support model to aid decision makers in selecting suitable agile partners is demonstrated in this paper. The block diagram of the selection as well as a hierarchical model is established and a set of Agile decision domains and attributes was identified. Based on questionnaire survey, the weights of all Agile dimensions, decision domains and attributes were determined. The reliability analysis reveals that all Agile attributes selected show a high level of internal consistency, thus they provide a validated base for the research.

Keywords: Partner selection, decision making method, Agile enterprise, AHP

1 INTRODUCTION

Agility describes a company that is able to change and adapt quickly to changing circumstances. In the words of one of the early proponents of the concept, "Agility is dynamic, context specific, aggressively change embracing, and growth oriented. It is not about improving efficiency, cutting costs, or battening down the business hatches to ride out fearsome competitive storms. It is about succeeding and about winning profits, market share and customers in the very centre of competitive storms that many companies now fear" (Goldman,1995). Increasingly Agile manufacturing is attracting greater attention from both academic and industrial communities. As a result, a number of programs are being conducted on relevant issues to propagate agile manufacturing concepts, build agile enterprise prototypes and realise an agile industry eventually.

The evaluation and selection of partners is very important to the success of an agile enterprise. In this study, the application of AHP in Agile partner selection is examined. As a result of past study carried out by one of the authors (Yusuf,et al 1999), there is opportunity to use attribute-dependent criteria to identify potential Agile companies that are deserving of further collaboration. Through easy-to-use software, expert judgement on key criteria in the Agile partner selection could be obtained and processed to provide scores for each company which are then used to rank companies in order of preference. Furthermore, knowledge acquired through AHP can be passed on to, and processed by, an Agile decision making system (Yusuf et al., 2000). An Agile decision making system comprises decision methods and database that incorporate external databases, heuristic decision principle obtained from experts, and other external information to yield even more powerful predictions for Agile partners.

This paper is organised as follows. First, the partner selection literature is reviewed. The methodology is then described, including the partner selection process, sampling process and partner selection model. The reliability of these Agile attributes are also examined and following that, the weights of each elements are calculated based on questionnaire survey.

2 LITERATURE REVIEW

Global competition is forcing corporations to look at their partnership map to reduce costs and time involved in the process. This is done periodically. The arrangement art of relationship management is now crucial to business success. A solid partnership helps corporations in gaining significant advantages over their global competitors. In the literature, partner selection research often starts at choosing criteria in particular organisation or industry. The literature review of this paper will investigate both criteria selection and decision making method selection which are the key issues in Agile partner selection.

2.1 Partner selection criteria

Many authors investigated how to choose decision criteria. For example, Angeles (2000) identified the trading partner selection criteria used by firms in a customer-supplier pair. By using factor analysis, the authors chose six critical factors that can be used as selection criteria: strategic commitment, trading partner flexibility, joint partnering, readiness, infrastructure, and communications. One way ANOVA and t-tests were used to test differences in the means of the responses of customer and supplier firms to the selection criteria.

Tatoglu (2000) examined partner selection criteria using a typology that distinguishes between partner-related and task-related selection criteria. The author investigated the strategic motivation for joint venture formation for a sample of Western partner firms. The relative importance of a set of strategic motives is identified and discussed with respect to the nationality of foreign partner. The research also identified the relationship between the relative importance of selection criteria and the nationality of foreign partner.

Al-Khalifa(1999) argued that the motivation for partner selection must be distinguished from that needed to enter into a joint venture, and the latter may be considered the "ends" dimension of motivation while the former is concerned with "means". The research recognised the importance to distinguish between "task related" factors and "partner related" factors in analysing the partner selection process. Based on a survey of 42 international joint ventures, respondents rank partner related factors as significantly more important than task factors in selecting a partner. The existence of partner related factors as a separate construct was also confirmed by the authors via factor analysis.

Talluri (1999) suggested that multi-organisational structures can be viewed as a solution for rapid introduction of products while maintaining high quality and minimal costs. One key issue in designing these new forms of organisations is the partner selection process. The business processes, owned by organisational partners, must be efficient both individually and as a collective group. The author proposed a two-phase quantitative framework to aid the decision making process in effectively selecting an efficient and a compatible set of partners. Phase 1 identifies efficient candidates for each type of business process (e.g. design, manufacturing, distribution, etc.) utilising data envelopment analysis. Phase 2 involves the execution of an integer goal-programming model to determine the best portfolio of efficient partners based on a number of compatibility objectives.

2.2 Partner selection method

Agile partner selection involves many factors including Agile dimensions, decision domains and Agile attributes. To consider these factors together is extending beyond the traditional method: optimising a single objective function over a set of feasible solutions (Marlow W.H., 1993). This leads to the multi-criterion decision analysis (MCDA). The MCDA methodology is made up of four steps: (1) identifying the decision problems, (2) structuring the preferences, (3) aggregating the alternative evaluations and (4) making recommendations. Traditional MCDA methods include the categorical method, weighted-point method, matrix approach, and multiple objective programming (MOP) such as goal programming. To elaborate, a categorical method rates companies on a number of equally-weighted factors and then allows the decision maker to evaluate the company with the total score. This method is simple and easily understandable, but it often introduces subjective error into the decision and oversimplifies the selection scenario by equally weighting all factors (Lowen R., 1999). Some authors suggested a matrix approach which evaluates and selects companies based on weighted scores obtained from a set of pre-determined benchmarks (Soukup,1987). To improve this matrix approach further, some authors suggested a performance matrix approach that can measure the company's performance under various unforeseen scenarios by estimating the probable deviations from the original goal that the company needs to obtain (Thompson,1990). Regardless of their strengths, none of these approaches can systematically measure both qualitative and quantitative factors and structure complex problems with a large number of criteria, attributes and alternatives.

From a review of the literature on criteria selection and decision making method selection, this paper build on the existing body of knowledge on the attributes of an agile organisation to use Analytic Hierarchy Process (AHP) as the tool to assist managers in selecting Agile partners. The most important reason to choose AHP was the simplicity of its procedures. In fact, Partovi et al.(1994) lists this simplicity as one of the benefits of using AHP. During the research it was also noticed that if the selection method is simple enough to be explained on different levels of the company's hierarchy, the resistance for using it reduces considerably. Secondly, the easiness of the use makes AHP much more probable than a method that requires higher education from the people who are using it. In fact, AHP is a popular management science tool that was developed more than two decades ago (Saaty, 1980). It enables managers to make potentially more effective decisions by structuring and evaluating the relative attractiveness of competing options for any type of managerial decision. Although AHP frequently has been used in a variety of decision scenarios, it has not been used in the area of Agile partner selection, even though other researchers have noted the usefulness of AHP in such contexts (Hokey,1994).

3 RESEARCH METHODOLOGY

3.1 Problem statement

Agility is an organisation's capacity to respond rapidly and effectively to unanticipated opportunities and to proactively develop solutions for potential needs. In order to establish and maintain this capability, manufacturing companies are forced to create agile business processes. Among them, one of the key issues is to select agile partners. The selection process involves many elements including different dimensions, decision domains and organisational attributes. Since the elements have systemic characteristics, they may be integrated into one model. Some authors (Gunasekaran,1999) (Sharifi,1999) have considered the potential need for such a model. Our study was to provide decision maker such a model to select suitable partners.

In the first phase of this study, based on literature review, the elements on different levels were identified to describe different aspects of the potential partners. In this phase a five level hierarchical structure was produced. This AHP-based structure can help decision maker get judgement easily. At the conclusion of this phase, the four main dimensions were identified: enriching the customer, co-operation, mastering change and uncertainty and leveraging the impact of people, information and technology. The ten decision domains with thirty-two relevant organisation attributes were also examined.

In the second phase of this study, the identified elements on each level of the hierarchical structure were included in a survey questionnaire designed for this study. The collected data were analysed with SPSS software package. The dominant features on each level of agility structure were identified based on the statistical analysis. The results of this phase worked out the weights of each element and that provided managers a comparative standard to make their own judgement.

The last phase is a simple example problem used to illustrate the proposed methodology.

3.2. The sample

The sample profile (Ren, J 2003) shows the number of employees at site, annual turnover and business category of the firms. The sample consisted of 105 companies in the UK. The table shows that companies with less than £30 million annual turnover represented about 60 percent of all respondents. Those with fewer than 500 employees accounted for 68.6 per cent of the total, indicating an industry structure dominated by SMEs. While respondents are well distributed across industrial sectors, there are two dominant groups: mechanical engineering, and computer/telecommunication. These two groups combined represent 44.7 per cent of the sample.

3.3 Data collection

In this research, we developed the survey instrument for the study by searching relevant literature to review the instruments and perspectives adopted by researchers who had studied similar issues. The questionnaire was pre-tested and pilot-tested. Pre-testing was carried out by discussing the questionnaire with colleagues working in the same area in the department. The pilot-testing was conducted via email and Fax. From the pre-tested and piloted-test results, we found that the initial questionnaire was too large, some questions were not clear, and the layout needed to be rearranged. The questionnaire was accordingly modified.

The hard copies of the questionnaire were mailed out to addresses taken from the European Business Directory. A covering letter which described the objectives of the research and procedures for completing the questionnaire, and a stamp addressed envelope were enclosed to facilitate returning of the questionnaires. A telephone reminder was made to some non-respondents three weeks after the questionnaires were mailed out. We targeted chief executive officers (CEOs) and senior managers to serve as respondents. Of the 1400 questionnaires mailed, 185 were returned with 105 valid for analysis. The overall response rate was therefore about 13.2 percent.

To make sure the questionnaire do meet our needs in a useful way, we use internal consistency reliability to determine the extent to which the elements in the questionnaire are related to each other. Actually, internal consistency reliability is a psychometric measure in assessing survey instruments and scale. It is the indicator of how well the different elements measure the same issues (Marks, 1995). High internal consistency of the attributes is necessary to ensure the accuracy or precision of the attribute. Cronbach's coefficient alpha(α) is most commonly used to assess the scale reliability. Normally the value of Cronbach's coefficient alpha(α) is between 0 and 1 and a higher level of α indicates a higher reliability of the scale. Lau(1999) stated that as a rule, alpha levels as low as 0.6 are acceptable for new scales. Meanwhile it would be difficult to justify a proposed indicator of a latent variable even in exploratory research if its reliability measures were less than 0.5. In this research (Ren, J, 2003), all α values range between 0.7327 and 0.9573 and thus all attributes show a relatively high level of internal consistency to the domains.

4 MODEL AND ANALYSIS

4.1 Partner Selection Process

Based on the literature review, a detailed partner selection process is proposed as shown in figure 1. The first step in the selection is the creation of a team, which has the managerial experiences needed. The team members are chosen from different levels of the organisation involved with the selection process to widen the points of views in the process.

After choosing the team and establishing the goal, the next two steps are to choose pertinent decision domains and attributes, and to establish hierarchy structure based on the company's strategy. At this stage, important guidelines have been suggested by Saaty(1982): representing the problem as thoroughly as possible; considering the environment surrounding the problem; identifying the issues or attributes that contribute to the solution; and clarifying the necessary participants associated with the problem.

In step 4, the priority weights of each alternatives are then determined through pair-wise comparison or other way like questionnaire survey. The weight of the corresponding higher level element is then used to weight the elements at the lower level in the hierarchy (composite weight). The procedure is repeated by moving downward along the hierarchy, computing the weight of each element at a particular level and using these to determine composite weights for succeeding levels. Then it is necessary to analyse comparative results that include weight analysis and consistency. Weight analysis can be used to assess the reliability of the weight. The priority weights of each element are the eigen-values in the corresponding eigen-vector of each matrix. The consistency of the data may be investigated during the analysis. The composite indices of the previous matrix represent the consistency of the data in the matrix with regard to the eigen-values.

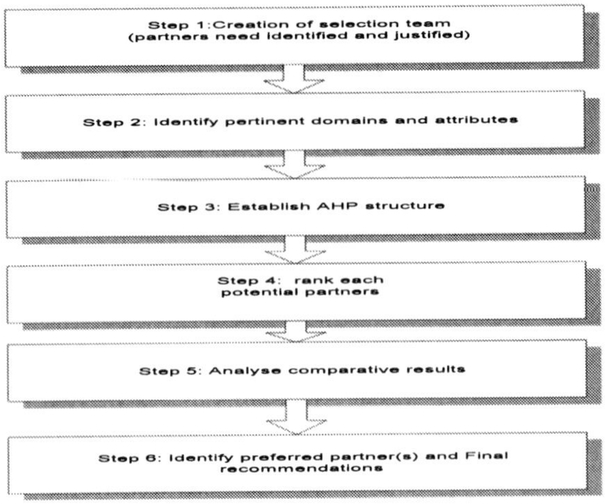

Figure 1 Agile partner solution process

In the final step of the proposed diagram, the partner that has the highest overall priority score is identified as the preferred one. The selection decision process ends when final recommendations are made concerning the most suitable partners.

4.2. Partner selection model

Based on the above analysis, a five level hierarchical model is devised (see Figure 2). The first level sets the main objective, here referred to as the agile partner selection. The main objective (i.e. referred to as the main considerations) is divided into four main dimensions or sub-objectives, which are enriching the customer (EC), co-operation (CO) , mastering change and uncertainty(MCU) and leveraging the impact of people, information and technology(LI). As shown in the figure 2, the first dimension is enriching the customer. This entails a quick understanding of the unique requirements of each individual customer and rapidly providing it. The second dimension entails co-operation (intra-organisational, inter-organisational co-operation such as supplier partnerships and virtual relationships with competing organisations) in order to enhance competitiveness. The third dimension utilises new organisational structure(s) to master change and uncertainty through techniques such as concurrent engineering and cross-functional teams. The fourth dimension leverages the impact of people, information and technology and recognises the importance of employees as a company asset, placing greater emphasis on education, training and empowerment. For further details of these four dimensions, see Goldman et al (1995).

The third level hierarchy includes ten decision domains and the fourth level consists of thirty two attributes(Yusuf,1999). In fact, to achieve agility in terms of the four dimensions, some decision domains with related attributes must be carefully identified. Team building, for example, is a key concept in manufacturing strategies and a critical decision domain. The associated attributes involve: *Empowered individuals working in teams; Cross functional teams; Teams across company borders* and *Decentralised decision making.* Empowerment enables workforce to make decision and work autonomously. That speeds up business processes and creates an environment in which innovation may thrive; Cross functional teams solves problems using a team of people from different business disciplines enabling better, broader and more robust solutions to be devised; Encouragement of teaming with other customers, even competitors allows companies to react quickly to market changes at low cost; Decentralised decision making speeds up operations, allows for workforce innovation, maintains information flow at a workable level. The ten decision domains and the associated attributes are show in table 1.

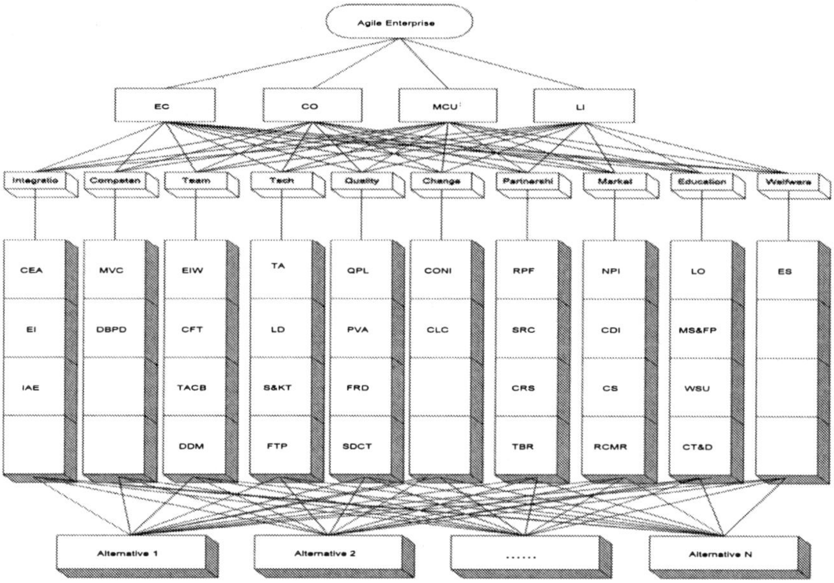

Figure2 Agile Partner Selection Model

Finally, at the fifth level of the hierarchy, there are some potential partners referred to as alternatives 1, 2, 3, …, and n, which must be assessed and compared.

4.3. Means and weights analysis

On the basis of developed hierarchical model shown in Figure 3, the elements on all levels of the model were suggested to which respondents indicated the level of importance. Five possible levels of importance were asked to be identified, including not important, slightly important, important, very important and absolutely important. Each of these importance levels was coded into SPSS as 1, 3, 5, 7, and 9 respectively. By calculating the means, the weights can be computed by the following formula:

$$W_i = \frac{M_i}{\sum_{j=1}^{n} M_j}$$

Where:

W_i = weight of the I th element;
M_i = mean of the I th element;
N = number of elements

Table 1 Domains and attributes (with Abbreviations used)

Decision domains	Attributes with abbreviation
Integration	Concurrent execution of activities(CEA) ,Enterprise integration(EI), Information accessible to employees(IAE);
Competence	Multi-venturing capabilities(MVC), Developed business practice difficult to Copy(DBPD);
Team building	Empowered individuals working in teams(EIW), Cross functional teams(CFT), Teams across company borders(TACB), Decentralised decision making(DDM);
Technology	Technology awareness(TA), Leadership in the use of current technology(LD), Skill and knowledge enhancing technologies(S&KT), Flexible production technology(FPT);
Quality	Quality over product life(QPL), Products with substantial value-addition(PVA), First-time right design(FRD), Short development cycle times(SDCT);
Change	Continuous improvement(CONI), Culture of change(CLC);
Partnership	Rapid partnership formation(RPF), Strategic relationship with customers(SRC), Close relationship with suppliers(CRS), Trust-based relationship with customers/suppliers(TBR);
Market	New product introduction(NPI), Customer-driven innovations(CDI), Customer satisfaction(CS), Response to changing market requirements(RCMR);
Education	Learning organisation(LO), Multi-skilled and flexible people(MS&FP), Workforce skill upgrade(WSU), Continuous training and development(CT&D);
Welfare	Employee satisfaction(ES).

Source: Yusuf et al (1999)

Dimensions

The mean values of the level of dimension are shown in Table 2. It can be seen from the table that the mean importance of the four dimensions were all greater than 5. That reflected that respondents thought all the four dimensions are important. It is not surprising that enriching the customer and industrial co-operation are highly ranked. A customer focus entails directing the efforts of the firm to meeting customer needs and wants. Recent research shows that enriching customer has been propelled by an increasingly competitive global business environment, accelerating technological developments which have shortened product life cycles and difficulty of many organisations in sustaining superior performance (Kwaku, 1998). Industrial co-operation is also a key issue in modern operational strategy. A single organisation is often may not be able to respond quickly to changing market requirements. Industrial co-operation based on core competencies of firms will help to improve the flexibility and responsiveness of organisations. Appropriate strategies/methods such as communication, training and education, and strategic alliances can be adopted for an effective co-ordination and integration of participating firms at different levels of co-operation (Gunasekaran, 1999).

Table 2 Means and weights for Agile dimensions

Dimension	Mean	Std.Deviation	weight
Enriching the customer	5.57	2.03	**0.2639**
Internal and external Co-operation	5.39	1.67	**0.2552**
Mastering the changing	5.04	1.84	**0.2387**
Leveraging the impact of people, information and technology	5.11	1.71	**0.2421**

Decision domains

The mean values of the level of decision domains are shown in Table 3 in rank order. As seen in the table, the mean values range from 5.027 to 6.257, ensuring our set of decision domains was comprehensive and meaningful. Four domains were identified as the dominant features: team building; quality, partnership and market. They took weights 0.1091, 0.1051, 0.1047 and 0.1025 respectively. Apparently, team building and partnership are highly co-operation related. Team building focuses on intral-organisational cooperation, and partnership focuses on inter-organisational co-operation. It is consistent with the above dimension analysis that market is strongly highlighted by respondents regarding enriching the customer.

Table 3 Means and weights for decision domains

Decision domains	Rank	Mean	Std.Deviation	weight
Team building	1	6.257	2.070	**0.1091**
Quality	2	6.037	1.837	**0.1052**
Partnership	3	6.007	1.852	**0.1047**
Market	4	5.882	1.845	**0.1025**
Competence	5	5.687	1.765	**0.0991**
Technology	6	5.687	1.887	**0.0991**
Integration	7	5.635	1.715	**0.0982**
Change	8	5.580	1.912	**0.0973**
Education	9	5.545	1.890	**0.0966**
Welfare	10	5.027	1.830	**0.0876**

Agile attributes

The thirty-two attributes are set in rank order in Table 4. The high-ranked group can be identified as: Customer satisfaction(CS), Technology awareness(TA), Leadership in the use of current technology(LD), New product introduction(NPI), Teams across company borders(TACB), Strategic relationship with customers(SRC), Skill and knowledge enhancing technologies(S&KT), Cross functional teams(CFT), Continuous improvement(CONI) and Trust-based relationship with customers/suppliers(TBR). The ten attributes are from six decision domains: team building, technology, change, partnership, market and education. It is interesting that none of quality related attributes are included in this set. The highest rank attribute regarding quality is Products with substantial value-addition(PVA), only stands the fourteenth position on the list. Perhaps quality should be treated collectively or individually to obtain a competitive advantage rather than divided to many attributes. Another reason is maybe dividing quality into some attributes is difficult since such dividing is often coloured by a particular managerial situation or problem(Curkovic S, 2000).

Table 4 Means and weights for agile attributes

Attributes	Rank	Mean	Std.D	weight	Attributes	Rank	Mean	Std.D	Weight
CS	1	6.34	1.97	**0.0370**	EI	17	5.30	1.43	**0.0309**
TA	2	6.16	1.71	**0.0359**	QPL	18	5.30	1.87	**0.0309**
LD	3	5.88	1.88	**0.0343**	CEA	19	5.27	1.77	**0.0307**
NPI	4	5.84	1.95	**0.0341**	CT&D	20	5.27	1.60	**0.0307**
TACB	5	5.77	1.66	**0.0336**	ES	21	5.27	1.51	**0.0307**
SRC	6	5.73	1.82	**0.0334**	CDI	22	5.25	1.86	**0.0306**
S&KT	7	5.70	1.78	**0.0332**	IAE	23	5.20	1.70	**0.0303**
CFT	8	5.66	1.60	**0.0330**	FRD	24	5.16	2.45	**0.0301**
CONI	9	5.59	1.88	**0.0326**	MS&FP	25	5.16	1.85	**0.0301**
TBR	10	5.59	1.91	**0.0326**	LO	26	5.09	1.73	**0.0297**
RCMR	11	5.52	1.74	**0.0322**	WSU	27	5.05	1.67	**0.0294**
CLC	12	5.45	1.95	**0.0318**	RPF	28	5.02	1.81	**0.0293**
MVC	13	5.41	1.75	**0.0315**	DBPD	29	4.70	2.06	**0.0274**
PVA	14	5.41	2.02	**0.0315**	SDCT	30	4.64	2.40	**0.0270**
CRS	15	5.38	1.51	**0.0314**	DDM	31	4.59	2.33	**0.0268**
FPT	16	5.36	2.44	**0.0312**	EIW	32	4.20	2.10	**0.0245**

Despite all of this, it is clear from the table that the high-ranked set of criteria reflects the respondents' concern with understanding the nature of the agility and ensuring adequate foundation to competitive bases. The understanding of the agility is thus the major concerns of the agile partner selection criteria. It may be conjectured that this set of selection criteria is the underlying motivation for agility formation.

A second group of selection criteria can be identified which are of intermediate rank: Response to changing market requirements(RCMR) , Culture of change(CLC), Multi-venturing capabilities(MVC), Products with substantial value-addition(PVA), Close relationship with suppliers(CRS), Flexible production technology (FPT), Enterprise integration EI, Quality over product life (QPL), Concurrent execution of activities (CEA), Continuous training and development (CT&D). This group reflects a concern with the operational requirement of the agile company. These operational concerns are clearly secondary.

4.4. Managerial implications

In this study, we have attempted to give empirical evidence to provide manufacturing managers a practical methodology for selecting suitable partners. A five-level hierarchical structure were produced here. It has clear that the structure can be used to help decision makers get judgement easily. The methodology suggests the need for identification for relative importance of each element on the different hierarchical levels. Linked to these is the need to choose an calculation instrument to meet the hierarchical structure. Thus AHP is highly recommended. However, different companies are under different competitive circumstances and thereby different important values(weights) could be given to each AHP elements. In order to give companies a guideline, we conducted a questionnaire in UK. The survey highlighted enriching the customer and industrial co-operation as the key agile dimensions. Also, team building; quality, partnership and market were examined as major decision domains. The significant contributions of high-ranked group of organisation attributes(i.e. Customer satisfaction(CS), technology awareness(TA)) were expected. However, the reason for an insignificant result for quality related attributes deserves further investigation.

In light of the empirical evidence produced in this study and that found in existing literature, we established the hierarchical structure of agile partner selection and worked out the weights of each AHP-based element. The outcomes of the research provide decision makers a practical tool when they have to choose partner. In fact, the hierarchical structure made the complex selection process easy to be understood; the established weight values gave a clear imagination of the relative importance of each selection criteria. As seen in the next section, a choice may be got easily using our proposed methodology. However, people could argue that the weight values or even the structure of the model may be different considering their own situation. Alternatively we will provide them a software package tool based on AHP. By using the tool decision makers can develop their own hierarchical structure and figure out the weights values.

5 CONCLUSION

The need for developing an Agile partner selection process is a continuous task. The selection model presented in this paper is aimed to aid decision makers in the turbulence competitive environment. The proposed methodology integrated various characteristics of Agility(Agile dimensions, decision domains and attributes) and their relationships. It has proven that the methodology will be beneficial in considering different characteristics on different levels. This technique has also proven useful for structuring the decision in the decision maker's mind. The proposed structure provides for simplification of a very difficulty decision with many dimensions. However, the large number of pair-wise comparison can be tedious for decision makers. In our example problem, 288 pair-wise comparison were required. If we choose five partner candidates, the pair-wise comparison will increase to 800. Due to time and effort required for a typical AHP process, an application of this methodology should be targeted at more strategies decision. Practically, the methodology envisaged in this research will not only helps in Agile partner selection but also in measuring different kind of partners and thus ensures protection of investment.

REFERENCES

1 Al-Khalifa A.K.,Peterson S.E., "The partner selection process in international joint ventures ",*European Journal of Marketing*, 33(1999), 106-108

2 Angeles R., Nath R., *"An empirical study of EDI trading partner selection criteria in customer-supplier relationships"* , *Information and Management*, 37(2000), 241-255

3 Curcovic S, Shawnee k, and Droge C(2000), "An empirical analysis of the competitive dimensions of quality performance in the automotive supply industry", *International Journal of Operations & Production Management*, Vol.20,pp.386-403.

4 Goldman,S.L., Nagel,R.N. and Preiss,K.(1995), *Agile Competitors and Virtual Organisations: Strategies for Enriching the Customer,* Van Nostrand Reinhold

5 Gunasekaran, A.(1999), "Agile Manufacturing :A framework for research and development", *International Journal of Production Economics*, Vol.62 , pp. 87-105.

6 Hokey,M(1994), "International supplier selection: A multi-attribute Utility approach", *International Journal of Physical Distribution & Logistic Management,* Vol.24,pp.24-33

7 Kwaku A, Satyendra S(1998), "Customer orientation and performance: a study of SMEs", *Management Decision,* Vol.36, pp.385-394.

8 Lau R.S.M.(1999), "Critical factors for achieving manufacturing flexibility", *International Journal of Operations&Production Management,* Vol.19, pp.328-341.

9 Lowen R.(1999), "Categorical Methods and Techniques in Fuzzy Mathematics", *Fuzzy Sets and Systems*, Vol. 104, No. 3, pp. 357-357(1).

10 Maks S. Litwin(1995), *How to measure survey reliability and validity,* Sage Publications.

11 Marlow W.H.,(1993), *Mathematics for Operations Research*, John Wiley&Son, New York, NY

12 Miller,C.N.,Roth, A.V.(1994), "A taxonomy of manufacturing strategies", *Management Science*, Vol.40, pp.285-304.

13 Partovi, F.Y. ,and Hopton, W.E.(1994) , "The analytic hierarchy process as applied to two types of inventory problems", *Production and Inventory Management Journal,* Vol.3, pp.13-19.

14 Ren, J (2003), *" Decision support method for agile enterprise design",* unpublished PhD thesis, the University of Exeter.

15 Satty,T.L.(1980),*The Analytic Hierarchy Process*, McGraw-Hill, New York, NY.

16 Saaty, T.L. (1982), *Decision Making for Leaders,* Lifetime Learning Publications, Beverly Hills,CA.

17 Sharifi, h. and Zhang, Z (1999), "A methodology for achieving agility in manufacturing organisations: An introduction", *International of production Economics*, Vol.62, pp.7-22

18 Soukup,W.R.(1987),"Supplier Selection Strategies", *Journal of Purchasing and Materials Management* ,Vol.23,pp.7-12

19 Partovi, F.Y.(1994), "Determining what to benchmark: an analytic hierarchy process approach", *International Journal of Operations and Production Management,* Vol.4 ,pp. 25-39.

20 Talluri S.,Baker R.C., Sarkis J.(1999), "A framework for designing efficient value chain networks - Derivations, meanings and uses", *International Journal of Production Economics,* Vol.62, pp. 133-144

21 Tatoglu E.,Western(2000), " Joint ventures in Turkey: strategic motives and partner selection criteria", *European Business Review*, Vol.12, pp.137-147

22 Thompson,K.N.(1990), "Supplier Profile Analysis", *Journal of Purchasing and Materials Management* ,Vol.26 , pp.11-18.

23 Yusuf, Y.Y., Ren,J, Gunasekaran,A,(2000) "An integrated framework for Agility-based decision making system", the 3th International conference intelligent manufacturing processes and system(MIT,USA) , pp.54-61.

24 Yusuf, Y.Y., Sarhadi, M. and Gunasekaran,A.(1999), "Agile Manufacturing: Concepts, Drivers and Assessment", *International Journal of Production Economics*, Vol.62 , pp.33-43.

Virtuality-based understanding scheme (VirUS) – a holistic typecast supporting research and behaviour-oriented management of virtual teams

C SCHMIDT
Department of Business Administration, HTW - University of Applied Sciences, Saarbrücken, Germany
B K TEMPLE
School of Engineering, Science, and Design, Glasgow Caldeonian University, UK
A McCREADY
Department of Business Information Management, Glasgow Caledonian University, UK
J NEWMAN
School of Computing and Mathematical Sciences, Glasgow Caledonian University, UK
S C KINZLER
Department of International Business, FH Aalen – University of Applied Sciences, Germany

ABSTRACT

This paper describes the background to a research project whose objective is to develop a Virtuality-based Understanding Scheme (VirUS) as a typecast of virtual teams.

In response to rapidly changing environments and developments in Information and Communication Technologies (ICT), organizations are increasingly using Virtual Teams (VTs). However, both managers and scientists have not yet agreed upon a common understanding of VTs. Classifications diverge as to what types of VTs exist and how degrees of virtualization in VTs can be identified. This creates difficulties for further research and managers striving to implement findings.

Furthermore, research on management of VTs lacks attention to potential individual dysfunctions arising out of virtual working environments. According to Türk [51] mismanagement of these dysfunctions can enhance Overstabilisation, Overcomplexity and Overregulation as corporate pathologies and result in higher dominance, social and psychic cost.

VirUS will serve as a basis for a projection of Türk's pathology concept. It involves an analysis of the relation between certain stages in VirUS and pathological states.

1 INTRODUCTION

There is a clear trend towards decentralized multicultural global corporate networks interacting largely through Internet technologies [45] but there is a danger that more dispersion will lead to increasing fragmentation. When coupled with a vocabulary that contains an increasing amount of jargon, such as the term "Virtual Team" (VT), businessmen are confronted with a changing working environment [30] bringing along both benefits and problems. They have to deal with new issues associated with the new complexities in order to decide and act for the strategic vitality of the company. Logically, this vitality is not only connected to economic elements. An organization with optimized balance sheets may be rated vital in an economic sense, while the state of its human capital lacks health in terms of social vitality. Therefore, monitoring the seemingly new phenomenon of "virtuality" in team-based environments is essential, because it might bring about major problems when individual dysfunctions, or pathologies, arise out of virtual structures and are disseminated among teams and departments even harming the organization as a whole [40].

There follows a review of interdisciplinary literature with the object of showing how the concepts of organizational pathologies, as considered in 1976 by Türk [51], have progressed very slowly in respect of virtual team behaviour. These organizational pathologies are explained using Türk's framework and are followed by a stocktaking of the literature on dysfunctional aspects in VTs and pathology-related managerial issues. Furthermore, the review summarizes the authors' views on how to classify and identify different degrees of virtualization in VTs. But first the terms "group", "team" and "virtuality" are considered, in order to further isolate and clarify the notion of VTs.

2 FROM GROUPS TO TEAMS

Organizations meet the growing demands of flexibility and efficiency by applying the concept of group working [8, 25]. Traditionally "groups" are defined as small entities who interact interdependently in a temporal context [16, 38, 39]. A group consists of two or more individuals with the condition that every person in the group can directly interact with other group members. Group communication towards accomplishing a common purpose is predominantly carried out face-to-face (F2F) [4, 19, 23, 32, 33]. Based on this, organizational work groups represent formal groups. Primarily initiated on the basis of an imposed decision, they are institutionalized and embedded into structural and technological corporate environments following corporate purposes. These purposes appear in the form of assignments following corporate goals.

While some authors use the terms "group" and "team" synonymously in an organizational context [11, 22, 29], others make the clear distinction, that a team is a group, but a group is not necessarily a team [6, 20, 31, 46]. Without further expanding on the discussion whether to use the terms "group" and "team" equally or not, the different approaches and their contribution to an in-depth discussion have to be acknowledged and valued. It has to be noted though that the authors' usage of the terms does not reflect a clear difference. Therefore, for the purpose of this paper, the terms are used in an equal sense.

3 GRASPING THE "VIRTUAL"

The phenomenon "virtuality" has become familiar with innovations achieved in ICT such as the breakthrough of the Internet. It has the flavour of a buzzword, because no commonly generated definition exists. As "virtuality" is often used to describe something that is not a concrete entity, establishing the meaning of the term seems to be a paradox. Therefore, it is helpful to start at its linguistic and historic roots.

Looking at its latin origin, "virtus" with the stem "vir" (lat.: man) denotes virtue and bravery. While many authors derive their working definitions using just this relation [29, 31, 37], the question arises as to how logical it is for bravery to be able to serve as a foundation to tap the meaning of "virtual" as it is described in dictionaries today. According to Sinclair [43: p. 1097] virtual means that "something has all the characteristics of a particular thing, although it is not formally recognized as being that thing". Wahrig [54: p. 803] describes "virtual" as "existent according to its force". In conclusion, the direct derivation from its latin roots lacks expressiveness. Contrary to that, the historical usage delivers an answer by closing etymological gaps. Many years ago, John Duns Scotus (1266-1308), a Scottish Franciscan friar, introduced terms to describe the reality of existence (realiter) as opposed to the possibility of existence (virtualiter). The heritage of Duns Scotus is carried on in informatics, where "virtual" is used to say that something has "the effect, but not the form" [11: p. 18], e.g. the "virtual cache". Based on its historical usage, Scholz [40] worked out a transdisciplinary and useful definition for both "virtual" and "virtuality". In his view, "virtual denotes the nature of an object, which is not existent in reality, but in potential. Thus, virtuality specifies a concrete object by its characteristics, which are not physically existent, but existent according to their potentiality." [40: p. 328]. The definition implies that a physically existing object is the prerequisite of virtuality. The virtualized object can be defined by its characteristics in comparison to its basis in reality. As Scholz does not define the nature of the real object as a basis for it to become virtual, Dahme and Raeithel [13] add a cohesion to individuals by stating that virtuality is an inner orientated form of reality, which is connected to the persons acting in an environment in a particular way. This is also reflected by Ecarnacao et al [15], who integrate humans as an essential element in the process of virtualization, which they refer to as the digital representation of a real environment.

Summarizing, virtuality denotes a state within the gradual process of virtualization, which can be described as the electronic or digital representation of real objects. This process is closely connected to technology and human beings. On the one hand, technology is the enabler for human beings to realize virtuality. On the other hand, it is the essential means for them to access and experience it in their perception.

4 THE CONCEPT OF VIRTUAL TEAMS

After explaining the notions of groups, teams and virtuality, an image of VTs can only be sketched. This lack of clarity is supported by the many diverse views on VTs often coinciding with definitions of other phenomena, such as virtual and networked organizations, teleworking or telecooperation.

From a systemic point of view a VT can be seen as a structural system which consists of elements who interrelate and are embedded into a process [1]. Elements of the VT are its members, its resources and activities. The state of a VT as a system is the number of relevant attributes. Combining the attributes authors in general agree that a VT is a system of various coworkers operating interdependently towards a common purpose or task. Team interaction and cooperation is predominantly carried out by electronic communication and collaboration applications and takes place at decentralized working locations. The working environment is digitally represented to a certain extent.

Based on this, VTs generally denote any kind of organizational team cooperation sharing the attributes mentioned above. Several authors support this idea by proceeding on the assumption that VT as a term is a genus itself for different classes of virtualized team or group operations

[14, 31], but apply different distinctive variables to classify its varieties. VTs in this notion embrace both intraorganizational varieties such as the virtual office or department and interorganizational ones, i.e. virtual organizations or virtual enterprises [12, 34, 47, 48]. On the contrary, some scholars use "VT" to specify a special composition of virtual organization design. As an example, Palmer and Speier [36] refer to a VT as an intraorganizational group constellation only.

5 TYPES AND DEGREES OF VIRTUALITY IN TEAM-BASED ENVIRONMENTS

There are different views about the starting point at which a team becomes virtual and what distinguishes it from traditional face-to-face settings. Team constellations which cross boundaries of time, space and organizational structure exist in most organizations. By juxtaposing the two extremes, a team solely operating face-to-face and a team exclusively working together electronically, it has to be noted that in business practice hybrid forms are prevalent [18, 44], namely those who mix up both personal encounters and virtual networking. Bearing in mind that co-workers might have a hard time meeting each other, although they work in the same building, the traditional team concept seems to underlie a transition. Parts of the working environment are shifted into the virtual, which implies that different varieties and degrees of virtualization in team constellations exist. Several scholars have presented possible classifications and approaches to separate the appearances of gradual virtual cooperation in team and organizational scopes. Rather than reiterate the approaches, there follows a list of categories that have been used to classify different constellations:

- Degree of time dispersion [31, 55], autonomy and hierarchy [3], proximity [31], isolation [29].
- Technology used [12, 14], Team members technical competency [7, 42, 53]
- Project duration [31], degree of completion [49, 50, 52], Complexity [29].
- Cultural differences [10]
- Nature of the task [3, 14]
- Interdependence ("team contact" vs. "team fusion") [56]

Bearing in mind the definition of virtuality, it can be said, that the more components of a team are expressed in the virtual, the more it is virtualized. Parallel to that, a classification is at least connected to the driving forces mentioned above, since other driving forces are conceivable (e.g. Frequency of F2F meetings, Frequency of technology usage,...). So far, literature offers no integrative concept that intertwines these two dimensions from a broader perspective.

6 ORGANIZATIONAL PATHOLOGIES ACCORDING TO TÜRK [51]

Organizations adopt managerial control and adaption mechanisms to substitute dynamic consensus and planning. These governance frameworks carried out by managers run the risk of neglecting the effects on the actors. The interplay between management strategies, non-situative structural arrangements of working environments and employees may cause pathologies[1]. These are reflected by behaviour and perception of individuals and have the

[1] A "pathology" denotes the suffering of an individual from organizational structures and governance strategies. The medical meaning "disease" is used in an allegoric way [51: p. 109].

potential to spread from the individual to the organizational level, when pathological patterns are embedded into the organizational culture. The organizational culture can be defined as "the implicit awareness of a business, which accrues from the behaviour of the members of the organization and, in return, governs the behaviour of individuals" [41: p. 779]. In this context, Türk [51] defined three basic pathological patterns, Overcomplexity (OC), Overregulation (OR) and Overstabilisation (OS). OC is mirrored in the overcharge of an individual's cognitive system. Management strategies promoting orientation are applied insufficiently or are completely neglected. The individual is confronted with high uncertainty, contradictive requirements and is not able to cope with an exceeding amount of alternatives and data to be processed. OC is reflected by structural overload, structural ambiguity or structural inconsistency. OR is marked by an exceeding reduction of complexity[2] and contingency[3]. It is characterized by a high simplicity, rigidity and repressivity. The available contacts and options an individual can operate in are reduced to a great extent. OS denotes that individuals perceive the social system they act in as fully independent and autonomous from themselves. They forget that the social system is taking into account their individual needs, which may result in an alienation from the product or the business as a whole.

As a consequence, these pathological patterns may result in higher costs and efforts to be burdened by both the organization and the employee. According to Türk these costs, exceeding the planned budget, are threefold and embrace dominance costs (DC), psychic costs (PC) and social costs (SC):

DC denote costs that the organization incurs in the light of known inadequacies because they have strategic goals to achieve. They appear either in form of increased effort or loss of yield. PC is associated with individual dysfunctions such as frustration, higher effort to overcome demotivation, anger, feelings of inferiority or alienation and stem from a negative development in terms of identity and orientation. SC refers to those efforts at the expense of an employee in terms of betraying his socio-emotional relations in order to maintain his working relations. Whereas OC hereby increases loads on positive socio-emotional relationships, OR is reflected in hostilities and the fact that the affected individual is prevented from building and fostering stable interaction relations with colleagues. OS has the most intense impact on the individual in terms of SC, since even private friendships are impeded or instable.

7 PATHOLOGIES IN VIRTUAL TEAMS: A MANAGERIAL APPROACH

The motivation to implement the VT concept often arises because of the various benefits associated with it. Team members are able to work any time any place since information can be distributed, retrieved and stored via technology. This reduces costs in terms of travelling, arranging meetings, hosting and facility maintenance [14, 31] and may result in shortened process cycles with better product and service outputs nearer on demand. Organizations have easier access to individual core competencies, knowledge and talent on a global scale while expanding their relationships and networks [17].

---

[2] The term "complexity" embraces the degrees of proximity, hierarchy and division of labour [51: pp. 44f.].
[3] "Contingency" denotes the spectrum of all possible options the environment offers for a specific scenario or systemic structure. Something is not what it is per se, it could be different according to other random influences and therefore is changeable [51: p. 18].

However, VTs harbour several negative aspects. As an example, an analysis by the German Ministry of Crime (BKA) revealed severe dysfunctions and inefficiencies after their teams were equipped and organized to work with modern communication tools across boundaries of space and time [9]. Although the study did not define the dysfunctions and its understanding of efficiency, it was the first time that a causal relation between virtualization and negative aspects was identified.

Most of the dysfunctions mentioned in the literature stem from the relation of proximity and collaboration underpinned by a study of Allen [5]. In his "50 feet rule" Allen ascertains that people are less likely to cooperate and communicate if they are more than 50 feet apart. Hollingshead [22] emphasizes this by spotting higher information suppression in computer-mediated groups (CMGs) compared to F2F groups. CMGs exchange less information which results in poorer outcomes. Furthermore, VTs made up of members with different cultural backgrounds have to overcome difficulties in understanding, the way of interaction and valuation [10]. Thus, VTs lack social integration [29]. As social integration is often referred to as group cohesion and studies revealed a positive relation between group cohesion and group performance [26, 35], it can be deduced that VTs are less effective in performance than F2F teams. Moreover, Haywood [21] perceives information overload as one of the main stress factors in terms of electronic media and its usage in VTs. In addition, Jarvenpaa and Leidner [24] identified low individual commitment, role overload, role ambiguity, absenteeism and social loafing.

Apart from these rather judgemental approaches to VTs, several authors focused on descriptive topics such as the role of communication and trust [24], transactive memory [18] or information richness and social presence through the use of groupware technology [14]. Managerial issues were covered to some extent but predominantly focused on processes, such as task and resource allocation. The recruiting of team members was highlighted as well as the importance of F2F meetings in the starting phase to create a clear vision and purpose [14, 29, 31]. The role of a VT manager is argued though. On the one hand some contributions cut down managerial functions to technical issues predominantly ensuring the proper infrastructure [37]: hence, the VT manager is seen as a coach or motivator presuming a higher self-responsibility of each team member [29]. On the other hand, Duarte and Snyder [14] broaden the spectrum of tasks of a VT manager by stating that "managing performance occurs at the team level and at the individual level" [14: p. 75]. While still ensuring autonomy to some extent, their ideal VT manager functions both as a coach and performance manager.

It is agreed that, if maintained poorly, the synergies of global connectivity and networking individual potential in a VT can not be experienced. Inexperience or ignorance concerning the infrastructure, the team members, their background and working scenario as well as the process is commonly understood to lead to failures, which result in dysfunctional states and higher costs.

However, the contributions predominantly contribute to general project management techniques on the group and structural level. What does not seem to have been the subject of recent scientific discussion relates to pathologies in relation to virtualization in a VT. This study would involve a generic view of the systemic elements of a VT as it relates to both the individual and the corporate environment.

8 CONCLUSION

As this paper illustrates, researchers have taken different approaches to grasp the new phenomenon of VTs. The review of the literature shows that there currently does not exist a common understanding of what a VT is, which stems from the diverse approaches to the terms, "group", "team" and "virtual". As a result, descriptions vary to a great extent as to what types of different VTs exist and how degrees of virtualization in team-based environments can be identified.

Furthermore, implications for the managerial level mainly concentrate on structural aspects in project management. Apparently, managerial issues concerning the individual that take into consideration possible dysfunctions arising out of the virtual working environment have been ignored. It would seem reasonable to suppose, that by not taking into account the team as the functional unit of every organization and trying to understand both the team's virtualization and its potential to harbour pathologies, the organization misses out on adapting to threatening factors to its stability.

Possible dysfunctions were included by introducing organizational pathologies according to Türk [51]. The model describes potential dysfunctions from a broader perspective. Negative aspects of VTs on the organizational, group, and individual level, which are mentioned in the literature to some extent, can be integrated by seeing them as symptoms of the pathologies OS, OR and OC. Given a latent pathological level for every organization, these pathologies result in a threefold contribution to costs as an extension of the efficiency concept, which the literature predominantly characterizes in terms of output variables.

The present literature review supports Adorno [2], who stated that, although it is known that employees may suffer from organizational structures (leading to higher effort and cost), the individual is predominantly excluded in organizational analysis.

By having identified gaps in current research, the following questions arise:

- How can VTs be classified in terms of their degree of virtualization and nature?
- Does a certain stage within that classification enhance the intensity of one of the mentioned pathologies?
- If so, how can managers assess their VT within a potential classification?
- Which managerial actions have to be carried out in order to minimize the risks evolving out of certain pathological constellations?

Yet, there still seems to exist a lack of an exhaustive approach to these challenging questions, which both ties together findings of different fields and provides practical implications for the managerial level. The aim of this research is therefore the creation of a typology of VTs and, on the basis of the typology, the projection of the pathology concept according to Türk [51] as an interim step and a basis for further considerations in behaviour-oriented management and software engineering.

REFERENCES

1 Ackhoff, R. L. (1971) "Towards a system of system concepts", *Man Sci*, Vol. 11, pp. 661-671.

2 Adorno, T. W. (1953) "Individuum und Organisation (Individual and Organization)", in: Neumark, F. ed. *Individuum und Organisation. Darmstädter Gespräch.* Darmstadt, pp. 21-35.

3 Ahuja, M. K. and Carley, K. M. (1998) "Network and Structure in Virtual Organisations", *Journal of Computer Mediated Communication* [online] 3(4) available from http://www.ascusc.org/jcmc/vol3/issue4/ahuja.html

4 Alioth, A. (1980) „Entwicklung und Einführung alternativer Arbeitsformen (Development and Implementation of alternative working structures)", in: Ulich, E. ed. *Schriften zur Arbeitspsychologie.* Band 27. Bern: Huber.

5 Allen, T. J. (1977) "*Managing the Flow of Technology: Technology Transfer and the Dissemination of Technological Information within the R&D Organization*", MIT Press.

6 Antoni, C.H. (1996) „*Teilautonome Arbeitsgruppen. Ein Königsweg zu mehr Produktivität und einer menschengerechten Arbeit? (Partial Autonomous Work Groups. A solution for higher productivity and to satisfy humans at work?)*", Weinheim: Beltz Psychologie.

7 Bosch-Sijtsema, P. M. (2002) "A Structure of roles within virtual organizations", *International Journal of Information Technology & Decision Making.* Vol. 1(3). London: World Scientific Publishing. pp. 371-384.

8 Boyett, J. H. and Conn, H. P. (1992) "*Workplace 2000: The Revolution Reshaping American Business*", Reprint ed. New York: Plume.

9 Brandt, H. (1990) „Technik im Dienste der Straftatenbekämpfung (Technology to reduce crime)", *BKA-Vortragsreihe zur Arbeitstagung des Bundeskriminalamtes.* Vol. 35. Wiesbaden. 7.-10. November 1989.

10 Carnevale, P. J. and Leung, K. (2003) "Cultural Dimensions of Negotiation", in: Hogg, M. A.; Scott Tindale, R. eds. *Blackwell Handbook of Social Psychology: Group Processes.* Oxford et al.: Blackwell Publishing. pp. 482-496.

11 Cohen, S. G. and Bailey, D. E. (1997) "What makes teams work: Group effectiveness research from the shop floor to the executive suite", *Journal of Management.* Vol. 23, pp. 239-290.

12 Davidow, W. H. and Malone, S. (1992) "*The virtual corporation. Structuring and revitalising the corporation for the 21st century*", New York: Harper Business.

13 Dahme, C. and Raeithel, A. (1997) „Ein tätigkeitstheoretischer Ansatz zur Entwicklung von brauchbarer Software (An activity-oriented theoretical approach to software engineering)", *Informatik-Spektrum.* Vol. 20. Berlin et al.: Springer. pp. 5-12.

14 Duarte, D. L. and Snyder, N. T. (2001) "*Mastering Virtual Teams: Strategies, tools and techniques that succeed*", 2nd ed. San Francisco: Jossey-Bass.

15 Ecarnacao, J. L., Dai, F., del Pino, A., Haase, H., Jacob, U., Unbescheiden, M. and Zachmann, G. (1997) "Grenzen der Virtualisierung (Boundaries of Virtualization)", in: Picot, A. ed. *Telekooperation und virtuelle Unternehmen.* Heidelberg: Decker. pp. 251 – 280.

16 Gebert, D. and Rosenstiel, L. v. (1981) „*Organisationspsychologie: Person und Organisation (Organizational Psychology: Person and Organization)*", Stuttgart et al.: Kohlhammer.

17 Goldman, S.L.; Nagel, R. N.; Preiss, K. (1995) „*Agile Competitors and Virtual Organizations. Strategies for Enriching the customer*", New York: Van Nostrand Reinhold.

18 Griffith, T. L.; Neale, M. A. (2001) "Information Processing in traditional, hybrid, and virtual Teams: From Nascent Knowlede to Transactive Memory", *Research in organizational behavior.* Vol. 23. Stamford: JAI Press. pp. 379-421.

19 Guzzo, R.; Dickson, M. W. (1996) "Teams in organizations: Recent Research on performance and effectiveness", *Annual Review of Psychology.* Vol. 47. pp. 307 – 338.

20 Guzzo, R.; Salas, E. (1995) "*Team Effectiveness and Decision Making in Organization*", San Francisco: Jossey-Bass.

21 Haywood, M. (1998) "*Managing Virtual Teams: Practical Techniques for High-Technology Project Managers*", Norwood, London: Artech House.

22 Hollingshead, A. B. (2003) "Communication Technologies, the Internet, and Group Research", in: Hogg, M. A.; Scott Tindale, R. eds. *Blackwell Handbook of Social Psychology: Group Processes*. Oxford et al.: Blackwell Publishing. pp. 557-573.

23 Hörschgen, H. (1992) „*Grundbegriffe der Betriebswirtschaftslehre (Basics of Business Administration)*" 3rd edition, Stuttgart: Schaeffer-Poeschel.

24 Jarvenpaa, S. L. and Leidner, D. E. (2000) "Communication and Trust in Global Virtual Teams", *Organization Science*. Vol. 10 (6). pp. 791-815.

25 Katzenbach, J.R. and Smith, D.K. (1993) "*The wisdom of teams: Creating the high-performance organization*", Boston: Harvard Business School Press.

26 Kelly, J. R. (2003) "Mood and Emotion in Groups", in: Hogg, M. A.; Scott Tindale, R. eds. *Blackwell Handbook of Social Psychology: Group Processes*. Oxford et al.: Blackwell Publishing. pp. 164-181.

27 Kets de Vries, M. F. R. and Miller, D. (1987) "*The Neurotic Organization. Diagnosing and Changing Counterproductive Styles of Management*", San Francisco, London: Jossey-Bass.

28 Kets de Vries, M. F. R. (1995) „Dysfunktionale Entscheidungsfindung und organisatorische Pathologien (Dysfunctional decision-making and organizational pathologies)", in: Scholz, C.; Zentes, J. eds.. *Strategisches Euro-Management*. Stuttgart: Schaeffer-Poeschel. pp. 247-258.

29 Konradt, U. and Hertel, G. (2002) "*Management Virtueller Teams (Management of virtual teams)*", Weinheim, Basel: Beltz.

30 Lipnack, J. (2003) "Virtual Teams: The Future is now", [online] available from http://www.linezine.com/7.2/articles/jlvtfin.htm

31 Lipnack, J.; Stamps, J. (2000) "*Virtual Teams: People Working across boundaries with Technology*", New York: John Wiley & Sons.

32 McGrath, J. E. (1984) "*Groups: Interaction and Performance*", Englewood Cliffs, NJ: Prentice Hall.

33 Mills, T. (1971) „*Soziologie der Gruppe (Group sociology)*" 3rd edition, Munich: Juventa.

34 Mowshowitz, A. (2002) "*Virtual Organization: Toward a Theory of Societal Transformation Stimulated by Information Technology*", Westport: Quorum Books.

35 Mullen, B. and Cooper, C. (1994) "The relation between group cohesiveness and performance: An integration", *Psychological Bulletin*. Vol. 115. pp. 210-227.

36 Palmer, J.W. and Speier, C. (1997) "A Typology of virtual organizations: An empirical study", in: Gupta, J. Ed. Proceedings of the Association for Information Systems 1997 Americas conference. Indianapolis, USA. 15-17 August 1997 [online] available from: http://hsb.baylor.edu/ramsower/ais.ac.97/papers/palm_spe.htm

37 Pindl, T. (2002) "*Führen und Coachen von virtuellen Netzwerken–unabhängig von Ort und Zeit (Leading and coaching of virtual networks–independent from location and time)*", Cologne: Deutscher Wirtschaftsdienst.

38 Rosenstiel, L.v. (1992) „*Grundlagen der Organisationspsychologie (Foundations of Organizational Psychology)*", Stuttgart: Schäffer - Poeschel.

39 Rosenstiel, L.v. (1995) „Kommunikation und Führung in Arbeitsgruppen (Communication and Leadership in Working Groups)", in: Schuler, H. ed. *Lehrbuch der Organisationspsychologie*. Bern: Huber.

40 Scholz, C. (2000a) „*Strategische Organisation-Multiperspektivität und Virtualität (Strategic Organisation-Multiple Perspectives and Virtuality)*" 2nd, rev. ed. Landsberg/Lech: Moderne Industrie. p. 252.

41 Scholz, C. (2000b) „*Personalmanagement: informationsorientierte und verhaltenstheoretische Grundlagen (Human Resource Management: information-oriented and behavior theoretic basics)2* 5th, rev. ed. Munich: Vahlen.

42 Shao, Y. P., Liao, S. Y. and Wang, H. Q. (1998) "A Model of Virtual Organisations", *Journal of Information Science*. Vol. 24 (5). pp. 305-312.

43 Sinclair, J. (Ed.) (1989) "*Pons Cobuild English learner´s dictionary*", London, Glasgow: Collins.

44 Snyder, B. (2003) "Workers Fear Cooperating in Virtual Teams May Make Them Obsolete" (online) available from http://www.gsb.stanford.edu/news/research/hr_virtualteams.shtml
45 Sommerlatte, T. (2003) „Trends–was bedeuten sie für unser Unternehmen? (Trends–what do they mean for our enterprises?)" [online] available from http://www.innovation-aktuell.de/kv0110.htm
46 Teufel, S., Sauter, C., Mühlherr, T. and Bauknecht, K. (1995) „*Computerunterstützung für die Gruppenarbeit (Computer support for group work)*", Bonn: Addison-Wesley.
47 Tjortjis, C., Dafoulas, G., Layzell, P.J. and Macaulay, L.A. (2002) "A Model for Selecting CSCW Technologies for Distributed Software Maintenance Teams in Virtual Organisations", *IEEE Workshop on Cooperative Supports for Distributed Software Engineering Processes (CSSE 02), Proceedings of the 26th Annual International Computer Software and Applications Conference (COMPSAC 2002).* Oxford, UK. 26–29 August 2002. pp. 1104-1108.
48 Travica, B. (1997) "The Design of the Virtual Organization: A Research Model", *Proceedings of the Association for information systems 1997 Americas conference* [online]. Indianapolis, USA. 15-17 August 1997. available from http://home.cc.umanitoba.ca/~btravica/pub_frameset.html
49 Tuckman, B. (1965) "Development Sequence in Small Groups", *Psychological Bulletin.* Vol. 63. pp. 384-399.
50 Tuckman, B. and Jensen, M.A. (1977) "Stages of Small Group Development Revisited", *Group and Organizational Studies.* Vol. 2. pp. 419-427.
51 Türk, K. (1976) "*Grundlagen einer Pathologie der Organisation (Principles of a pathology of an organization)*". Stuttgart: Enke.
52 Van Wijk, J., Geurts, D. and Bultje, R. (1999) "7 Steps to Virtuality: Understanding the Virtual Organisation Processes before designing ICT-Support", *Proceedings of Objects, Components and the Virtual Enterprise, An interdisciplinary workshop at OOPSLA 98* [online]. Vancouver, Canada. 18-22 October 1998. available from http://www.cs.tcd.ie/Virtues/ocve98/proceedings/005.html
53 Venkatraman N. and Henderson J. C. (1996) "*The Architecture of Virtual Organizing: Leveraging Three Interdependent Vectors*", Discussion Paper. Boston: Systems Research Center, Boston University, School of Management.
54 Wahrig, G. (1985) "*Fremdwörter-Lexikon (Dictionary of foreign loan words)*", Munich: Bertelsmann.
55 Wigand, R.T., Picot, A.and Reichwald, R. (1997) "*Information, Organization and Management -Expanding Markets and Corporate Boundaries*", Chichester: Wiley.
56 Wojda, F. (2000) "*Innovative Organisationsformen – Neue Entwicklungen in der Unternehmensorganisation (Innovative forms of organization–new developments in organizational design)*", Stuttgart: Schäffer-Poeschel.. Stuttgart: Schaeffer-Poeschel.

Web-based Modelling and Simulation

A virtual environment approach for planning implant placement

T ITO
Department of Mechanical Engineering, University of Tokushima, Japan

ABSTRACT

This paper presents the studies on tailor-made design and production of guide templates for osseointegrated implants, which is under progress as a joint research between Department of Mechanical Engineering and School of Dentistry at the University of Tokushima. A guide template is used to support the accurate operation to insert fixture of implants into the appropriate site in terms of position and direction as designed before surgical operation.

The procedure in the project includes several basic steps: digital bone modelling based on medical data, planning of implant placement, designing of guide templates to support the operation, implant simulation based on the model, evaluation on physical model using rapid prototype techniques, production of guide templates based on the design.

This paper focuses on the design issue of implant placement planning in a virtual environment to perform a tailor-made implant treatment. The placement should be designed based on the consideration on where, in which direction, and how deep the fixture should be inserted. This study uses a virtual environment where a designer can feel the geometry of bone with bone density to make an appropriate planning of implant placement in an intuitive manner. For surgical operation, the tailor-made guide template is used to perform an accurate operation.

The paper also describes some of the basic issues regarding design and production of tailor-made guide template. What the project is pursuing is not the technology of standardized template production, but the technology of tailor-made design of templates and its production towards tailor-made implant treatment which satisfies each patient based on individual needs.

1 INTRODUCTION

Nowadays, oral rehabilitation based on osseointegrated implants is becoming a popular and highly successful treatment not only in Japan but also worldwide. Furthermore, new technologies, such as Novum system based on a high level of standardization, have made it possible to achieve immediately fixed prostheses based on osseointegrated implants, which could dramatically reduces the total treatment period for implant.

In the mean time, tailor-made approaches are under study based on the progressing computer technology and advancing medical image processing. Starting from computer tomography data or CT scan data of patients, 3 dimensional bone models, which are almost identical to real ones in terms of shape, can be digitally prepared in high quality level. These models make it possible for dental doctors to study on each patient from various points of views, not only from physical shape of bone but also from bone conditions. The placement should be designed based on the consideration on where, in which direction, and how deep the fixture should be inserted.

This study uses a virtual environment where a designer can feel the geometry of bone with bone density to make an appropriate planning of implant placement in an intuitive manner, of which result can also be reviewed using some commercial software. The paper covers the whole procedures to set up this virtual environment. The procedure in the project includes several basic steps: digital bone modelling based on medical data, planning of implant placement, designing of guide template to support the operation, implant simulation based on the model, evaluation on physical model using rapid prototype techniques, production of guide templates based on the design. The paper also addresses a rapid prototyping approach to tailor-made design and production of guide template, which would make it possible to accurately place implants in terms of direction and depth based on the planning prepared in the virtual environment.

2 BASIC PROCEDURES FOR IMPLANT TREATMENT

Surgical operation for implant treatment is conducted based on the principle procedures as follows: (1) to make a hole into the bone to place a fixture, (2) to place a fixture under sterile conditions, (3) to stitch up the cut, (4) to keep it until enough osseointegration, (5) to connect abutment upon the fixture, and (6) to build superstructure on the abutment such as crown structure.

Implant placement planning which is conducted prior to surgical operation plays a very critical role for appropriate implant treatment. However, implant placement planning requires various knowledge and experience to consider the condition of bone as well as the geometrical shape. Even if the placement plan has completed in an appropriate manner, surgical operation to satisfy the design requires the high level of skill.

This research focuses on two critical issues in this implant treatment. One is design issue for appropriate implant placement planning and the other one is implementation issue for its actual operation. Figure 1 shows a digital image which represents that 6 sets of fixture with abutment are placed on lower jaw bone, which shows the ideal result of implant placement based on the appropriate placement plan. To design this ideal placement is not an easy task so that computer-supported tools are often used.

The planning is designed using digital bone model, considering various points of view such as bone density, bone shape, etc. In this case, some commercial software helps dental doctor to decide which fixture should be appropriate, and then to consider where, in which direction, how deep each fixture should be placed. Interactive software for predictable treatment planning of dental implants is available on the market.

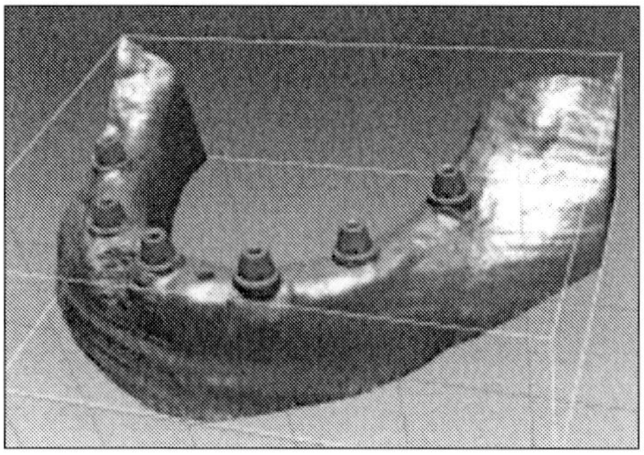

Fig. 1 Ideal implant placement in virtual environment

Although placement planning is supported by the software, digital image interaction is the only interface provided to the users by the software. This research is to create an environment which provide more interactive interface for intuitive interaction in placement design, where the user can not only view the digital 3D model, but also can feel the condition of bone so that the user should make the design more intuitively.

When the placement planning is appropriately completed, it is a dental doctor who performs the operation based on the plan. In order to make sure that the operation should be performed precisely, a guide template is normally used in the operation.

3 OVERALL PROCEDURE OF TAILOR-MADE IMPLANT TREATMENT

Considering the basic procedure of implant treatment mentioned in the previous section, this section presents overall procedure of tailor-made implant treatment based on a virtual environment approach, which are under study for our research in a medical application area. The procedure of this approach includes (1) digital bone modelling, (2) implant placement planning, (3) geometric designing of guide templates, (4)simulation of implant operation, (5)manufacturing and evaluation on physical model using rapid prototype techniques, (6)surgical operation using the guide template based on the design. This section briefly describes the overall procedure of tailor-made implant treatment.

3.1 Digital bone modelling

For tailor-made implant treatment, digital model of the bone should be created precisely. This study uses medical data for the modelling, such as CT scan data, and creates digital bone model, from which preparation of physical rapid prototype is also available. The left model in Figure 2 is an example of digital bone, which was based on medical data, or CT scan data of a patient, and the right one in Figure 2 is a physically rapid-prototyped model based on the digital model. Although this figure shows physical shape of the bone, the model can also include additional information such as bone density, which provides very critical information to determine the appropriate implant placement planning.

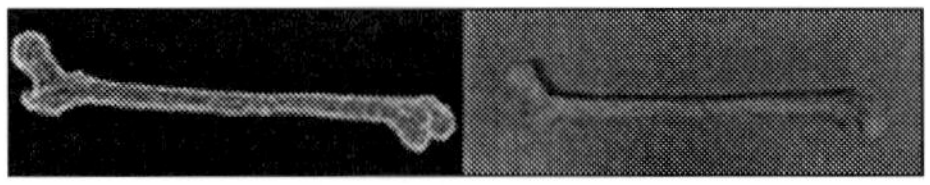

Fig. 2 Digital model and its physical model

3.2 Implant placement planning

Using the technique of digital modelling, jaw bone of a patient for implant treatment is prepared. The model not only represents the actual physical shape of bone of a patient, but it also contains the material condition information on the bone, or bone density. Even if bone tissue exists around the portion to be implanted, it is not always possible to insert a fixture if bone density is low and very fragile. Using the digital bone model in a virtual environment as well as a physical prototype model as show in Figure 3, implant planning is studied and designed by dental doctors. This phase is one of the main focuses of this paper.

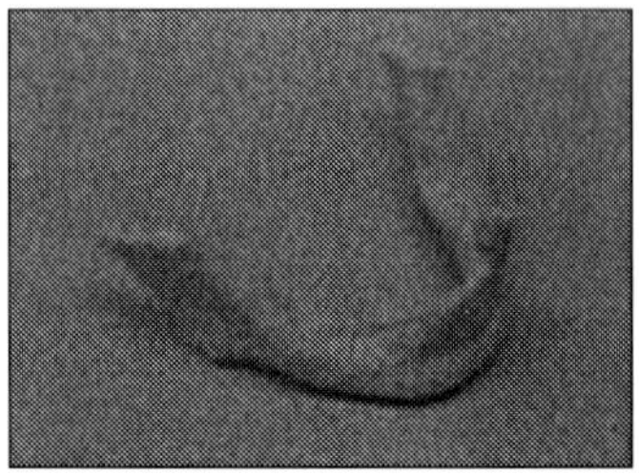

Fig. 3 Example of jaw bone model

3.3 Guide template designing

Once the implant placement is designed, surgical operation is carried out based on the design. Even if the design is completely made, it is not possible to carry out the operation precisely by manual operation without any supporting device. To make it possible, a guide template is normally used to make a hole to insert a fixture into bone. Some general guide templates have been used so far and one solution might be to use one of those general templates. Moreover, to prepare a guide template in a timely manner, one of these templates is normally used. For tailor-made implant treatment, however, these general templates are not suitable in terms of accuracy and some other reasons. In our research, guide template can also be designed in a virtual environment. The more detail will be presented in the next section.

3.4 Implant operation simulation

Using the guide template, implant operation is simulated physically as well as virtually. Design of guide template is evaluated as well as a training environment can be provided in this simulation. If anything wrong is found, designing of guide template may be reconsidered or implant placement plan may be restudied. In any case, preparation for implant treatment is made as much as possible before the actual surgical operation, which would secure the tailor-made treatment.

3.5 Guide template manufacturing

Once the guide template is designed as a 3D CAD model, a rapid prototyped physical model can easily be prepared to evaluate its physical geometry in addition to the evaluation in a virtual environment. Then CAM data can also be generated to be processed in CNC machining to obtain the tailor-made guide template. Although mechanical processing is very difficult and there remain many issues to be solved in guide template manufacturing, some of our results will be presented in the following section.

3.6 Implant operation

Using the guide template, implant operation is carried out based on the placement plan, of which clinical results and its evaluation is out of the scope of this paper and will be reported elsewhere.

4 IMPLANT PLACEMENT DEVICE WITH HAPTIC FUNCTION

This section covers the implant placement plan as mentioned in 3.2. Considering appropriate treatment for each patient, dental doctor studies to decides which fixture should be appropriate, and then decides where, in which direction, and how deep each fixture should be placed. Some commercial software makes it possible to support this planning through user-friendly interaction with graphical user interface. However, intuitive interaction based on feeling and touching, which is common in our lives, is not available in such software.

To support the placement planning in a more intuitive manner, an implant placement device with haptic function is under study in our research. Figure 4 shows a schematic view of the device composed of some arms attached with positioning point for vertical axis movement. The arms provide free movement on horizontal directions, whereas pointing tip provides vertical movement with haptic function, which is provided by the motor-drive with rotary encoder. The pointing tip moves up and down with forced feedback based on the hardness of bone, which enables the user to feel the condition of bone in an intuitive manner when designing the position of fixture on the bone. Interaction is not provided by the commercial software, but it will enhance the accuracy and efficiency of implant placement planning.

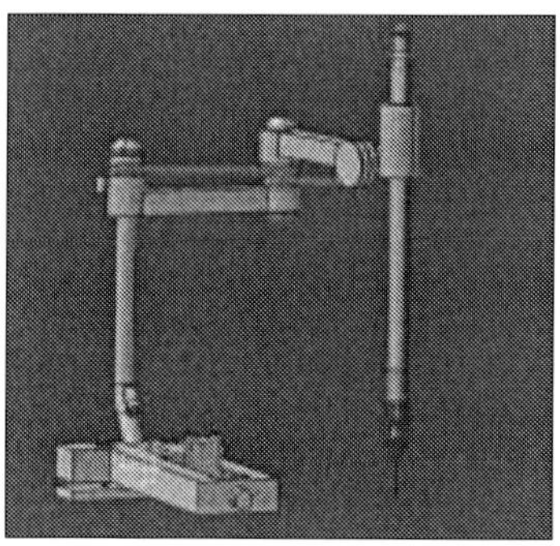

Fig. 4 Schematic view of implant placement device with haptic function

5 PROCEDURE OF TAILOT-MADE GUIDE TEMPLATE

The procedure of our approach includes digital bone modelling using reverse engineering process based on medical data, geometric designing of guide templates based on the digital bone model, implant simulation using the model, fabrication and evaluation on physical model using rapid prototype techniques, production of guide templates based on the design. This section presents the procedure of digital modelling.

The first step in out procedure is to create digital data of jaw bone from CT scan data, of which rapid prototype model. Considering bone density and thickness, placement of fixtures is designed. Once these fixtures are implanted, primary bar is installed on top of them, then secondary bar is connected and denture is installed on it. For this reason, fixture placement determines the overall quality and reliability of implant treatment.

Guide template to be put on the top of jaw bone plays a very critical role to support the accurate operation based on the implant placement plan. The guide template is placed on the upper part of the jaw bone, which means that the contact surface of guide template to jaw bone should be exactly identical to the shape of jaw bone.

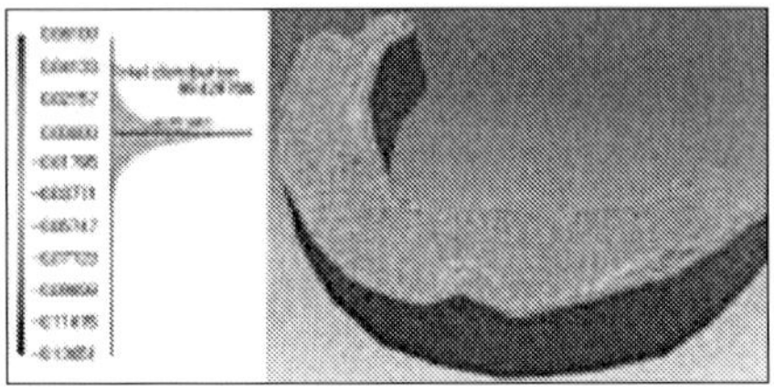

Fig. 5 Accuracy analysis of contact surface

Figure 5 shows contact surface of guide template designed by the digital jaw bone. Accuracy of the contact surface determines the quality of guide template. To evaluate the accuracy of contact surface, the designed surface and the jaw bone surface in the original CT data were compared as shown in Figure 5. The comparison data in this example shows the total distribution of p>96%, which is a very high accuracy. Using the contact surface data, guide template is designed.

6 ENGINEERING ISSUES FOR GUIDE TEMPLATE

The guide template is designed in terms of both digital point of view and physical point of view. When the quality of the designed surface shape satisfies the requirements, overall geometrical shape of guide template is designed as shown in Figure 6. This example shows 6 guide holes to support drilling path and one centre hole to calibrate the guide template to jaw bone.

The guide template is also evaluated in terms of physical point of view. The upper object in Figure 7 shows a rapid prototype model (M1) based on digital design. Thanks to the rapid prototype technology, any complicated shape of object can be created in a precise way, which make it possible to evaluate it physically, some of which might be difficult to evaluate only in digital environment.

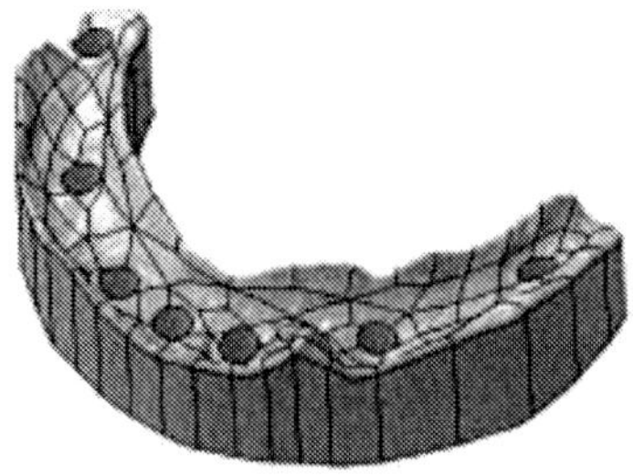

Fig. 6 Design of guide template with drill-holes

The bottom one in Figure 7 shows a metal model (M2), which was manufactured by mechanical processing with CNC-machine based on the digital model. Rapid prototyping makes it possible to manufacture any complicated shape, whereas mechanical processing does not always provide the same manufacturability. In this case, M2 is almost identical to M1, but the study raised the several critical issues. For example, how to shorten production lead time, how to make bottom narrow shape, non-perpendicular hole, how to improve quality, and so no.

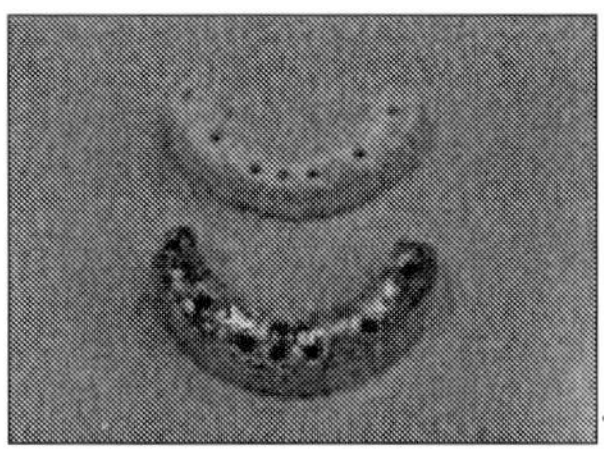

Fig. 7 Examples of guide template

7 CONCLUDING REMARKS

This paper presented an approach of virtual environment for planning of implant placement for osseointegrated implants. To support the accurate operation based on the implant placement, the paper also presented tailor-made implant treatment based on tailor-made design and production for guide templates for surgical operation.

Digital planning of implant placement can be available in some software. However, intuitive interaction, which is common in our lives, is not available in such software. Haptic interaction proposed in this research does not only provide a more suitable environment to study the bone condition of a patient for tailor-made treatment. But it also provide a virtual environment where the patient in treatment himself/herself can feel the bone condition and much clearly understand the implant treatment which is going to be given, which will surely contribute to enhance the informed consent in surgical treatment.

Tailor-made design of the templates and its production plays a very critical role for this purpose. Therefore, to improve the quality of design and production of guide template enhance the reliability of treatment. Experiments for mechanical processing on metal materials raised various kinds of open issues as mentioned in the previous section, which should be studied using advanced technologies, such as virtual manufacturing techniques, micro-laser processing, collaboration technologies, and so on.

On one hand, further detailed studies on each phase are required, but on the other hand, we came to the step where sample guide templates are available through mechanical processing on metal materials. Although the quality of the sample is not high enough, and further efforts are required to meet the actual requirements, but we have a feeling that out approach is feasible and is in the right direction towards the tailor-made implant treatment, which is the ultimate goal of this research.

8 ACKKNOWLEGEMENT

This research was supported in part by the Grants-in-Aid for Scientific Research in 2003-2004 (Scientific Research (C) 15500347) from the Japan Society of Promotion of Science. The author would like to thank Prof. Teisuke Sato of Department of Mechanical Engineering, and Prof. Tetsuo Ichikawa, Dr. Yoritoki Tomotake, DDS. Keiko Shibutani of School of Dentistry of the University of Tokushima. The author would also like to thank the members of Collaborative Engineering Laboratory at the University of Tokushima.

9 REFERENCES

1. P. I. Branemark, P. Engstrand, L. O. Ohrnell, et al, "Branemark Novum: A new treatment concept for rehabilitation of the edentulous mandible. Preliminary result from a prospective clinical follow-up study", Clin. Implant Dent. Res., 1, 2-16, 1999.

2. P. Engstrand, U. Nannmark, L. Martensson, I. Galeus, P.I. Branemark, "Branemark Novum: Prosthodontic and dental laboratory procedures for fabrication of a fixed prosthesis on the day of surgery", Int. Journal of Prosthodont., 14, 303-309, 2001.

3. T. Ito, "Collaborative implant design using the virtual manufacturing approach", INT. J. Computer Integrated Manufacturing, 16, 7-8, 541-545, 2003.

4. T.Ito and T.Sato, "Virtual manufacturing approach to collaborative design and production for hard-tissue implants", 10th International Conference on Human-Computer Interaction, June 22-27, Crete, Greece, 2, 934-938, 2003.

6. SIM/Plant, URL: http://www.materialise.com/simplant/main1_ENG.html.

5. Y. Tomotake, O. Ishida, H. Kanitani and T. Ichikawa, "FuzzyImmediate Implant-supported oral rehabilitation using a photocurable plastic skull model and laser welding. A technical note on the screw-retained type: part 1", The International Journal of Prosthodontics, 15, 3, 303-306, 2002.

Robustness and reliability of manufacturing a pulley in a virtual environment

X LIU, M DANIELS, and **B SHIRVANI**
Technology Innovation Centre, University of Central England in Birmingham, UK

ABSTRACT

Manufacturing like metal work, will have changed significantly in the next five years. Except for Research and Development, more and more companies are transferring their manufacturing operations to low labour cost countries. Virtual engineering plays a vital role in this process. However, virtual engineering technology has to be robust and reliable. This paper describes the reliability of manufacturing in a virtual environment by comparing the results from simulation and experiment of a pulley forming process.

Manufacturers are concerned with the process efficiency, localized buckling and spring back. These characteristics are observed in the simulation process and methods to minimize their effect are investigated. For understanding this operation and verifying the numerical models, experiments which were designed to show the development of the deformation in workpiece have been conducted. The numerical models were built to represent the same conditions as the tests, and the results have shown good correlation between them. The results obtained have shown that designers who can be in different locations can see the manufacturing process in a digital fashion.

Keywords: Simulation, Metal Forming, Finite Element Analysis.

1 INTRODUCTION

For reasons of cost and weight, light gauge sheet is used wherever possible. Therefore, in many cases the process is required to gather the metal into defined areas. This is achieved by superimposing axial loads on top of radial loads using a pressure-controlled tailstock as shown in Fig. 1. Whilst the headstock-mounted tooling is fixed, that part held on the tailstock can be powered axially under controlled pressure. This pressure is governed by the width of the workpiece which changes during the forming process. Very high axial forces are generated whilst profiling the V-section. These range from 50 to 700 kN depending on the work rate, material properties and material thickness.

Figure 1: The forming machine with finished workpiece

Two-groove pulley forming tests have been carried out on the material (BS1149 CR1) in the pre-rolling operation using a Leico LEIFIELD 8631 forming machine [Dierig, H. et al. (1997)]. The machine is suitable for producing different types of pulleys, as well as other hollow, rotationally symmetrical parts, such as splined components. The blanks may be deep-drawn or plain discs with a raised centre. In this way, a high degree of flexibility is gained, both in terms of the shape of the finished part as well as the design of the hub in the blank.

Material handling is enhanced by an automatic loader. The high degree of flexibility described above and operator-friendly level of automation ensure economic production of small to large batches, from single unit to volume production.

Radial forming processes lend themselves to the production of small, medium and large batches, from one off to series production. By combining several operations in one setting, additional savings can be achieved over and above those resulting from

- low energy consumption,

- high material yield, and
- work hardening, resulting in
- improved serviceability.

2 TECHNICAL APPROACH

2.1 Finite Element Modelling

Ring forming is a specialized process typically used to manufacture parts with revolved geometries such as pulleys. The 3-D rolling setup includes an internally mounted idle roller and two continuously rotating driver rollers which also move horizontally towards the workpiece. A tailstock is used to put pressure on the top of the workpiece. The pressure on the workpiece provided by a hydraulic pump is controlled by the forming machine automatically preventing the workpiece from collapsing. The spindle ensures that the rollers and the workpiece synchronise. Fig. 2 shows the setup of the pulley forming system.

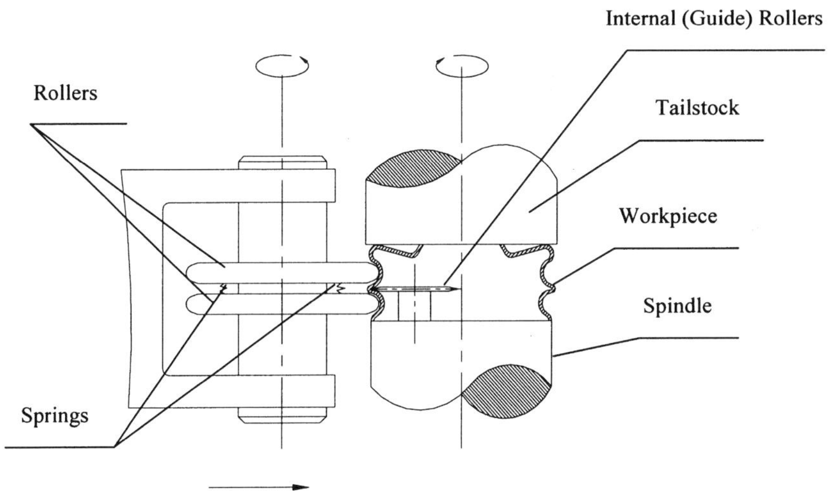

Figure 2: Pulley forming system setup

The numerical model is based on the following facts. The driver rollers have a diameter of 209 mm, and the guide roller has a diameter of 100 mm. The workpiece has an initial diameter of 200 mm and a thickness of 2 mm. The internal and driver rollers are arranged horizontally and are in contact with the inner and outer surfaces of the workpiece, respectively. The internal roller is rotated around its stationary axis; while the driver rollers rotate at 600 r.p.m and move horizontally to the workpiece at a specified feed rate.

In the current paper, several simplifications have been made based on the experiments. The tailstock and the internal roller are replaced by defining the movements of the nodes of the workpiece which the tailstock and the internal roller contact during the forming operation.

This will dramatically reduce the computing time because use of these boundary conditions not only avoids the contact algorithms but also reduces the number of equations.

The methodology used in this work firstly requires the nodes which are on the workpiece and also contact with the tailstock to be identified. They are then grouped as a node set which are defined with the same displacements as the tailstock. Secondly, the horizontal degrees of freedom of the nodes which are on the workpiece and contact with the internal roller are restrained. After these two treatments the tailstock and internal roller will not appear in the model. Experiments have been carried out in order to validate the above simplifications.

By making use of the data obtained from the tests a finite element model was created to represent the real operation. This allows the pulley forming process to be considered in greater detail than would otherwise have been possible. Through the correlation of finite element predictions with measured results the ability of the LS-DYNA package to simulate real behaviour will be evaluated. After the validation, the input parameters can be changed to achieve different objectives.

The following assumptions have been made for creating the FE model:

1. The focus is on the deformation of the wall of the workpiece. The top part does not affect the deformation as it is separated from the whole cup (the area inside the dashed lines was not included in the model, as shown in Fig. 3).
2. The two rollers (see Fig. 2) are connected with springs, which are replaced by the definition of the movements of the rollers. That is, the paths of the two rollers can be defined by monotonous curves.
3. The internal roller (also see Fig. 2) has point contact with the workpiece and the tailstock has line contact with the workpiece.
4. The thickness of workpiece is uniform.

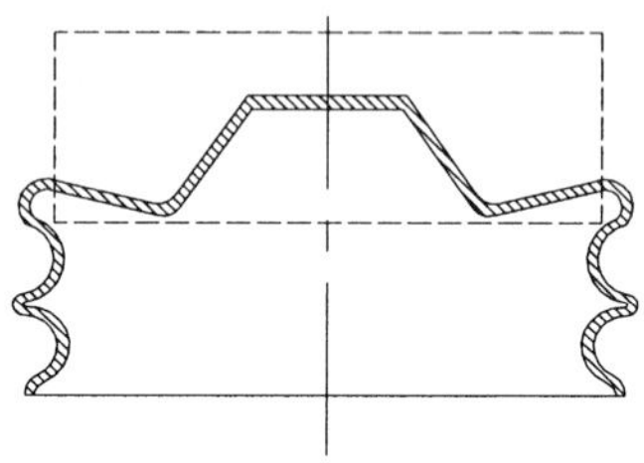

Figure 3: The real pulley geometry.

Materials

Poisson's ratio is 0.292, Young's modulus is 210.0E+03 N/mm^2 and the mass density is 7.85E-09 tonne/mm^3. The following table gives the mechanical properties.

Table 1: Mechanical property of the investigated materials

	YS (MPa)	UTS (MPa)	E (%)
Typical value	192	319	44

Shell Element

Three shell element types were used in this work, namely the Hughes-Liu shell, the Belytschko-Tsay shell and the Belytschko-Wong-Chiang shell. For formability analysis only, the default shell type Belytschko-Tsay shell element needs less CPU time than the other two types. Three integration points are adequate for pulley forming analysis, but when a springback analysis is to be performed five points or seven points are used [Hallquist, J. (1998)].

Contact

Penalty-based contact interfaces are used for this simulation [Maker, B. N. and Zhu, X. (2000)]. The thickness of the workpiece is included in the contact calculation. The workpiece is meshed with 5459 shell elements and the upper and lower rollers each with 1616 linear shell elements as shown in Fig. 4.

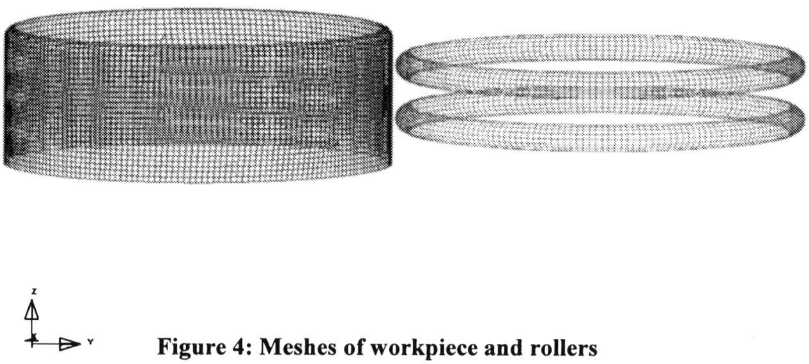

Figure 4: Meshes of workpiece and rollers

2.2 Experimental Validation

There are three objectives for conducting the experiments:

1. providing data to define the movements of the forming tools for the finite element model, including displacements and velocities,
2. understanding the effects of the pulley forming operation on the flow of material, and
3. validating the finite element model.

There are many variables which can be used to validate the simulation. In this work the height of the workpiece and the diameters of the grooves after the forming operation are used to assess the reliability of the finite element modelling.

Fig. 5 shows clearly the development of the deformations of the workpiece both from simulation and experiment. Table 2 shows a good match between them.

Figure 5: The comparison of the simulating and testing

The variables used for model verification are the two groove diameters, D_1, D_2, the radii of the groove cross-section, R_1, R_2, and the distance between the centres of R_1, R_2, and the bottom of the workpiece, H_1, H_2 as shown in Fig. 6.

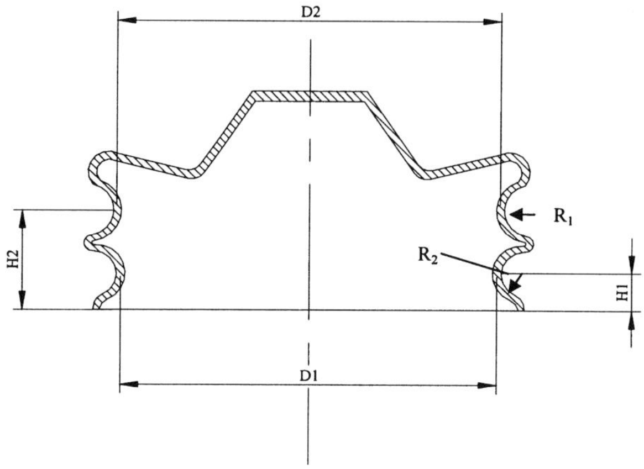

Figure 6: The variables for verification

The final results of the simulation are the measures shown in Table 2, where

$$\text{discrepancy} = \frac{(\text{Value from experiment - Value from simulation})}{\text{Value from experiment}}$$

The radii of the grooves are from the positions of three nodes in the grooves with the same x and y coordinates.

Table 2: Results from experiment and simulation (mm)

	Test	FEA	Discrepancy (%)
Height change	17.8	17.75	0.28
Diameter 1	176.18	175.867	0.177
Diameter 2	176.69	177.323	0.358
H_1	25.90	24.364	5.93
H_2	53.35	52.961	0.793
R_1	7.0	6.95	0.714
R_2	7.0	6.85	2.14

Table 2 shows the discrepancies between experimental and calculated displacements are less than 1% in the diameters, less than 6 % in the heights and radii. Therefore, this FEA model is satisfactory under the assumed conditions.

3 CONCLUSIONS

The flexibility and power of the finite element analysis technique has been demonstrated by its successful application in this research to a pulley forming process. This permits the incorporation of seemingly complicated process parameters in a most effective way and makes simulating manufacturing processes possible.

A virtual manufacturing environment has been created using LS-DYNA simulation. The results have demonstrated the power of finite element simulation for analysing sheet metal forming applications related to potential forming defects such as buckling and springback. It can be used either during the forming design stage or as a trouble shooting tool in the production phase.

The basic idea is that, when pulley forming conditions are given, including the dimensions of workpiece and forming tools, the material properties of workpiece and the feed distance of the forming tools, the results of this operation can be simulated on a computer. The results can be published on the web and shared by designers in different locations in the world.

REFERENCES:

1. Dierig, H. et al. (1997) "The Profiling of Pulleys and Multi-functional Components" (2nd Edition), translated by Pollitt, D.H., Leico GmbH & Co., Ahlen.
2. Hallquist, J. (1998) *"LS-DYNA Theoretical Manual"*, Livermore Software Technology Corporation.
3. Maker, B. N. and Zhu, X. (2000) "*Input Parameters for Metal Forming Simulation Using LS-DYNA*", 6th International LS-DYNA Users Conference, Michigan. 12.1-12.12.

Development of an integrated simulation tool for flexible machining lines

Q WANG and **P G MAROPOULOS**
School of Engineering, University of Durham, UK

ABSTRACT

Changes within the manufacturing environments are forcing automobile manufacturers to perform at new levels of speed, flexibility and inter-company collaboration. To support these changes a prototype multi-component flexible machining line has been developed that allows manufacturers to balance lines by allocating operations according to machine specifications. An integrated computer software tool is developed to enable the responsive design, operation and cost effective configuration of multi-component flexible machining lines. The software will calculate operating cycle times and generate a respective machine layout. Discrete event simulation models which link this software will also provide a dynamic perspective of the balanced production line. The new software will allow users to analyse throughput, work-in-process, resource utilisation and buffer sizes.

1 INTRODUCTION

The Technology Foresight Programme has identified that UK manufacturing productivity remains below that achieved in most industrial nations. Within the automotive component industry productivity represents a particular problem since there is a high level of over-capacity (1) and increasing customer demands for both lower prices (2) and greater levels of service. To overcome these limitations a flexible machining line has been developed in order for UK automotive suppliers to maintain competitive productivity levels.

A major constraint to the use of this technology is the lack of proficient processes for the responsive design of flexible machining lines (FM/cLs). A responsive design process is essential to enable frequent design reconfiguration when product volume and variation changes. The principal aims of this research are to develop an integrated computer-based simulation/optimisation toolkit that can significantly improve the FM/cL design process making possible the responsive design of efficient and cost effective FM/cLs and ensuring rapid FM/cL implementation times.

2 SYSTEM DESIGN

The multi-component flexible machining line has been designed to reduce lead time and cost, and to improve quality when machining dissimilar parts. Therefore, much of the machining is performed using cellular CNC machining centres, linked together by manual or robotic work handling. The aim of cellular manufacturing (3) is to group parts with similar processing needs into part families, and machines that meet these needs into machine cells.

Since the processing sequence of parts can differ, the location of machines seriously affects the material handling requirement. A traditional case would be to set up a separate line for each part. Though this would reduce the material handling cost, it would increase the idle time of the machines, and it would not be cost effective. Separate lines for individual components could reduce the material handling requirement, but conversely would increase machine idle time, reducing the cost efficiency of the machine.

An alternative would be to consider a line setup for a number of parts grouped together on the basis of the similarity in their processing sequences. This would increase the utilisation of the machines and reduce the material handling cost.

A framework of the system design and its overall sequence of activities are shown in Figure 1. The main tasks are database design, development of optimisation routines, fitness function calculation and development of procedures for controlling operational planning aspects. In addition, each of these main tasks will require identification of the data inputs/outputs between other tasks. During the development work, assumptions will need to be made and verified with automobile manufacturers.

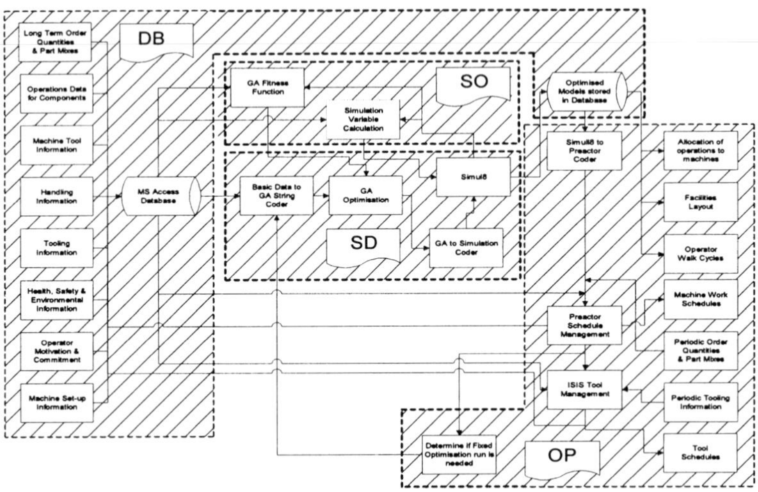

Fig. 1 Framework of system design for flexible machining lines
DD: database design, SD: system design, SO: system optimisation, OP: operational planning

3 DESIGN IMPLEMENTATION

3.1 Design guidelines

Design assumptions:

1. Within the Lead Time the line will produce a single batch of basically similar parts.
2. The line will be laid out in U shape, but where a number of machines are needed to achieve the Takt time, these machines will be doubled up.

What we try to achieve are:

1. The use of Genetic Algorithm to allocate operations to machines taking into account their capabilities and cost.
2. The use of Genetic Algorithm to allocate machines to walk cycles.
3. Modelling of the line within Simul8 and automatic generation of models by XML.
4. Basic walk cycle simulation.

3.2 Design process

3.2.1 Genetic algorithm

A genetic algorithm (GA) (4, 5) is a randomised parallel search method based on evolution.

In a classical GA, the system parameters are presented as a binary string. Borrowing from biology, an encoded parameter can be thought of as a gene, where the parameters values are the genes alleles. The string produced by the concatenation of all the encoded parameters forms a genotype. Each genotype specifies an individual, which is in turn a member of a population. An initial population of individuals, each represented by a randomly generated genotype, is created and the GA starts to evolve solutions from this initial population (6). The three basic genetic operators: selection, crossover, and mutation perform the search. Genotypic strings can be thought of as points in a search space of possible solutions. They specify a phenotype, which can be assigned a fitness value. Fitness values affect selection and in this way guide the research.

Selection of a string depends on its fitness relative to that of other strings in the population. The genetic search process is iterative: evaluating, selecting, and recombining strings in the population during each iteration (generation) until reaching some termination condition. The basic algorithm, where P (t) is the population of strings at generation t, is shown below:

```
Initialise P (t)
Evaluate P (t)
While (termination condition not satisfied) do
Select P (t+1) from P(t)
Recombine P (t+1)
Evaluate P (t+1)
t =t+1
```

3.2.2 Databases design

Figure 1 described a framework of system design for the flexible machining lines. Within such system and essentially the information drivers, are the databases. These have been separated into four areas, each with its own inputs into the GA string. The first of these databases contains part routing information (Figure 2). Within this database each of the process steps required for manufacturing are listed using a designated operation number. Each of the process steps has a description which links it to the class of the processes required to compete the operation. Each of the operations requires a set-up time and a process time indicating non-value adding and value adding time respectively.

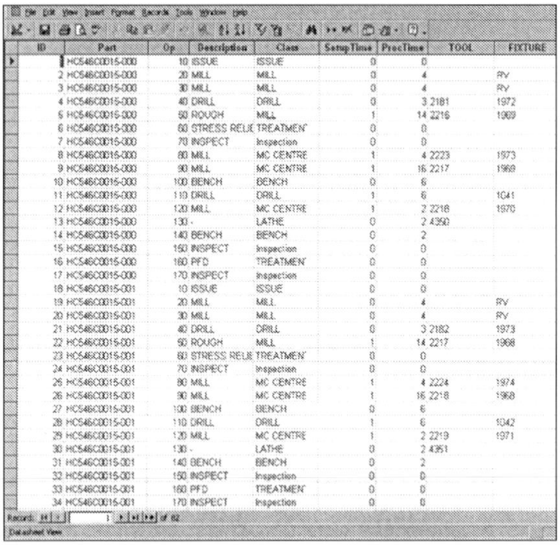

ID	Part	Op	Description	Class	SetupTime	ProcTime	TOOL	FIXTURE
1	HC546C0015-000	10	ISSUE	ISSUE	0	0		
2	HC546C0015-000	20	MILL	MILL	0	4		RV
3	HC546C0015-000	30	MILL	MILL	0	4		RV
4	HC546C0015-000	40	DRILL	DRILL	0	3	2181	1972
5	HC546C0015-000	50	ROUGH	MILL	1	14	2216	1969
6	HC546C0015-000	60	STRESS RELIE	TREATMEN'	0	0		
7	HC546C0015-000	70	INSPECT	Inspection	0	0		
8	HC546C0015-000	80	MILL	MC CENTRE	1	4	2223	1973
9	HC546C0015-000	90	MILL	MC CENTRE	1	16	2217	1969
10	HC546C0015-000	100	BENCH	BENCH	0	6		
11	HC546C0015-000	110	DRILL	DRILL	1	6		1041
12	HC546C0015-000	120	MILL	MC CENTRE	1	2	2218	1970
13	HC546C0015-000	130	-	LATHE	0	2	4350	
14	HC546C0015-000	140	BENCH	BENCH	0	2		
15	HC546C0015-000	150	INSPECT	Inspection	0	0		
16	HC546C0015-000	160	PFD	TREATMEN'	0	0		
17	HC546C0015-000	170	INSPECT	Inspection	0	0		
18	HC546C0015-001	10	ISSUE	ISSUE	0	0		
19	HC546C0015-001	20	MILL	MILL	0	4		RV
20	HC546C0015-001	30	MILL	MILL	0	4		RV
21	HC546C0015-001	40	DRILL	DRILL	0	3	2182	1973
22	HC546C0015-001	50	ROUGH	MILL	1	14	2217	1968
23	HC546C0015-001	60	STRESS RELIE	TREATMEN'	0	0		
24	HC546C0015-001	70	INSPECT	Inspection	0	0		
25	HC546C0015-001	80	MILL	MC CENTRE	1	4	2224	1974
26	HC546C0015-001	90	MILL	MC CENTRE	1	16	2218	1968
27	HC546C0015-001	100	BENCH	BENCH	0	6		
28	HC546C0015-001	110	DRILL	DRILL	1	6		1042
29	HC546C0015-001	120	MILL	MC CENTRE	1	2	2219	1971
30	HC546C0015-001	130	-	LATHE	0	2	4351	
31	HC546C0015-001	140	BENCH	BENCH	0	2		
32	HC546C0015-001	150	INSPECT	Inspection	0	0		
33	HC546C0015-001	160	PFD	TREATMEN'	0	0		
34	HC546C0015-001	170	INSPECT	Inspection	0	0		

Fig. 2 Part routing database

The process classification database shown in Figure 3 is used to establish the range of process that can be carried out on the parts. In order for the generation process to be completed successfully, all operation classifications should be identified and entered into this area.

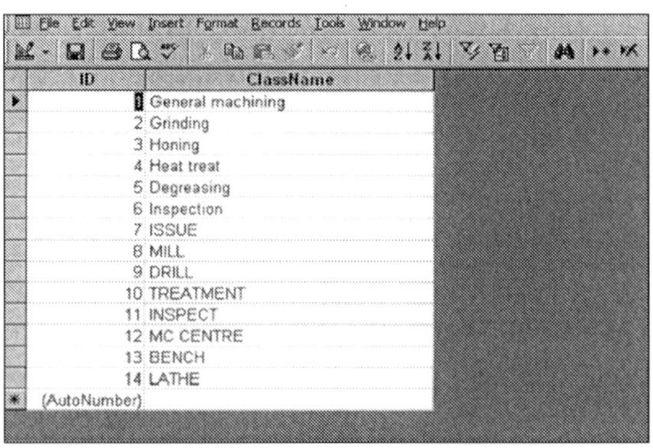

ID	ClassName
1	General machining
2	Grinding
3	Honing
4	Heat treat
5	Degreasing
6	Inspection
7	ISSUE
8	MILL
9	DRILL
10	TREATMENT
11	INSPECT
12	MC CENTRE
13	BENCH
14	LATHE
(AutoNumber)	

Fig. 3 Process classification database

The part quantity database (Figure 4) is where the demand input data is stored. The ID column contains a unique identifier is used to indicate the sequence and ties the parts table to the routing database. It can be seen that part HC546C0015-000 starts with the unique identifier 1 while Part HC546C0015-001 has the unique identifier 18, this correspond to the position in the part routing table where part HC546C0015-001 starts.

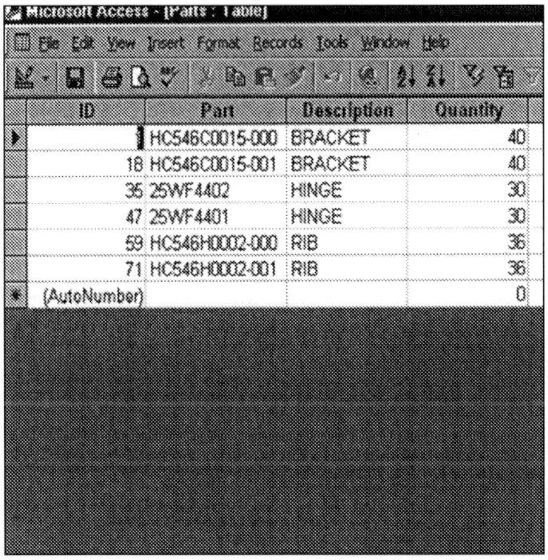

ID	Part	Description	Quantity
1	HC546C0015-000	BRACKET	40
18	HC546C0015-001	BRACKET	40
35	25WF4402	HINGE	30
47	25WF4401	HINGE	30
59	HC546H0002-000	RIB	36
71	HC546H0002-001	RIB	36
(AutoNumber)			0

Fig. 4 Part/quantity database

The machine/process relationship database (Figure 5) is the area where the operation is linked to machines and also allows the selection of different machines with different process capabilities. Within this database, machine set-up time is entered. Although the process time of a selected machine may not be required to complete the task, the set-up time between the parts could have a significant influence on the overall effectiveness of the optimisation process.

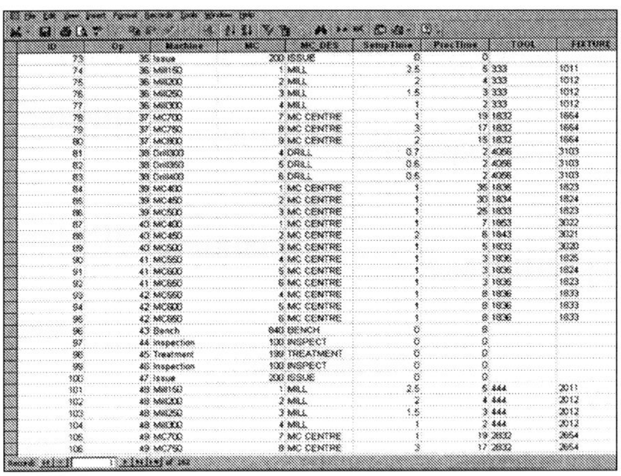

ID	Op	Machine	MC	MC_DES	SetupTime	ProcTime	TOOL	FIXTURE
73	35	Issue	200	ISSUE	0	0		
74	36	Mill150	1	MILL	2.5	5	333	1011
75	36	Mill200	2	MILL	2	4	333	1012
76	36	Mill250	3	MILL	1.5	3	333	1012
77	36	Mill300	4	MILL	1	2	333	1012
78	37	MC700	7	MC CENTRE	1	19	1832	1864
79	37	MC750	8	MC CENTRE	3	17	1832	1864
80	37	MC800	9	MC CENTRE	2	15	1832	1864
81	38	Drill300	4	DRILL	0.7	2	4056	3103
82	38	Drill350	5	DRILL	0.6	2	4056	3103
83	38	Drill400	6	DRILL	0.5	2	4056	3103
84	39	MC400	1	MC CENTRE	1	35	1836	1823
85	39	MC450	2	MC CENTRE	1	30	1834	1824
86	39	MC500	3	MC CENTRE	1	25	1833	1823
87	40	MC400	1	MC CENTRE	1	7	1853	3032
88	40	MC450	2	MC CENTRE	2	6	1843	3031
89	40	MC500	3	MC CENTRE	1	5	1833	3030
90	41	MC550	4	MC CENTRE	1	3	1836	1825
91	41	MC600	5	MC CENTRE	1	3	1836	1824
92	41	MC650	6	MC CENTRE	1	3	1836	1823
93	42	MC550	4	MC CENTRE	1	8	1836	1833
94	42	MC600	5	MC CENTRE	1	8	1836	1833
95	42	MC650	6	MC CENTRE	1	8	1836	1833
96	43	Bench	840	BENCH	0	8		
97	44	Inspection	100	INSPECT	0	0		
98	45	Treatment	199	TREATMENT	0	0		
99	46	Inspection	100	INSPECT	0	0		
100	47	Issue	200	ISSUE	0	0		
101	48	Mill150	1	MILL	2.5	5	444	2011
102	48	Mill200	2	MILL	2	4	444	2012
103	48	Mill250	3	MILL	1.5	3	444	2012
104	48	Mill300	4	MILL	1	2	444	2012
105	49	MC700	7	MC CENTRE	1	19	2832	2854
106	49	MC750	8	MC CENTRE	3	17	2832	2854

Fig. 5 Machine/process relationship database

3.2.3 Simulation models

After the selection process (7) had been completed the generated data is transferred to Simul8 and the model created (8).

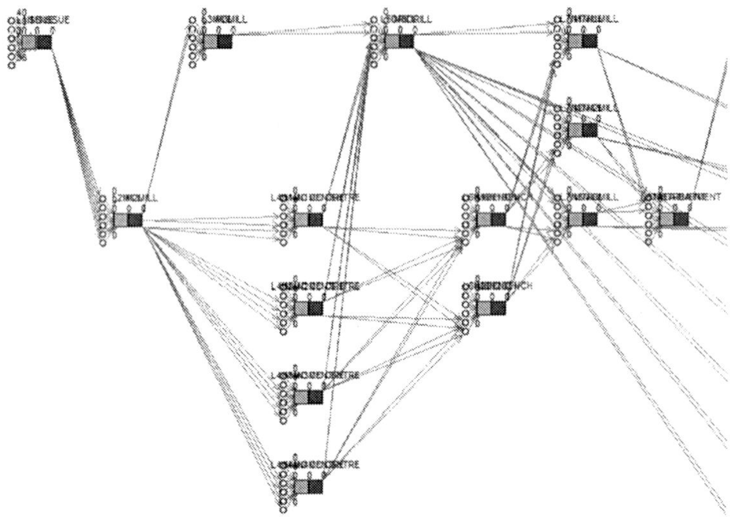

Fig. 6 Initial simul8 model

It can be seen in Figure 6 that parts were introduced on the left at the work station, where the process of introduction of the raw material was handled and then passed on to the machine that would process the operation.

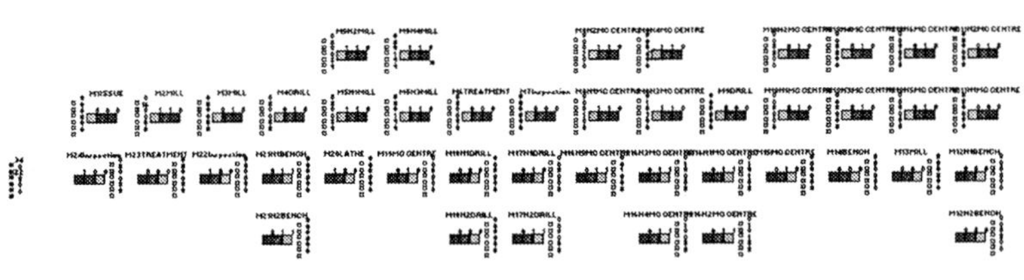

Fig. 7 'U' shaped cell

The model was then refined to reflect the desired 'U' shaped cell (Figure 7) which is normally the most efficient cell layout within an agile manufacturing environment. The 'U' shape is especially useful for the design of the operator walk cycles. As shown before in Figure 6, the system has grouped the machines that were designated to the operator. In this model there was a flow of parts but no walk cycles were shown or included, however, the walk cycle information was used to create the cell into the 'U' shape.

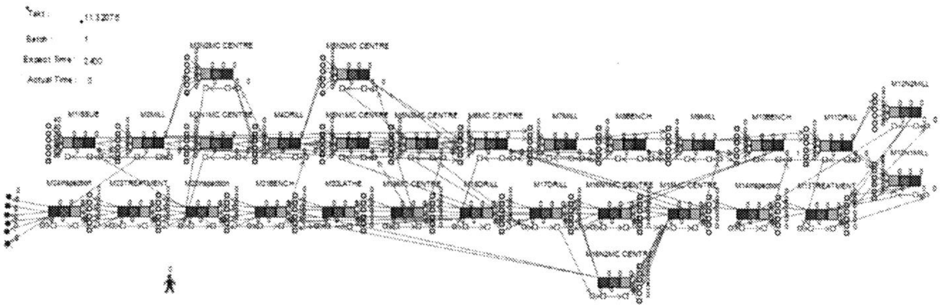

Fig. 8 Six parts model with operator walk cycles

It can be seen in Figure 8 that the cell has been designed to process six part types and the machines are grouped into the 'U' shape. The figure of the man is a designated resource that does not affect the running of the model but is used to tie the operator and the machine together.

Figure 9 shows a detailed representation of the machine, operator load and unload station. The work centre consists of six work storage areas, which are represented by the blue circles. The machine is made up of two work centres and a storage bin. The work centres are used to represent the load and unload areas and link the part to the next process. It was decided to use a storage bin as the actual process area to allow for the use of pallets and multiple process/storage devices. The process time is represented by the wait time in the bin.

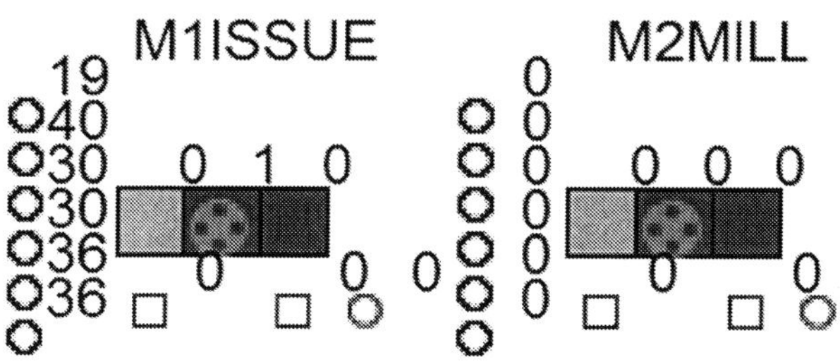

Fig. 9 Work station

The operators are controlled by the two red squares, which are work centres and the red circle, which is a storage bin. The red work centres are the areas where the operator performs the load and unload operation and the storage bin is the area where the operator waits to allow for slack in the system. If an operator arrives at a work station before the part is ready to be loaded and/or unloaded, he waits in the storage area and is released to perform the task when required.

4 CONCLUSIONS

The automotive industry has used discrete event simulation for many years to investigate the capabilities of the different manufacturing systems involved in building automobiles. In this paper, simulation of machining areas allows manufacturers to investigate the impact that new equipment, conveyor, and production schedules during concept, design, and build phases of a product life cycle. Machining models investigate the impact of synchronous machining areas separated by conveyors and the impact on system performance. The system can be evaluated by linking machine characteristics, simulation variable and Genetic Algorithm.

REFERENCES

1 Naughton, K. (1999) "The global six", *Business Week*, January, pp 16 - 21.

2 Wooding, P. (1997) "We know what we have to do. We know where we have to do it. General Motors' Chairman Jack Smith spells out the way forward to its suppliers of the year", *Automotive Sourcing*, Vol 5, No 4, pp 36-52.

3 Irani, S.A. (1999) "*Handbook of cellular manufacturing systems*", Wiley-Interscience, 1999.

4 Davis, K. (1996) "*Handbook of genetic algorithm*", International Thomson Computer Press, London.

5 Michalewicz, Z. (1994) "*Genetic algorithms + data structure = evolution program*", Springer-Verlag, New York.

6 Kado, K., Ross, P. and Corne, D. (1995) "A study of genetic algorithm hybrids for facility layout problems", *Proceedings of the 6th International Conference on Genetic Algorithms*, pp 204 - 213.

7 Paul, R. J. and Chanev, T. S. (1998) "Simulation optimisation using a genetic algorithm", *Simulation Practice and Theory*, Vol 6, pp 601 - 611.

8 Robinson, S. (1993) "The application of computer simulation in manufacturing", *Integrated Manufacturing Systems*, Vol. 4, pp 18-23.

Web-based Java analysis software for engineering applications of axisymmetric plane stress problems

X L YOU
Consultant, Bournemouth, UK

ABSTRACT

Web-based Java analysis software for engineering applications of axisymmetric plane stress problems is developed in this paper. In order for researchers and engineers to apply the most advanced analysis method, a simple, efficient and accurate numerical method is proposed to predict deformations and stresses of axisymmetric plane stress problems. Java language and Hyper Text Markup Language (HTML) are used to produce applets of the developed numerical method and static Web pages, respectively. A numerical example is presented to demonstrate the application of the developed numerical method and Java analysis software.
Keywords: Web-based Java analysis software, axisymmetric plane stress, elastic and elastic-plastic, numerical approach

1 INTRODUCTION

Axisymmetric plane stress problems can be found in many engineering applications such as high speed gears, fly wheels, turbine rotors which are usually simplified as rotating disks and circular disks subjected to internal or external pressure or temperature changes. The elastic and elastic-plastic analyses of these engineering components have become an important issue in solid mechanics and engineering applications. Developing Web-based analysis software is important for researchers and engineers to tackle these problems using advanced analysis techniques through Internet.

Up to now, engineering analysis software is mainly designed for a single user. Accompanying the quick development of computer science, the Internet and Web applications provide a new opportunity to share and distribute information. Two obvious advantages of Web applications are: they overcome the limits of user's hardware and software, and offer a platform for Web collaboration.

There have been some research studies on Web-based design and manufacture, Web-based modelling and simulation, and on-line learning and education.

For Web-based design and manufacture, Szykman et al. presented a language for the modelling of engineering design artefacts and described the implementation of prototype tool suite [1]. Cheng et al. developed an engineering approach for implementing Web-based design and manufacturing support systems and addressed design information integration, remote execution of the systems, use of Java programming, client-server architecture, open computing and user interface design, and the human-computer interaction [2]. Lan et al. proposed a novel integrated system of rapid product development based on rapid prototyping and established a network service system which offers better support for the rapid product development of small and medium sized enterprises [3]. Zeng et al. presented a Web-base CAD system installed on the server with which the designer can build a three-dimensional model of products through the Web browser of any system [4]. Xue and Xu discussed a new approach for Web-based collaborative concurrent design in which systems, product libraries and product databases for modelling different product development life-cycle aspects are distributed at different locations that are linked through the Web [5]. Qin et al. proposed a Web-based conceptual design prototype system which interpret on-line sketches as 2D and 3D geometry, extract 3D hierarchical configurations, allow editing of component behaviors, and produce VRML-based behavioural simulations for design verification and Web-based application [6]. Zhang et al. gave a Web-based intelligent forging process plan and investigated the related key technologies which lead to the development of a prototype system [7].

For Web-based modelling and simulation, Salisbury et al. introduced a government owned, browser-based simulation visualizer capable of displaying simulated entities to any number of distributed sites [8]. Zhang et al. presented a framework for modeling the quality of Web-based information systems [9]. Chen and Lu created a completely Web-based, platform-independent, power system simulation package with various analyses distributed in a clustered environment [10]. Choi et al. proposed a Web-based scalable deformable model by simulating deformation of soft tissues as a successive force propagation process which avoids laborious formulation of stiffness matrices in conventional mass-spring models [11]. Huang explored a generic framework in which a dynamic environmental modeling process can be interactively visualized, controlled and tuned on the Internet by using Java 3D-based hybrid Internet computation [12].

For on-line learning and education, MacDonald et al. presented the demand-driven learning model (DDLM) which provides an explicit statement of a high-quality standard of Web-based learning (WBL) [13]. In the work of Papanikolaou et al., learners' knowledge level and individual traits are used as valuable information to represent learners' current state and personalise the educational system accordingly, in order to facilitate learners to achieve their personal learning goals and objectives [14]. Yen and Li presented a Web-based learning and instructional system for Pneumatics. The system includes course material, remote data acquisition modules, and a pneumatic laboratory set [15].

Compared to the above work, few publications can be found on Web-based engineering analysis software. Manuel et al. described how to take advantage of Internet services and object technology to solve 2D partial differential equations in a distributed manner [16].

In order to develop Web-based analysis software of axisymmetric plane stress problems, advanced analysis methods should be proposed. Here we concisely introduce some existing work on rotating disks and circular disks under external pressure and issues to be addressed.

Elastic analysis of isotropic rotating disks with constant thickness can be found in some books of elasticity such as the book by Timoshenko and Goodier [17]. Elastic thermal analysis of rotating disks of varying thickness and density was performed by Gurushankar [18]. Sterner et al. presented a unified numerical approach to investigate rotating disks including turbine rotors which have arbitrarily variable thickness and density [19]. Adopting Tresca's yield criterion, Gamer determined elastic-plastic deformations and stresses in rotating disks made of perfectly plastic materials [20]. Later on, Gamer extended his work from perfectly plastic to linear strain-hardening. In [21], he discussed deformations of rotating solid disks with constant thickness. In [22], he examined elastic-plastic stress distribution in the rotating annulus and in the annulus under external pressure. Using the same linear strain-hardening material model but considering the thickness variation, Güven gave analytical expressions of elastic-plastic deformations and stresses of rotating disks with some specially varying thickness and density [23]. He also obtained elastic-plastic stress distributions in circular disks under external pressure with some specially variable thickness functions [24]. More recently, Eraslan and Orcan employed the same Tresca's yield condition and linear strain-hardening stress-plastic strain relationship to investigate rotating disks which have more specially variable thickness functions [25].

For practical engineering materials, their stress-strain relation is nonlinear with obvious strain-hardening. The existing studies into perfectly plastic rotating disks did not consider the strain-hardening effects of the materials and those on linear strain-hardening cannot describe the nonlinearity of material stress-strain relationship realistically. In addition, all the existing work only tackled the cases of constant thickness or specially variable thickness. Clearly, improvements should be made to cope with the nonlinearity of material behaviours and arbitrarily variable thickness, density and temperature.

In this paper, we will develop a simple, efficient and effective numerical method which can deal with realistic stress-strain relations of materials and any variations of thickness, density and temperature. Based on this advanced analysis method, a Web-based Java software system will be developed.

2 OVERVIEW OF NUMERICAL METHOD

The proposed numerical method used to develop the Web-based Java analysis software can be derived from the basic equations of axisymmetric plane stress problems. For elastic analysis, these basic equations consist of strain-displacement and stress-strain relationships, Hooke's law, and deformation compatibility and equilibrium equations.

For axisymmetric problems, there is a radial displacement only. The relation between strains and radial displacement is called geometric equation which can be written as

$$\begin{aligned} \varepsilon_r &= \frac{du}{dr} \\ \varepsilon_\theta &= \frac{u}{r} \end{aligned} \tag{1}$$

where ε_r and ε_θ are the radial and circumferential strains, u is the radial displacement, and r is the radial coordinate.

The deformation compatibility equation can be achieved from the geometric equation (1) which has the form of

$$r\frac{d\varepsilon_\theta}{dr}+\varepsilon_\theta-\varepsilon_r=0 \tag{2}$$

Assuming the variation of the temperature along the radial direction is $T(r)$ and the thermal expansion coefficient of the material is α, and taking into account the thermal deformation and the assumption of plane stress, the strain-stress relation for elastic analysis reads

$$\begin{aligned}\varepsilon_r&=\frac{1}{E}(\sigma_r-\nu\sigma_\theta)+\alpha T(r)\\ \varepsilon_\theta&=\frac{1}{E}(\sigma_\theta-\nu\sigma_r)+\alpha T(r)\end{aligned} \tag{3}$$

where σ_r and σ_θ are the radial and circumferential stresses, $T(r)$ is the temperature change, and E and ν are Young's modulus and Poisson's ratio of the material.

For axisymmetric plane stress problems, there are no shear stresses. The equilibrium equation is given by

$$\frac{d}{dr}(hr\sigma_r)-h\sigma_\theta+h\rho\omega^2r^2=0 \tag{4}$$

where h is the thickness of rotating disks or pressurized circular disks, ω is the angular velocity of rotation, and ρ is the density of the material.

The governing equation for elastic problems can be derived from the above equations which can be expressed as

$$\begin{aligned}\frac{d^2\sigma_r}{dr^2}=-\frac{1}{r}\Bigg\{&(3+\frac{r}{h}\frac{dh}{dr})\frac{d\sigma_r}{dr}+\frac{1}{h^2}\left\{r\left[h\frac{d^2h}{dr^2}-\left(\frac{dh}{dr}\right)^2\right]+(2+\nu)h\frac{dh}{dr}\right\}\sigma_r\\ &+\left[(3+\nu)\rho+r\frac{d\rho}{dr}\right]\omega^2r+E\alpha\frac{dT}{dr}\Bigg\}\end{aligned} \tag{5}$$

For elastic-plastic analysis, the total strains consist of the elastic, plastic and thermal strains. Therefore, Eq. (3) should be changed into

$$\begin{aligned}\varepsilon_r&=\frac{1}{E}(\sigma_r-\nu\sigma_\theta)+\frac{\varepsilon_e^p}{\sigma_e}(\sigma_r-\frac{1}{2}\sigma_\theta)+\alpha T(r)\\ \varepsilon_\theta&=\frac{1}{E}(\sigma_\theta-\nu\sigma_r)+\frac{\varepsilon_e^p}{\sigma_e}(\sigma_\theta-\frac{1}{2}\sigma_r)+\alpha T(r)\end{aligned} \tag{6}$$

where σ_e and ε_e^p are the equivalent stress and equivalent plastic strain, respectively. The equivalent stress is represented by

$$\sigma_e=\sqrt{\sigma_r^2-\sigma_r\sigma_\theta+\sigma_\theta^2} \tag{7}$$

From tensile tests, we can determine the relationship between uniaxial plastic strain and stress which can be formulated as

$$\varepsilon^p=f(\sigma) \tag{8}$$

where ε^p and σ are the plastic strain and stress from the tensile tests.

The same relation as that of Eq. (8) exists between the equivalent plastic strain and equivalent stress. Therefore, the equivalent plastic strain can be determined by

$$\varepsilon_e^p = f(\sigma_e) \tag{9}$$

The governing equation for elastic-plastic analysis can also be derived from the above equations whose mathematical expression is

$$\frac{d^2\sigma_r}{dr^2} = -\left\{\left[rF_1 + (3+\frac{r}{h}\frac{dh}{dr})(\frac{1}{E}+\frac{f}{\sigma_e})\right]\frac{d\sigma_r}{dr} + \left\{(0.5+\frac{r}{h}\frac{dh}{dr})F_1 + \frac{1}{Eh}\left\{(2+\nu)\frac{dh}{dr}+\frac{r}{h}\left[h\frac{d^2h}{dr^2}-\left(\frac{dh}{dr}\right)^2\right]\right\} + \frac{f}{h\sigma_e}\left\{2.5\frac{dh}{dr}+\frac{r}{h}\left[h\frac{d^2h}{dr^2}-\left(\frac{dh}{dr}\right)^2\right]\right\}\right\}\sigma_r + \left(rF_1 + \frac{3+\nu}{E}+\frac{7f}{2\sigma_e}\right)\rho\omega^2 r + (\frac{1}{E}+\frac{f}{\sigma_e})\frac{d\rho}{dr}\omega^2r^2 + \alpha\frac{dT}{dr}\right\} \Bigg/ \left[r(\frac{1}{E}+\frac{f}{\sigma_e}) + (0.5+\frac{r}{h}\frac{dh}{dr})F_2\sigma_r + rF_2\frac{d\sigma_r}{dr} + F_2\rho\omega^2r^2\right] \tag{10}$$

where

$$F_1 = \frac{{df}/{d\sigma_e} - f}{2\sigma_e^3}\left\{-\left[\left(\frac{r}{h}\frac{dh}{dr}-1\right)\sigma_r + r\frac{d\sigma_r}{dr}+\rho\omega^2r^2\right]\frac{d\sigma_r}{dr} + \left[\left(1+2\frac{r}{h}\frac{dh}{dr}\right)\sigma_r + 2r\frac{d\sigma_r}{dr} + 2\rho\omega^2r^2\right]\left\{\left\{\frac{1}{h}\frac{dh}{dr}+\frac{r}{h^2}\left[h\frac{d^2h}{dr^2}-\left(\frac{dh}{dr}\right)^2\right]\right\}\sigma_r + \left(2+\frac{r}{h}\frac{dh}{dr}\right)\frac{d\sigma_r}{dr} + \frac{d\rho}{dr}\omega^2r^2 + 2\rho\omega^2 r\right\}\right\} \tag{11}$$

$$F_2 = \frac{{df}/{d\sigma_e} - f}{2\sigma_e^3} r\left[\left(1+2\frac{r}{h}\frac{dh}{dr}\right)\sigma_r + 2r\frac{d\sigma_r}{dr} + 2\rho\omega^2r^2\right] \tag{12}$$

and

$$\sigma_e = \left\{-\sigma_r\left[\frac{r}{h}\frac{dh}{dr}\sigma_r + r\frac{d\sigma_r}{dr}+\rho\omega^2r^2\right] + \left[\left(1+\frac{r}{h}\frac{dh}{dr}\right)\sigma_r + r\frac{d\sigma_r}{dr}+\rho\omega^2r^2\right]^2\right\}^{\frac{1}{2}} \tag{13}$$

The boundary conditions for different problems have different forms. For the rotating solid disks, the boundary conditions are

$$\begin{aligned} r &= 0 \quad \sigma_r = \sigma_\theta \\ r &= b \quad \sigma_r = 0 \end{aligned} \tag{14}$$

where b is the radius of the rotating solid disks.

For rotating hollow disks, the above boundary conditions become

$$\begin{aligned} r &= a \quad \sigma_r = 0 \\ r &= b \quad \sigma_r = 0 \end{aligned} \tag{15}$$

where a and b are the inner and outer radii of the rotating hollow disks, respectively.

For circular disks subjected to internal and external pressure, the radial stress at inner and outer radii should be equal to the externally applied pressure, i. e.,

$$\begin{aligned} r &= a \quad \sigma_r = -p_i \\ r &= b \quad \sigma_r = -p_o \end{aligned} \tag{16}$$

where p_i and p_o are the internal and external pressure, respectively.

Substituting the first of Eq. (14) into the equilibrium equation (4), the first derivate of the radial stress at the axis of rotating disks is determined by

$$\frac{d\sigma_r}{dr} = -\frac{1}{h}\frac{dh}{dr}\sigma_r \tag{17}$$

3 SOFTWARE SYSTEM IMPLEMENTATION

The system architecture is depicted in Fig. 1. It consists of distributed Web browsers and a Web server. The Preprocessor and Postprocessor run on the distributed Web browsers by remote clients. The Solver, ImageGenerator and Database (DB) run on the Web server. The Web browsers and Web server are connected together through Internet.

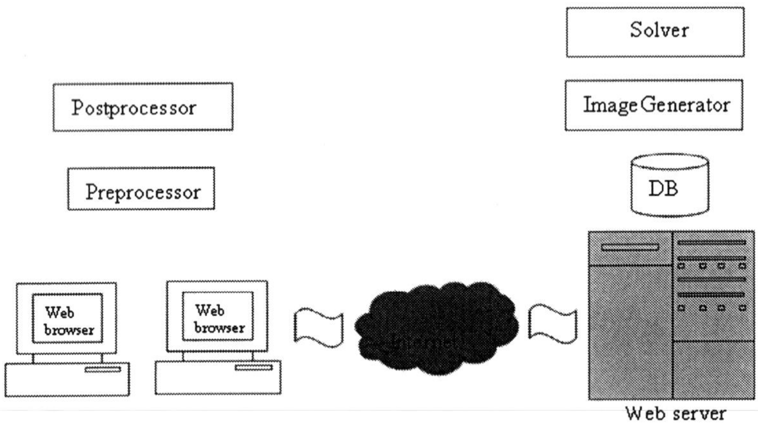

Fig. 1 System architecture

Java language is used to generate applets of the developed numerical method. The Hyper Text Markup Language (HTML) is used to produce static Web pages. A user-friendly Web-based interface is created for users to input the information required by the analysis, submit the information as a batch job to the server for numerical calculations and image generation, and send the images and output data back to the clients.

The clients can input data and define the output of the calculated results from the Preprocessor. The input information includes geometry size, material properties, loads and total intervals. The geometry size consists of the thickness and internal and external radii of the disks. The material properties have Young's modulus, Poisson's ratio, density, thermal expansion coefficient and stress-strain data. The loads on the disks are rotating angular velocity, internal and outer pressures, and temperature.

Postprecessor is used to generate output requirements of the calculated results. The clients can list undeformed geometry sizes and calculated radial displacement. They can also draw undeformed and deformed shapes of disks. The obtained strains are classified into total stains, elastic strains and plastic strains, and the stresses are represented in terms of stress components and equivalent stresses. Once again, the clients can request lists or plots of these quantities.

After the above preprocessing and postprecessing are completed, a batch job file is submitted to Web server to start the resolution of the problem.

According to the total intervals given by the clients, the Solver uniformly divides the disks into N equal intervals along the radial direction and obtains a small distance increment $\Delta r = \frac{b}{N}$ for solid disks or $\Delta r = \frac{b-a}{N}$ for hollow disks.

The calculation starts from the axis of solid disks or the inner radius of hollow disks. For solid disks, a guessed value of the radial stress at the axis of disks is assumed. Then, its first derivative is obtained from Eq. (17). For hollow disks, the radial stress at the inner radius is known. Therefore, only the value of the first derivative of the radial stress at this position is specified.

After the radial stress and its first derivative become known, the circumferential and equivalent stresses are determined from Eqs. (4) and (7), respectively. The stress state is identified according to the equivalent stress. If the equivalent stress at the axis of solid disks or inner radius of hollow disks is not larger than the yield stress, Eq. (5) is used. Otherwise, Eq. (10) is adopted. These two equations can be unified as

$$\frac{d^2\sigma_r}{dr^2} = g\left(r,\ \sigma_r,\ \frac{d\sigma_r}{dr}\right) \tag{18}$$

Having known the values of the radial stress, and its first derivatives at the axis or inner radius of the disks, the following equations are used to determine the radial stress and its first derivative at the next position

$$\begin{aligned}
&(\frac{d\sigma_r}{dr})_{i+1} = (\frac{d\sigma_r}{dr})_i + \frac{\Delta r}{6}(K_1 + 2K_2 + 2K_3 + K_4) \\
&(\sigma_r)_{i+1} = (\sigma_r)_i + \Delta r\left[(\frac{d\sigma_r}{dr})_i + \frac{\Delta r}{6}(K_1 + K_2 + K_3)\right] \\
&(i = 0,1,2,3\cdots,N)
\end{aligned} \tag{19}$$

where

$$\begin{aligned}
K_1 &= g\left(r_i,\ (\sigma_r)_i,\ (\frac{d\sigma_r}{dr})_i\right) \\
K_2 &= g\left(r_i + \frac{\Delta r}{2},\ (\sigma_r)_i + \frac{\Delta r}{2}(\frac{d\sigma_r}{dr})_i,\ (\frac{d\sigma_r}{dr})_i + \frac{\Delta r}{2}K_1\right) \\
K_3 &= g\left(r_i + \frac{\Delta r}{2},\ (\sigma_r)_i + \frac{\Delta r}{2}(\frac{d\sigma_r}{dr})_i + \frac{\Delta r^2}{4}K_1,\ (\frac{d\sigma_r}{dr})_i + \frac{\Delta r}{2}K_2\right) \\
K_4 &= g\left(r_i + \Delta r,\ (\sigma_r)_i + \Delta r\,(\frac{d\sigma_r}{dr})_i + \frac{\Delta r^2}{2}K_2,\ (\frac{d\sigma_r}{dr})_i + \Delta r K_3\right)
\end{aligned} \tag{20}$$

and the subscript $i = 0$ indicates those at the axis or inner radius of disks, and $i = N$ stands for those at the outer radius of the disks.

The above process is repeated until reaching the outer radius. Then, the obtained radial stress is compared to the boundary condition of radial stress at this position. If the required precision is satisfied, the calculation is completed. Otherwise, a new value of the radial stress at the axis of solid disks or its first derivative at the inner radius of hollow disks is assumed, and new iteration begins.

According to the requests of clients for output, the ImageGenerator draws undeformed and deformed shapes of disks, and creates the plots of stresses and strains from the calculated results. Then, it sends these image files back to the clients.

The input data and obtained results are stored in the Database. Each time when a new calculation is to be carried out, the Database will be searched to check whether the same problem has been tackled. If yes, the calculation will not be performed and the calculated results previously stored in Database are directly sent back to the clients. In doing so, the clients can fully share the same calculated results and avoid the waste of computer resources.

4 APPLICATIONS

Using the above-described Web-based Java software system, in this section, we investigate elastic-plastic deformations and stresses of a rotating hollow disk of constant thickness.

The geometric size, physical properties, stress-strain relationship and loads are shown in Fig. 2. The total intervals are taken to be 100. Using the plot output to draw the curves of radial and circumferential stresses against the radial coordinate, the calculated results were obtained by the above-presented Java software system and depicted in Fig. 2.

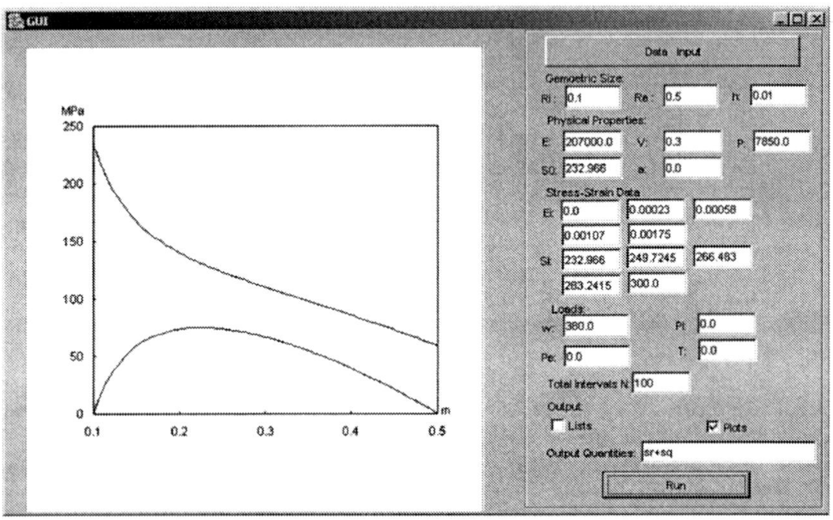

Fig. 2 User interface of elastic-plastic analysis of axisymmetric plane stress problems

5 CONCLUSIONS

A Web-based Java software system has been developed to tackle elastic and elastic-plastic analyses of axisymmetric plane stress problems. It is based on a new, simple, efficient and effective numerical method. The governing equation of the numerical method was derived from the basic equations of axisymmetric plane stress problems in elasticity and flow theory in plasticity. It was solved with the algorithm of boundary-value problems. The system was implemented through a Web browser and Internet by using Java language and Hyper Text Markup Language (HTML). The preprocessing and postprocessing of axisymmetric plane stress problems are completed by Web browsers. And the resolution, image generation and data storage for the problems are carried out by a Web server. The calculated results of the numerical example indicate the validity of the software system.

REFERENCES

1 Szykman, S., Racz, J., Bochenek, C. and Sriram, R. D. (2000) "A Web-based system for design artifact modelling", *Design Studies*, 21(2), 145-165.

2 Cheng, K., Pan, P. Y. and Harrision, D. K. (2001) "Web-based design and manufacturing support systems: implementation perspectives", *International Journal of Computer Integrated Manufacturing*, 14(1), 14-27.

3 Zeng, J., Chen, W. and Ding, Q. (2003) "A Web-based CAD system", *Journal of Materials Processing Technology*, 139, 229-232.

4 Lan, H., Ding, Y., Hong, J., Huang, H. and Lu, B. (2003) "A Web-based manufacturing service system for rapid product development", *Computers in Industry*, 54(1), 51-67.

5 Xue, D. and Xu, Y. (2003) "Web-based distributed system and database modeling for concurrent design", *Computer-Aided Design*, 35(5), 433-452.

6 Qin, S. F., Harrison, R., West, A. A., Jordanov, I. N. and Wright, D. K. (2003) "A framework of Web-based conceptual design", *Computers in Industry*, 50(2), 153-164.

7 Zhang, J., Peng, Y. and Ruan, X. (2004) "A Web-based cold forging process generation system", *Journal of Materials Processing Technology*, 145, 1-6.

8 Salisbury, C. F., Farr, S. D. and Moore, J. A. (1999) "Web-based simulation visualization using Java3D", in 1999 *Winter Simulation Conference Proceedings (WSC)*, Dec 5-Dec 8 1999, Phoenix, AZ, USA, Vol. 2, pp. 1425-1429.

9 Zhang, Y., Zhu, H., Greenwood, S. and Huo, Q. (2001) "Quality modelling for Web-based information systems", 8th Workshop on Future Trends of Distributed Computing Systems (FTDCS'01), Oct 31-Nov 2 2001, Bologna, Proceedings of the *IEEE Computer Society Workshop on Future Trends of Distributed Computing Systems*, pp. 41-47.

10 Chen, S. and Lu, F. Y. (2002) "Web-based simulations of power systems", *IEEE Computer Applications in Power*, 15(1), 35-40.

11 Choi, K. S., Sun, H., Heng, P. A. and Cheng, J. C. Y. (2003) "A scalable force propagation approach for Web-based deformable simulation of soft tissues", *Proceedings of the 8th International Conference on 3D Web Technology*, Mar 9-12 2003, Saint Malo, France, Web3D Symposium Proceedings, pp. 185-193.

12 Huang, B. (2003) "Web-based dynamic and interactive environmental visualization", *Computers, Environment and Urban Systems*, 27, 623-636.

13 MacDonald, C. J., Stodel, E. J., Farres, L. G., Breithaupt, K. and Gabriel, M. A. (2001) "The demand-driven learning model: A framework for Web-based learning", *The Internet and Higher Education*, 4(1), 9-30.

14 Papanikolaou, K. A., Grigoriadou, M., Magoulas, G. D. and Kornilakis, H. (2002) "Towards new forms of knowledge communication: the adaptive dimension of a Web-based learning environment", *Computers & Education*, 39(4), 333-360.
15 Yen, C. and Li, W.-J. (2003) "Web-based learning and instruction support system for pneumatics", *Computers & Education*, 41(2), 107-120.
16 Manuel, A., Juan, D. L. and Hans, V. (2001) "Web-based simulation of systems described by partial differential equations", *Proceedings of the 2001 Winter Simulation Conference*, Dec 9-12 2001, Arlington, VA, Winter Simulation Conference Proceedings, Vol. 1, pp. 629-636.
17 Timoshenko, S. P. and Goodier, J. N. (1970) "*Theory of elasticity*", New York: McGraw-Hill.
18 Gurushankar, G. V. (1975) "Thermal stresses in a rotating, nonhomogeneous, anisotropic disk of varying thickness and density", *Journal of Strain Analysis*, 10(3), 137-142.
19 Sterner, S. C., Saigal, S., Kistler, W. and Dietrich, D. E. (1994) "A unified numerical approach for the analysis of rotating disks including turbine rotors", *International Journal of Solids and Structures*, 31, 269-277.
20 Gamer, U. (1983) "Tresca's yield condition and the rotating disk", *Journal of Applied Mechanics*, Transactions of American Society of Mechanical Engineers, 50, 676-678.
21 Gamer, U. (1984) "Elastic-plastic deformation of the rotating solid disk", *Ingenieur-Archiv*, 54, 345-354.
22 Gamer, U. (1984) "The elastic-plastic stress distribution in the rotating annulus and in the annulus under external pressure", *Zeitschrift für Angewandte Mathematik und Mechanik*, 64, 124-128.
23 Güven, U. (1992) "Elastic-plastic stresses in a rotating annular disk of variable thickness and variable density", *International Journal of Mechanical Sciences*, 34(2), 133-138.
24 Güven, U. (1998) "Stress distribution in linear hardening annular disk of variable thickness subjected to external pressure", *International Journal of Mechanical Science*, 40(6), 589-601.
25 Eraslan, A. N. and Orcan, Y. (2002) "On the rotating elastic-plastic solid disks of variable thickness having concave profiles", *International Journal of Mechanical Sciences*, 44, 1445-1466.

Remote Control, Condition Monitoring, and Tele-robotics

Development of a web-based condition monitoring and control system for CNC milling processes

K BLLAU, K CHENG, M WHITBREAD, and **G E TAYLOR**
Faculty of Information and Technology, Leeds Metropolitan University, UK

ABSTRACT

This paper describes the methods for the design of web based condition monitoring and remote control system. The integration of hardware system consists of sensors, network cards and server with software system. The benefits of monitoring different process parameters such as vibration, temperature, spindle speed, motor power consumption, fluid pressure, cutting force and feed rate, may include the detection of the machine instability, the surface integrity, the too wear and breakage. The developed system can be implemented on wide range of manufacturing applications. The core of the system is based on different approaches. These approaches have been investigated and used to develop one system for the monitoring and control purposes.

Keywords; *Internet based condition monitoring, remote control, CNC milling process*

1 INTRODUCTION

Researchers and developers have introduced different techniques for condition monitoring, such as using wavelet transforms and fuzzy techniques for real time condition monitoring [1]. Intelligent remote monitoring and diagnosis of manufacturing process has been developed by using an integrated approach of neural networks and rough sets [2]. Huang and Chen introduced a neural network based tool breakage monitoring system for end milling operations [3]. Lee explored remote machinery condition monitoring using wireless technology and Internet [4]. Moreover, research on using feed motor current signatures has been undertaken to detect the tool flute breakage.

The survey above reflects the importance of monitoring and detecting unusual signatures to prevent damage of machine tools and components, which becomes crucial within intense global competition environment. A survey of machine tool manufacturers in the United State clearly indicates the distinct need for new technology features to be built into machine tools for e- manufacturing, fault diagnosis and maintenance purposes through on–line monitoring and remote control functions [2].

In past view years, Internet usages has been increased sharply with remarkable development in communication technology, networking and web-based programming. These advantages can be integrated and applied in the system/tool development for web-based condition monitoring and control. The aim of this study is to develop a web- based condition monitoring and control system for CNC milling processes. The system should be low cost and industrial feasible especially for manufacturing [5].

2 THE SYSTEM DESIGN CONCEPTS

Internet, networking technology and web-based programming present wide a range of technological basis that can be used to develop a web-based condition monitoring & control tool. But before we start the development we should clarify the requirements and the system specifications.

What we will monitor?
Monitoring systems measure the conditions of the machine tools and the process itself [6]. In this developed system we focus on monitoring specific parameters that affect both the machine itself and the cutting process as well. Factors such as cutting force, temperature, vibrations, power consume and fluid pressure are crucial, and monitoring them provides an early indicator to what could happen in real time working environment.

What we will control?
Control is one of the most complicated processes, because there are a variety of machine control systems designed by different adventures. The variety of control systems make the process of interfacing the different hardware and software system very complicated and usually there should be a middle ware system.

There are a couple of design and programming tools for prototyping the system, such as Adobe Photoshop used for design of the user interface, FrontPage used to design the web-pages and link all together. PHP functions are also programmed by using PHP expert editor and all of these functions have been tested through the same software. In order to test the whole system an Apache server has been set up and a number of trails have been carried out to test and validate the prototype system.

3 THE SYSTEM ARCHITECTURE

Fig 1 shows architecture of the system development which consists of three main phases. Each phase is designed for implementing specific functions and activities.

3.1 Phase one

The phase one is designed to cover the interfacing activities with either different sensors or the machine control system itself. Currently we found that dealing with different sensors is easier than dealing with the machine control system. In fact there are many different control systems, which, make the processes of interfacing the machine control system with this developed system hard and long process.

Phase one is where the data start to be collected from different sensors and pass it to the MySQL server through a data logger, and then handled by the Apache server. The handling step is approached by the implementation of PHP codes. This means that the phrase target has been approached and the rest is a loop of the same step within certain time. For example, if we set up that every 10 seconds we need to collect data from sensors, then the server will send the request to the phase one to send the data every 10 seconds.

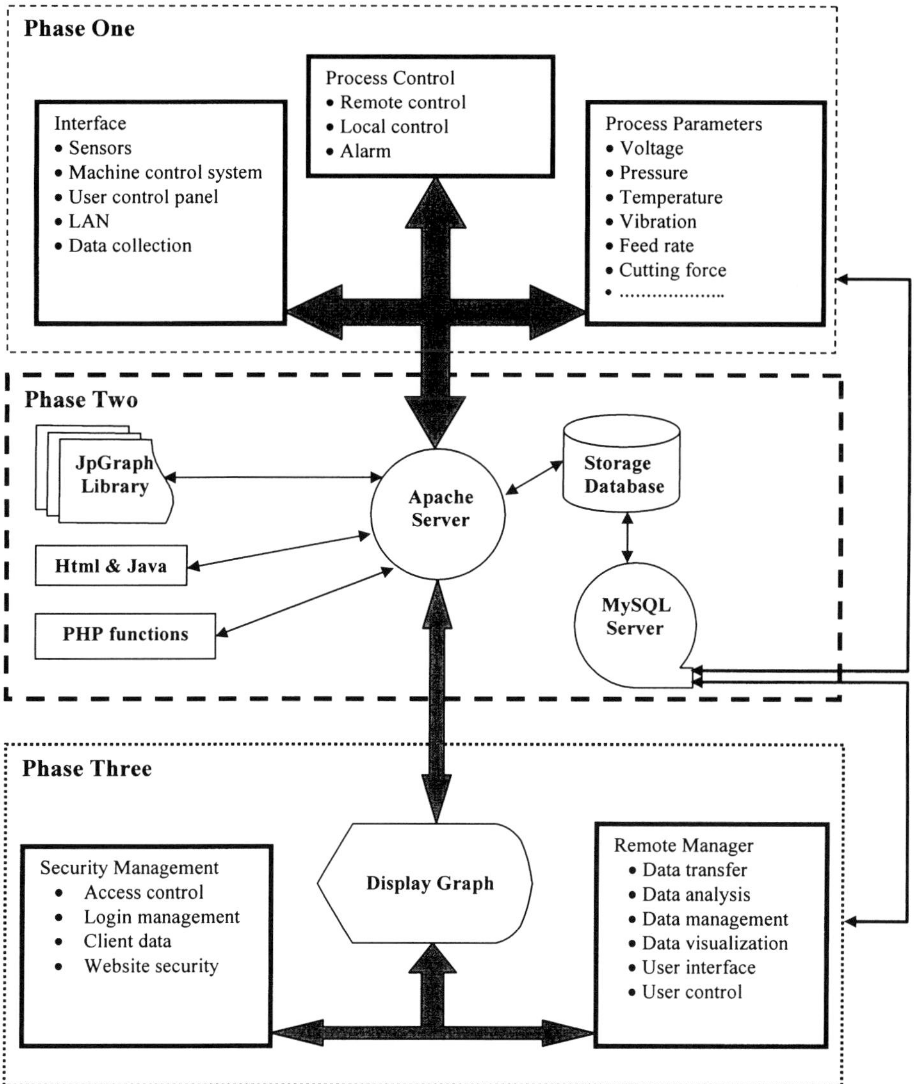

Fig. 1 Architecture of the developed system

3.2 Phase two

Phase two is the server side, which is designed to process the sent data from phase one. It basically contains an Apache server supporting PHP scripts, Java scripts and MYSQL. These Internet technologies have been used to develop different PHP functions and Java codes, each of these PHP code files has a specific function. The phase two functions start when the request of data is sent to the phase one, then the process of processing these data will follow specific steps.

The first step will send these data to the database where they will be stored. The second step is these data have to be processed by the PHP functions in order to read them then present them as visual reports (Graphs).

The mechanism for condition-monitoring is based on a comparison process between the data stored in database. ***dbconnect.php*** is a file contains a PHP code used to connect to the database (see Fig. 2). Later these data will be processed by PHP functions and determine whether the process is stable and within the working limits. Fig. 2 shows an example of these functions.

```
<?php

$hostname_connect = "localhost";

$database_connect = "master database";

$username_connect = "root";

$password_connect = "";

$connect = mysql_pconnect($hostname_connect, $username_connect,

$password_connect) or die(mysql_error());

?>
```

Fig. 2 A PHP function known as *dbconnect.php* used to connect to the database

3.3 Phase three

The final phase concerns with the data display on the client browser. The client will find different types of visual reports that help to understand the general condition of the machine connected to the system. Each monitored parameter has provided with it is own visual report and specific upper and lower limit zones have been clearly shown on the client browser for easy monitoring as shown in Fig. 3.

4 JPGRAPH LIBRARY AND PHP PROGRAMMING FOR VISUAL REPORTS

The JpGraph library makes it easy to plot graphs with minimum codes and able to create complex graphs, which require a very fine grade of control [7]. With this library the data can be plotted and reported visually. The method used to design this part is by implementing PHP programming language with JpGraph library onto the server.

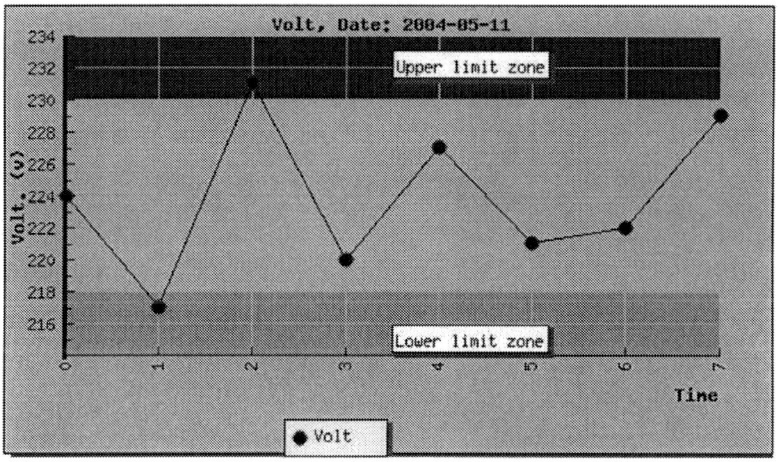

Fig. 3 A graph generated by JpGraph library

5 PHP DECISION FUNCTIONS

The system is based on a number of functions designed to make decisions including whether the machine condition is within the limits or not. The increase in temperature, pressure or vibrations will be reflected on the plotted graph. Fig 4 illustrates one of these functions that monitoring the temperature limits, and same idea has been used to design other functions for monitoring other parameters.

```
<?php

if ( $Temp1 >= 55 or $Temp1 <=22 ) {

echo " Machine temperature within the limits"

}

else { echo " Machine temperature out of the limits"

} {/if}

?>
```

Fig. 4 A PHP decision function used to monitor the machine temperature

6 USER INTERFACE AND CONTROL PANEL

The system provides with an easy use control panel to make the monitoring process visually very intuitive and simple. Different softwares are used to design the panel and link it with different PHP functions. These functions are integrated with indicators in order to provide the client with different information reflecting the machine condition. Fig. 5 is a snapshot of the main page designed to give access to any client around the world to any machine linked with this system. So any user can easily go through different pages from where he or she can find out different information regarding the machine condition.

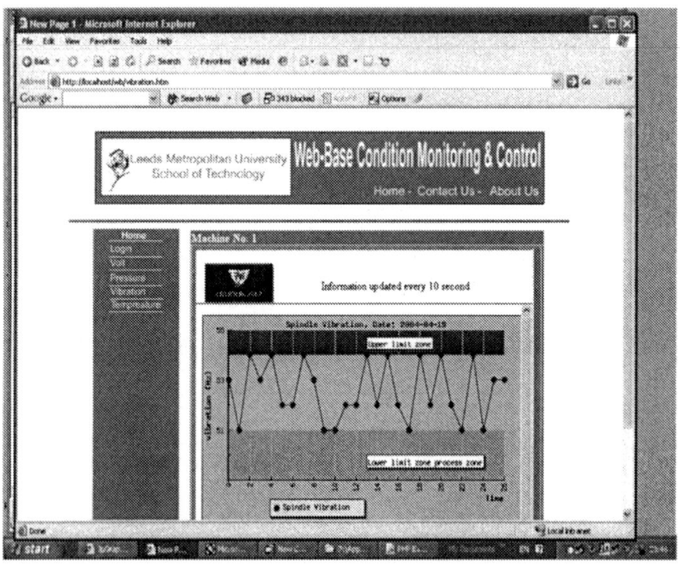

Fig. 5 Snapshot of the main page

7 SYSTEM IMPLEMENTATION

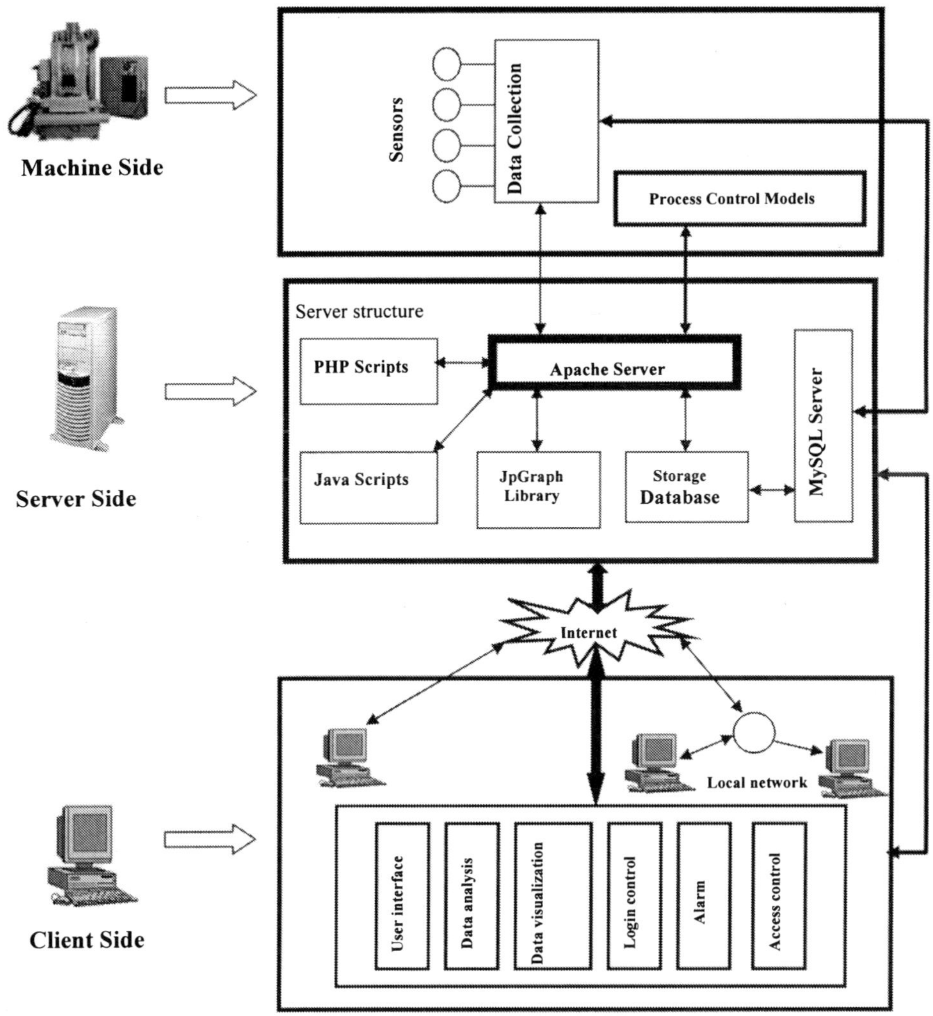

Fig. 6 The developed system structure

Figure 6 illustrates the whole structure of the system which consists of three parts including machine side, server side and client side. These three parts are interfaced to each other by software and hardware system.

8 EXPERIMENTAL TRAILS

In order to evaluate the developed system, we have implemented PHP codes to generate random data within specific limits for each parameter (see Fig 7). This step should cover the phrase one as shown in Fig. 1. It works as different readings have been generated from sensors.

```
<?php
srand ((double) microtime( )*1000000);
$temp = rand(27,36 );
$volt = rand(217,231 );
$vibration = rand(44,75 );
$pressure = rand(51,54 );
$cutforce = rand(44,75 );
<?
```

Fig. 7 PHP code generating random data for a specific parameter

The resulted data are stored into the database, then these data will be handled by the PHP functions. All resulted data have been plotted on graphs and presented on visual reports. The PHP codes proof that they have handled the data as designed and presented them in such creative way. Furthermore the upper and lower limits and the optimum zone are clearly plotted , which means any operator can easily pick up any unusual points out of the limits.

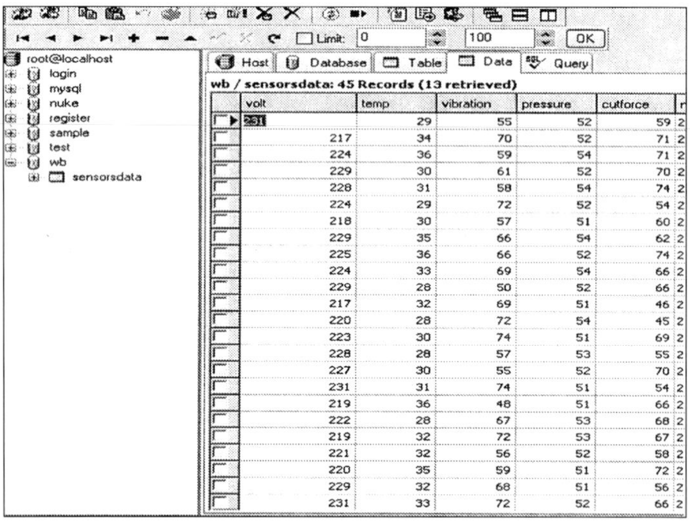

volt	temp	vibration	pressure	cutforce	r
231	29	55	52	59	2
217	34	70	52	71	2
224	36	59	54	71	2
229	30	61	52	70	2
228	31	58	54	74	2
224	29	72	52	54	2
218	30	57	51	60	2
229	35	66	54	62	2
225	36	66	52	74	2
224	33	69	54	66	2
229	28	50	52	66	2
217	32	69	51	46	2
220	28	72	54	45	2
223	30	74	51	69	2
228	28	57	53	55	2
227	30	55	52	70	2
231	31	74	51	54	2
219	36	48	51	66	2
222	28	67	53	68	2
219	32	72	53	67	2
221	32	56	52	58	2
220	35	59	51	72	2
229	32	68	51	56	2
231	33	72	52	66	2

Fig. 8 Data generated randomly every 1 second and stored in the database

The system has been automatically tested by refreshing the pages at every desired time. It proves that even though we set up the refreshing time in a very short time (every 1 second or less), the system can pickup the data generated and then store and plot them accordingly.

9 LINKING UP WITH A BRIDGEPORT VMC 500 CNC MACHINE

Fig. 9 shows an application scenario of the system developed of the Bridgeport VMC 500 CNC milling centre.

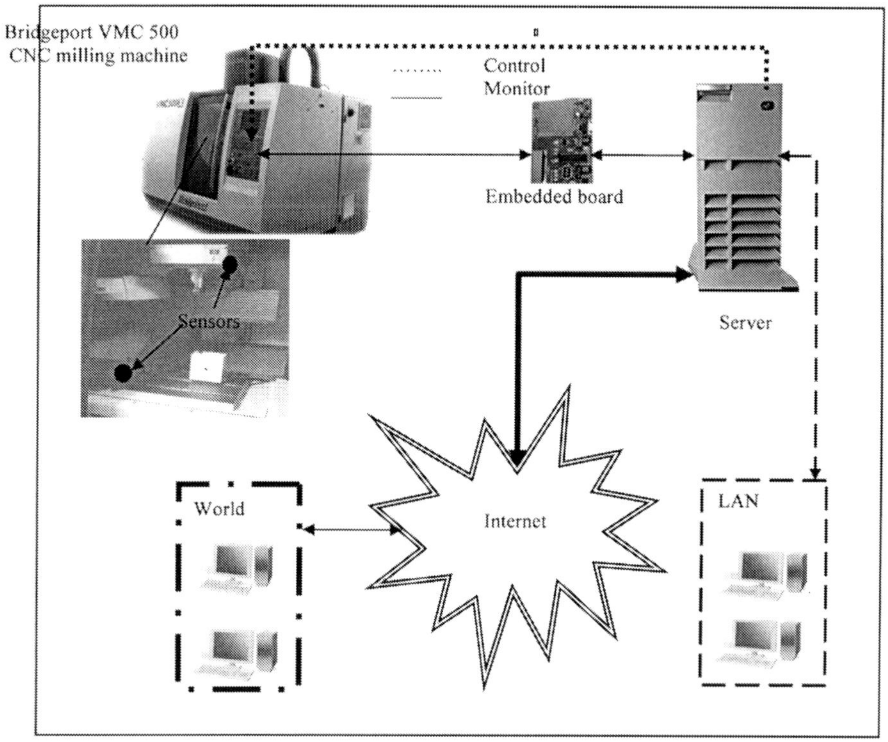

Fig. 9 The developed system interfaced with Bridgeport VMC 500

10 CONCLUDING REMARKS

Implementing the system as presented will make the monitoring process become more effective and easier. Information can be accessed any time and used as historical information for future usage, which provides the client the flexibility and ability to monitor the machines from anywhere in the world. Furthermore, the advantage of reporting the machine condition visually through graphs gives any operator ability to pick up any crucial values out of the upper or lower limit. Currently the system designed is implemented mainly with CNC milling machines, but it is possible to use as a generic tool for remote monitoring and control of other machines.

REFERENCES

1. Xiaoli Li, Shiu Kit Tso (2000) "Real – time tool condition monitoring using wavelet transforms and fuzzy techniques", *IEEE transaction on system*, Vol. 30, N0. 3,pp 352 – 357.
2. Tung – Hsu (Tony) Hou, Wang-Lin and Li Lin (2003) "An intelligent remote monitoring and diagnosis of manufacturing process using an integrated approach of neural networks and rough sets", *Journal of intelligent manufacturing*, Vol. 14, pp 239-253.
3. Huang P. T. and Chen J. C. (2000) "Neural network based tool breakage monitoring system for end milling operations", *Journal of industrial technology*, Vol.16, number 2
4. Lee L. D., Jr. "Buddy", "Remote machinery condition monitoring using wireless technology and internet", http://www.reliabilityweb.com, accessed 12/02/2004.
5. Altun Z. G., Topaloglu U. M. and Saygin A. V. (2002) "Web-based system, monitoring and control", university of Cukurova, departement of Physics, Turkey
6. Park, J., Pasek, Z.J., Birla, S., Yansong, S., Koren, Y., Shin, K.G., and Ulsoy, A.G. (1995) "An open architecture testbed for real-time monitoring and control of machining processes", *American control conference*, Seattle, Washington, pp. 200-204.
7. JpGraph - OO Graph Library for PHP - what is JP Graph? http://www.aditus.nu/jpgraph/, accessed 12/02/2004.
8. Altintas, Y. and Munasinghe, W.K. (1996) "Modular CNC design for intelligent machining, part 2 - modular integration of sensor based milling process monitoring and control tasks", *ASME Journal of manufacturing science and engineering*, Vol. 118, No. 4, pp. 514-521.
9. Du, R., Zhang, B., Hungerford, W., and Pryor, T. (1993) "Tool conditioning monitoring and compensation in finish turning using optical sensor, proceedings of symposium on mechatronics", *ASME Winter annual meeting*, pp. 245-251.
10. Koren, Y., Pasek, Z.J., Ulsoy, A.G., and Benchetrit, U. (1996) "Real-time control architectures for system performance", *Annals of the CIRP*, Vol. 45, No. 1, pp. 377-380.
11. Hosmon J. H. and Fritz D., "Connecting Your legacy CNC machine tool to the Internet", http://www.mmsonline.com, accessed 12/02/2004.
12. Longanbach, D.M. and Kurfess, T.R. (1998) "Real-time measurement for an internal grinding system", *Transactions of NAMRI/SME*, Vol. 26, pp.317-322.
13. Albert M., "Feedback from a CNC in real time" http://www.mmsonline.com, accessed 12/02/2004.
14. Rober, S.J. and Shin, Y.C. (1995) "Modelling and control of CNC machining using a PC-based open architecture controller", *Mechatronics*, Vol. 5, No. 4, pp. 401-420.

Remote process monitoring through Internet

X CHEN, T LIMCHIMCHOL, and **S HUSSAIN**
School of Mechanical, Materials, Manufacturing Engineering, and Management, University of Nottingham, UK
G LIU
School of Electrical Engineering, Glamorgan University, UK

ABSTRACT

Condition monitoring is a very important aspect of automated manufacturing processes. Any malfunction of a machining process will deteriorate production quality. More importantly, a malfunction may cause catastrophic accident if there is not a proper monitoring system in place. Development of the internet technology provides an excellent opportunity for implementing a remote monitoring system to machining processes. This paper reviews the requirement of machining process monitoring. It is realised the responsive speed for monitoring different performance parameters would be very different. It is also noticed that speed of internet communication varies tremendous. Therefore an internet based machine monitoring system should have great flexibility to cope such environmental variation.

1 INTRODUCTION

Over the last decade, the internet technology has been used extensively in many areas of education, science and also business. The major advantage this technology is that enables the communication between different sides of the world. Online process monitoring is very important for an effective production. By using internet as the communication infrastructure, internet based condition monitoring would provide an early warning of tool damage or process degradation.

The machining condition monitoring technique is a method to monitor a machining operation and tool quality during machining by interpreting the meaning of sensed signals from the machine [1]. The sensors attached to the machine tool provide useful information that can be stored in a computer. The common data extracted from a machining process are force, vibration, acoustic emission, power and moment. These data will provide information to describe characteristics of the tool conditions [2]. By analysing these data, operator may take actions accordingly in order to avoid the failure of process. It is noticed that the information extracted by indirect methods often involves influences from different factors such as tool wear, materials properties and electrical power fluctuation etc.. For process analysis, a precise tool wear monitoring is essential. Therefore, direct tool wear vision check is still important to industry.

Development of internet technology provides a low cost remote communication tool which may be used for remote process monitoring. The benefit of using internet is that people can access remote plant and monitor manufacturing processes quickly by connecting a PC to the internet from any part of the world. A manager or an expert can easily access the information of production processes at different sites. In this way, any tool failures and problems in the manufacturing plant can be analyzed and solved effectively. This research intends to evaluate the remote tool condition monitoring using internet technology.

2 ARCHITECTURE OF A CLIENT/SERVER SYSTEM FOR INTERNET BASED CONDITION MONITORING

The internet based condition monitoring is a monitoring system which monitors processes and tool quality by presenting process data in a dynamic graphical interface through a web site. The structure of the system is composed of a database sever and a web sever as shown in figure 1.

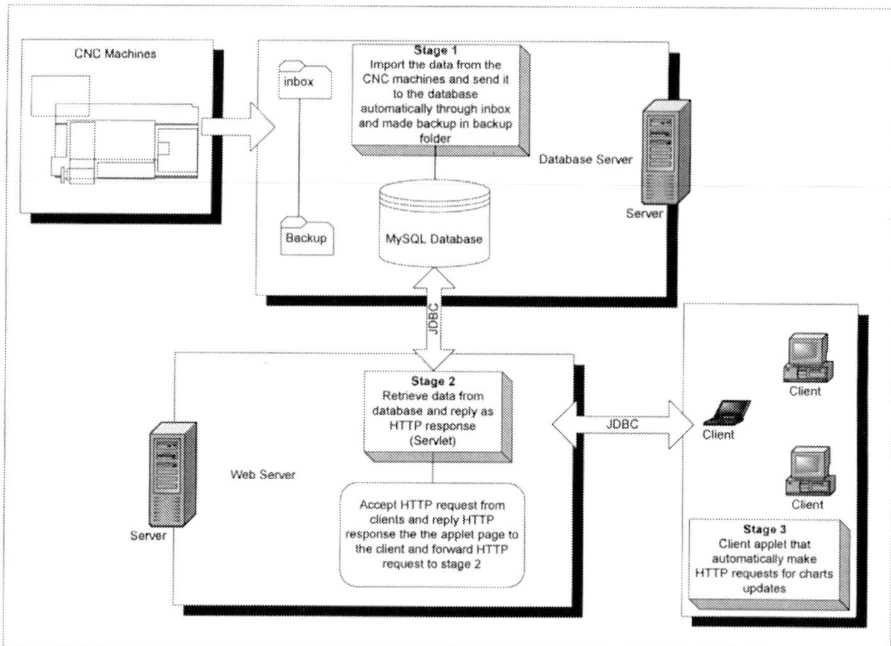

Figure 1 Internet based condition monitoring system architecture diagram

The database sever is set up to allow machining process data to be stored. The web server is integrated into the monitoring system, which can be accessed via a normal web browser that is Java enabled. Therefore a user with a Java supported web browser is able to access the data from any internet portal. A Java platform is setup to combine the web server and the database server in order to compile the source and running program. By installing Java 2 Platform, Standard Edition 1.4.2, the platform provides a compiler that programmers can use it to compile all kinds of Java programs. It also provides an interpreter that can be used to run Java Applications.

Due to its reliable performance and availability, MySQL are selected for database server programming. MySQL is a multi-user SQL relational database server. It can run on UNIX, windows, Mac and OS/2 platforms. The SQL (Structured Query Language) is a high-level, declarative database language used for defining and querying relational databases [3]. Three basic functions are implemented in organising the database.

(a) Data maintenance (add, remove);
(b) Data retrieval (querying);
(c) Data control (create, monitoring, restricting access).

As soon as the data from CNC machine generate a file in the monitoring folder, a program written in Java transfers the data file automatically to the database server and makes another copy to the backup folder. The data in the MySQL database are available to the web sever and will be overwritten automatically after they reach the maximum memory space allowance. The MySQL database can also read the data from the back up folder to review the processes.

As illustrated in figure 1, the data of monitored machining process parameters can be transferred to a folder (Inbox) in the database server system. Once a data file created in the monitoring folder, the data will be automatically stored into the MySQL database and be readily available for internet services. The data table was designed and created by MS ACCESS and then export through Open Database Connectivity (ODBC) to MySQL. When a client requests data from the web server, web server will retrieve the data from the database server through JDBC:ODBC connection and then send it to the client. The JDBC (Java Database Connectivity) is a Java application program interface (API) developed by JavaSoft for executing SQL statements. The JDBC API acts as a bridge connection between Java programs and wide ranges of databases. In this application, the JDBC defines Java classes to represent database connections, SQL statements, etc. and allows a programmer to connect database, issue SQL statements and process the results through Java programming. Basically the JDBC is an interface for Java applet and it's applications to connect to the databases [4, 5]. The connection between java program and MySQL database are established by JDBC:ODBC via MyODBC driver.

The web server is basically intermediary between the contents of the database and the client's browser. When the web server receives a request for a set of information, the server will load the information from the database and send it to the user via the internet. In this study Apache Tomcat 5.0 is used for the web server programming. Once Tomcat is executed, the Java Servlet can then be processed. The JSP files in the program are running under Tomcat web server which is waiting for data request from clients. The JSP is one type of servlet program, which uses with tomcat in order to connect with the web server. With java applet, the system publishes dynamic graphs on the webpage to continuously demonstrate the signals from the CNC machine.

The web based condition monitoring system is designed to collect and assess process condition signals for a remote access. Users can access the system and all of the information it managed. Java applet which connects to MySQL database via ODBC by using MyODBC driver will transfer the data from the web server as requested to the client. The service can be obtained at anytime and from anywhere, through a standard Java enabled internet browser.

As an internet application, it is important that the web sever can provide information that is able to be represented at different computer platforms. Java applet is one of available technologies for such purpose. The advantage of Java applet is that a small set of program code can be downloaded from the internet and function properly within a Java enabled web browser. Such browser normally has a Java virtual machine which enables the same code to run on different machine platforms.

For security reason, the system is protected by encrypted connection to the database server. Only authorised user will be able to view the process from web pages.

3 DISCUSSION ON APPLICATIONS

In this study a web-based condition monitoring program was developed to monitor different processes with different conditions. A framework of the monitoring system is designed to enable a number of the signals from the process to be stored and to be accessed via the internet. So forces, power, torque, vibration and acoustic emission of machining processes can be inspected remotely. Different parameters of the processes are monitored by using different sensors with an integrated data logging system. During the machining process, the data of monitored parameters are written into a file in a monitoring folder. The file is then transferred to the database server and to be available for the web server to provide data for remote clients.

A grinding process and a milling process were monitored as examples. Figure 2 is a screen shoot of grinding force signals monitored via internet connection. The variation of grinding force in four grinding cycles is presented. The workpiece of mild steel is ground by an Al_2O_3 wheel. The change of grinding forces indicates the wheel wear during grinding. Different parameters can be checked by selecting different options shown on the screen.

Figure 3 shows the cutting force signals of a milling process, where mild steel is cut by a high-speed steel milling tool. In order to see the effect of each cutting tooth, the scanning speed of the screen can be changed. In figure 3, the changes of cutting force with each cutting edge can be seen clearly.

In most cases, the speed of the data transmission is satisfactory and reliable. However in the case of heavy internet information traffic, the data transfer may delay. In this project historical machining data are retraceable by an established backup folder. Therefore the data will not be lost due to the speed of internet.

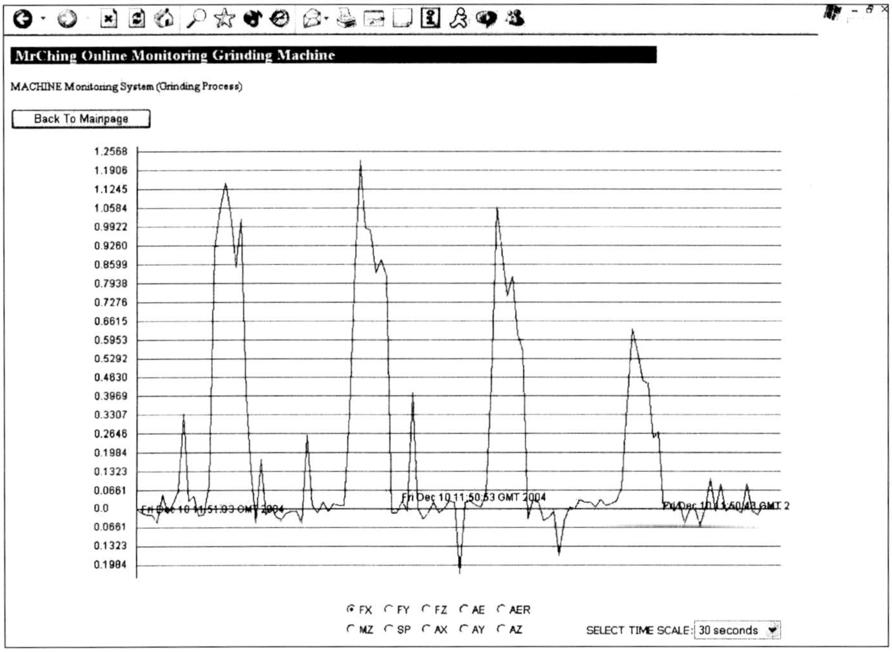

Figure 2 Horizontal grinding force in four grinding cycles

For a tool wear vision monitoring, it seems to be a problem if the internet traffic is heavy. Large image file transfer takes time. For example, the cutting edge wear of a milling cutter can be observed under a microscope. The images can be digitalised and stored in a computer. For cutting tool wear monitoring, it would be sufficient enough to take images of tool wear after each cutting pass. So the image refresh-rate need not too high. In most cases, the image capture time interval allows computer to calculate wear area and refresh the images on the webpage. Figure 4 is a screen shot of a tool wear image monitoring through internet. The image is refreshed every three seconds. By using PHP script, the images are sent frame by frame to MySQL and web server enabling client to see them. In the mean time, the tool wear area is calculated by a digital image processing. The acquired images can be stored in the MySQL database for future use.

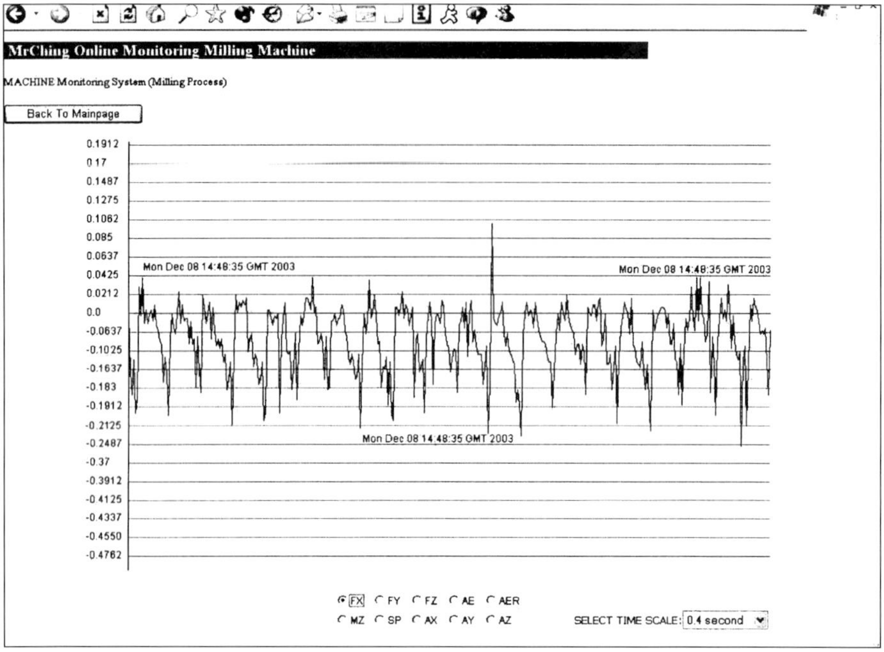

Figure 3 Horizontal milling forces in two revolutions of milling cutting

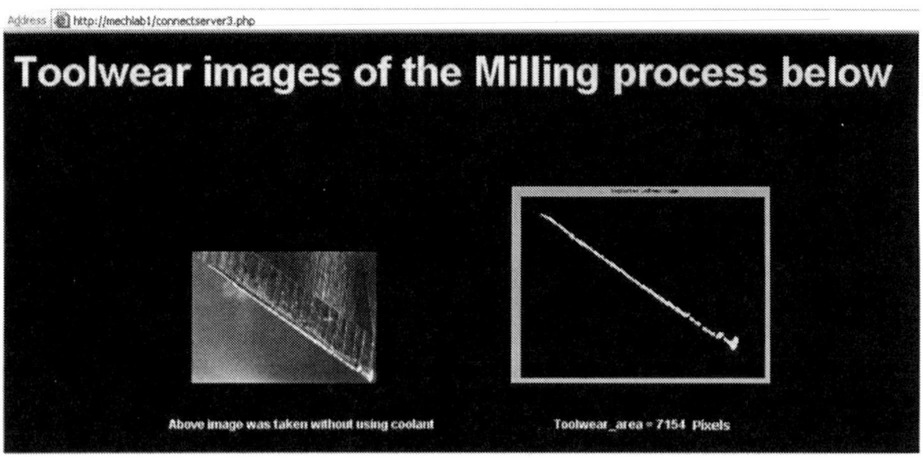

Figure 4. Tool wear image monitoring through internet

Benefits gained from the web based online condition monitoring system may be

- Online tracking process information,
- Fast response to process problems,
- Breakdown avoidance,
- Improvement of continuous operation,
- Resources sharing,
- Provided information services between manufacturing plant,
- Improvement of communication between human and the machines,
- Increase process knowledge from the historical data obtained,
- Enhancement of remote decision making.

Along with the advances of data transmission technologies, the internet-based process monitoring may realise its maximum benefits for manufacturing industry.

4. CONCLUSION

Web based processing monitoring device with Java technology is a state of art information technologies to ensure consistent production quality. With Java applets and a database system, the machining data can be accessed via internet. The machining process performance can be visualised by graphical plotting at a remote place. Both numeric data and graphical images can be viewed and accessed via the Internet. By using this system, an automated data diagnosis and classification can be integrated into a web based condition monitoring system. In this paper, it has demonstrated that the performance of a machining process can be monitored and assessed from anywhere in the world. Further information of productivity and results of diagnoses may be shared among partners at different locations. The development of a remote diagnostic system should provide manufacturers and users with greater flexibility for conducting manufacturing activities

REFERENCES

1 Errki Jantunen, 2002, "A summary of methods applied to tool condition monitoring in drilling" *Int. J. Mach. Tools & Manufact.*, vol. 42, pp. 997-1010.

2 P. W. Prickett, C. Johns, 1999, "An overview of approaches to end milling tool monitoring" *Int. J. Mach. Tools & Manufact*, vol 39, pp. 105-122.

3 Mehmet Kuntalp, Orkun Akar, 2004 "A simple and low-cost inter-based teleconsultation system that could effectively solve the health care access problems in underserved areas of developing countries" , *Computer Methods and Programs in Biomedicine.*

4 N. Zingirian, M. Maresca, S. Nalin, 1999, " efficiency of standard software architectures for Java-based access to remote databases" *Future Generation Computer System*, vol. 15, pp. 417-424.

5 Sameer S. Pradhan and Wilfred V. Huang, 1998, "Virtual manufacturing information system using Java and JDBC", *Computers Ind. Engng*, vol. 35, pp. 255-258.

Application of mobile agents in multiple online robots

L CRAGG and **H HU**
Department of Computer Science, University of Essex, Colchester, UK

ABSTRACT

Mobile agents provide the functionality of all other distributed computing paradigms in a unified environment and a natural design philosophy for distributed computing systems. These properties can provide multiple robot system developers with a wide range of options for initial development and future extension of their system and an intuitive and robust design environment. In this paper we present two examples to show how by taking advantage of these properties the adaptability and fault tolerance of a multi-robot architecture (the ALLIANCE architecture) can be extended in multiple online robots.

1 INTRODUCTION

ALLIANCE was developed for fault tolerant control in multiple robot systems. It was originally developed using static code and one-way radio frequency message passing to allow inter-robot communication (1). ALLIANCE has been shown to be effective (2) however real-world applications continue to demand advances in fault-tolerant architectures.

Online robots offer us the opportunity to extend fault tolerant multi-robot architectures like ALLIANCE because through their use of Internet and networking functionality they allow us to develop systems using modern distributed computing paradigms which offer us new design options. One such modern distributed computing paradigm is mobile agents. In this paper we use mobile agents to implement ALLIANCE in multiple online robots in order to show the functionality such an implementation can provide.

The rest of this paper is organised as follows. Section 2 outlines background information on ALLIANCE, online robots and distributed computing paradigms. Section 3 presents a detailed description of mobile agents. In Section 4, details of our implementation environment and architecture are introduced. Section 5 shows experiments conducted using the proposed environment and architecture. Experimental results and analyses are given in Section 6. Finally, conclusions and future work are summarised in Section 7.

2 BACKGROUND

2.1 The ALLIANCE architecture

Fig. 1 shows a simplified diagram of ALLIANCE. Its main components are Motivational Behaviours (MB's) and Behaviour Sets (BS's). MB's receive input from Inter-Robot Communication (IC) and sensors, and are each linked to an associated BS. BS's receive input from sensors and provide output to robot actuators. BS's are independent robot activities which may contain sub-tasks. MB's determine whether their associated BS is selected at run-time based on the robots motivation to perform them. Motivation is calculated using a mathematical formula which takes into account the behaviour of other robots, the behaviour of the host robot, and sensory information from the environment. Please see (1) for a detailed description.

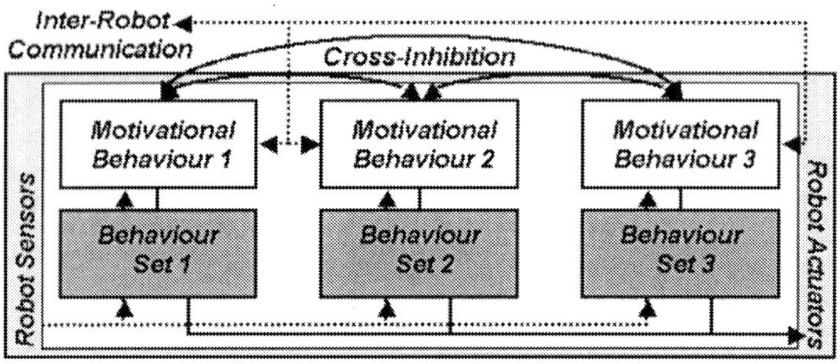

Fig. 1 The ALLIANCE architecture

2.2 Online robots

Online robot systems use Internet software protocols (such as TCP/IP or UDP/IP) for communication and Internet communication infrastructure as transmission hardware. Online robot systems have been developed using single (3) and multiple robots (4). Using widespread, inexpensive Internet technology allows this type of robot system to be widely accessed (e.g. via standard web browsers) and cost effective.

2.3 Distributed computing paradigms

A number of distributed computing paradigms exist with which distributed computing systems such as multiple robot systems can be structured. These include:

2.3.1 Client/server

In this paradigm (Fig. 2a) a server has knowledge and resources to perform a task. A client sends a message to the server requesting task execution. The server performs the task using its knowledge and resources and returns a result to the client.

2.3.2 Remote computation

In this paradigm (Fig. 2b) a client has knowledge while a server has resources to perform a task. For the client to obtain the result to a task, it must send its knowledge to the server. The server uses this knowledge and its own resources to perform the task before returning a result to the client.

2.3.3 Code on demand

In this paradigm (Fig. 2c) a client has resources while a server contains knowledge to perform a task. For the client to obtain a result to a task, it must send a message to the server to request knowledge. The client uses the server's knowledge and its own resources to perform the task and obtain a result.

2.3.4 Mobile agents

In this paradigm (Fig. 2d) an agent server is host to a mobile agent which contains knowledge and results from previous tasks, but not the resources it requires to perform a new task. Another agent server contains the required resources. Instead of forwarding knowledge to the new agent server, the mobile agent moves to access the resource. When the mobile agent has completed its task it can remain at the new server or move to another agent server carrying its knowledge and task results.

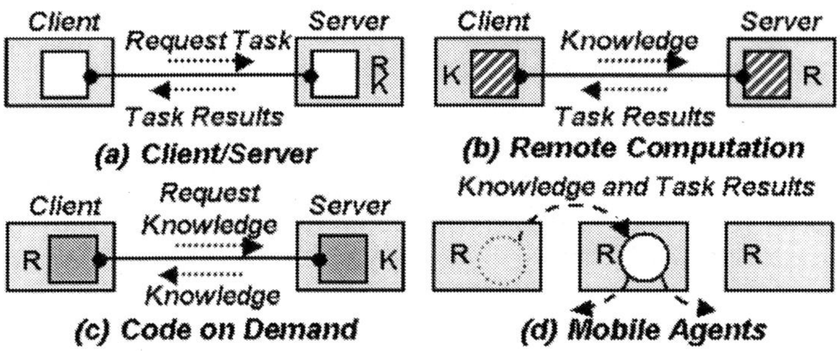

Fig. 2 Distributed computing paradigms

3 MOBILE AGENTS

In the previous section we compared activity in a variety of distributed computing paradigms. In this section we examine mobile agents in more detail.

3.1 Execution environment

Mobile agents exist within an execution environment, provided by multiple agent servers (Fig. 3). Agent servers provide areas known as places in which an agent can reside and interface with functionality provided by the server and host computer. Most agent servers and mobile agents are written in Java which is an interpreted computer language and can therefore execute on a variety of heterogeneous computer platforms.

Multiple host computers providing a distributed (but unified) mobile agent execution environment each host a single agent server which can be uniquely identified using the hosts IP address. Agent servers may however contain multiple places, between which multiple mobile agents can simultaneously migrate. Basic mobile agents are lightweight computing components, because many of the functions which they execute are provided by agent servers in a function library e.g. functions for agent creation, destruction, cloning, migration and fault tolerant storage to persistent media such as a computer hard drive. Function libraries allow developers to concentrate specifically on the wider purpose of the mobile agents in their system rather than on their basic execution requirements.

Most agent servers make extensive use of multi-threading in order to execute agent activity. Mobile agents usually run in an independent thread, allowing multiple agents to execute concurrently. Mobile agent developers can also make use of multithreading within individual agents to allow simultaneous activity to be executed e.g. agent communication while also conducting high level planning etc. Mobile agents can make use of many communication mechanisms including sockets, RMI and CORBA. Although remote communication using these mechanisms is possible, migration and local communication are more often employed in mobile agent systems because this helps to reduce network load in comparison to other distributed computing paradigms which must all communicate remotely across networks.

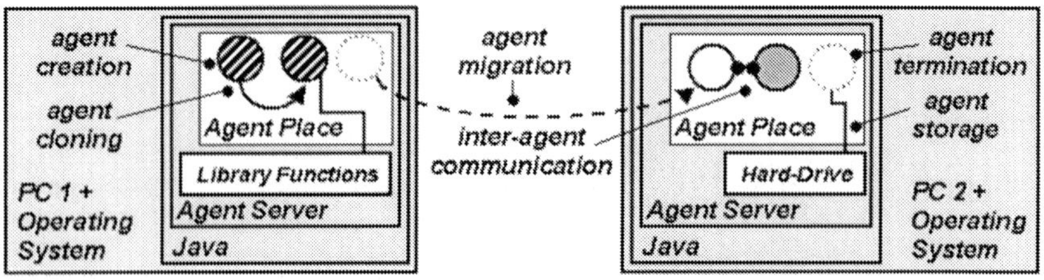

Fig. 3 Mobile agent execution environment

3.2 Characteristics and functionality

Mobile agents are goal driven and autonomous; they can be reactive, adaptable, dynamic, temporally continuous, operate asynchronously, communicate, and learn. Because they can provide all of these characteristics and also move from place to place, mobile agents can provide the functionality found in all other distributed computing architectures combined. In addition they provide a more natural design philosophy for distributed computing systems when compared to other distributed computing paradigms i.e. traditional client/server based architectures require more complex interconnected networks of heterogeneous computing nodes than mobile agents. Using mobile agents we do not need to consider network connections and consider all distributed computing nodes as offering equivalent functionality. This allows the development of systems which are more likely to be logically sound/robust.

Mobile agents can be used to produce intelligent, dynamic, fault tolerant and scalable distributed systems and reduce network load. This has allowed them to be applied in a range of application areas (5) including computational outsourcing, dynamic network management, load balancing, personalised user presence, software deployment, intelligent data, temporary applications, application of dynamic protocols and intelligent remote action (6).

4 IMPLEMENTATION

This section describes how we have implemented ALLIANCE in a mobile agent environment, starting initially with a description of the hardware/software employed.

4.1 Hardware/software

Our research hardware/software implementation is shown in Fig. 4. It includes two PC's, one running Windows^{XP} and the other Linux. The PC-Windows^{XP} contains Java 1.4, Grasshopper 2 (GH2) Agent Platform (7) (to host mobile agents), a Timing Server (to time experiments) and Apache Web Server v2.0 (to serve Java mobile agent .class files to GH2). The PC-Linux contains SSH (to allow remote login to multiple robots), SFTP (to allow downloading of files to multiple robots), and a Timing Server. The implementation also includes three ActivMedia Pioneer 2DX mobile robots (8) each running Linux and network accessible via wireless LAN. Each robot contains Java 1.3, GH2 and ActivMedia's Aria robot control software.

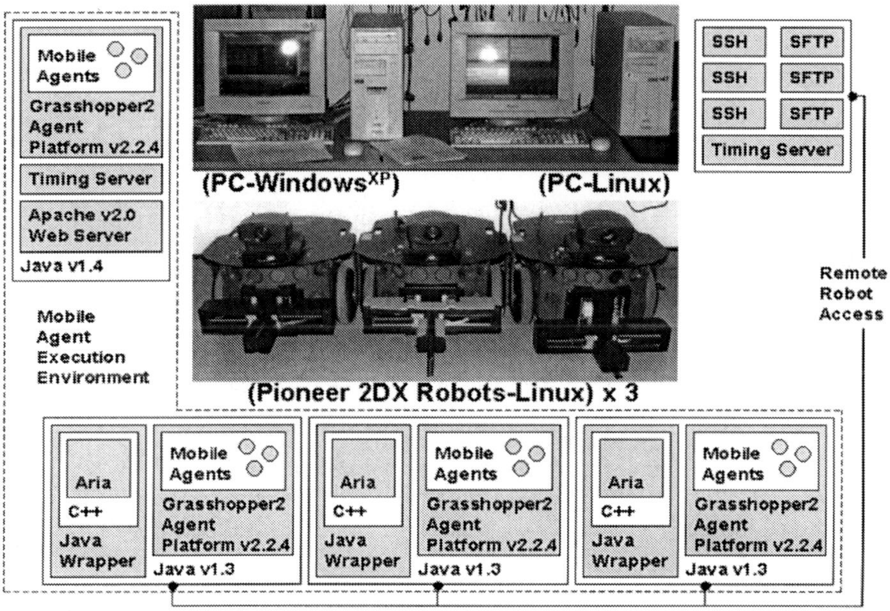

Fig. 4 Mobile robot/mobile agent implementation

GH2 is a free OMG MASIF compliant Java based mobile agent server which provides an execution environment and function library for Java based mobile agents. The function library provides functions for the creation, destruction, cloning, migration and persistent storage of agents. A mobile agent developer overrides a live() function in which they incorporate their mobile agent's run-time behaviour. Apache Web Server is used to store agent Java .class files. The GH2 agent servers obtain .class files from this central repository.

ActivMedia's Aria robot control software allows various levels of control to be executed on ActivMedia Pioneer 2DX robots, it is written in C++ but provided with a Java Wrapper which allows Aria commands to be written directly in our Java based mobile agents. The Timing Servers are multi-threaded Java based servers which listen on known ports and register the time of any connection. Because it is not possible to precisely synchronise the internal clock of distributed computers we employed a single timing server during each experiment to generate consistent times for experiment duration calculation.

4.2 Architecture

Fig. 5 shows our ALLIANCE implementation with the main element of an ALLIANCE Control Agent (ACA), a mobile agent which engages in three main activities: Calculation of MB's, execution of BS's, and IC. To conduct IC, calculate MB's and execute BS's concurrently, we implement these activities as subclasses of the thread class. We start each thread in the ACA's live() function allowing them to execute simultaneously. Using synchronised methods these classes are able to access shared variables within the main body of the ACA allowing coordinated behaviour. As these activities operate independently the main agent thread is free to interact with the server and other agents. The MB thread calculates a motivation value for each BS every 100ms. The BS contains code for the execution of any BS selected by MB's in the MB thread and executes every 100ms.

For IC the IC thread creates and interacts with a second form of mobile agent, an ALLIANCE Communication Agent (AComm). At each cycle of the IC thread i.e. every 4 seconds the IC thread on each robot creates a single AComm. This agent is created with a list of all known robot addresses, the ID of the creating robot and the behaviour of the creating robot. Upon creation this agent automatically creates a copy of itself on all the other robots in its list and then removes itself from the creating robot.

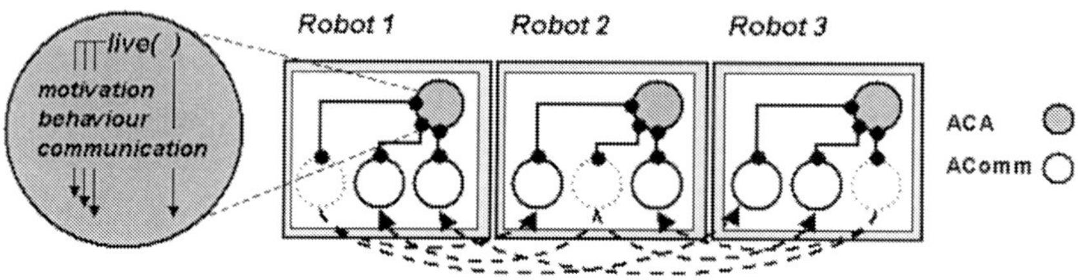

After creating the AComm the IC thread uses the agent server to locate and interact locally with any AComm's which have been copied to this robot from other robots. It obtains the ID information and current behaviour of any AComm's currently residing on the robot, thereby obtaining information on the activity of all other communicating teammates. After providing the local IC thread with their information payload AComm's remove themselves from the robot host in order to prevent a build-up of agent numbers and resource use. The IC conducted in this way allows the MB thread to be provided with relevant information with which to calculate motivation values for behaviour selection.

5 EXPERIMENTS

Using the implementation shown in Fig. 4 and 5, two sets of experiments were conducted. These are described in the following sections.

5.1 Adaptability experiments

Adaptability experiments were designed to investigate how mobile agents might affect the adaptability characteristics (i.e. the ability to update or adapt system code at run-time) of an ALLIANCE controlled multiple robot system. One set-up used an ALLIANCE mobile agent implementation (in which mobile agents are employed to allow existing ACA's to be dynamically updated by new ACA's at run-time), whilst the second investigated the updating of a traditional static code ALLIANCE implementation (using remote robot login (SSH) and file transfer using (SFTP)).

Fig. 6a shows mobile agent adaptability experiment activities. Initially existing ACA's are used to control multiple robots via Robot Agents (RA's) (1). A GH2 platform is then started which automatically creates a new ACA (2). The new ACA clones itself to all robots, at which clones communicate with existing ACA's causing them to remove themselves from the robot (3). The new ACA's then connect to the robot via RAs before communicating with a timing server (4). Timing of an experiment trial begins on the creation of the GH2 platform in step (2) and ends when the last new ACA communicates with the timing server in step (4).

Fig. 6b shows static code adaptability experiment activities. Initially existing Alliance Control Programs (ACP) are used to control multiple Robots (1). SFTP is then used to download new ACP's (2) and SSH used to stop existing ACP's (3). SSH is then used to start the new ACP's which connect to their robot and communicate with a timing server (4). Timing of an experiment trial begins on the creation of the first SFTP connection to robots in step (2) and ends when the last new ACP communicates with the timing server in step (4).

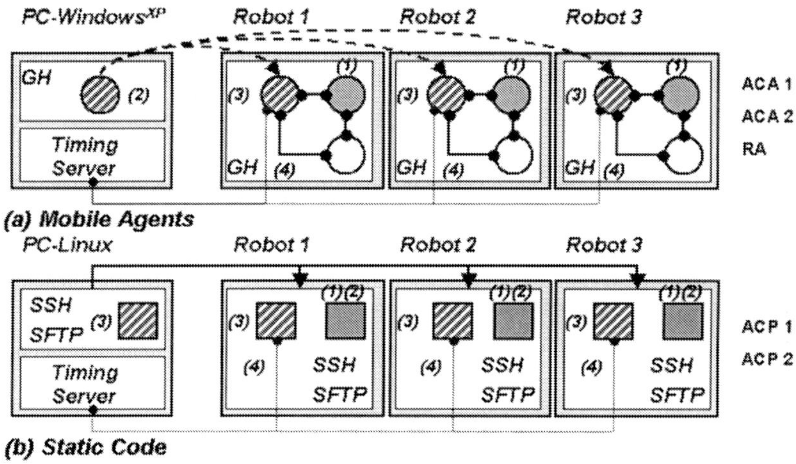

Fig. 6 Adaptability experiments

5.2 Fault tolerance experiments

Fault tolerance experiments were designed to investigate how mobile agents might affect the fault tolerance characteristics (i.e. the ability of a system to remain operational in the presence of faults) of an ALLIANCE controlled multiple robot system. One set-up used an ALLIANCE mobile agent implementation (in which an existing ACA is able to autonomously migrate from a failing robot to a secondary robot in order to maintain mission level data, before migrating on to a new replacement robot), whilst the second investigated a traditional static code ALLIANCE implementation (in which a failing robot loses mission level data when it fails and is replaced using a new robot with no previous mission experience).

Fig. 7a shows mobile agent fault tolerance experiment activities. Initially an ACA executing on Robot 1 detects a terminal fault and communicates with a timing server (1). This ACA then migrates to Robot 2 (2) and waits until a Robot Start Up Agent (RSUA) communicates with it to show that a new Robot 1 has been started (3). The ACA waiting on Robot 2 then migrates to the New Robot 1 connects to the robot via an RA and communicates with the timing server once more (4). Timing of an experiment trial begins on the connection of ACA to the timing server in step (1) and ends on ACA connection to the timing server in step (4).

Fig. 7b shows static code fault tolerance experiment activities. Initially an ACP executing on Robot 1 detects a terminal fault and communicates with a timing server (1). SFTP is then used to download a new ACP to a new Robot 1 (2). SSH is then used to start this new ACP (3). The new ACP connects to the new Robot 1 and then communicates with the timing server (4). Timing of an experiment trial begins on the connection of ACP to timing server in step (1) and ends on ACP connection to the timing server in step (4).

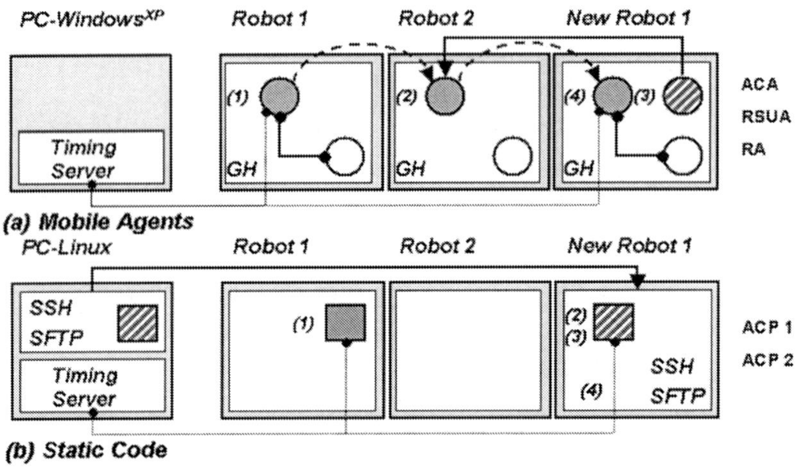

Fig. 7 Fault tolerance experiments

6 EXPERIMENTAL RESULTS AND ANALYSIS

6.1 Adaptability experiments

Adaptability experiment results are shown in Fig. 8 and Table 1. The adaptability experiments show that it is considerably quicker and easier to update and adapt the control software on multiple robots using mobile agents in comparison to a traditional method using static code. In the experiments scripts were used wherever possible in the static code experiments to remove human performance factors and to speed up and automate the updating process. Despite this mobile agents updated code much more quickly.

In addition the times for mobile agents include the time required to start an agent server at the PC-Windows^{XP}, while in practice such a server would normally be in operation at all times. The mobile agent performance is an improvement over traditional methods because agent creation occurs in parallel on all robots, while in the static code implementation activities occur in series. Due to this parallel execution, as robot team numbers increase, the mobile agent to static code time ratio should widen i.e. mobile agents relative to static code should be ever more effective as robot number increases.

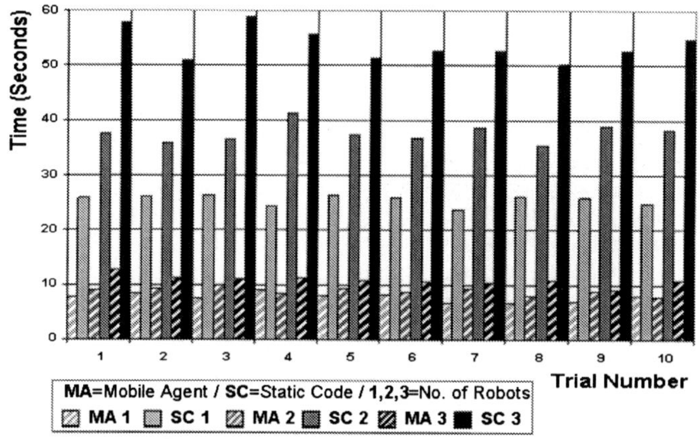

Fig. 8 Adaptability experiment results chart

Table 1: Adaptability experiment results table

Type\Trial	1	2	3	4	5	6	7	8	9	10
MA 1	7.8	8.4	7.6	8.9	8.0	8.1	6.6	6.6	6.9	7.8
SC 1	25.9	26.1	26.4	24.3	26.2	26.0	23.7	26.2	25.8	24.8
MA 2	9.0	9.2	10.0	8.4	9.2	8.5	9.2	8.0	8.7	7.8
SC 2	37.5	35.8	36.5	41.3	37.4	36.8	38.6	35.4	38.8	38.2
MA 3	12.6	11.2	11.0	11.3	10.9	10.5	10.4	10.8	8.9	10.7
SC 3	58.0	51.0	59.0	55.8	51.3	52.7	52.7	50.2	52.6	54.8

6.2 Fault tolerance experiments

Fault tolerance experiment results are shown in Fig. 9 and Table 2. The fault tolerance experiments have shown that using mobile agents it is possible to retain the state and data of a failing robot. However it took more time (approximately 7-8 seconds more) using mobile agents to start up a new replacement robot in comparison to the static code implementation. Although this process took longer using mobile agents, the traditional static code implementation lost robot state and data when the robot failed.

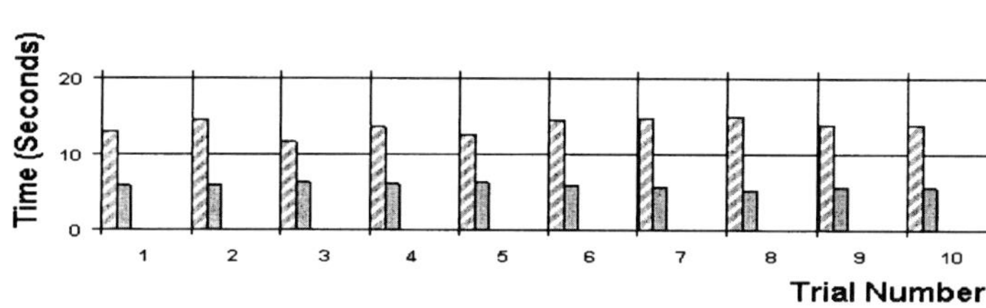

Fig. 9 Fault tolerance experiment results chart

Table 2: Fault tolerance experiment results table

Type\Trial	1	2	3	4	5	6	7	8	9	10
MA 1	13.0	14.5	11.7	13.6	12.6	14.7	14.8	15.0	13.9	13.9
SC 1	5.8	5.8	6.3	6.0	6.2	5.9	5.7	5.3	5.7	5.6

7 CONCLUSIONS AND FUTURE WORK

In this paper we have examined an implementation of the ALLIANCE architecture in networked robots using mobile agents in which we extend adaptability and fault tolerance. These functional extensions arise as a result of the implementation of ALLIANCE in the mobile agent environment and not through its architectural adaptation. These extensions could equally be achieved with other distributed mobile robot control architectures. Mobile agents offer many opportunities for the extension of functionality in multiple robot systems, because they unify the functionality found in all other distributed systems, and provide a natural design philosophy for distributed systems such as multiple robot architectures (based upon an equal allocation of basic functionality throughout all mobile agent server enabled system components). In conjunction these characteristics can allow multiple robot system developers to employ mobile agents as a tool to develop robust cohesive architectures in which system structure can be easily extended by making use of in built functionality available in the mobile agent environment. We are using this implementation of ALLIANCE as the control selection mechanism of a multiple tele/autonomous robot architecture (9) in which we aim to make further use of the beneficial functionality which mobile agents offer.

REFERENCES

1. Parker, L. E. (1994), "ALLIANCE: An Architecture for Fault Tolerant, Cooperative Control of Heterogeneous Mobile Robots", *Proc. of the IEEE/RSJ/GI Int. Conference on Intelligent Robots and Systems* (IROS '94), September 1994, pages 776-783.
2. Parker, L. E. (2001), "Evaluating Success in Autonomous Multi-Robot Teams: Experiences from ALLIANCE Architecture Implementations", *Journal of Theoretical and Experimental Artificial Intelligence*, 13, 2001, pages 95-98.
3. Yu, L., Tsui, P.W., Zhou, Q. and Hu, H. (2001), "A Web-based Tele-robotic System for Research and Education at Essex", *Proc. of IEEE/ASME International Conference on Advanced Intelligent Mechatronics*, Como, Italy, 8-11 July 2001, pages 284-289.
4. Tsui, P.W. and Hu, H. (2002), "A Framework for Multi-Robot Foraging over the Internet", *Proceedings of IEEE International Conference on Industrial Technology*, 2002.
5. Milojicic, D. et al. "Mobile Agent Applications", *IEEE Concurrency*, July-September 1999, pages 80-90.
6. Camarinha-Matos, L. and Vieira, W. "Intelligent Mobile Agents in Elderly Care", *Journal of Robotics and Autonomous Systems,* 27 (1999), pages 59-75.
7. IKV++ (2003), "Grasshopper 2 Homepage", World Wide Web, http://www.grasshopper.de/, IKV++ Technologies AG, 2003.
8. ActivMedia (2003), "Software, Documentation and Technical Support", World Wide Web, http://robots.activmedia.com/techhome.html, 2003.
9. Cragg. L. and Hu, H. (2003), "Application of MAs to Robust Tele-operation of Internet Robots in Nuclear Decommissioning", *Proc. of IEEE International Conference on Industrial Technology*, Maribor, Slovenia, 10-12 December 2003, pages 1214-1219.

Pump asset operation and performance monitoring over the Internet

K J DARBYSHIRE and **G WATERWORTH**
Faculty of Information and Technology, Leeds Metropolitan University, UK

ABSTRACT

This paper describes the development of a web-based database to provide hydraulic engineers with secure real-time pump characteristics. The aim is to monitor power, flow and pressure produced by individual pumps at fifteen-minute intervals.

The pump data is compiled and stored on a secure web-based server using java server pages (JSP) and server query language (SQL) statements. Pump characteristic curves can be modelled from this data, which illustrates which pump or combination of pumps is the most efficient and cost effective.

The results illustrate that the recording of pump characteristics over the Internet provides an efficient method of pump asset operation and performance monitoring. The stored data may be used in the development of a sophisticated scheduling scheme or to provide historical information for water demand prediction.

1 INTRODUCTION

Throughout the water industry many methodologies have been adopted to enable significant energy cost reduction incurred by Water Pumping Station (WPS). Refurbishing worn equipment, using highly efficient motors, employing the most efficient pumps and ensuring that water is pumped through the smallest hydraulic head will make substantial savings. Over the last few years, Internet has been broadly used to circulate information around the world. Using existing technologies of Hypertext Mark-up Language (HTML) and Java / JSP, with a remote SQL web based server, power, flow and pressure produced by individual pumps can be monitored over the Internet, allowing pump operators to remotely select efficient pump combinations. To simulate pump characteristics produced from the data loggers, a web enabled front end is used to interact with the database. The Java Database Connectivity

(JDBC) Application-Programming Interface (API) allows the user to interact with the SQL database server, enabling Microsoft Excel to extract flow, pressure and power readings every fifteen minutes. Using direct data access (DDA) the spreadsheet can be imported into MATLAB, enabling pump efficiency and system curves pump be modelled.

2 WATER SUPPLY SYSTEM

Figure 1 illustrates a typical high-lift water pumping station consisting of five-fixed speed pumps connected in parallel via a 38-inch water main. Each pump extracts clean water from a nearby water storage tank and supplies treated water to the consumer. Pumps are individually monitored using Micro-Controller Processors (MCP), which log the pumps parameters flow, power and pressure on the delivery side of the pump [1]. This raw data is transmitted to the 'pump control room' through remote telemetry units (RTU) or outstations [2]. Supervisory Control and Data Acquisition (SCADA) enables automatic control of pump operation (on / off) and archives the pump data. Figure 2 depicts the data transfer between the RTU and SCADA.

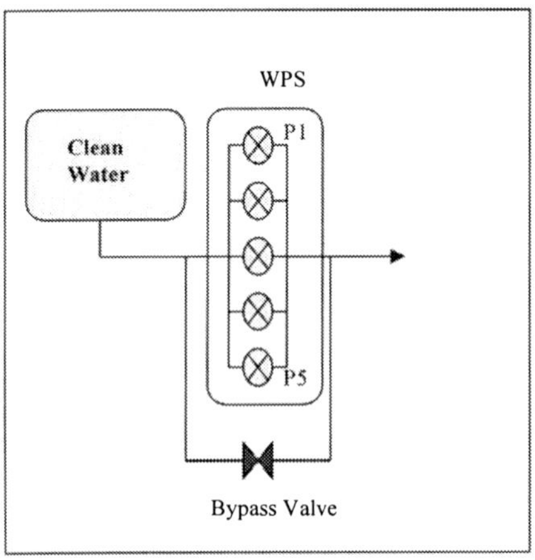

Figure 1. Water Supply System

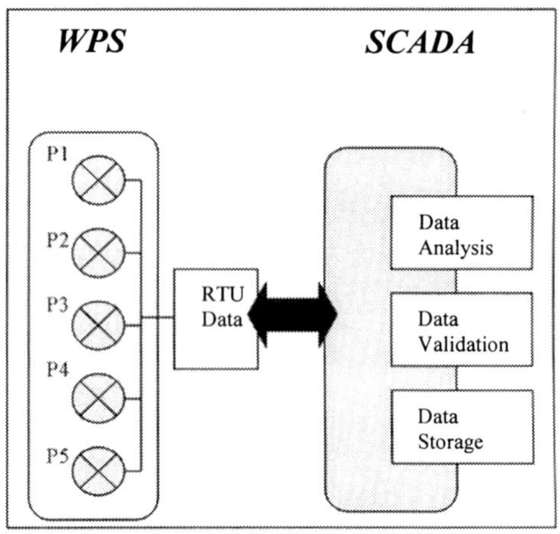

Figure 2. Data Transfer Using RTU's

3 WEB TECHNOLOGIES

JSP is a web scripting technology similar to Microsoft Active Server Pages (ASP) and Netscape Server Side Java Script (SSJS) [3]. JSP is platform independent, and can easily evolve in complexity by adding other Java components such as JavaBeans [4]. JSP is a presentation / representation layer enabling users to mix static HTML content with server side scripting to produce a dynamic output. The database is implemented as a classic '2-tier' client / server, which is illustrated in Figure 3. JDBC allows the user to interact with the remote database, which is isolated on the server side, through SQL statements sent over a TCP / IP network.

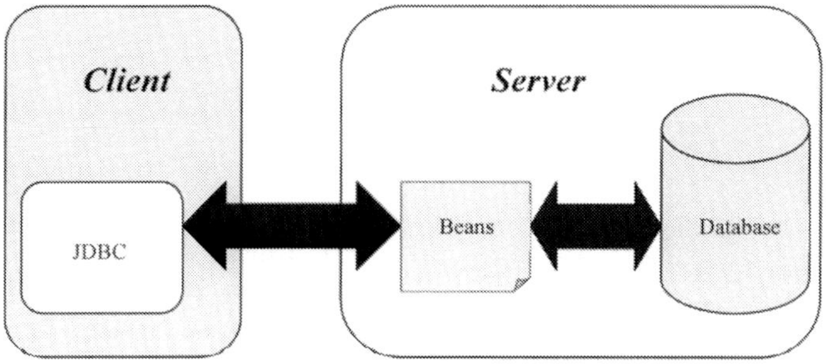

Figure 3. Data Transfer Using RTU's

4 RESULTS

The results illustrate the integration of a data source on the server-side enabling pump characteristics to be store and calculated to derive pump efficiency curves [5]. The front end of the database consists of add, delete and select pump data through Graphical User Interface (GUI). Figures 4 illustrates the 'select data' GUI. Alternatively a GUI administration client, such as My SQL Control Center (MySQLCC) [6] or MSQL Front the user may construct quires, view server status information and amend / delete files as illustrated in Figure 5.

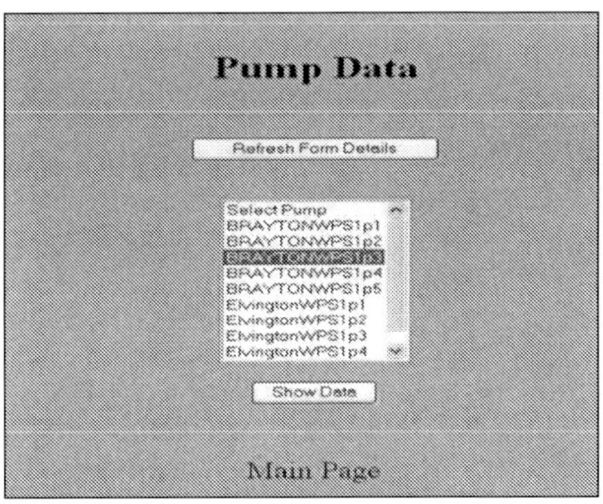

Figure 4. Pump Selection Data

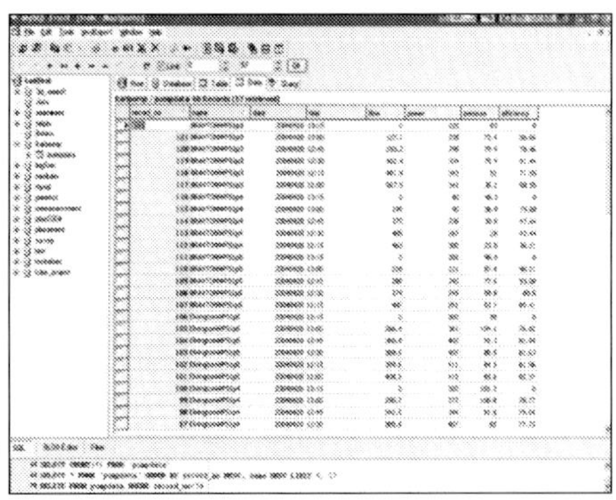

Figure 5. MSQL Front Server Entries

If the user processes a request, the pump characteristics are entered and sent to the server, which are collected by a servlet under the form of a 'request' [7]. From these parameters, the servlet connects to the database, extracts the pump parameters and calculates the efficiency of the pump. The user is then faced with a request formatted web page, depicted in Figure 6.

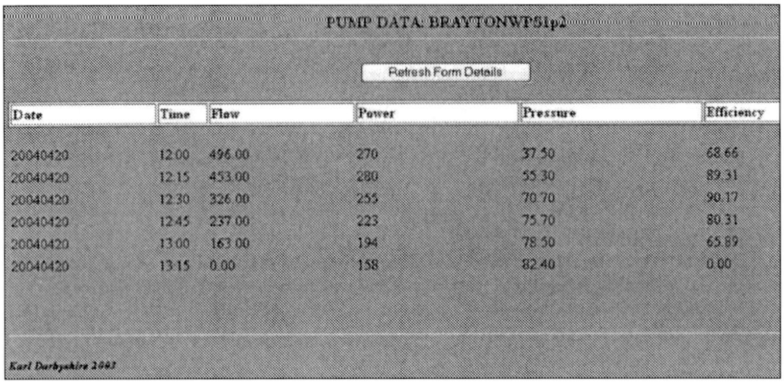
PUMP DATA: BRAYTONWPS1p2

Refresh Form Details

Date	Time	Flow	Power	Pressure	Efficiency
20040420	12:00	496.00	270	37.50	68.66
20040420	12:15	453.00	280	55.30	89.31
20040420	12:30	326.00	255	70.70	90.17
20040420	12:45	237.00	223	75.70	80.31
20040420	13:00	163.00	194	78.50	65.89
20040420	13:15	0.00	158	82.40	0.00

Karl Darbyshire 2003

Figure 6. Request Table Data from Server

Matlab is used to model pump efficiency curves from the achieved data. Through the application of hydraulic equations [8], and reference to Euler's equation, individual pump efficiencies are calculated aiding ideal pump selection. Figure 7 illustrates head, flow and efficiency curves.

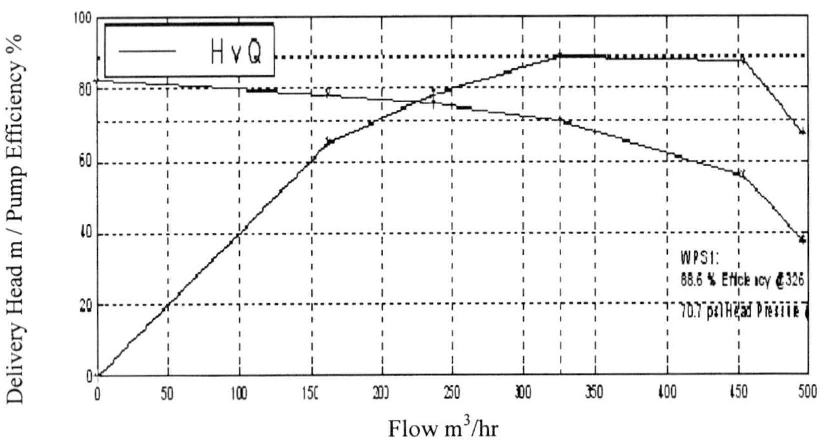

Figure 7. Matlab Pump Efficiency Curves

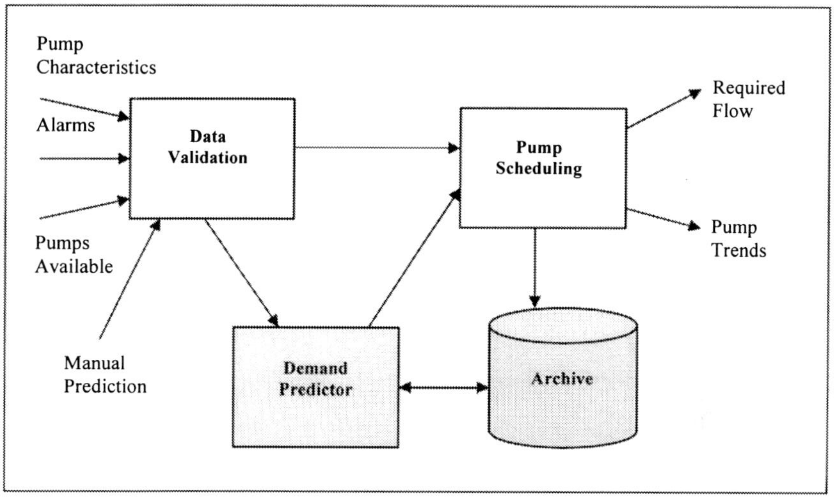

Figure 8. Data Flow of Decision Support System

5 CONCLUSION

This paper has proposed a cost effective method of pump asset operation and performance monitoring over the Internet. The combined web methodologies of Java, JSP and SQL technologies provide an efficient and accurate process of secure remote logging of pump data. The results illustrate that the web server can remotely accessed pump characteristics to log, view or delete pump data. Through the aid of Matlab, pump efficiency curves have been derived to enable pump engineers to identify inefficient pumps and shortfalls. Through operating pumps at their maximum efficiency, with respect to flow and power consumption, potential energy cost reductions are achievable. The stored data may be used in the development of a sophisticated scheduling scheme [9] or to provide historical information for water demand prediction. A logical extension of this work would be to fully automate the system to create an online decision support system. This aspect will bring potential savings in manpower costs due to the system being operated according to instructions generated by the decision support system rather than an experience pump operator. The tasks required to enable such a system are data collection validation and storage, demand prediction, pump scheduling, controlling the pumps, monitoring the results of the pumping and condition of equipment and raising alarms or forwarding SMS and email messages using G3 technology [10]. Figure 8 illustrates how data is communicated between tasks.

REFERENCES

1. Kang, B. L et al. (2000) "Distributed Measurement and Control Based on the IEEE 1451 Smart Transducer Interface Standards", *IEEE Transactions on Instrumentation and Measurement.* Vol 49, No 3, June.
2. Reynolds, L (2004) "Developments in Control in the Water Industry", *IEE Computing and Control Engineering.* February/March 2004. Pages 39-43.
3. Kuslich, J. J (1999) *Introduction to JavaServer Pages; Serve-Side Scripting the Java Way.* [Internet] Available from: <http://developer.netscape.com/viewsource/kuslich_jsp/kushlish_jsp.html#bio/> [Accessed 6 Feb 2004]
4. van der Liden, P (2002) Just Java. *Java Series Fifth Edition* Sun Microsystems. The Sun Microsystems Press.
5. Darbyshire, K. J (2003*) Pump Data Management* [Internet] Available from: <http://care.ies.lmu.ac.uk/karlpump/index.jsp> [Accessed 6 Feb 2004]
6. MySQLCC [Internet] Available from: <http://www.mysql.com/products/msqlcc/index.html> [Accessed 6 Feb 2004]
7. Brogden, B. Minnick, C (2001) *Java Developers Guide to E-Commerce with XML and JSP.* 2001 Sybex Inc.
8. Forlags, E. P (1981). *The British Pump Market 181-82.* Heffers Printers Ltd, Sweden.
9. Darbyshire, K. J. Waterworth, G. (2002) "Optimal Scheduling Utilizing Factorial Analysis and Constrained Optimisation", *Advances in Manufacturing Technology XVI.* Leeds Metropolitan University UK, Edited by Cheng, K. Webb, D. 10-12 September 2002. Professional Engineering Publishing Limited UK.
10. Mustard, S. Harrison, S. (2004) "Future Requirements for Remote Telemetry Units", *IEE Computing and Control Engineering.* February/March 2004. Pages 45-47.

Distributed hardware architecture and intelligent sensors for mobile robots

A DILSHAD
Intelligent Robotics Laboratory, Nanyang Technological University, Singapore

ABSTRACT

Mobile robots and unmanned autonomous vehicles need sufficient sensors and real time information on the environment throughout the mission. This paper describes major constrains related to the mission of an unmanned autonomous vehicle or a team of mobile robots and suggests a few solutions for each of them. The paper proposes a system of intelligent sensors that is controlled in a distributed environment via wired fieldbus for an autonomous vehicle and via wireless field bus architecture for a group of mobile robots. In the proposed system each device has low cost computing power installed in it, making each device a 'smart' device throughout the work-cell of robotic system. Each device will be able to execute simple functions on its own such as diagnostic, control, and maintenance functions as well as providing bi-directional communication capabilities to communicate with the neighboring robots.

1. INTRODUCTION

Unmanned autonomous vehicles and mobile robots can significantly reduce the manpower requirements and risk to humans for hazardous tasks. The robots and UAVs can perform very effectively in the areas where humans cannot reach or in dangerous environments. Unlike human controlled robots, autonomous robots require significant amount of intelligence to carry out different tasks and operations independently that are mainly dependent on sensory information.

Most of the control strategies for mobile robots focus more on sensing and acting [1] [2]. Many biological systems provide motivation for this. For example, [3] introduces the concept of the *Umwelt* that describes the set of environmental triggers to which an animal responds. Similarly, robots may be configured to respond instinctively to a certain set of features. This forms the basis for Brooks' [1] [4] subsumption architecture.

For a good autonomous robot it is necessary to equip it with state of the art sensors to provide highly reliable and real-time sensory information that can meet the requirements of specified task. On the other hand the robot is also subjected to some other constrains that create a trade-off in providing required intelligence onboard. These constrains are discussed in next heading.

When a team of these kinds of robots performs a mission, usually they talk or communicate with each other to allocate tasks and to specify different actions to one another. It is proposed that the sensory intelligence can be distributed to the different team members of robot squad and a fieldbus approach can be used to provide communication architecture.

2. CONSTRAINS FOR MOBILE ROBOTS

An autonomous robot performs its mission under various constrains that limit its operational capabilities as compared to conventional robots. Some major physical constrains are weight, size and power or fuel, while major technological constrains are processing capabilities, inter communication dependencies, task allocation and carrying out different operations to achieve the goal. These constrains are also closely related to each other for example increase in fuel or power will also increase the size and weight of the robot that consume more power in order to move and operate. Mobile robots usually consist of many motors and actuators, different distributed sensing units and processing power. For example *Herbert* and *Genghis* [5] are two robots in MIT's mobile robots laboratory built with subsumption architecture, *Herbert* has a 24-processors distributed, loosely coupled, onboard computer. They are 8-bit CMOS microprocessors consume low electrical power and communicate by slow serial interfaces (10 packets per second, each packet 24 bits wide). It had 30 infrared proximity sensors, an onboard manipulator with a number of simple sensors attached to the hand and a laser light stripping system to collect three dimensional depth data. To compute high performance vision algorithm four onboard processors are linked in a chain. The 2nd robot *Genghis* is a 1 Kg, six-legged robot that can walk successfully over rough environment. It has 12 motors, 12 force sensors, 6 pyroelectric sensors, one inclinometer and 2 whiskers to carryout its tasks.

For a complete and dynamic autonomous robotic system we need a comprehensive set of sensors, adequate computing power and memory, reliable communication architecture and well-designed software routines for planning, task allocation and other computations. All these requirements must be fulfilled within limited size, weight and available fuel or power. For a team of similar mobile robots we can use intelligent sensor system with field bus architecture to provide highly reliable, accurate and real time sensed information to the team so that the robots can negotiate each other for taking actions and allocating task.

3. INTELLIGENT SENSOR SYSTEM

The idea of intelligent sensor system for a team of mobile robots is that the sensor system of each robot can directly communicate with all sensors of other robots and then by using sensor integration and fusion techniques for example using [6] [7] a highly reliable and accurate result can be obtained that will be applied to every member of the team. This approach can provide many advantages for example in a team of 'N' mobile robots, each robot has 'M' number of sensors, by using proposed approach we may turnoff some redundant sensors to save power and in case of any fault or failure a similar unit can be turned-on to provide same level of reliability.

To illustrate this concept let we have a squad of 5 autonomous mobile robots that made an autonomous system of robots, the task of squad is to find, enter in a number of different warehouses and investigate that whether they contain tranquilizer / drugs, explosive material or food items. Each robot consists of for example 20 different sensors, 8 of which are to control internal motion and remaining 12 used for navigation, chemical examinations and other investigations. Robots can find and enter in any warehouse simultaneously altogether to investigate and the collective information needs to be fuse to get maximum reliable conclusion. According to the reliability statistics to get highly reliable conclusion it is required to fuse maximum of 3 member's data. The conclusion is then stored in each robot's memory. This approach can be used throughout the mission to navigate, eliminate hurdles, chemical investigations and so on. Another advantage is that if a specific sensor of a member robot failed, the robot can continue its mission by using others information. For a critical mission in which weight, size and power constrains are much tight (very limited), we can distribute the sensors over different members to senses different attributes for example some members can only listen and some can only view then by using sensor fusion a complete perception for the team can be obtained.

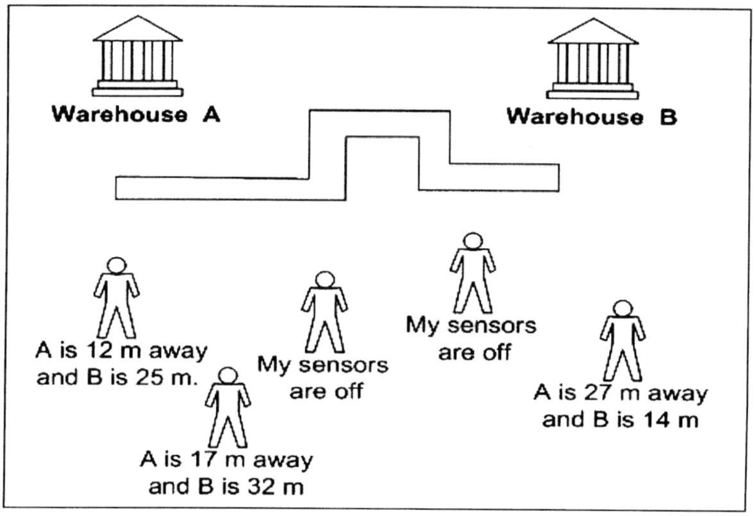

Fig. 1. Individually sensed information will be fused to get overall conclusion. Note that navigation sensors of some robots are turned off that will not affect overall conclusion.

4. PLANNING AND TASK ALLOCATION

The coordination of a group of mobile robots performing a mission is very essential especially for planning and task allocation. To complete the mission in dynamic environment, a good planning is required so that robots can complete the mission in minimum time duration while interacting in real-time as discussed in a dynamic mission planning for mobile robots [8].

Robots must allocate different tasks to each other by negotiation and other means and also determine the sequence of different actions that will be carried out by each member robot. In a dynamic and changing environment these allocations are also dynamic and can be changed as and when required. One good strategy of task allocation is suggested in [9] in which a team of mobile robots use free market architecture that is a novel economic approach in which robot allows to bid and negotiate to carry out the task while trying to maximize their personal profile. Another approach is proposed in [10] in which the overall task allocation problem is formulated using Mixed-integer Linear Programming (MILP).

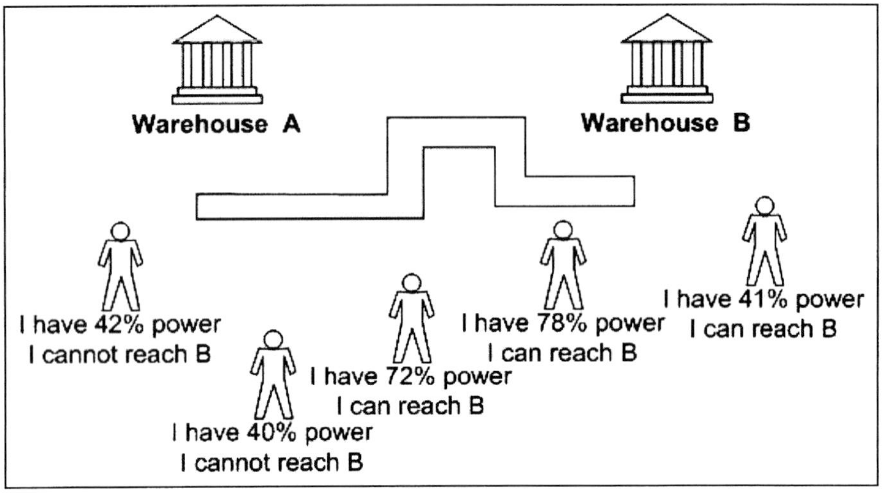

Fig. 2. Negotiation process for task allocation, robots negotiating each other to go inside warehouse B. It is clear that to complete mission first 3, 4 & 5 will go in B and then after further negotiation again three of them will go in A.

5. COMMUNICATION

Reliable communication is essential for the success in the mission of mobile robots team because all the decisions, planning and task allocation are carried out by intercommunication of the robots in real time, in other words the mission is highly communication-sensitive. Wireless communication significantly decreases the reliability by increasing communication failure rate. In [11] a method is presented for multi robot communication-sensitive reconnaissance. The approach utilizes collections of precompiled vector fields in parallel to coordinate a team of robots in a manner that is responsive to communication failure. In [13] an interesting approach is presented in which an adaptive communication protocol is used for cooperating mobile robots. The robots are initially given a fixed but un-interpreted vocabulary for communication. In attempting to perform their tasks, the robots learn a private communication language and as state of affairs change, the robot adapt their language to allow continued success at their tasks.

5.1 Robot communication using existing technology

Robot communication can be carried out through existing communication technology like satellite or mobile communication in which robots can communicate with each other via satellite or mobile phone network. This approach is not always reliable, in satellite communication computational and transmission delays can cause less accurate reporting in many cases. Satellite lock can also occur near high-rise buildings and under the bridges. Mobile and satellite based communication is also not available in many covered areas and indoor surroundings. On the other hand these types of communications are rather expensive specially to develop for small and medium enterprises. These problems can be eliminated through proposed fieldbus communication architecture as discussed below.

5.2 Fieldbus architecture for robot communication

Fieldbus is a term that describes a new digital communications network used in the industry to link isolated field devices, such as controllers, transducers, actuators and sensors. Each field device has low cost computing power installed in it, making each device a 'smart' device. Each device will be able to execute simple functions on its own such as diagnostic, control, and maintenance functions as well as providing bi-directional communication capabilities. This approach can be used for mobile robots if we can implement the fieldbus architecture as wireless fieldbus.

In [13] a radio protocol for low power wireless sensor networks is described in which a special medium access control protocol meets the requirements in terms of energy saving and application layer functionality. This work and many other related preliminary works provide enough motivation to investigate the functionality of the fieldbus as wireless field bus because the fieldbus is a highly reliable communication architecture that is currently in the industry. Using radio waves as communication media between robots can easily solve the problems of satellite communication and mobile networks. There is a wide range of radio technologies already available for use in different fields of telecommunications. However, they are not designed for robotics or industrial applications or on industrial standards and still no comprehensive solutions are available to integrate them appropriately into state of the art industrial communication architectures.

Radio based wireless fieldbus can also provide consistent link between base station and every mobile robot that can be used to transfer real-time information about the mission to the base including large multimedia, real-time voice and low-resolution digital video sequences for many applications. Strategically important control signals can also be transferred from base to the mobile robots to respond in extremely unexpected situations, moreover, it can be used to design and develop fault tolerant robotic systems. The main objective of the wireless fieldbus system is to support both the real-time requirements of robots control mechanism as well as a user-defined QoS for the multimedia communications. Wireless transmissions of data and video images can be performed by a radio modem based on RAKE spread spectrum technology. This technology allows taking advantage of propagation phenomena such as multiple paths.

In the wireless shared medium, all information transmitted in a fieldbus could be easy captured by any wireless device even if it is not involved in the fieldbus segment. Therefore, in order to provide security and authentication mechanism, the wireless fieldbus must offer an option to encrypt data to be transmitted.

Wireless fieldbus can also support TCP/IP applications. With the integration of the TCP/IP protocol stack into the fieldbus communication stack, control data and multimedia data can be efficiently supported by the same network infrastructure.

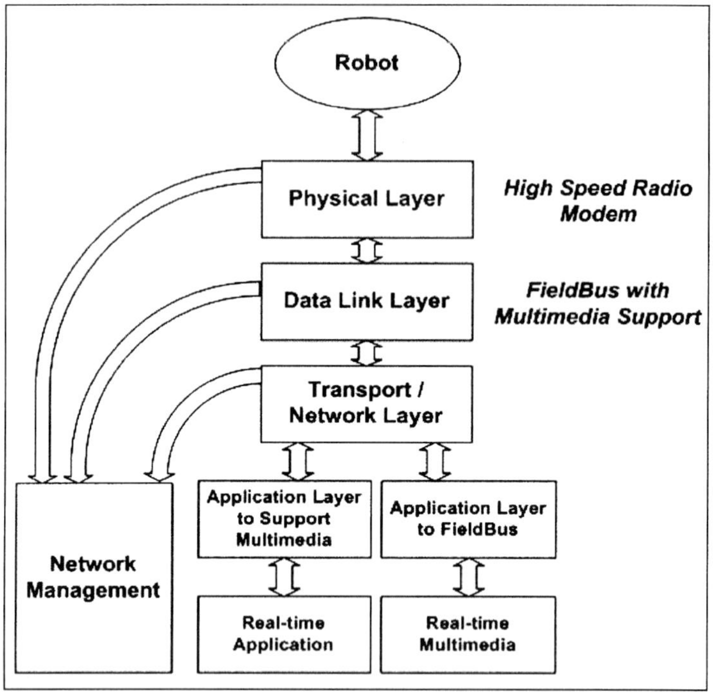

Fig. 3. Proposed wireless fieldbus architecture to provide reliable data transmission with real-time behaviour, security, flexibility and multimedia support

6. CONCLUSION

Mission of a team of autonomous mobile robots is subjected to a number of physical and technical constrains that are closely related to each other and create a trade-off in selection of sensors, computational capabilities and required electrical power. Using the proposed distributed intelligent sensors system in which the overall perception of the team is based on individual perceptions of some member robots can reduce the problems. Different types of sensors can also be deployed on different member robots in order to reduce size, weight and power requirements. Any sensor integration and fusion technique can be used to fuse these sensors to get overall outcome that will apply to every member robot. On the basis of sensed data, robots then negotiate for planning and task allocation. In order to provide reliable data transmission, proposed fieldbus architecture can be used for inter-robot communication that is an essential requirement of multi robot system. The proposed wireless fieldbus architecture can provide reliable data transmission with real-time behavior, security, flexibility and multimedia support. The latest micro-scale technology can also be used to face the size, weight and power constrains as [14] [15] introduced the idea of building complete robots out of silicon on a VLSI fabrication line.

Using proposed system of intelligent sensors and wireless fieldbus architecture, mobile robots can be applied more effectively in many civil and defense applications like surveys, security, surveillance & reconnaissance, target acquisition, medical re-supply, border petrol, maritime security, crowd control search and rescue and environment monitoring as presented in [16].

REFERENCES

1 Rodney A. Brooks, "A robust layered control system for a mobile robot", *IEEE J. Robotics and Automation,* RA-2, April (1986), pp.14-23.

2 Steels, L. "Building Agents with Autonomous Behaviour Systems", The `artificial life' route to `artificial intelligence'. Building situated embodied agents. *Lawrence Erlbaum Associates, New Haven. 1994.*

3 Von Uexkull, J., "A stroll through the worlds on animals and men." in Instinctive Behaviour, ed. K. Lashley. International University Press.

4 Rodney A. Brooks, "A robot that walks: Emergent behavior form a carefully evolved network" *Neural Computation,* 1(2) (Summer 1989), pp. 253-262.

5 Rodney A. Brooks., "Elephants Don't Play Chess", *Robotics and Autonomous Systems,* Vol. 6, 1990, pp. 3-15.

6 Takamasa, K., "Improved Sensor Selection Technique by Integrating Sensor Fusion in Robot Position Estimation", *Journal of Intelligent and Robotic Systems* 29: 2000, pp.79–92.

7 Hakyoung Chung, Lauro Ojeda, Johann Borenstein, "Sensor fusion for Mobile Robot Dead-reckoning With a Precision-calibrated Fiber Optic Gyroscope", *IEEEInt. Conf. on Robotics and Automation, Seoul, Korea,* May 21-26, 2001, pp. 3588-3593.

8 Brumitt, B. L. and Stentz, A., "Dynamic Mission Planning for Multiple Mobile Robots", *Proc. of the IEEE Int. Conf. on Robotics and Automation,* No. 3, 1996, pp. 2396-2401.

9 Dias, M. and Stentz, A., "A free market architecture for distributed control of a multirobot system," in *Proc. of the 6th Int. Conf. on Intel. Auto. Systems,* July 2000.

10 Alighanbari, M., Kuwata, Y. and How, J. P., "Coordination and Control of Multiple UAVs with Timing Constraints and Loitering," Proceedings of the *IEEE American Control Conference,* June 2003.

11 Wagner, A.R. and Arkin, R.C., *Multi-Robot Communication-Sensitive Reconnaissance, College of Computing Georgia Institute of Tech.* Atlanta, USA.

12 Yanco H. and Stein, L.A., An adaptive communication protocol for cooperating mobile robots, *Proc. of 2nd Int. Conf. on from animals to animats 2 : simulation of adaptive behavior,* p.478-485, August 1993, Honolulu, Hawaii, USA.

13 Stein, G. and Kabitzsch, K., "A Radio Protocol for Low Power Wireless Sensor Networks. *Dresden University of Technology,* Dresden, Germany (http://www.iai.inf.tu-dresden.de/tis/forschung/publikationen/FET2003b.pdf)

14 Flynn, A.M., "Gnat robots (and how they will change robotics)", *IEEE Micro Robots and Teleoperators Workshop,*Hyannis, MA (November 1987).

15 Flynn, A.M., "Gnat robots: A low-intelligence, low-cost approach", *IEEE Solid State Sensor and Actuator Workshop,* Hilton Head, SC (June 1988).

16 Mullens, K.D., Pacis, E.B., Stancliff, S.B., Burmeister, A.B., Denewiler, T.A., Bruch, M.H. and Everett, H.R. "An Automated UAV Mission System," *AUVSI Unmanned Systems in International Security 2003 (USIS 03), London, England,* September 9-12, 2003.

Assessment of the possibility to remotely monitor and control a pH process – an experimental study

N N KARNACHI, G WATERWORTH, and **K J DARBYSHIRE**
Faculty of Information and Engineering Systems, Leeds Metropolitan University, UK

ABSTRACT

The control of pH is important in many processes including wastewater treatment, chemical processes and biological processes. This paper considers a model reference non-linear control scheme. The method is tested using a 7-litre continuously stirred tank reactor to neutralise a strong acid using a strong alkaline solution. The method is first realised using a simulation of the process. Subsequently, it is demonstrated on an experimental rig using real-time control. Furthermore, the process is monitored remotely and controlled with a software using e-Technology. Experimental results confirm that a robust control and remote monitoring of the process is achievable.

1 INTRODUCTION

In many processes, pH neutralisation is a very fast and simple reaction. In terms of practical control, it is recognised as a difficult control problem (1), (2), (3). The difficulties arise from high process nonlinearity (the process gain can change tens or hundreds of times over a small pH range) and from changes in the pH characteristics due to changes in influent concentration. Various techniques have been developed to control process pH. Young and Rao (4) presented a variable structure controller ("sliding mode control for a neutralisation process") involving strong acids and bases. Parrish and Brosilow (5) used non-linear inferential control in a simple simulated neutralisation process, using static estimation of the concentration of a single monoprotic weak acid. Kulkarni et al. (6) presented non-linear internal model control for a simulated system of sodium hydroxide (NaOH) and hydrochloric acid (HCl). Li et al. (7) and Li and Biegler (8) presented non-linear feedback methods for a simulated neutralisation process. In the present work, a non-linear controller design is implemented. It uses a design procedure presented by Jayadeva et al. (9). The controller is implemented practically on a 7-litre reactor.

2 THEORY

In the present work, the design of a robust non-linear controller is introduced. It considers a model reference controller developed by Jayadeva et al. (10). The method is taken originally from a paper by Yuocef -Toumi and Ito (11). The control scheme is non-linear model reference (12).

2.1 Controller Design

Consider a single input and single output (SISO) state variable system of the form

$$x_1 = f_1(x_1, x_2, \ldots x_n) + g_1(x_1, x_2, \ldots, x_n)u + d_1(x, t)$$

$$\cdot \quad \cdot \quad \cdot \quad \cdot \quad \cdot \quad \cdot \quad \cdot \quad \cdot \quad \cdot$$

$$x_n = f_n(x_1, x_2, \ldots, x_n) + g_n(x_1, x_2, \ldots, x_n)u + d_n(x, t) \qquad [1]$$

$$y = c_1 x_1 + c_2 x_2 \ldots + c_n x_n \qquad [2]$$

where, u is a scalar manipulative input, $x_1, x_2, \ldots x_n$ are the states and y is a scalar output. f_i and g_i are nonlinear functions of state variables. $d_1, d_2, \ldots d_n$ represent general disturbances. The output variable y is a linear function of the state variable. $c_1, c_2, \ldots, c_n$ are constant scalars. Yuocef-Toumi and Ito presented a robust nonlinear feedback controller design for a general nonlinear multi-input state variable system, from which a least square solution for the manipulative variable was obtained. The method is applied to the specific form of Equation [1] and [2] to obtain an exact solution for the manipulative variable. Equation [1] and [2] can be written in vector form as

$$\dot{x} = f + gu + d \qquad [3]$$

$$y = cx \qquad [4]$$

where the vectors f and g are functions of x, and c is a constant row vector. Let us assume the reference model in the scalar output y_m is given by

$$\dot{y}_m = \lambda_m y_m + b_m r \qquad [5]$$

where, λ_m is the eigen value of reference model. The scalar e is defined as the difference between reference value and the process output. Therefore,

$$e = y_m - y \qquad [6]$$

The control objective is to force the error to vanish with a desired dynamics:

$$\dot{e} = \lambda e \qquad [7]$$

Where, λ is the eigen value for the error system. By combining Equations [3] – [7] we obtain the equation that governs the error dynamics. Therefore,

$$\dot{e} = \dot{y}_m - \dot{y} \quad [8]$$
$$= \lambda_m y_m + b_m r - cf - cgu - cd$$

as $\lambda_m e = \lambda_m (y_m - y)$ *therefore,*

$$= \lambda_m e + (\lambda_m y + b_m r - cf - cgu - cd) \quad [9]$$

It is possible to determine the manipulative variable u in Equation [9] such that

$$(\lambda_m y + b_m r - cf - cgu - cd) = ke \quad [10]$$

From which we have the manipulative variable

$$u = (cg)^{-1} [b_m r - cf - cd - ke + \lambda_m y] \quad [11]$$

Therefore Equation [9] becomes

$$\dot{e} = (\lambda_m + k) e$$
$$\dot{e} = \lambda e \quad [12]$$

Where, k is a scalar error feedback gain. The error system eigen value λ can be assigned arbitrarily through proper choice of the error feedback gain k. The control law in Equation [11] is used to calculate u in order to get the desired error dynamics (10).

Now consider the application of the above control design to the model for the pH process described by McAvoy et al. (13). The process consists of a strong acid flowing into a constant volume tank which is thoroughly mixed with a strong base. The feed flow rate of the base is to be controlled in such a way as to produce a neutral outlet from the tank. The equation describing this process is given by

$$\dot{x} = a_1 x - a_3 u (x + a_2) + a_1 D \quad [13]$$

Where, x is the deviation from neutrality. Note that, x and the pH value y, are related by the non-linear equation:

$$x(t) = 10^{-y(t)} - 10^{y(t)} K_w \quad [14]$$

Where K_w = water equilibrium constant = 10^{-14}, $a_1 = F_1/V$, F_1 is the acid flow in litres and V the volume of the mixing tank; a_2 = C_{base} = concentration of base; a_3 = $1/V$ are constant parameters; u = F_2, is the manipulative variable, base flow control in litres; D = C_{acid} = concentration of acid = the disturbance variable. It is to be noted that Equation [14] is valid for the strong acid / strong base case only. For the general case, there are two model equations [1], [3].

Now, comparing Equation [13] with [3], we have,

$$f(x) = -a_1 x; \quad g(x) = -a_3 (x + a_2); \quad d(t) = a_1 D$$

And the output equation,

$$h(x,y) = x + 10^{y-14} - 10^{-y} = 0 \qquad [15]$$

The control objective is to keep pH, $y(t) = 7$ = constant in the presence of disturbances occurring in the process in general, making $y(t)$ follow a given reference trajectory. In the control design, the output equation is a linear function of the state variables. But, Equation [15] is a non-linear implicit output equation. Hence, for this nonlinear process, the controller design procedure requires to be suitably modified. Therefore we apply the following partial differentiation identity to Equation [15]:

$$\frac{\partial h}{\partial y}\dot{y} + \frac{\partial h}{\partial x}\dot{x} = 0 \qquad [16]$$

hence

$$\dot{y} = -\left\{ \frac{\partial h}{\partial x} \Big/ \frac{\partial h}{\partial y} \right\} \dot{x} \qquad [17]$$

Using Equations [3], [6], [8] and [17] we get:

$$\dot{e} = \lambda_m y_m + b_m r + [(\partial h/\partial x)/(\partial h/\partial y)][f(x) + g(x)u + d(t)] \qquad [18]$$
$$= \lambda_m y_m + b_m r + J\,[f(x) + g(x)u + d(t)] \qquad [19]$$

$$where\ J = (\partial h/\partial x)/(\partial h/\partial y) \qquad [20]$$

If we make,

$$\lambda_m y_m + b_m r + J\,[f(x) + g(x)u + d(t)] = ke \qquad [21]$$

then the control law is calculated as:

$$u = -(Jg)^{-1}(\lambda_m y + b_m r - ke + Jf + Jd) \qquad [22]$$

Equation [19] becomes:

$$\dot{e} = (\lambda_m + k)e \qquad [23]$$
$$= \lambda e \qquad [24]$$

It is to be noted that since the disturbance term $d(t)$ appears in the control law, it is essentially a combined feedback-feedforward control action (10). The expression for the control law of Equation [22] in terms of the plant variable y only is given by

$$u(t)=\frac{[(10^{y-14}+10^{-y})(2.303)(\lambda_m r+b_m r-ke)-a_1(-10^{y-14}+10^{-y})+a_1 D]}{[a_3(-10^{y-14}+10^{-y})+a_3 a_2]} \quad [25]$$

3 EXPERIMENTAL SET UP

Figure 1 shows the experimental set up for the pH neutralisation system. The process stream (influent) consists of a diluted strong acid (HCl) and the titrating stream is a more concentrated strong base (NaOH). Table 1 consists of typical operating conditions. The process stream is fed through two feed tanks, and a 3-way valve is placed in the feed line, which allows switching between two different feed concentrations. A remote control peristaltic pump (RM pump) is used
to control the flow rate of the titrating stream. The volume of the reactor vessel is kept constant at 5-litres with an over flow system. An agitator is used to ensure proper mixing. The pH of the influent, the pH of the mixture in Continuously Stirred Tank Reactor (CSTR) and the influent flow are measured by a data acquisition system (National Instruments E series I/O card and a PC with LabVIEW Instrumentation package). The control objective is to maintain the pH value at the set point = 7. The control output is calculated according to the non-linear model reference control law (Equation [25]). The digital output is converted to an analogue output, and the signal is transmitted to a remote control peristaltic pump (RC pump) that controls the base flow rate. The sampling time for the measurements is 0.1 of a second and the control law is executed at approximately the same time (considering the time taken for control computation by the package and the operating system).

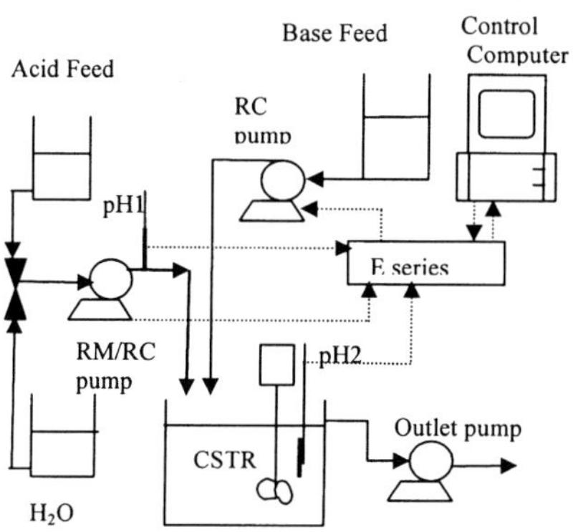

Fig 1 Laboratory set-up of the pH Neutralisation Process

Table 1- Typical operating conditions for the pH neutralisation process

Parameters	Values
Acid Flow (F_1)	Variable
Base Flow (u)	Manipulative
variable	
Conc. Of acid (D)	0.01M – 0.005M
Conc. Of base (a_2)	0.2M

4 SIMULATIONS

A continuous time simulation of the controller was undertaken to confirm the results obtained by Jayadeva et al. (10). Open loop response of the process was considered for a disturbance of

100% reduction in the concentration of influent by diluting the influent and effective reducing the concentration from 0.01M to 0.005M at 1.4seconds. Figure 2a shows the open loop response of the plant. With the non-linear model reference controller, the closed loop response of the plant is as shown in Figure 2b. The corresponding change in the control action is a shown in Figure 2c.

The above simulations assumed steady conditions while the above disturbance occurred. Therefore the controller was tested for a change in operating point by simulating the process at different initial condition. Figure 2d is the simulation response of the plant for a change in the operating point from pH 7 to pH 3 along with the disturbance in concentration at 1.4 seconds. The controller responds robustly to both the disturbances as a glitch can be observed at 1.4 second interval in Figure 2d. The controller was also successfully tested for disturbances both in the flow of the influent and concentration. Finally, the continuous controller is studied with sampled input and output signals before practical implementation. Hence zero order holds are applied to model this effect on the continuous process. The Simulink model incorporates the change in disturbance with respect to time. The effect due to the change in influent flow was also studied with slight modification in the model.

Analysis was done for the allowable sampling time for real-time implementation with Zero–order hold at both input source and output sampling.

5 REAL TIME IMPLEMENTATION OF THE CONTROLLER

The pH sensor is assumed to be linear and the temperature is assumed to be constant (1). pH sensors have very high source impedance and it is therefore necessary to use a high input impedance buffer amplifier. A low pass filter is used to reject AC mains 50Hz.

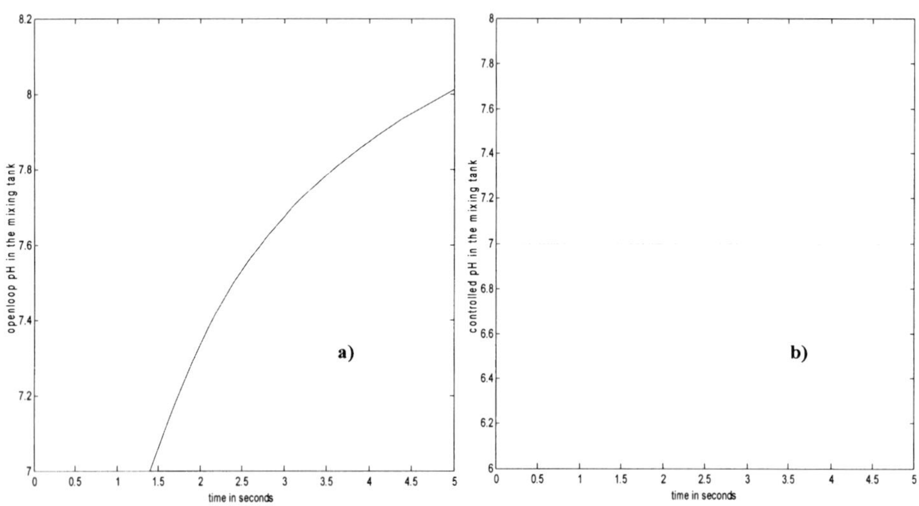

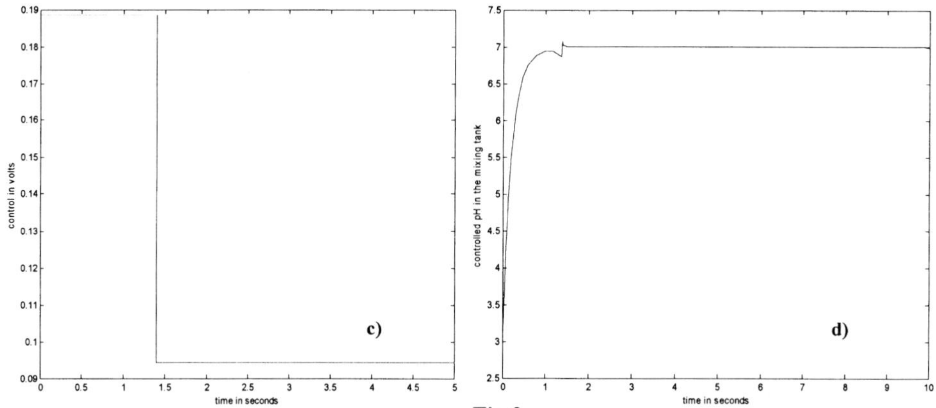

Fig 2

a) **Step response simulation of the plant for a disturbance of 100% reduction in influent concentration.**
b) **Controlled pH value for the same disturbance at 1.4 seconds as in a).**
c) **Controller in action for the disturbance of 100% reduction in the influent concentration.**
d) **Controlled pH for a simulation of the controller at different operating point along with the disturbance at 1.4 seconds as in a)**

Differential analogue input mode is preferred to single channel analogue input of the I/O card for sensor signal feed, as Common Mode Rejection Ratio (CMRR) is very high in this mode. The control signal range for the pump so that it responds linearly is 0-10 volts. The concentration of the solutions is accordingly chosen considering the constraint.

5.1 Software Platform

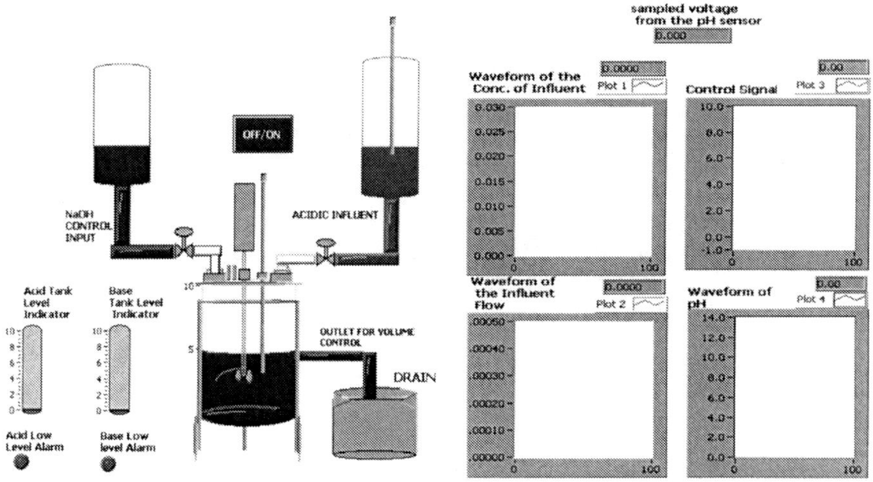

Fig 3 Pictorial view of the front panel of a VI to control the pH process

LabVIEW, a real-time virtual instrument package, is used to implement the control strategy. This is a development environment based on a graphical programming language. A LabVIEW file is called Virtual Instrument (VI). Each VI consists of two items: a block diagram (Figure 4) and a front panel (Figure 3). The front panel is the graphical user interface where the user can enter commands, values, and manipulate objects such as knobs and buttons to control the application. The front panel is also where results from the execution of the application are presented through indicators, charts, and many other graphical displays. We can drop objects in the front panel, which in turn create terminals in the block diagram. The data can be wired in or out of these objects, and pass it back and forth between the front panel and the code in the block diagram. The easy accessibility of Matlab code within the LabVIEW environment is utilised for complete implementation and data retrieval and remote monitoring of hazardous processes. Figure 4 partly shows the VI diagram of the program for 0.1of a second sampling delay.

While remotely monitoring the process, front panel of the VI can be programmed to generate specific alarms such as one for low level of control reagent. The process can be completely shut through the Front panel in case of emergencies. The alkalinity of the mixture in the CSTR can be gauged by using logical operatives to indicate this by changing the colour of the solution in the front panel along with graphical representation. Figure 5 further illustrates the possibility of monitoring and controlling the process using the internet.

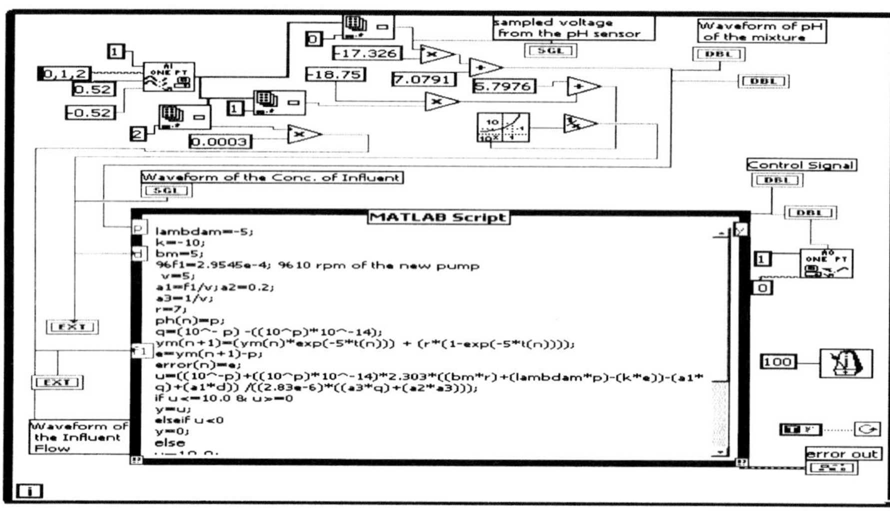

Fig 4 Pictorial VI block diagram to control and monitor the pH process

6 ANALYSIS

The controller is tested for the most common of the disturbances, which are the changes in the flow of the influent and the concentration. The experiment is conducted approximately for 3 minutes with the change in concentration after 1 minute (Figure 6) and then the change in flow after 2 minutes. The controller robustly responds for the disturbances with no apparent change in the pH of the mixture. It was inferred during simulation that, the maximum sampling time can be 0.2 second. But the response of the plant was not as quite the continuous time response (Figure 2). One of the reasons is the limitation of the control pump. The instantaneous control values were noted to be the same as the continuous time values.

7 CONCLUSIONS

This study is a part of the research to propose a non-linear adaptive control scheme for pH control of wastewater and implement it on an industrial scale for a water company in the United Kingdom. A lot of research is only simulation based, understandably, due to many factors such as cost etc. Therefore, the importance of real-time implementation has also been

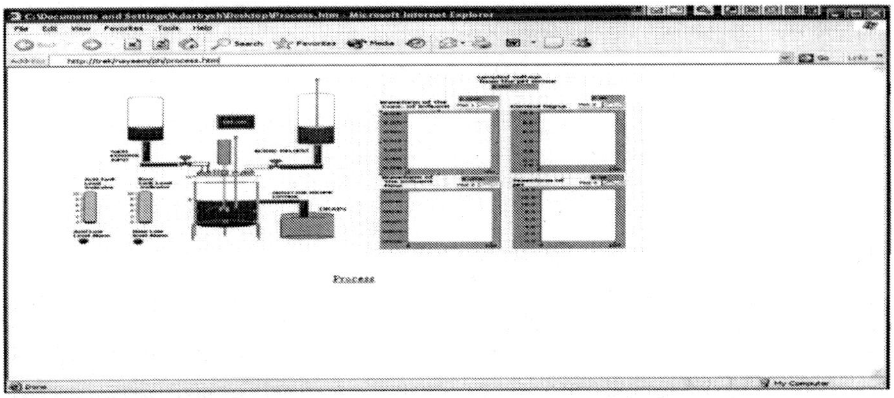

Fig 5 Browser window to monitor and control the process using the internet

emphasised in this study. This study has opened doors for further investigation into simulation, real-time implementation and remote monitoring. As this study aimed at exploring requirements to liaise software with hardware, the experimentation has been successful in doing so.

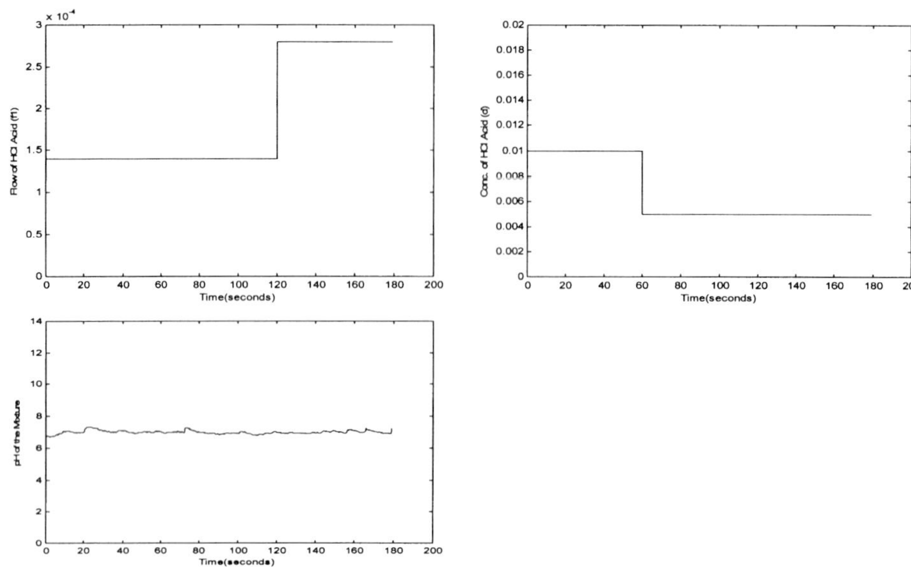

Fig 6 pH response of the mixture in the CSTR for the disturbances in the influent flow (after 2mins and concentrations (after1min)

REFERENCES

1. Shinskey. F.G. 1973. "pH and pION Control in Process and Waste Streams". John Wiley & Sons, New York, USA.
2. Pishvaie, M. R. and M. Shahrokhi. 2000. "pH Control using Non-linear Multiple Models, Switching and Tuning Approach." *Ind. Eng. Chem. Res.*, 39, 1311.
3. Wright, R. A., M. Soroush and C. Kravaris. 1991. "Strong acid equivalent control of pH processes: An experimental study." *Ind. Eng. Chem. Res.*, 30, 2437.
4. Young, G. E. and S. Rao. 1986. "Robust control of a Non-linear Process with System Uncertainty and Delay using Variable Structure," *Proc. Am. Control Conf., 1210.*
5. Parrish, J. R. and C. B. Brosilow. 1988. "Non-linear Inferential Control." *AIChE Journal*, 34, 633.
6. Kulkarni, B. D., S. S. Tambe, N. V. Shukla and P. B. Deshpande. 1991. "Non-linear pH Control." *Chem. Eng. Sci.*, 46, 995.
7. Li, W. C., L. T. Biegler, C. G. Economou, and M. A. Morari. 1990a. "Constrained Pseudo-Newton Control Strategy for Non-linear Systems." *Comput. Chem. Engg.,* 14, 451.
8. Li, W. C. and L. T. Biegler. 1990b. "Newton-type Controllers for Constrained Non-linear Processes with Uncertainty." *Ind. Eng. Chem. Res.*, 29, 1647.
9. Jayadeva, B. J., M. Chidambaram and K. P. Madhavan. 1990b. "Robust Control of Batch Reactors." *Chem. Engng. Commun.* 87, 195.

10. Jayadeva, B. J., Y. S. N. M. Rao, M. Chidambaram and K. P. Madhavan. 1990a. "Nonlinear Controller for A pH Process." *Computers chem. Engng.*, 14(8), 917.
11. Yuocef-Toumi, K. and O. Ito. 1987. "Controller Design for Systems with Unknown Nonlinear Dynamics." Proc. Am. Control Conf. Minneapolis, 836.
12. Karnachi, N. N. and Waterworth, G. October, 2003. "Non-linear Model Reference Control of a pH Process: An Experimental Study." 15th European Simulation Symposium, Delft, The Netherlands, 430.
13. McAvoy, T. J., E. Hsu and S. Lowenthal. 1972. "Dynamics of pH in a Controlled Stirred Tank Reactor." *Ind. Eng. Chem. Process Des. Dev.*, 11(1), 68.

Virtual reality telerobotic system

U KARTOUN, H STERN, and **Y EDAN**
Department of Industrial Engineering and Management, Ben-Gurion University of the Negev, Be'er-Sheeva, Israel

ABSTRACT

This paper describes a telerobotic system operated through a virtual reality (VR) interface. A least squares method is used to find the transformation mapping, from the virtual to real environments. Results revealed an average transformation error of 3mm. The system was tested for the task of planning minimum time shaking trajectories to discharge the contents of a suspicious package onto a workstation platform. Performance times to carry out the task directly through the VR interface showed rapid learning, reaching standard time (288 seconds) within 7 to 8 trials - exhibiting a learning rate of 0.79.

1 INTRODUCTION

Teleoperation is used when a task has to be performed in a hostile, unsafe, inaccessible or remote environment (1). A telerobot is defined as a robot controlled at a distance by a human operator (HO) regardless of the degree of robot autonomy. Telerobotic devices are typically developed for situations or environments that are too dangerous, uncomfortable, limiting, repetitive, or costly for humans to perform (2). Some applications include: underwater (3), space (4), resource industry (5) and medical (6)(7).

Examples for using graphical models to allow users to control robots off-line and practice control techniques can be found in the RobotToy research (8), the KhepOnTheWeb project (9) and the WITS (Web Interface for Telescience) project (10). NASA developed WITS for controlling remote vehicles on planets such as Mars and Saturn. In the Tele-Garden project (11), users can tend a garden that contains live plants through a graphical representation of the environment. The University of Western Australia's Telerobot experiment (12) provides Internet control of an industrial ASEA IRB-6 robot arm. The PumaPaint project (13) is a website allowing users to control a PUMA-760 robot equipped with a parallel-fingered gripper, to perform painting tasks on an easel.

The virtual reality telerobotic system, described in this paper allows a HO to: (a) perform off-line path planning by manipulating an object in a VR robotic scene, and (b) perform on-line control by indirectly controlling the real robot through manipulation of its VR representation in real-time. The available commands for control are the manipulator coordinates (x, y, z) and open / close gripper. When a command is chosen, the VR model is initially updated. After conversion, the joint angles are sent to the real robot for execution.

To demonstrate the utility of the system, we focus on the task of bag shaking. The usual method for bomb squad personnel is to blow up a suspicious bag and any explosives contained therein. However, if the bag contains chemical, biological or radiological canisters, this method can lead to disastrous results. Furthermore, the "blow-up" method also destroys important clues such as fingerprints, type of explosive, detonators and other signatures of use in subsequent forensic analysis. Extraction of the bags contents using telerobotics, which avoids these problems, is the subject addressed here. In the on-line control mode suggested here, the user controls the robot to perform the task or develops a plan off-line before downloading it to the robot's controller for execution. For off-line bag shaking, the HO chooses a sequence of spatial locations (including inter point speeds) in the VR model to define a shaking trajectory. The trajectory is then downloaded to the real robot for execution.

In this paper we report on experiments for on-line control. The system architecture is presented in Section 2. Before carrying out the on-line control experiments it is necessary to calibrate the VR-telerobotic system. Calibration experiments are described in section 3. In section 4 we report on user experiments, using on-line control for the bag lifting and shaking task. The paper ends with conclusions and directions for future work in Section 5.

2 SYSTEM ARCHITECTURE

The proposed virtual reality (VR) telerobotic system contains a human operator (HO), VR web-based control interface, internet access method, a remote server, a robot and its controller, and visual sensory feedback (Fig. 1).

2.1 User and control interface

The HO requires interacting with the remote robot through a control interface (Fig 2). The interface developed includes a five degree of freedom articulated robotic arm model of the "CRS A255" robot (14), two views of the real robot, a checkerboard on a table, and a world coordinate diagram that shows the *x*, *y* and *z* directions in the 3D scene. In addition, overall and close views of the robot site are displayed on-line.

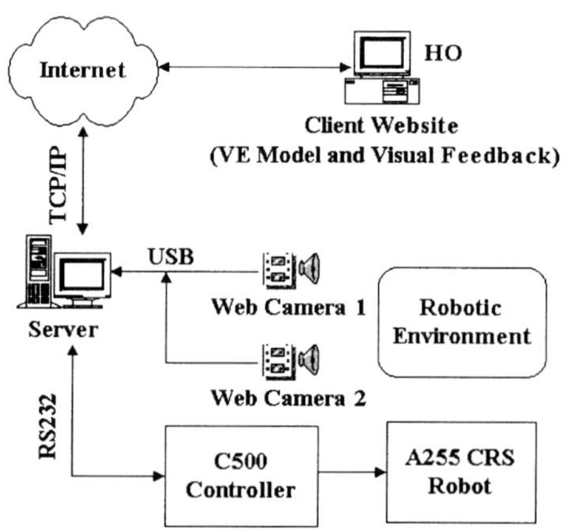

Figure 1. System architecture layout

The system has six different operational stages controlled through predefined control panels. These are: changing the real robots speed, showing a 3D-grid (Fig. 3) that contains spatial locations which the robot gripper moves to when selected, selecting the viewing aspect of the VR model, planning shaking policies, planning off-line paths for downloading to the real robot, and on-line simultaneous control (in real-time) of the VR and real robots.

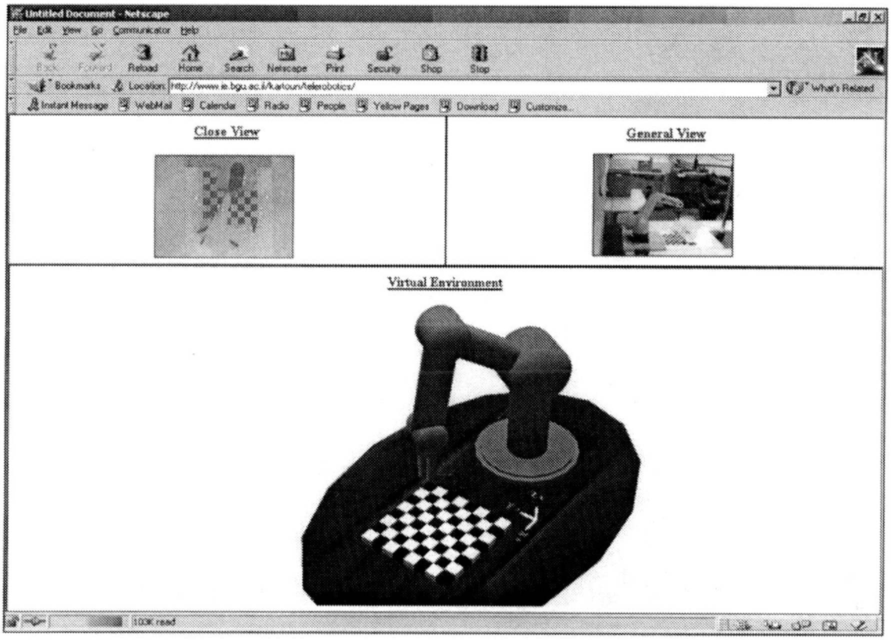

Figure 2. Web-based interface (camera views, and virtual environment)

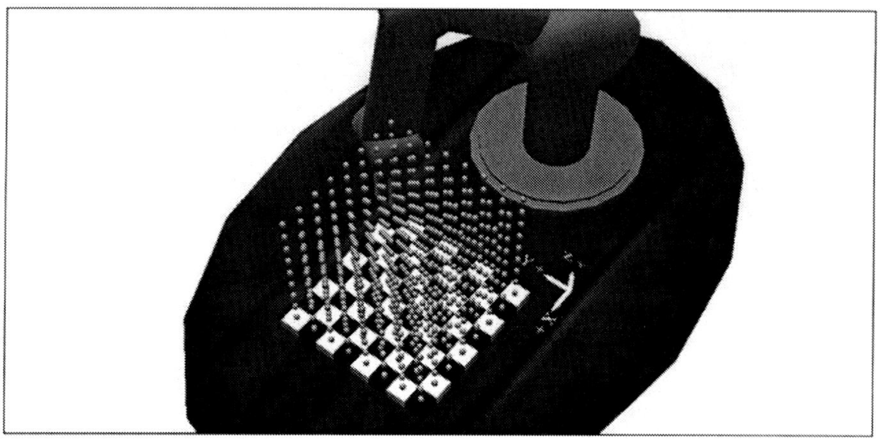

Figure 3. 3D-grid

2.2 Virtual environment

The virtual environment (VE) model was built and developed using "3D-Studio-Max" (15) and "Alice" (16) softwares. "Alice", a rapid prototyping software for creating interactive computer graphics applications (17)(18)(19)(20)(21), was chosen to be the VR software. It is designed to enable rapid development and prototyping of interactive graphics applications and uses "Python" (22)(23) as the language for writing its scripts.

2.3 Communication

The HO communicates with the server, connected to a robotic arm (Fig. 4) through a web-browser. Commands sent from the VE client are transmitted through TCP/IP to the server that extracts them and updates the real robot.

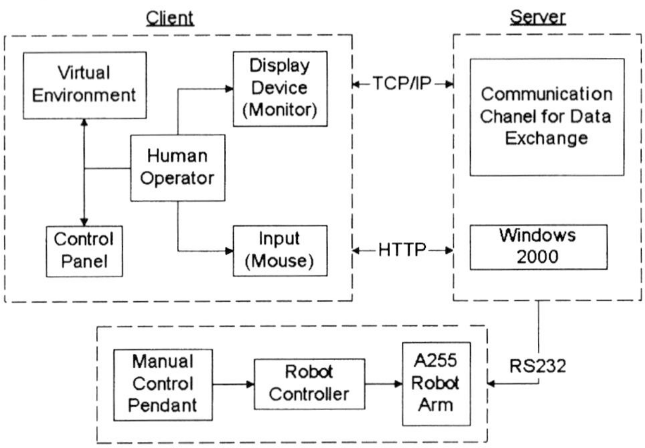

Figure 4. Client-server communication architecture

2.4 Sensory feedback, robot, and controller

The remote server is connected to the robot controller, and two USB web-cameras (Fig. 1), which constantly send updated images to the client for observation of the environment. Images are in 24 bit colour and of size 240X180 pixels which appear as close-up and overall views (Fig. 2) in the client browser interface. The "A255" robot system consists of robot arm and controller. The robot arm is equipped with a special gripper capable of sliding under bags.

3 SYSTEM CALIBRATION

3.1 Kinematics

The inverse kinematics (IK) equations were solved using a closed form analytical solution (24)(25)(26). It has the benefit of being an exact solution and very fast to calculate. For the VR robot, an IK algorithm was implemented to determine the joint angles required to reach an end point location by supplying the (x, y, z) coordinates. The side and top views of the VR robotic chain are shown in Fig. 5. The distance x' is obtained from the projection of the shoulder and the elbow links onto the X-Y plane. The x and the y values are the horizontal and vertical values of the robot gripper position relative to the robot base-point, P_B. The z coordinate is taken vertical to the platform plane and measured from the base of L_1.

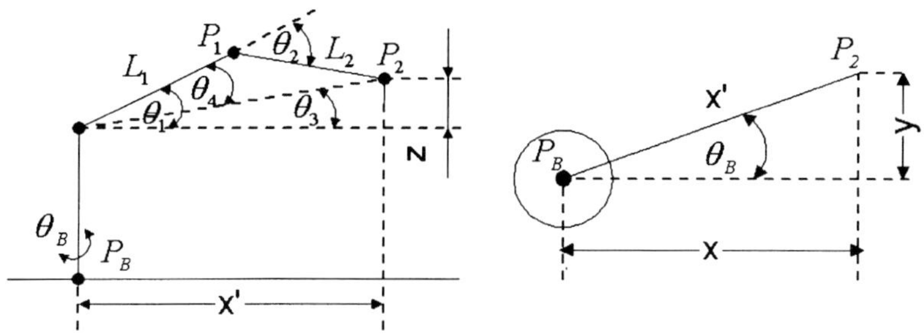

(a) Side view **(b) Top view**

Figure 5. "A255" robotic chain

Using trigonometric identities yields:

$$\theta_1 = \arctan\left(\frac{z(L_1 + L_2\cos(\theta_2)) + yL_2\sin(\theta_2)}{zL_2\sin(\theta_2) - y(L_1 + L_2\cos(\theta_2))}\right), \quad [1]$$

$$\theta_2 = \arccos\left(\frac{y^2 + z^2 - L^2{}_1 - L^2{}_2}{2L_1L_2}\right), \quad [2]$$

$$\theta_3 = \arctan(z/x'), \quad [3]$$

$$\theta_4 = \arctan\left(\frac{L_2\sin(\theta_2)}{L_1 + L_2\cos(\theta_2)}\right), \quad [4]$$

$$\theta_B = \arctan(y/x), \quad [5]$$

$$x' = \sqrt{x^2 + y^2}. \quad [6]$$

Now, given any end point position (x, y, z), the joint angles required to reach it are determined by θ_B, θ_1, and θ_2 using [1] through [6].

3.2 Transformation matrix

A transformation matrix, providing a 1 to 1 mapping between the VR and real robots is estimated from corresponding pairs of 32 intersection points on the checkerboard appearing in the VR and the real environments. The estimate is based on the method of aligning a pair of shapes, which uses a least-squares approximation (27). Given two similar shapes, x_1 (the VR environment) and x_2 (the real environment), a rotation θ, a scale s, and a translation (t_x, t_y) are calculated. The transformation equation using the 3x3 transformation matrix is shown in equation [7]. The points $(x_{vr(I,j)}, y_{vr(I,j)})$ and $(x_{r(I,j)}, y_{r(I,j)})$ are in the VR scene and in the real environments, respectively.

$$\begin{pmatrix} s\cos\theta & -s\sin\theta & t_x \\ s\sin\theta & s\cos\theta & t_y \\ 0 & 0 & 1 \end{pmatrix}\begin{pmatrix} x_{vr(i,j)} \\ y_{vr(i,j)} \\ 1 \end{pmatrix} = \begin{pmatrix} x_{r(i,j)} \\ y_{r(i,j)} \\ 1 \end{pmatrix} \quad [7]$$

3.3 Transformation accuracy

Given the calibrated transformation matrix an experiment to determine the transformation error was performed. The experiment (Fig. 6) starts with controlling the VR robot arm through the VR interface. The coordinates of 32 points (x_{vr}, y_{vr}), selected from checkerboard intersections, were set as test points. These points were inserted into [7] to obtain (x_r, y_r) which were sent to the real robot.

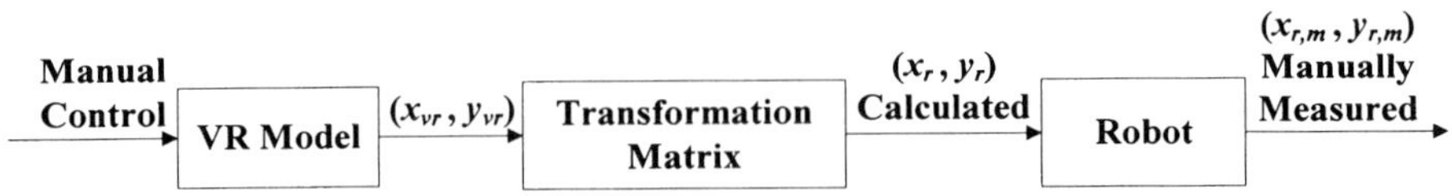

Figure 6. Calibration experiment flow chart

A pen inserted into the real robot's gripper marked its controlled position. The coordinates of the robot's pen on the checkerboard intersection points $(x_{r,m}, y_{r,m})$ in the real environment were measured manually by a ruler. The average transformation error between the robot's pen positions, and the desired points in the real checkerboard was found to be 3mm.

4 SYSTEM TEST USING VR ON-LINE CONTROL

The system was tested using on-line control through the VR interface for the task of shaking out the contents of a plastic bag. The HO views the images of the robotic workspace in the client browser, and commands the robot by selecting points in the 3D VR scene. Using the calibrated mapping of points in the VR scene to the real workspace, the robot is controlled to carry out the task. A view of the experiment to empty the contents of a plastic bag onto the platform is shown in Fig. 7. It is assumed that the bag contains ten identical electronic components known in advance. Five inexperienced operators performed ten identical experiments, and performance times (the amount of time it takes to extract all the objects from the bag) were recorded. The learning curve of task completion time was reduced quickly reaching standard time (288 seconds) after 7 to 8 trials (Fig. 8).

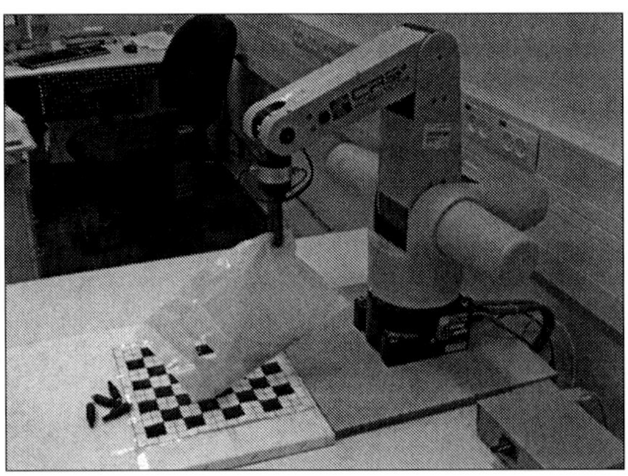

Fig 7 (a) Overall view

Fig 7 (b) Close view

Figure 7. "A255" robot, plastic bag, platform and electronic components

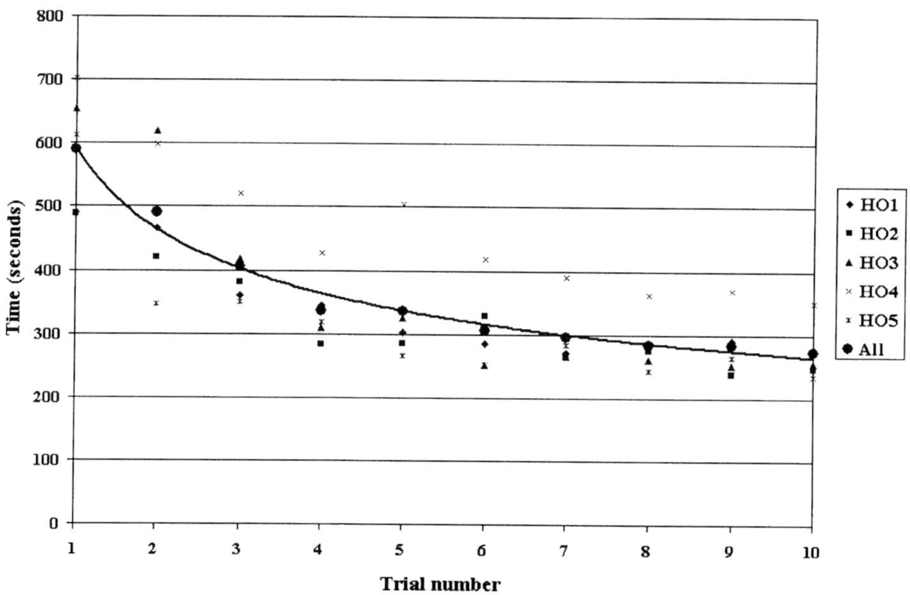

Figure 8. Learning curve experimental results

5 CONCLUSIONS AND FUTURE WORK

This paper describes the design, implementation and testing of a real-time VR-telerobotic web-based system. Visual feedback arrived to the human interface via two independent web-cameras and a transformation matrix mapped the VE to the real one was calculated to calibrate the system. The system allows a human operator to: (a) perform off-line path planning by manipulating an object in a VR robotic scene, (b) perform on-line control by indirectly controlling the real robot through manipulation of its VR representation in real-time. A least squares method is used to find the orientation, rotation and translation of 1 to 1 mapping between the virtual and real environments. Results revealed an average transformation error of 3mm. The system was tested for the task of planning minimum time shaking trajectories to discharge the contents of a suspicious package onto the workstation platform. Performance times to carry out the task directly through the VR interface showed rapid learning, reaching standard time (288 seconds) within 7 to 8 trials. This fast learning rate was calculated as 0.79. A future extension is to use visual feedback from the cameras for creating a second transformation mapping the real environment to the virtual one. Errors for both transformations will be calculated and the real robot will perform corrections in order to reduce them. Possible extension of the VR graphical display of the robotic environment presented here is to use visual input (*e.g.*, data gloves, movement trackers, hand gestures) and output (*e.g.*, head mounted displays, shutter glasses) devices. Future research will include development of a cooperative human-robot learning system for remote robotic operations using a VR interface (28).

ACKNOWLEDGMENTS

This project was partially supported by the Ministry of Defense MAFAT Grant No. 1102, and the Paul Ivanier Center for Robotics Research and Production Management, Ben-Gurion University of the Negev.

REFERENCES

1. J. Bukchin, R. Luquer, and A. Shtub, "Learning in tele-operations", *IIE Transactions*, vol. 34, no. 3, pp. 245-252, 2002.
2. N. Durlach and S. N. Mavor, "Virtual Reality: Scientific and Technological Challenges", *National Academy Press*, Washington D.C., 1995.
3. L. Hsu, R. Costa, F. Lizarralde, and J. Soares, "Passive arm based dynamic positioning system for remotely operated underwater vehicles", *IEEE International Conference on Robotics and Automation*, vol. 1, pp. 407-412, 1999.
4. G. Hirzinger, B. Brunner, J. Dietrich, and J. Heindl, "Sensor-based space robotics-ROTEX and its telerobotic features", *IEEE Transactions on Robotics and Automation*, vol. 9, no. 5, pp. 649-663, 1993.
5. A. Goldenberg, J. Wiercienski, P. Kuzan, C. Szymczyk, R.G. Fenton, and B. Shaver, "A remote manipulator for forestry operation", *IEEE Transactions on Robotics and Automation*, vol. 11, no. 2, pp. 185-197, 1995.
6. D. Kwon, K. Y. Woo, and H. S. Cho, "Haptic control of the master hand controller for a microsurgical telerobot system", *IEEE International Conference on Robotics and Automation*, pp. 1722-1727, 1999.
7. D. Sorid and S. K. Moore, "The virtual surgeon", *IEEE Spectrum*, pp. 26-39, 2000.

8. http://robotoy.elec.uow.edu.au/
9. O. Michel, P. Saucy, and F. Mondada, "KhepOnTheWeb: an experimental demonstrator in telerobotics and virtual reality", *Proceedings of the International Conference on Virtual Systems and Multimedia, IEEE VSMM'97*, pp. 90-98, 1997.
10. P. G. Backes, K. S. Tso, and G. K. Tharp, "The web interface for telescience", Presence, *MIT Press*, vol. 8, pp. 531-539, 1999.
11. K. Goldberg, J. Santarromana, G. Bekey, S. Gentner, R. Morris, J. Wiegley, and E. Berger, "The Telegarden", *Proceedings of ACM SIGGRAPH*, 1995.
12. K. Talyor and J. Trevelyan, "A telerobot on the world wide web", *National Conference of the Australian Robot Association*, 1995.
13. M. Stein, "Interactive internet artistry, painting on the world wide web with the PumaPaint project", *IEEE Robotics and Automation Magazine*, vol. 7, no. 1, pp. 28-32, 2000.
14. "A255" Robot System, *CRS Robotics Human Scale Solutions*, 1998.
15. http://www.discreet.com/
16. http://www.alice.org/
17. M. J. Conway, R. Pausch, R. Gossweiler, and T. Burnette, "Alice: a rapid prototyping system for building virtual environments", *Adjunct Proceedings of ACM CHI'94 Human Factors in Computing Systems Conference*, vol. 2, pp. 295-296, 1994.
18. R. Pausch, T. Burnette, A. C. Capehart, M. Conway, D. Cosgrove, R. DeLine, J. Durbin, R. Gossweiler, S. Koga, and J. White, "A brief architectural overview of Alice, a rapid prototyping system for virtual reality", *IEEE Computer Graphics and Applications*, 1995.
19. M. J. Conway, "Alice: easy-to-learn 3D scripting for Novices", *Ph.D. Thesis, Faculty of the School of Engineering and Applied Science at the University of Virginia*, 1997.
20. S. Cooper, W. Dann, and R. Pausch, "Alice: a 3-D tool for introductory programming concepts", *Proceedings of the Fifth Anual CCSC Northeastern Conference*, pp. 107-116, 2000.
21. W. Dann, S. Cooper, and R. Pausch, "Making the connection: programming with animated small world", *Fifth Annual SIGCSE/SIGCUE Conference on Innovation and Technology in Computer Science Education (ITiCSE)*, pp. 41-44, 2000.
22. http://www.python.org/
23. M. Lutz, Programming Python, *O'Reilly and Associates*, 2001.
24. J. Lander, "Oh my god, I inverted kine", *Game Developer Magazine*, pp. 9-14, 1998.
25. J. J. Craig, Introduction to Robotics: Mechanics and Control, *Addison-Wesley*, 1989.
26. P. J. McKerrow, Introduction to Robotics, *Addison-Wesley*, 1991.
27. T. F. Cootes, C. J. Taylor, D. H. Cooper, and J. Graham, "Training models of shape from sets of examples", *Proceedings of the British Machine Vision Conference*, pp. 9-18, 1992.
28. U. Kartoun, "A human-robot collaborative learning system using a virtual reality telerobotic interface", *Ph.D. Thesis Proposal, Department of Industrial Engineering and Management at the Ben-Gurion University of the Negev*, Israel, 2003, Available: http://www.ie.bgu.ac.il/kartoun/site/articles/Uri_Kartoun_PhD_Proposal_Final_October_2003.pdf

Conditions for through-process analysis of product surface in steel manufacture

F D VAN DEN BURGH, S DUYSTERS, C A A JONKER, G E N KUIPER, P T M VAN LIESHOUT, M P TIMMERMANS, and **H T PLOEGAERT**
CRDT-MIE Department, Corus Research, Ijmuiden, The Netherlands
M B DENYS
CRDT-RMS Department, Corus Research, Ijmuiden, The Netherlands
S G THORNTON
CRDT-MIE Department, Corus Research, Teesside, UK

ABSTRACT

In the steel industry, there is a trend to equip an increasing number of production lines with camera-based inspection systems to monitor the surface quality of the steel. These systems are dedicated to automatically detect and classify surface defects. When camera systems are used in different processing lines where the steel subsequently passes through, it becomes possible to track the surface quality through the chain. Linking through process data enables fast root cause analysis and optimisation of manufacturing practice to minimize the defect level.

The evaluation of the changes in surface condition through the process is hampered by problems in data synchronisation, changes in product geometry (scaling), parts of the product that are taken out of the process chain (slit edges, cut off heads or tails) and delays in the processing chain (e.g. due to maintenance of a production line). This paper highlights the conditions and encountered problems in integrated process analysis in an industrial environment.

1 INTRODUCTION

Two important markets for Corus Netherlands are the automotive and the packaging sectors. Car manufacturers use steel for the structural frame, chassis (the "car body"), engine, wheels, etc. A common product delivered to the automotive industry is galvanised strip steel, i.e. flat material of 1.80 – 2.20 m wide, 0.6-1.6 mm thick and 1 – 2 km long, that is wound onto coils of typically 20 tonnes. In packaging, steel is used to manufacture cans for food, beer and beverage as well as aerosol spray containers. A common product type supplied to this market are coils of tin coated steel, ~ 1 m wide, ~ 0.2 mm thick, and typically 6 km long.

Galvanised strip steel is used to make external body parts like roofs, doors and bonnets. Obviously, these external parts require steel surfaces that are absolutely free from surface defects like scratches, roll imprints, pits and oxide stripes. Similarly, the packaging industry requires their products to be absolutely defect-free. Critical material produced at the steel manufacturer is therefore inspected on its surface quality.

In the past, the task of visual inspection was mostly done by manual inspection by the operators or inspectors. Evidently, the change in manning of a manufacturing line, the changing degree of attentiveness of the operator and variations in perception of the relative importance of errors introduce a substantial level of subjectivity into this procedure. The advent of fast cameras, high bandwidth data transport and increased computational power has provided the possibility to fulfil automated inspection of the surface by computerized camera-based vision systems. These vision systems, or Surface Inspection Systems (SIS) perform image capture, image processing and automated defect detection and classification. Operators have usually the possibility to view the statistics of defect occurrence over a given production period and they can retrieve the images of individual defects, which are stored as Regions Of Interest (ROI).

Generally speaking, SIS's are installed at the end of a production line, but in certain cases, these systems inspect the incoming product. Since the manufacturing of galvanised steel is done in subsequent production phases, like hot rolling, pickling, cold rolling, annealing and galvanising, each of which line can be equipped with one of more SIS's, this creates the possibility to study the evolution of the surface quality through the production chain.

Various conditions and events during production in an industrial environment however interfere when establishing a connection between corresponding data. In the next sections, we will focus on the factors that trouble data synchronisation and propose directions for solutions.

2 INTEGRATED STEEL PRODUCTION

To obtain a proper understanding of the through-process conditions in the manufacturing of steel, a concise introduction into this topic is given below. Fig. 1 shows schematically the product flow through the principal processes needed to produce metal-plated steel from the raw materials, iron ore and coal. Table 1 further details these production steps and specifies

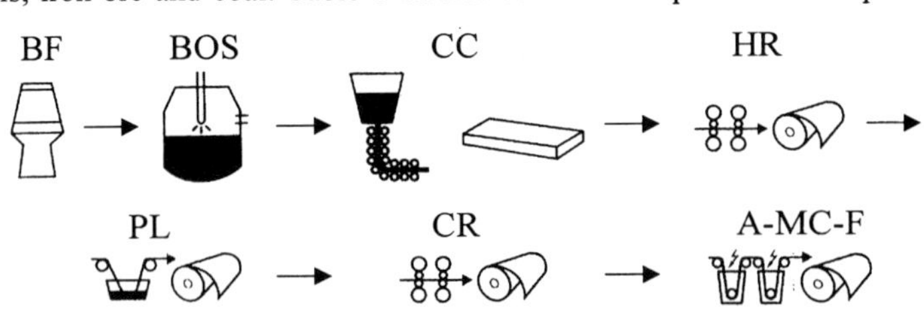

Fig. 1 Schematic drawing of the most important production phases in steel manufacturing, see also table 1.

Table 1 Overview of steel production process.

Production Process	*Description of process*	*Unit & type of semi–finished product*
BF = Blast Furnace	Extract iron from ore (at ~ 1500 °C)	torpedo car holding ~ 300 tonnes of liquid hot metal
BOS = Basic Oxygen Steelmaking	Reduce Carbon content and add alloying elements (at ~ 1600 °C)	ladles (~ 300 tonnes) of liquid steel
CC = Continuous Caster	Cast steel in a mould and cut in blocks	Rectangular solid steel blocks of dimensions 6-12 m x 1-2 m x 0.2 m ("slabs")
HR = Hot rolling	Reduce thickness & define microstructure (at 600 – 1200 °C)	Coils of 6 – 30 tonnes of 1-2 m wide, 2-7 mm thick and 100-2000 m long strip of hot rolled steel
PL = Pickle line	Pickle steel to remove oxides	Coils of pickled steel, similar in size as those leaving the HR process
CR = Cold rolling	Reduce thickness & orient microstructure (at 20 – 120 °C)	Coils of typically 20 tonnes of 1-2 m wide, 0.2 – 3 mm thick and up to ~10 km long strip of cold rolled steel
A-MC-F = Anneal & Application of metal coating & "Finish"	Heat treatment (anneal) to ensure formability of steel, Apply zinc or tin layer to ensure corrosion resistance, Define surface roughness & flatness profile in a "finish" step	Zinc or tin coated steel as final product. Also, Corus applies organic coatings to steel.

the type of output of each of the factories. After pre-processing, the raw materials are fed to a blast furnace (BF), in which the iron oxide is reduced to obtain liquid iron at a temperature of about 1500 °C. The resulting liquid hot metal is transported in torpedo cars to a so-called steelmaking converter. Here the overabundant carbon is removed by oxidation using an oxygen lance in an exothermic reaction and alloying elements are added to the resulting melt to yield liquid steel. This process is known as "Basic Oxygen Steelmaking (BOS)".

Next, the liquid steel is continuously cast (CC) through a water-cooled mould at about 1-2 m/min. After solidification, the steel is cut in parts of typically 12 m long, called "slabs". The slabs are transported to the hot strip mill, where they are reheated to ~1200°C to facilitate plastic deformation by hot rolling (HR). In a couple of reduction steps, the thickness of the slab is reduced from 22 cm to finally 3-7 mm by rolling. The thickness reduction is accompanied by a length elongation from 12 m to ~700 m. At the end of the hot strip mill, the steel is actively cooled down from ~900°C to below 600°C. The exact temperature evolution in time in this temperature domain largely defines the final microstructure of the steel strip.

During coiling at the end of hot strip mill, iron oxide is formed at the surface; this must be removed before further cold rolling, otherwise the oxide is impressed into the steel surface. This task is performed in the pickle line (PL) using acid. The pickled steel is then further reduced to customer specified thickness in the cold mill. During the cold rolling (CR) process the surface quality and ductility are largely defined also.

After the cold rolling mill, the steel is annealed (A) to obtain a ductile material. Next, for the purpose of corrosion protection, a metallic coating (MC) is applied. For the automotive market, the applied metal is a zinc layer, while for the packaging market, a tin layer is applied in electrolytic baths. Finally, the product is submitted to some finishing (F) operations, to deliver steel strip with the required surface roughness and shape, which is then shipped to the customer.

3 SURFACE INSPECTION SYSTEMS

3.1 Application of surface inspection systems in the steel industry

Surface Inspection Systems (SIS's) are typically installed at the end of the production lines HR, PL, CR and A-MC-F. These commercially available systems monitor the surface by optical, camera-based inspection. The raw image data is real-time analysed. Regions that exhibit anomalies are selected and the "defects" within these Regions of Interests (ROI) are characterised on several features like length, width, area, peak and mean intensity, intensity distribution quantified as first and second moments in width and length. Based on classification rules, the defects are attributed to a certain defect class. Classification rules are established in separate "training" periods in which an 'administrator' calibrates the system by adjusting settings like discrimination levels and by classifying manually the detected defects. The SIS's typically yield output in the form of raw images of defects (the ROIs), defect maps, and coil statistics.

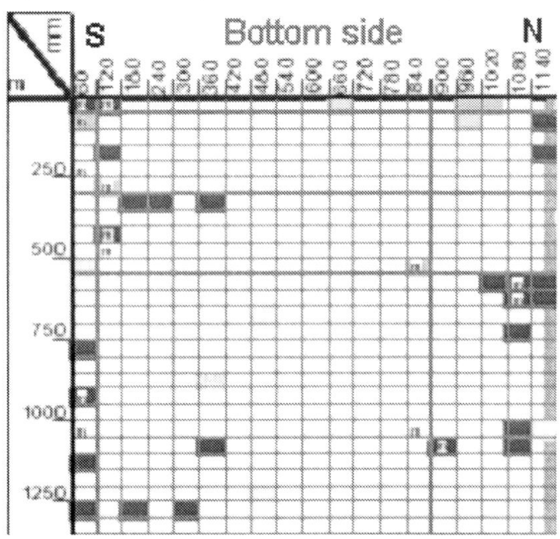

Fig. 2 Illustration of a defect map (partial). Horizontal is the position over the width, vertical the position along the length of the steel strip.

Fig. 2 presents an example of a defect map. The chequered rectangle represents the steel surface of (a part of) the coil. Defects are mapped on it as coloured squares, where different colours are used for different defect types.

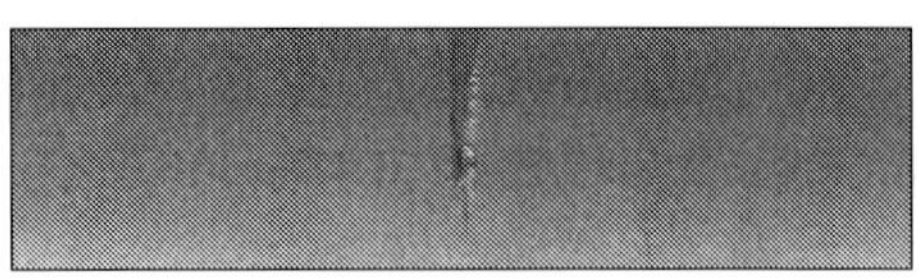

Fig. 3 Image of a defect.

The ROIs, of which Fig. 3 shows an example, enable the operator to check if the system correctly performs the defect classification. Also, if a "new" type of defect appears, the system will attribute these defects to a "remainder" class. By inspecting the ROIs of these new defect types, the operator can define a new defect class based on their similarity in appearance.

3.2 Need for SIS data integrity checks

In practice, the defects or the background might exhibit (gradual) changes in appearance. The current SIS systems have only a limited ability to take these gradual trends into account, and require therefore regular re-calibration, which causes gaps in the monitoring of the surface quality. The limited robustness of SIS's against those gradual changes is one of the arguments to link and compare data from multiple SIS systems to verify the data integrity. A second reason for realisation of this data integrity check finds its origin in the condition that within a business, SIS systems tend to have a single owner who applies continuous efforts to optimise his equipment, having limited communication with the owners and users of the other systems at different manufacturing lines. This results in locally different calibrations, different names for the same defect types, or, adversely and even worse, the same names for different defect types; thus differences in interpretation in results can easily arise. Although this is partly an organisational problem, which can be relieved by regular meetings between operators and the introduction of common procedures, the technical solution to couple and compare defect maps obtained from various SIS's helps in the identification of potential problems and deviations in classification rules, and supports efforts to achieve consistency.

3.3 Logistic operations on semi-finished products

Producing "overlays" of defect maps that originate from the same piece of material, but obtained at different process stages, is however far from straightforward, because of the many different intermediate operations which may be applied to the steel (semi-finished) product. Tables 2, 3 and 4 show which factors potentially intervene at the different levels. Table 2 lists the operations on the level of the entire unit of semi-finished product (see 3rd column of table 1). Semi-finished products can be sold or rejected ("deleted" from further processing), but can also be purchased ("created" and "inserted" in production process). In general, transport of the semi-finished product introduces a time delay, which varies from unit to unit, and the order of downstream processing is not necessarily consistent with the upstream processing order ("permutation").

Table 2 Operations on semi-finished products due to logistics

Operation	*Explanation / Example of reason:*
Delay	Transport, maintenance of downstream production line
Permutation	Change in priority of delivery or processing order
Create (& insert)	Purchased semi-finished product from external supplier
Delete (& reorder)	Shipment of semi-finished product to customer

3.4 Operations on semi-finished products during processing

Table 3 details operations which play a role during the downstream processing. Most obvious is the change in dimensions of the product, notably its length, during the operations HR, CR and, to a lesser extent, A-MC-F. The coiling and subsequent de-coiling at the start of the next process introduces evidently a switch between head and tail of the strip. More hidden is the interchange that can occur between the top and bottom face of the strip as coiling and de-coiling happen in both clockwise and counter clockwise directions. The swaps between top & bottom and head & tail can be mathematically described as a rotation operation.

Even more hidden are pieces of the material that are "deleted" from further processing, because of the slitting of edges or removal of bad parts, which are usually the head and the tail, but occasionally a central part. Rarely, a bad or even broken central part is replaced by another piece of strip. This corresponds to a create & insert operation on the material.

Table 3 Operations on the material due to processing.

Operation	*Explanation / Example of reason:*
Scaling ("Resize")	Dimensions length, width and thickness are rescaled by milling & tensioning operations
Rotation	Coiling and de-coiling induce changes in head and tail, and can cause a mutation of top and down face
Delete (& Translate)	Cutting of head, tail and edges. Sometimes, a bad quality intermediate part is removed and both ends are rejoined
Create (& Insert)	Intersection of a new piece (rare)

3.5 Operations on the level of individual defects

Finally, considering the defect map, several different "operations" can also be distinguished on the level of the individual classified defects. These are listed in Table 4. If a defect is detected, which an upstream positioned SIS had not observed, it can be considered as being created since the previous inspection station. Oppositely, if the upstream SIS identified a defect at a given location, but the next SIS does not observe it, the defect can be considered as being deleted from the defect map.

In both situations, it can be the case that a defect is actually present at the surface, but the SIS fails to observe it. Alternatively, the defects can have really disappeared or appeared during the process. To make a clear distinction and to avoid confusion of these cases, it is helpful to introduce the terms "true positive", "true negative", "false positive" and "false negative", see Table 5. Here the word "true" ("false") refers to the presence (absence) of a defect in real world, and the word "positive" ("negative") denotes that the SIS has (not) detected a defect. A reliable SIS has a low score for "false positive" and "true negative", and a high score for "true positive".

Table 4 Operation on the defect level.

Operation on defect:	*Explanation / Example of reason:*
Create	New defect is created during process, or the defect has not been detected by previous SIS ("true negative")
Delete ("Erase")	A previously defected defect has become invisible or has disappeared, or it has been wrongly left undetected ("true negative"), or it has been wrongly identified as a defect by the previous SIS ("false positive")
Change apparent class	Defect has been attributed to another class

Table 5 Logic table of defect presence in real and SIS-observed worlds

		Defect detected by SIS?	
		positive	negative
Defect present in real world?	true	*true positive*	*true negative*
	false	*false positive*	*false negative*

Besides the creation and deletion of defects, whether in the real world due to the modification of the surface by the undergone process, or in the virtual defect map due to true negative or false positive identification, it can also happen that the two subsequent SIS systems correctly detect a defect, but they classify it differently. Hence, the same terminology as used in table 5 can be adopted for the defect *class* (whereas table 5 originally was dedicated to the status of the defect *presence*).

3.6 Transformation of maps & defect mappings

The surface area A_1 at the location of camera inspection system SIS-1 can be transformed to the surface area A_2 from the second SIS-2, by the following set of operations:

$$A_2^{'} = \prod_j Q_j A_1 \qquad (1)$$

where the symbols Q_j represent the operations listed in tables 2 and 3. A_2' is a best estimate for A_2. Some of the operations Q_j are not commuting, and hence need to be carried out in order of their occurrence. Commonly, each unit of the semi-finished product has an identification code, by which the data can be linked unit-wise. This saves the effort of taking into account the operations of table 2.

Suppose we want to predict the defect map at the position of SIS-2 on the basis of the defect map M_1 obtained at SIS-1. In analogy to equation 1, a best estimate M_2' for the predicted defect map at the location of SIS-2 can be given as:

$$M_2^{'} = \prod_j Q_j M_1 \qquad (2)$$

The relation between the predicted defect map M_2' and the actual defect map M_2 is given by the operations on defect level which have been listed in table 4. Denoting these operations by T_m, with the suffix m denoting the different kind of operations, we have thus:

$$M_2 = T_m M_2^{'} \qquad (3)$$

The retrieval of the collection of operators T_m from the defect maps M_1 and M_2 is essential to establish a quality control loop on individual defect level. The correlation of the T_m operators obtained at many coils over time gives insight into the performance of the individual SIS systems, and, more importantly, of the production installations.

A complication in establishing the correct mapping is the fact that operations are not always systematically logged, in particular the operations listed in table 3. The practical approach is therefore to use as much as possible knowledge on the operations Q_j (tables 2 and 3) to conduct the surface area transformation to obtain a best estimate M_2'. Then, starting from this best estimate M_2', and the actual defect map M_2, adjust the transformations Q_j to obtain matching maps. Algorithms to perform this task are e.g. Self Organising Map (SOM) methods, correlation algorithms, statistical techniques, etc. It is still a challenge to develop a robust algorithm or series of algorithms that can performs this task in an automated and sufficiently reliable way. Finally, after matching the defect maps, the defect operators T_m can be extracted.

4 RESULTS

The current status of through-process defect analysis is mainly based on coil statistics. Besides, through-process defect mapping occasionally occurs on a manual basis; due to the mentioned lack of universal and robust algorithms, this procedure is not yet implemented as an automated task. An alternative 'non-rigorous' approach to deal with the uncertainty associated with the mathematical manipulation of these data is to adopt a knowledge engineering approach together with the resident SIS experts. In the process of viewing and investigating numerous defect maps, over time these experts formulate heuristic rules which guide their decisions about the implied state of the product. Through interview and application of knowledge elicitation techniques, it is often possible to extract these 'rules of thumb' such that they can be incorporated into the data processing operation. Thus a form of intelligent filter is applied to make an automated assessment of the defect distribution, expressed in terms which make it of immediate value to upstream and downstream processes. This first step in integrating SIS's serves to replace routine data aggregation freeing up technical time for more value adding tasks. The increased visibility and speed of feedback also serves to stimulate user demand for further developments such as automated analytical routines triggered by specific key observations. Such generated interest serves to support efforts to implement improved tracking systems which in turn support the rigorous approach described earlier.

An example of current practice of through-process defect analysis is given in Fig. 4, which is a screen shot of a client-server based application that runs within Corus. The features of this viewing application were determined from interviews carried out with technical personnel at each stage of the process to ascertain the current approach to data handling. The total number of defects, or, optionally, the defect density of a user selectable series of defect types can be viewed over a user configurable time span. With this application plant technical people can easily observe trends. By ticking or un-ticking checkboxes, the product grades can be selected that have to be taken into account. Because of the delays (see table 2) due to e.g. transport of semi-finished products, a product can be cast in week 1, hot rolled in week 2 and pickled in week 4. Hence, different views have been defined for the different factories, which allow the user to view the defect statistics according to cast week, hot roll week and pickle week. As such, production technical people have direct feedback on the quality performance of the material leaving their own plant.

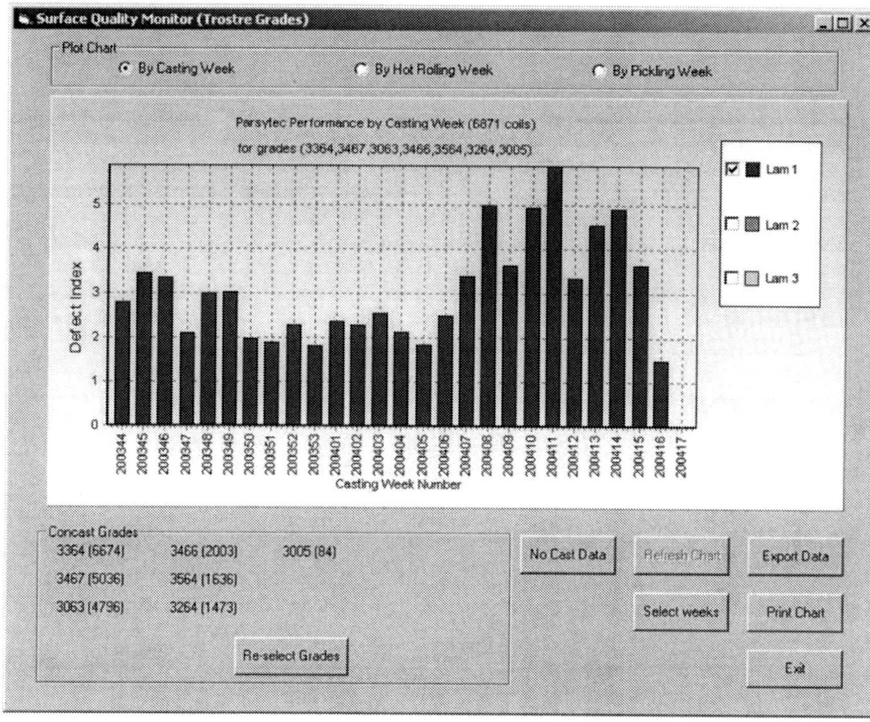

Fig. 4 Screen shot of the through-process defect monitor.

In addition to the access to overall performance indicators, a click on a bar in the chart yields an associated table with the defect statistics of each individual coil that resides in the selection for the specified week. Subsets of through process records for specific periods of processing or for selections of steel grades can be extracted into data files for further analysis in end-user tools such as spreadsheets. Indicative defect maps can also be viewed for coils with defect levels above a certain threshold.

5 CONCLUSIONS & OUTLOOK

In the previous sections it has become clear that a structured comparison of surface defect maps obtained at various stages of the production process of steel strip, offers many advantages. The areas of benefit are for instance:

- Check the reliability of SIS system, e.g. on the following characteristics:
    - Detection performance
    - Classification performance
    - Trends in detection and classification performance. For example, systematic deviations over the width of the strip or over time can elucidate deficiencies in illumination conditions or focusing, but these can also be related to (changes in) processing.

- Verify and optimise process performance:
  - Determination of which defects appear or disappear during a given process
  - Identification of trends in the creation or deletion of defects. For example, replacement of a tool can introduce a sudden increase in the creation of certain defects. An automatic alarm generated by a through process analysis system offers the possibility to signal the operators of such an anomaly and urge them to take action. Fast reaction to suddenly occurring quality problems saves costs. Defect densities can also be studied in relation to width and length position of the coil, as well as top and bottom sides.
  - When trends in the creation or deletion of defects can be correlated to machine settings, the way is opened to process improvements. Applications linking SIS data to process parameters are demonstrated in e.g. (1, 2).

- Reroute material:
  - Based on previously obtained statistical relationships between surface state early in the process and the surface quality of the finished product, operators may decide to reroute products halfway the production process, acting on information received from inspection results at an early production stage. This will finally result in less material rejections and a higher quality product.

Past work has been carried out mainly on the level of coil statistics, which has resulted in through-process availability of this information to the plant technical personnel. In the near future, information on individual defect maps will be online available to all people concerned.

A knowledge-based approach is currently being applied to support through process utilisation of data from different SIS's. This incorporates heuristic approaches as applied by SIS experts to cope with incomplete knowledge of product manipulation through the various processes. This also serves to energise the organisation in pursuing of more rigorous systematic approaches.

The next step, automated analysis of through-process defect maps and correlation of this information to process parameters, is currently in the phase of academic development. Robustness and universal applicability of these mapping procedures is currently a major concern for their use within the industrial production environment like Corus.

6 REFERENCES

1 Peters, H., Heckenthaler, T. and Holzknecht, N. (16-20 June 2003) "Optimisation of flat product quality by intelligent data exploitation", *Proceedings of the 3rd European Rolling Conference. METEC*, Düsseldorf, Germany, pp317-322

2 Keplinger, S. (16-20 June 2003) "New Efficient methods for quality control in steel production", *Proceedings of the 3rd European Rolling Conference. METEC*, Düsseldorf, Germany, pp323-326

e-Business, e-Commerce, and e-Services

Location-based services – a vehicle for IT&T convergence

K MICHAEL
School of Information Technology and Computer Science, University of Wollongong, Australia

ABSTRACT

Location-based services (LBS), more than any other mobile commerce application area has served to bring together information technology and telecommunications (IT&T) industries. While much has been written on the potential of LBS, literature on how it is a catalyst for digital convergence is scant. This paper identifies and explores the various levels of converging technologies in mobile commerce by using three LBS case studies. Through literal replication the findings indicate that IT&T technologies are converging at the infrastructure, appliance and application level. It is predicted that mCommerce applications will increasingly rely on industry convergence to achieve their desired outcomes.

1 INTRODUCTION

Location-Based Services (LBS) is a branch of m-Commerce that has revolutionised the way people communicate with others or gather timely information based on a given geographic location. Everything living and non-living has a location on the earth's surface, a longitude and latitude coordinate that can be used to provide a subscriber with a wide range of value added services (VAS). Subscribers can use their mobile phone, personal digital assistant (PDA) or laptop to find information relating to their current location. Typical LBS consumer applications include roadside assistance, who is nearest, where is, and personal navigation. LBS business applications differ in their focus and many are linked to core business challenges such as optimising supply chain management (SCM) and enhancing customer relationship management (CRM). Some of the more prominent LBS business applications include: fleet management (incorporating vehicle navigation), property asset tracking (via air, ship and road) and field service personnel management (i.e. people monitoring). The emergency services sector in the United States (US) was responsible for driving the first pin-point location service, demonstrating to the world the potentially life-saving functionality of the technology. As of October 2003, the Federal Trade Commission (FTC) enforced that wireless operators provide the Automatic Location Identification (ALI) of a caller to the emergency dispatcher. ALI standards designate that more than two-thirds of emergency calls received require the location of the individual to be accurate to within 50 metres, and 95 per cent of calls to within 150 metres. The technology is available for potential mass market deployment, how feasible it is however is a separate issue altogether. This paper provides an overview of the devices, applications and technologies used by three companies that offer LBS applications. The overall aim is to show the current state of development in leading edge LBS product innovations and to demonstrate that LBS have served to bring together information technology and telecommunications (IT&T) industries. The first section of the paper reviews previous literature and develops an analytical framework for the investigation; second each LBS product innovation will be examined; and third a discussion on the high-level effects LBS has had on IT&T convergence ensues.

2 LITERATURE REVIEW

2.1 Who, what, when, where & wi-fi?

The evolution of mobile location-based services has been well documented in a paper by Rao and Minakakis (1). This article summarises the platforms, technologies and standards of mobile LBS and does well to differentiate between the various techniques that can be used to determine an accurate location of an object or individual. These techniques include: cell identifier (cell ID), global positioning systems (GPS), assisted global positioning system (aGPS), and the broadband satellite network. Zeimpekis et al. (2) go into more explicit detail about each of these and identify a whole range of indoor and outdoor positioning techniques categorising these into "self positioning" and "remote positioning". It should also be noted that location technologies can be classified as either handset-based or network-based. Cousins and Varshney (3) provide a brief overview of the location framework required for mobile location services whereas Varshney (4) goes into greater depth for each element in the framework. Balatseng and Hanrahan (5) specifically use the Global System for Mobile (GSM) to describe the logical architecture required to support mobile station positioning. Maass (6) can be credited with an implementation-level paper on location-aware mobile applications based on directory services. Varshney's (4) paper however stands out from the rest of the literature in that he makes the important connection between the type of service offering and the level of accuracy required. He also includes the wireless LAN (wi-fi) network in the location management architecture, instituting radio frequency identification (RFID) as a significant technology embedded in the LBS framework.

In terms of target markets for LBS, Rao and Minakakis (1) identify three target markets including the consumer, niche consumer/ business, and industrial/ corporate. Cousins and Varshney (3) also separate state-driven applications from those that are business driven which is important when discussing the overall capabilities (present and future) of LBS (7). Typical services specified by most authors range from mapping, directory services, shopping, alerting, SCM, CRM, intelligent transportation, emergency and e-health. These can be applied in any given scenario- Business-to-Consumer (B2C), Business-to-Business (B2B) and even Citizen-to-Government (C2G) relationships. Interestingly the work of Burak and Sharon (8) on FriendZone is among the few analysing usage of a single LBS commercial application. The distinction between push and pull services is also important (4). The FriendZone service is a 'push' mode of operation allowing a subscriber to locate friends and acquaintances nearby, whereas checking on the next movie showing closest to a location is an example of a 'pull' mode of operation. Some of the more common revenue business models for LBS services include the traditional subscription-based model, pay-per-view, micropayments and application service provider (ASP) facilitator (1).

2.2 The gap in the literature

The gap in the literature is two-fold. First, a paper needs to be written showcasing cutting edge LBS product innovations that reveal the current state of development. A lot of sensational material exists in the popular media about what is possible with LBS but a candid view of billable applications that are being offered now is required. Second, a look at how LBS is spurring on convergence at various levels within IT&T needs to be demonstrated. Traditional telephone companies are no longer the typical service providers (SPs). New business models are changing the rules of engagement between established companies and new entrants who are looking for niche markets. The definite move toward a packet-based solution using Internet Protocol (IP) is also blurring the line between the once easily identifiable carrier-grade applications and enterprise-level offerings. The need to reduce the time-to-market (TTM) for opportune LBS was exemplified during the SARS outbreak in 2003. Hong Kong mobile telephone operator, Sunday, rapidly developed and launched an application that warned subscribers via short message service (SMS) about buildings with confirmed or suspected SARS cases within approximately one kilometre radius of their location.

3 METHODOLOGY

The research approach for this paper is exploratory. Multiple case studies will be used to gather evidence to satisfy the two main objectives stated above. The main unit of analysis is the product innovation, and the sub-unit of analysis is the LBS technology used to implement that product innovation. Three US companies have been chosen for this study, each with billable LBS market applications. AT&T Wireless (www.attwireless.com), Wherify Wireless (www.wherifywireless.com) and Applied Digital Solutions (www.adsx.com) offer product innovations that represent the diverse ways that LBS applications can be implemented. The case study protocol is composed of the following questions: What is the product innovation? What are the LBS applications the company can support? When were the company's LBS services officially launched? Who is the target market? What kind of device(s) is/ are being used by the subscriber? What are the subscriber pricing plans (i.e. connection, monthly, usage fees)? Is it a carrier-grade or enterprise-level application? What is the level of accuracy when locating a subscriber? What do the LBS services require in terms of IT&T? It is the latter question that pertains to showing that LBS is a catalyst to IT&T convergence. In citing Kampas, Chen (9) provides a high-level framework for possible convergence at three separate layers occurring at the infrastructure, appliance and application levels. Chen also describes the notion of "colliding industries" including the communication, electronics, computing and information/ entertainment sectors.

The data gathered by the researcher will be drawn completely from information provided on the company web sites published between the period of April 2002 and April 2004. The online documentation reviewed will typically include: company background, product briefs, application user guides, technical specifications and press releases. In this manner, the method of investigation can be considered wholly e-research (10). External validity is ensured given that the companies are registered on the New York Stock Exchange and must provide factual content to their present and potential subscriber base. The possibility of researcher bias is minimised in this paper given its intent is not to prove that one service is better than another, but to document the current state of development.

4 CASE STUDIES

4.1 Product innovations

4.1.1 The versatile mMode

AT&T Wireless was the first mobile carrier to launch m-Commerce applications in the US in July 2001. Following the success of NTT Docomo's i-mode and c-mode in Japan, mMode provided a value-added data-centric package to AT&T's voice and SMS basic plans. Subscribers to mMode can use numerous devices to communicate including IP-enabled phones, PDAs, handhelds and even vertical devices such as the Panasonic Toughbook and Microslate Sidearm. The service is carrier-grade and is based on a GSM network architecture that uses new network elements, namely the Gateway Mobile Location Centre (GMLC), Serving Mobile Location Centre (SMLC), and the Location Measurement Unit (LMU). AT&T Wireless is now rolling out the general packet radio service (GPRS) network and EDGE technology, increasing bandwidth by targeting specific coverage areas as demand increases and it becomes economically justifiable to do so. The accuracy of the specific location-based applications is dependent upon the general location of the mobile transmission tower most recently contacted by the customer's device. For example, the IP device could be right next to a tower or some fifteen kilometres away. In metropolitan areas the accuracy is greater given the number of base transceiver stations is higher than in less urbanised areas.

4.1.2 The wrist-worn GPS Personal Locator

mMode's location identification is not pin-point such as in the Wherify Personal Locator solution that is based on a combination of GPS satellites and code division multiple access (CDMA) PCS network triangulation methods. The Personal Locator wrist-worn device is accurate within 30 metres of the wearer, possibly even as close as a metre. The GPS device can be controlled by both the subscriber and individual wearer, allowing the parent subscriber to track the wearer, and for the wearer to alert the parent subscriber and/or location centre headquarters in case of an emergency. Coverage is available throughout the US given the GPS capability but is dependent on the PCS network coverage footprint. The Wherify frequently-asked-questions (FAQs) page (11) states: "[i]f a GPS signal is received, but the Locator is outside the digital wireless coverage area or does not receive a digital wireless signal, no location report will be provided. If the Locator receives a digital wireless signal, but no GPS signal is available, a CDMA tower-based location report will be available for emergencies." On December 30th 2003, Wherify unveiled its new GPS Universal Locator Phone which is targeted at all age groups of both the consumer and business market.

4.1.3 The VeriChip implant

While mMode requires the subscriber to *carry* a device, and the Personal Locator requires an individual to *wear* a device, VeriChip is radical in that it requires the subscriber to be *implanted* with a microchip (see table 1 for a comparison list of attributes). The campaign to *Get Chipped* was launched in early 2003, and the first person to do so formally was implanted in September of that year. The chipping procedure only lasts a few minutes. There are a number of Veri centres where the procedure can take place in the US and internationally. There is even a high-tech *ChipMobile* bus fully equipped to perform the implant procedure, 'on the road'. Applied Digital Solutions (ADSX) initially invested heavily in another product they called the Digital Angel in 2002, which resembled the Personal Locator solution but aimed at a broader market base than just children. The Digital Angel wristwatch was more slim-line but required the user to carry an additional wallet with battery power. While remnants of the Digital Angel web site are still operational today, it is the VeriChip which has become the flagship product of the VeriChip Corporation (a subsidiary of ADSX). About the size of a grain of rice, the VeriChip is the world's first subdermal radio-frequency identification (RFID) microchip. According to an ADSX press release (12): "[t]he standard location of the microchip is in the triceps area between the elbow and the shoulder of the right arm." In theory an implantee could be identified in a wi-fi network, such as in a workplace or university campus. Whereas GPS has limitations in-building locations due to construction materials used, RFID thrives in a local area network (LAN) setting, allowing walkways and door entries to act as scanners. RF energy from the scanner triggers the dormant VeriChip and in turn sends out a signal containing the unique verification number. The exchange of data is transparent and seamless in the case of RFID, there is no need to physically stop to verify a biometric feature- the network is ubiquitous. In another scenario, an individual could be identified by the RFID implant, giving emergency services access to the implantee's medical data and history that could be potentially life-saving. Unlike other fixed services, m-Commerce applications grant the subscriber access to services twenty-four hours a day, seven days a week. In the case of the VeriChip it is not only "always on" but "ever-present" inside the body of the subscriber. Unlike physical biometric attributes, the VeriChip is inconspicuous to the naked eye.

Table 1 LBS product innovations and their attributes

Company Offering LBS	**AT&T Wireless**	**Wherify Wireless**	**Applied Digital Solutions**
Product Innovation	mMode	Personal Locator * new product: GPS Universal Locator Phone	VeriChip * old product: Digital Angel
Launch Date	July 2001	October 2001	September 2003 (1st person implanted)
Device Used	- IP-enabled Phones - PDAs & Handhelds - Vertical Devices	GPS Wristwatch	RF/ID Chip Implant
Network Type	GSM/GPRS * 2003 EDGE began to be rolled out	aGPS + CDMA 1900 MHz PCS network	RFID * old product: aGPS + GSM or CDMA
Bandwidth	* future 2 Mbps	(n/a) based on alerts	* dependent
Coverage	millions of customers in > than 6,500 cities	national with greater accuracy when locator in PCS footprint	local area network (LAN), personal area network (PAN), wi-fi
Type	Carrier-grade	Carrier-class * blurry/ Enterprise	Enterprise-level * dependent
Accuracy	General location (typically < 10 miles)	1-30 metres	* 1-30 metres

4.2 LBS applications

4.2.1 "My mMode: this time it's personal"

mMode is heavily oriented towards the consumer market, although AT&T Wireless also offer package deals to business users specifically for the purposes of email (plus attachments), web access, and remote access. mMode was marketed as the beginning of mLife, next generation services that 'one could not live without' (13). Among its mCommerce suite that includes news, music and finance services are a number of LBS solutions (a list of these can be found in table 2). mMode's LBS applications are diverse- everything from a mobile traffic report to directions 'to the nearest' and find people nearby (14). Some of the more creative LBS are chat and date, and travel and dining. There are four plans subscribers can choose from including: mini, mega, max and ultra. The plans are charged monthly ranging from $2.99 to $19.99 USD and include a limited megabytes (MB) download. Additional usage fees are charged at between 2c and 0.6c per extra kilobyte (KB) received or sent, dependent on the plan. These fees do not include voice calls and SMS. The mMode service is bundled allowing the subscriber maximum personalisation to choose from any application they require. The myMode web site allows the subscriber to customise their preferences and settings.

4.2.2 Personal Locator "Just For Kids"

In contrast to AT&T Wireless, Wherify strategically chose to enter the market with a niche LBS application for a Personal Locator *Just For Kids*, specifically targeted at parents of children between the age of four and twelve. The device previously cost $399 USD but was recently slashed for a "back to school special" to $199. Monthly plans for the LBS application range from an average of $19.95 to $44.95 dependent on the plan chosen (liberty, independence or freedom). There is a one-time activation fee of $35 USD plus usage fees related to additional page requests above the included locates, additional operator assistance calls and subsequent emergency calls. Wherify makes it clear that it is looking to diversify to other niche applications including Alzheimer's and law enforcement, even though the Locator for Kids is the only marketable application demoed on the web site at the present time (15).

Table 2- Present and future LBS applications as stated on the company web site

Product	Applications
mMode	Weather & Traffic: Mobile Traffic, AccuweatherAlerts, My Snow Report What's Nearby: 10Best, Find People Nearby, Find Things Nearby, Go2 Directions, Go2 Directory, White Pages, Yellow Pages Chat & Date: Match Mobile, Upoc Communities Travel & Dining: 10Best, 10Best Dining, 10Best Nightlife, Go2 Dining, Go2 Travel, Sabre Virtually There, Vindigo
Personal Locator	Locator For Kids, Corrections (Monitoring Parolees), Law Enforcement, Executive Security, Auto Recovery, Vehicle and Cargo Tracking, Fleet Management, Mobile Workforce Communication, Property Asset Tracking, Personal Electronics, Pet Care, Child Safety, Parental Supervision, Personal Protection, Alzheimer's and Memory Loss, Supervision, Animal Identification
VeriChip	VeriPay: Payment Technology for Cash and Credit Transactions VeriMed: Healthcare-related Information that is Patient-Supplied VeriGuard: Versatile Secure Access Technology Corrections: Offender Monitoring (Released on Parole, Probation or Pre-trial)

4.2.3 "Get Chipped" with VeriChip: "technology that cares"

There is little information on the ADSX web site about the pricing of the VeriChip, however it is stated that the global VeriChip subscriber (GVS) registry subscription fee is $9.95 USD monthly. There is a cost for the implant medical procedure as well, although this is not provided. In 2002 the first one hundred pre-registered persons were granted a $50 USD discount on the chipping procedure (16). The pricing for the new VeriPay and VeriGuard services has yet to be published on the WWW and probably will not be given these are typically targeting business-to-business-to-consumer (B2B2C) solutions which are highly complex in design. The "Trusted Traveller" and residential security programs (i.e., prisoners serving their sentence from home) are two examples of VeriGuard LBS applications. One desirable feature of VeriGuard is that it could operate in conjunction with other auto-ID technologies like smart cards and biometrics, rendering customer legacy systems reusable.

4.3 Information technology and telecommunications (IT&T) requirements

4.3.1 mMode: how does it work?

Using the "find people nearby" service, the GSM/ GPRS network works as follows to determine a subscriber's approximate location. An application request is made by a subscriber. The application server subsequently makes a location request to the gateway mobile location centre (GMLC). The GMLC in turn queries the home location register (HLR) and then contacts the appropriate mobile switching centre (MSC). Another location request is generated to identify the base station controller (BSC) where the mobile is currently using the serving mobile location centre (SMLC). The BSC then can use the location measurement unit (LMU) alongside the appropriate base transceiver stations (BTS) to determine the location of the subscriber by using the uplink time distance of arrival (UTDOA). The location information is then sent back via the above-mentioned pieces of hardware/ software until the message reaches the application server and a response is given to the subscriber. The AT&T Wireless web site provides an excellent facility to aid external developers of mobile solutions (17). Freely available for download are whitepapers, style guides, software development kits (SDK), programming guides, sample code and emulators. In table 3 can be found the major building blocks of the mMode technical solution.

Table 3 The mMode building blocks

Network	Technologies & Platforms	Devices
GSM/GPRS/EDGE	AT&T Wireless Communication Manager	Internet-enabled Phones
Migration- CDPD (Cellular Digital Packet Data) to GPRS	BlackBerry Handhelds	PDAs and Handhelds
Network Evolution	e-Wallet Billing Platform	Modems
Building Wireless Solutions	Wireless Java™ (J2ME)	Vertical Devices
Network Security	Markup Languages	Interoperability
	Messaging: SMS, SMPP, Multimedia Message Service (MMS), Wireless Application Protocol (WAP) Push, e-mail	
	mModeSM, Palm, PocketPC Smartphone, WAP 2.0	

AT&T Wireless differs significantly from Wherify and Applied Digital Solutions, given it owns much of its network infrastructure. AT&T Wireless also has a large existing customer base that is used to an excellent quality of service (QoS) and certain level of post sales support. Launching LBS applications nation-wide with potentially tens-of-thousands of new subscribers joining daily, requires equipment that can handle data traffic levels and systems that have been thoroughly tested for faults. mMode contains diverse LBS services- ensuring that each of these works properly and is interoperable with a range of media devices is a labour-intensive activity which is one reason why they have decided to outsource as well.

4.3.2 Personal Locator: all the bits and pieces

Wherify's location service centre (LSC) is at the heart of its current and pending product innovations. A carrier-class server and software hub, the LSC manages and presents location-based information. Unlike mMode, Wherify utilises wireless data and aGPS. Consider the following scenario where a parent wants to be reassured that their child made it to school alright after missing the bus. The parent requests a location report via the Internet using a Microsoft IE browser (or ringing the toll-free telephone number). The LSC contacts the child's Personal Locator via the PCS network (if within the footprint), and then downloads the current GPS data and requests a location. Using the data from the LSC, the device that is identified by an electronic serial number (ESN), finds the closest satellite and then computes the longitude and latitude coordinates of the child's location. The Personal Locator then communicates location information to the LSC and the LSC generates a location report for the parent via the Internet. The whole process from request to report takes about sixty seconds. The parent is able to look at the report visually on a scalable map which shows streets and other feature points in a vector or aerial view, using geographic information systems (GIS) capabilities. Each report requested by the parent is logged in the customer's event file database for billing and subscriber profiling. The location database includes a time stamp along with the longitude/ latitude coordinates. The wearer's profile is also stored including: age, gender, height, weight and features.

Wherify make no secret of their technology partners. They include an impressive list of companies: SiRF who provide the GPS chipset that is integrated into the Personal Locator based on a-GPS; Qualcomm for the CDMA chipset; Baldwin Hackett & Meeks who are applications developers, Conexant who provide the RF board; Advanced Micro Systems who specialise in flash memory; Compaq for the server technology; Intrado for emergency communications; and GlobeXplorer Online for the component of aerial photography. Security firewalls are paramount in the Personal Locator system as is redundancy and fault tolerance. During an emergency situation for instance, the LSC is even able to interact with public safety answering points (PSAP) through Wherify's emergency operation service. There are customer care representatives available 24x7x365.

4.3.3 VeriChip made very easy

The least complex of the three case studies in terms of technology requirements is the VeriChip. RFID networks are usually small in scale when compared to nation-wide or global networks. They include the following components: the RFID transponder, a reader that captures information, an antenna that transmits information, and a computer which interprets or manipulates the information gathered. In the case of VeriChip, there is a requirement that each subscriber registers their personal details (and other relevant information they desire) on the GVS database. At this stage all the transponders issued by VeriChip are passive but it is likely that active transponders will be issued in the future, despite the fact that they require onboard battery power to operate internal electronics. When an individual passes an associated scanner, information is read and sent to the computer via an antenna. Dependent on the application, a log may be retained or the implantee's location updated a predefined number of times in a set period. Given global standards are an issue for debate in RFID, proprietary systems are used.

5. DISCUSSION

5.1 Defining Convergence

Convergence means different things to different people and is usually loosely applied to denote the coming together of two distinct technologies, i.e. the merging of several products into a single good. The *2003 Penguin Concise Dictionary* states that *convergence* is a "jargon term" and gives examples of the merging of the television (TV) and computer, or telephone and computer, or TV and WWW. To anyone who has studied technological trajectories at any length, convergence is far from being a jargon term, but a well-constituted concept in the field of innovation (18, 19, 20). Terms like "digital convergence", "technological convergence", "application convergence" and "industry convergence" have been used interchangeably in some instances, and in others each has carried a loaded meaning. For example, Covell (18) states: "(d)igital convergence is the merging of these improved computing capabilities, new digital multimedia technologies and content, and new digital communications technologies. This combination of computing power and functionality, digital networked interconnectedness, and multimedia capability enables new forms of human interaction, collaboration, and information sharing." Greenstein and Khanna (20) on the other hand, distinguish between "convergence in substitutes" and "convergence in complements". The distinction of these 'kinds' of convergence finally puts an end to the debate over usage. Convergence thus can occur at any level of detail, in any part of the subsystem.

5.2 LBS: a catalyst for IT&T convergence

Throughout this paper, technologies at the appliance, application and infrastructure level have been shown for each of the LBS cases. What can be seen is a coming together of what were once somewhat unrelated technologies. Most obvious perhaps is the convergence of wireless capabilities and the Internet as depicted in the mMode case. For example, IP-based phones can already receive voice, text and multimedia. And as for the vertical devices mentioned, many of these are converged technologies in themselves (e.g. the wireless PDA that is also a phone and MPEG3 player). In the case of Wherify, the traditional wristwatch has now been turned into a Personal Locator with the aid of a GPS chipset. And chip implants have found there way under the skin of human beings to converge with living tissue- chips once as big as bricks, now smaller in size than a grain of rice.

Yet it is not only at the device level that convergence is occurring. A whole suite of new applications are being created using content from syndicates, once considered to be unrelated. The Yellow Pages directory for instance, used to "find the nearest", or "the best 10 nightlife" locations as well as providing "shopping discount alerts". And geographic information systems once used for computer-aided design (CAD), now used to visually represent the geographic location trail of a child, using high resolution aerial photography once synonymous with superior defence intelligence systems. There are even applications like VeriPay that are forecasted to change the way that humans interact with other technologies like automatic teller machines (ATMs). Who needs to carry a card at all? Applications once used solely for businesses purposes, now permeating the consumer market given their cross-functional nature. At the infrastructure level also, multiple network technologies are being used in tandem to locate subscribers including PCS with aGPS. Another example provided, was the VeriGuard system that will have the capability to incorporate other automatic identification (auto-ID) reader equipment belonging to smart card and biometrics. Even at the protocol level, the very essence of traditional voice calls will be packetised, i.e. voice will be data. It is obvious through the evidence provided in this paper that *convergence in complements* is occurring, given the products are working better together than separately (20). LBS has shown itself to also involve a diverse range of businesses from vertical and horizontal industries- from independent software vendors (ISVs) developing the applications, to third party suppliers building enabling technologies and platforms, to industry bodies setting the appropriate standards for communications, to marketing consultants invited to develop and spearhead brand awareness campaigns. LBS brings not only the industries but the technologies to increasingly work together to form larger and larger systems (20).

6 CONCLUSION

Location-based services are pulling together a vast array of digital technologies like never before. The convergence between technologies is a cultural-changing force. Miniaturisation in design in particular is allowing for once separate technologies to be fused. From handset phones to smart watches to implants, the more invasive the technologies are becoming, the greater the precision for locating the subscriber or wearer or implantee. The question now, that all this technology can be used in an integrated fashion, is how far will entrepreneurs take LBS in the future? How many different players can become involved in offering LBS specifically before the state of affairs becomes too cluttered and confused? Do content providers reach mutually exclusive agreements with service providers (SPs) so that there is minimal conflict of interest? And if so, does this not limit the number of SPs to a few large players that can actually deliver LBS? And how many different types of LBS can one service provider practically offer? Looking at the dilemma from another perspective- will consumers require subscription to mMode, the Personal Locator and the VeriChip solution and carry with them a PDA, wear a GPS watch and be implanted with a chip, to circumvent a variety of limitations of each technology? Or are future directions set on a trajectory of even greater convergence proportions between all of the technologies discussed in this paper. For instance, will one device be able to cater for the needs at each level of accuracy- global, national, regional, local and in-building or will service providers amalgamate their networks to offer super-LBS services from satellite-based to network-based to LAN-based and PAN-based. Whatever the outcome, we are surely entering into a period where pervasive computing will become a dominant force in the way we live, work, and interact with one another.

ACKNOWLEDGEMENTS

The author is currently involved in collaborative work with Nortel Networks on the theme of the Mobile Location Centre (MLC).

REFERENCES

1 **Rao, B.** and **Minakakis, L.** (2003) EVOLUTION of Mobile Location-Based Services. *Communications of the ACM*. Vol 46, No 12, pp. 61-65.

2 **Zeimpekis, V., et al.** (2003) A Taxonomy of Indoor and Outdoor Positioning Techniques for Mobile Location Services. *Journal of ACM SIGecom Exchanges*. Vol 3, No 4, pp. 19-27.

3 **Cousins, K.** and **Varshney, U.** (2001) A Product Location Framework for Mobile Commerce Environment. Proc. ACM 1st International Conference on Mobile Commerce, pp. 43-47.

4 **Varshney, U.** (2003) Location Management for Mobile Commerce Applications in Wireless Internet Environment. *ACM Transactions on Internet Technology*. Vol 3, No 3, pp. 236-255.

5 **Balatseng, O. E.** and **Hanrahan, H. E.** (2002) MS Positioning for the Support of Mobile Location Services. [http://www.ee.wits.ac.za/~comms/output/satnac02/balatseng.doc, 2004].

6 **Maass, H.** (1998) Location-aware Mobile Applications Based on Directory Services. *Mobile Networks and Applications*. Vol 3, pp. 157-173.

7 **Deitel, H. M., et al.** (2001) e-Business and e-Commerce for Managers. Prentice Hall, New Jersey, pp. 168-170.

8 **Burak, A.** and **Sharon, T.** (2003) Analysing Usage of Location Based Services. CHI 2003: New Horizons, April 5-10, Florida, USA, pp. 970-971.

9 **Chen, S.** (2001) Strategic Management of e-Business, John Wiley and Sons, New York, pp. 5-7.

10 **Anderson, T.** and **Kanuka, H.** (2003) E-research: methods, strategies, and issues. Allyn and Bacon, Boston.

11 **Wherify** (2004) Frequently Asked Questions. Wherify Wireless. [http://www.wherifywireless.com/faq.asp, Last Accessed: 15 April 2004].

12 **ADSX** (21 November 2003) Applied Digital Solutions. CEO Announces "VeriPay™" Secure Subdermal Solution for Payment and Credit Transactions at ID World 2003 in Paris. Applied Digital Solutions. [http://www.adsx.com/news/2003/112103.html, Last Accessed: 15 April 2004].

13 **McDonough, B.** (17 April 2002) AT&T Wireless Pushes mLife with mMode. CIO Today. [http://cio-today.newsfactor.com/perl/story/17307.html, Last Accessed: 6 April 2004].

14 **AT&T** (2003) Feature and Services User Guide. AT&T Wireless. [http://www.attwireless.com/personal/features/mmode/mmode_guide.jhtml, Last Accessed: 15 April 2004], pp. 1-39.

15 **Wherify** (2003) Wherify Wireless GPS Locator For Kids. Wherify Wireless. [http://www.wherifywireless.com/prod_watches.htm, Last Accessed: 15 April 2004], pp. 1-120.

16 **ADSX** (2003) Implantable Personal Verification Systems. Applied Digital Solutions. [http://www.adsx.com/prodservpart/verichip.html, Last Accessed 15 April 2004], pp. 1-2.

17 **AT&T** (2003) Developer Tools. AT&T Wireless. [http://www.attwireless.com/developer/tools/, Last Accessed: 15 April 2004].

18 **Covell, A.** (2000) Digital Convergence: how the merging of computers. *Communications, and multimedia is transforming our lives*. Aegis Publishing Group, Rhode Island, pp. 14.

19 **Baldwin, T.F., et al.** (1996) CONVERGENCE: integrating media, information & communication. Sage Publications, California, pp. 209.

20 **Greenstein, S.** and **Khanna, T.** (1997) What Does Industry Convergence Mean? *in D.B. Yoffie (ed.), Competing in the Age of Digital Convergence*. Harvard Business School Press, USA, pp. 204.

The international material data system – e-business readiness in the automobile supply chain?

I POLLOK
EDS Operations Services GmbH, Rüsselsheim, Germany
B K TEMPLE and **D K HARRISON**
School of Engineering, Science, and Design, Glasgow Caledonian University, UK
D A EDGAR
Caledonian Business School, Glasgow Caledonian University, UK
S C KINZLER
Department of International Business, F A Aalen – University of Applied Sciences, Aalen, Germany

ABSTRACT

The internet application International Material Data System (IMDS) was developed in order to globally collect material data, which will later be needed by the OEMs for documenting the accordance with the EU Directive 2000/53/EC concerning End-of-life vehicle management. This Directive demands 95% recycling of the mass of a car by 2015.

For successful data collection it is important that all tiers within this chain act efficiently and in accord. The bottom-up principle of the application implies that the lowest tier must be the first to provide information. If information of one of the suppliers in the chain is provided late, the whole process is delayed.

As the whole automobile supply chain is involved three influential facts could be detected: firstly, the IT infrastructure should guarantee a good internet access; secondly, the departments within a company need to be aware of the data demand and have to cooperate in order to support the data flow and thirdly, the relationship between the companies has to be based on co-operation, too.

1 INTRODUCTION

The internet application International Material Data System (IMDS) was developed in order to globally collect material data, which will later be needed by the OEMs for documenting the accordance with the EU Directive 2000/53/EC concerning End-of-life vehicle management. This Directive demands 95% recycling of the mass of a car by 2015. Germany's automobile industry voluntarily committed to earlier completion of this task initiated by a consortium of eight automobile manufacturers - Audi, BMW, Daimler-Chrysler, Ford, Opel, Porsche, Volkswagen and Volvo - in association with the IT service provider, Electronic Data Systems (EDS). The system was implemented in 2000. Subsequently, eight other car manufacturers have also joined and the participation of more companies is being negotiated.

The application - initiated by the OEMs - relies on a bottom-up approach to data entry. The lowest tier in the supply chain ("tier n") must be the first to provide the data on their products and pass them to the next tier supplier ("tier n-1") who ordered the product. They provide information on the product they have made from the “tier n” supplier. The information passes up the supply chain and every OEM receives a complete account of the chemical characteristics of their car for recycling purposes. If information of one of the suppliers is provided late, the whole process is delayed as the well-developed information flow is an essential feature for a supply chain which acts under high competitive pressure (1).

The success of IMDS is based on three critical issues as the whole automobile supply chain is involved: firstly, every supplier must have access to good IT equipment; secondly, there must be a good inter-departmental co-operation in order to provide accurate data and thirdly, there must be a good working relationship between the different tiers. These three critical factors are thought to be preconditions for a successful e-business solution and consequently, the response to IMDS is an indicator of the ability to pass on the data through the electronic network.

As the automobile industry is forced to innovate (2) and is characterized by large fixed costs and overcapacity (3) its supply chain is a good example where collaboration can create higher value (4). In addition, electronic processes can enable greater efficiencies (5). IMDS was chosen for analysis as it is an example for an e-business project covering the entire automobile supply chain. The analysis concentrates on the progress that has been made in the use of e-business techniques.

This work provides evidence that there are considerable differences between the different company sizes concerning the use of e-business applications and the IT equipment. The level of “computerisation” within companies was found far from complete, especially in the smaller companies on whose information the whole chain depends.

Although the automobile industry’s supply chain is well established, the paper provides findings that the electronic measures for efficiency improvement have been less successfully applied and derives some recommendations for further implications.

2 METHODOLOGY

Two online questionnaires were designed with a time delay of a year. The first part of the two questionnaires asked for IMDS application features like design, performance, services etc. The second part of both questionnaires contained general questions about the different company sizes but also specific questions on the intra-company infrastructure and IT applications.

The first survey was put online for a month in December 2001 and delivered 981 valid entries. The percentage of missing questions varies from about 6% to 7%. This leaves an average number of about 920 entries for analysis, which can be taken as representative population of about 17,000 users.

The second survey was put online for the same time period in February 2002 and delivered 2,418 responses. The percentage of missing questions varies from about 9% to 10%. This leaves an average number of 2,180 entries for analysis, which is taken as representative population of about 24,000 users.

The analysis is divided into three parts according to the above mentioned three preconditions for fast data delivery within IMDS. The following question from the first questionnaire was chosen for analysing the existence of electronic equipment and the IT possibilities within the companies:

Question 1: What is the percentage of PCs/work place in your company?

The second precondition - good inter-departmental cooperation – can be tested with further two questions from the first questionnaire which aimed at the commitment to integrating company systems and thus to improve inter-departmental information flow as well as the information about new means of procurement:

Question 2: Does your company use Enterprise Resource Planning software, e.g. SAP, BAAN?

Question 3: Does your purchasing department use e-procurement?

Both questions aimed at finding out whether employees are familiar with possible e-business applications in their company.

The third precondition is a good working relationship between the different supplier tiers. For checking the existing relationship, the following two questions from the second questionnaire were analysed:

Question 4: Which of the following words would best describe the relationship to your automobile industry customers?

Question 5: According to your opinion, how are suppliers for new products selected in your company?

The focus of both questions is the relationship between the suppliers and customers on the different tiers.

3 FINDINGS

The findings for each question are grouped in accordance with the size definitions of the Chamber of Commerce in Heilbronn/Germany. Most of the companies in the two questionnaires have either 100-500 or more than 1,000 employees. This picture mainly corresponds to the findings of Dahlhoff et al. (6) but delivers a more global view on the automobile industry structure as Dahlhoff analysed the German structure of this branch of industry. For better comparison the responses to the sets of questions the company sizes in each category are set to 100% in each question.

Table 1 Company size compared with Dahlhoff et al.

Number of employees	Present work (questionnaire 1)	Present work (questionnaire 2)	Dahlhoff et al
less than 100	17%	20%	8%
101 to 500	39%	41%	26%
501 to 1000	12%	12%	12%
more than 1000	32%	27%	54%

3.1 IT infrastructure

One indicator for the first precondition - the extent to which the company is committed to electronic equipment - is the relative number of PCs:

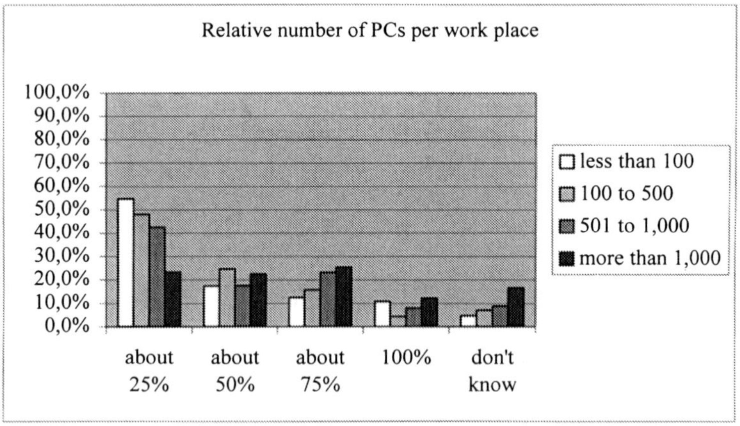

Fig. 1 Relative number of PCs per workplace

Figure 1 clearly shows an increase in the supply of PC workstations with company size. This is due to the fact that small companies are generally on the lower tiers whose contact with the OEM is indirect. Therefore, they are under less pressure to equip and in general, they have lower budgets for regular investment into IT equipment and networks.

3.2 Inter-departmental cooperation

Rungtusanatham et al. (7) points out that Enterprise Resource Planning (ERP) can enable a company to capture valuable information throughout and beyond its boundaries. ERP systems are considered “a key starting point for leveraging new Internet technologies” (8). Therefore, the use of an ERP application in the companies was questioned in the survey. The responses are shown in Figure 2 in relative percentages in order to compare the company categories.

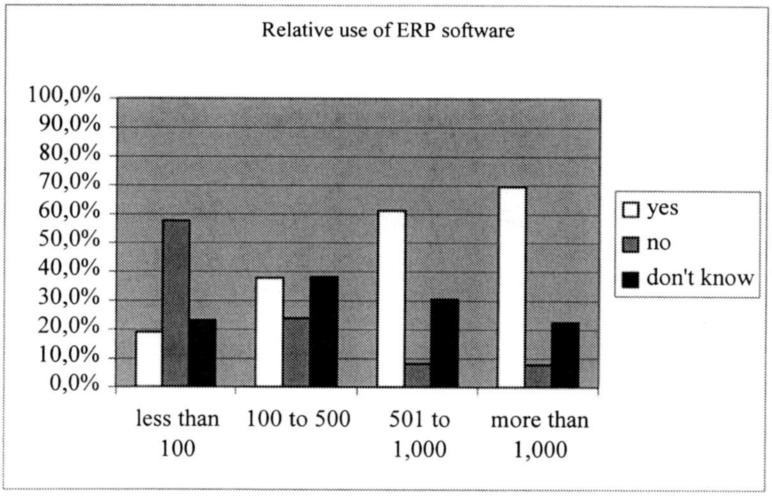

Fig. 2 Relative use of ERP software

About half of the companies report the use of ERP software. From Figure 2 it can be concluded that the use of ERP software increases with the size of the company. It might have been expected that the level of "don't know" responses increases with company size as it does for IT equipment. In contrast to the question about IT equipment of the entire company, most employees would be expected to use an ERP system every day and hence to know about its existence. Especially in companies having between 100 to 500 employees there was a major uncertainty. In small companies the clear "no" (nearly 58%) leads to the assumption that an ERP solution is not implemented in this company category because it is too cost-intense.

Taking all responses to this question into consideration, there is a large number in the "don't know" category. This might be due to the fact that only two ERP systems (SAP and Baan) were given as examples and that employees in some cases are not aware that they are working with an ERP system.

The second question on e-procurement delivered the following responses:

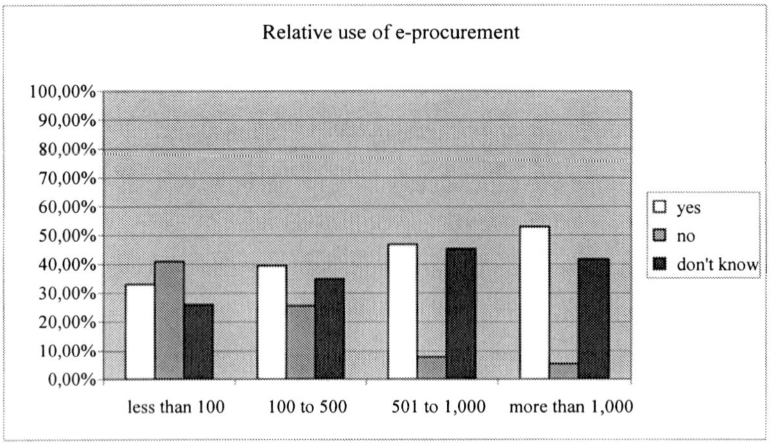

Fig. 3 Relative use of e-procurement

This picture resembles Figure 2. The main differences are the even higher number on the "don't know" category and a lower overall average use of e-procurement. Small companies do not, in general, use means of e-procurement. In the remaining categories, however, there is a severe lack of information correlating with the company size since most of the responses are "don't know". This might suggest that the responding department is not aware of the e-business activities within the company.

Could the responses be influenced by who responded? In view of the nature and anonymity of the of the questionnaire, the respondent could have worked in one of several departments in the company. The only employee who could be guaranteed to answer with authority would be a purchasing officer. Thus the large number of "don't knows" could merely be attributable to ignorance through "compartmentalization". It is well known that increasing size can often lead to this effect and hence to a certain ignorance of other parts of the company. The above findings certainly suggest that something of this nature pertained here in so far as the large percentage of "don't know" responses can only result from an increasing localisation. Nevertheless, the result gives cause for concern as the nature of the trading conditions can influence product design and manufacturing philosophy.

3.3 Working relationship between the different tiers

The question about the relationship to the customers was answered by the companies of the different tiers as follows:

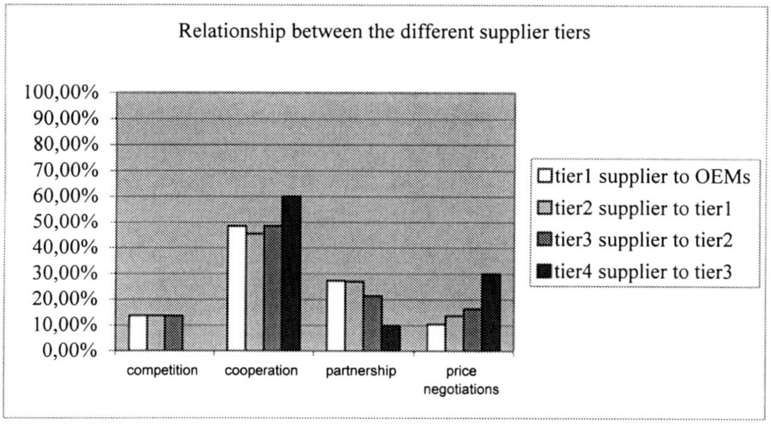

Fig. 4 Relative relationship to the customers of different tiers

Only the larger tiers seem to characterise their relationship to their customers as cooperation- and partnership-driven. This is not surprising as they work closely with the OEMs and therefore have a better position in negotiations. Although cooperation was also very important for the smaller company categories, price negotiation was ranked the second term after cooperation. Two factors influence this position: Firstly, cooperation is ranked highest because smaller companies do have a closer, personal contact and interest in satisfying their customers' needs. Secondly, the lower tiers are closer to the raw materials industry where prices for commodities are easier to compare e.g. for steel.

However, all answers might have been biased to a certain extent by the fact that the questionnaire was positioned in an OEM application where the suppliers should answer as it might be expected.

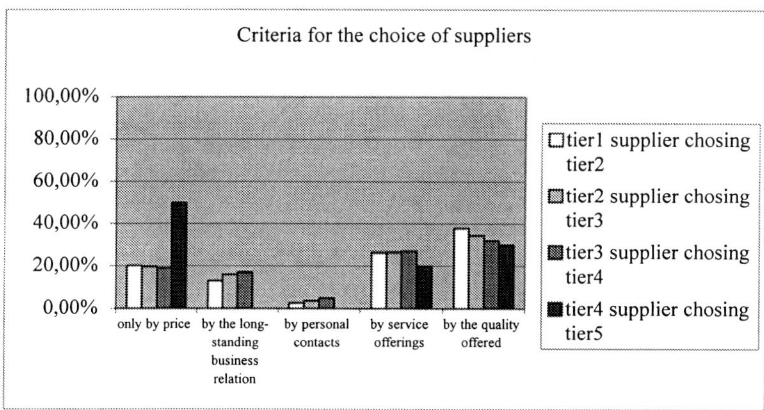

Fig. 5 Relative relationship to the suppliers of different tiers

The same picture can be drawn for the criteria by which suppliers are chosen. For the small companies price is the main driver whereas for larger companies the quality is the most important factor.

4 CONCLUSIONS

The common IT solution for recycling data collection was developed according to a legal requirement for the automobile industry, but both the information and the data exchange necessary for the IMDS project has been shown to differ in companies of different sizes. The findings from this survey support the hypothesis that even in the automobile industry, which usually is very innovative and trend-setting in the use of new technologies, the supply chain infra-structure is not yet fully implemented.

Small and medium-sized companies represent about 60% of the automobile industry according to the two surveys. Concluding from the findings those two company categories are not well-equipped regarding their IT infrastructure and e-business commitment. Especially at the lower tiers representing the smaller company categories the price is the most important argument. Hence the small companies are more likely to compete by price and they also chose their suppliers by price only. On the other hand, smaller companies seem to have the advantage of better information flow but they are unwilling to invest in cost-intense IT solutions like ERP software that could improve supply chain performance.

The larger company categories are better equipped and better positioned towards their customers. The major obstacle in this company category seems to be the lack of inter-departmental cooperation which stems from the traditional view of detached departments and leads to faulty intra- and inter-company communication.

Additionally, the second questionnaire has revealed a low level of awareness of e-business activities beyond the department boundaries. We suggest that this lack of awareness offers an obstacle to a companies efficient operation and by implication, to the efficient use of IMDS.

If the lack of awareness of e-business importance could be overcome and the relative high level of commitment of the employees would be used by the small and medium-sized companies a very large part of the automobile industry could improve its performance and speed up the information flow by effectively using information technology. Larger companies need to improve the interdepartmental communication and to make a more efficient use of their existing IT infrastructure in order to guarantee the data provision for e-business projects like IMDS.

REFERENCES

1 Childerhouse, P. et al., (2003), "Information flow in automotive supply chains – present industrial practice", *Industrial Management & Data Systems,* Vol. 103, No. 3, pp. 147

2 Dyer, J., (2000), "Collaborative Advantage: Winning Through Extended Enterprise Supplier Networks", *Oxford University Press*, ISBN 0-19-513068-5, p. 6

3 PricewaterhouseCoopers, (2001), "Supplier Survival. Survival in the Modern Automotive Supply Chain", pp. 1

http://www.pwcglobal.com/Extweb/pwcpublications.nsf/4bd5f76b48e282738525662b00739e22/4efba3d2e36509cc85256bde006bec7b/$FILE/supplier_survival_for_web.pdf

4 Sahay, B., (2003), "Supply Chain Collaboration: the key to value creation", *Work Study,* Vol. 52, No. 2, ISSN 0043-8022, pp. 76-83.

5 Prasad, S. and Sounderpandian, J., (2003), "Factors influencing global supply chain efficiency: implications for information systems", *Supply Chain Management: An International Journal,* Vol. 8, No. 3, ISSN 1359-8546, pp. 241-250.

6 Dahlhoff, H., Dudenhöffer, F. and Densing, C., (2000), "Marketing und E-Commerce – Ein neues Bild der Automobilzulieferer entsteht (Marketing and E-Commerce – A new picture of automobile suppliers emerges)", VDA and Centre of Automotive Research, Druckerei Henrich GmbH, ISSN D946-0179, pp. 3.

7 Rungtusanatham, M. et al., (2003), "Supply-chain linkages and operational performance. A resource-based-view perspective" *International Journal of Operations & Production Management*. Vol. 23, No. 9, pp. 1091

8 PA Consulting Group, (2002), "Ready to profit from the Potential of e-business? A study of e-maturity in the European automotive industry 2001-2002", *PA Knowledge Limited,* pp. 11

A staged e-business development model

J VARLEY and **O MOLLOY**
IT Department, National University of Ireland, Galway, Ireland

ABSTRACT

Within this paper we present a staged e-Business development model and toolset. The model allows organisations to assess their current eBusiness capabilities, and advise them on the steps to take in areas of technology, management, and business operations to progress to the following stage in the model. The approach taken was to design five separate stages of e-Business capability, from having order visibility, to a stage of eCollaboration. Organisations are assessed similar to a Capability Maturity Modelling (CMM) approach [1], and our framework then used to progress their e-Business initiatives to higher stages in the development model.

1 BACKGROUND

1.1 SMEs and e-Business

e-Business is commerce conducted in cyberspace. It is the execution of real-time business processes with the assistance of Internet technologies. In essence, it is business on the Internet - but it also includes all the processes involved in business. In the US ECommerce Sales in the fourth quarter of 2003 were 17.2 billion, a 25 % increase from 2002 [2], this rapid growth of e-Business and Business-to-Business communication has provided new opportunities for today's Small to Medium Sized Enterprises (SMEs). B2B trading capabilities can make SME's services and products more attractive to large enterprises. They can strive to attain a competitive advantage and reach out to new markets previously untapped. They can aim to streamline and improve current business processes and integrate their systems with that of both customers and suppliers to improve supply chain management, helping to reduce manual supply chain management costs, reduce inventory levels and increase customer satisfaction levels provide to a higher level of service.

1.1 Motivations and benefits of a Staged eBusiness Development Model

The vast majority of work in the B2B integration area has been undertaken by large enterprises. These development initiatives have typically resulted in large complex software solutions that are both difficult to implement and highly expensive. Many SMEs are unable undertake a projects of this scale for reasons such as lack of qualified staff and lack of capital [3]. Our Staged e-Business Development Model will aid SMEs in integrating with each other and with large enterprises solutions. The model will include CMM like assessment in all areas of e-Business with our customisable implementation framework and guidelines for implementation. CMM has already been used in the area of eSourcing [4], we continue this work into an overall eBusiness area and augment it with the implementation framework. There will be no highly expensive software or hardware requirements and the solution will be highly flexible. These factors combined should ease the worries of the SMEs with regard to undertaking an e-Business initiative and will provide a clear path of where to start and how to progress.

2 STAGES OF THE DEVELOPMENT MODEL

Level 1. Auditing and Strategy Definition.
Level 2. Order Visibility and Business to Customer Transactions.
Level 3. Business-to-Business trading.
Level 4. Publishing of Standards
Level 5. e-Collaboration.

2.1 Level 1

2.1.1 Technology Audit

It is necessary to audit an organisations IT infrastructure and establish the inventory of IT infrastructure items, their configurations and the applications and information that the IT infrastructure supports. The basic technology audit will cover the areas of the computer hardware available, the specifications of desktops and servers, the telecommunication Infrastructure, Network specifics such as ISDN, fibre optic cabling and transmission, Internet Services providers, and broadband considerations, and the existence of a business web site. Also to be included in the audit should be the area of in house IT skills - particularly valuable would be skills in Network support / maintenance / back-ups, Intranet development and support, Web content development. It is also important to audit the applications the organisation uses, from the operating system deployed to the use of ERP and MRP systems, and what interfaces are available to these. A complete and detailed audit will mean implementation of eBusiness projects will be a great deal easier. Knowing what areas should be outsourced is vital, and reduces any wasted internal effort on aspects of the implementation that require expertise in specific areas.

2.1.2 e-Business Strategy Definition

Before any decisions are made or e-Business projects are started it is necessary to create an e-Business and IT strategy. These goal of which should be to preserve and leverage existing business processes and technology while developing e-Business initiatives, improving overall performance and providing competitive advantage. There are a number of steps to creating an effective e-Business strategy.

- Review current business processes. It is essential to have a clear and complete view of the business so as to analyse what areas have potential for development.
- Identify business problems or efficiencies that could be solved by applying ICT in a new way.
- Identify projects to undertake. Anything from training staff to fully utilise current systems to an initial e-Business step of setting up an e-Commerce website
- Analyse costs and benefits of the projects and prioritise them

- Clearly set up times budgets and responsibilities for projects to be undertaken and manage these projects ensuring as little time and cost increases as possible.

The e-Business strategy has to have support from top management down to the lower level workers, who will have to train for, use and maintain new systems and processes.

2.2 Level 2

2.2.1 Order Visibility

This level provides customers with the ability to query their current order status, it requires no altering of legacy systems just the creation a front end dashboard which can create order queries which can then be processed against the back–end system, using a Business Event and Activity Monitor (BEAM), returning the required information to the customer (See Fig.1).

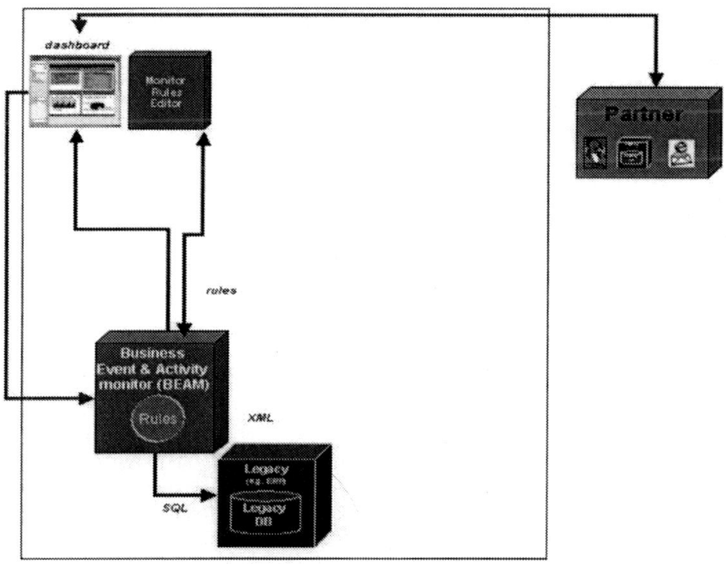

Fig. 1 Order Visibility

2.2.2 Providing Business to Customer transactions

The next technology hurdle on the road to eBusiness success is setting up an eCommerce website to allow Business to Customer transactions. The critical issues here are creating an online product catalogue, dealing with customer queries, security issues and handling payment. Any online catalogue must be kept synchronised with whatever Enterprise Resource Planning (ERP) or Backend system is in use within the business. EnableSME's technology framework provides a module for this purpose (Fig.2). As the following architecture shows the B2C module can be implemented alongside any ERP or Backend with it only being necessary to write one mapping to integrate with new systems.

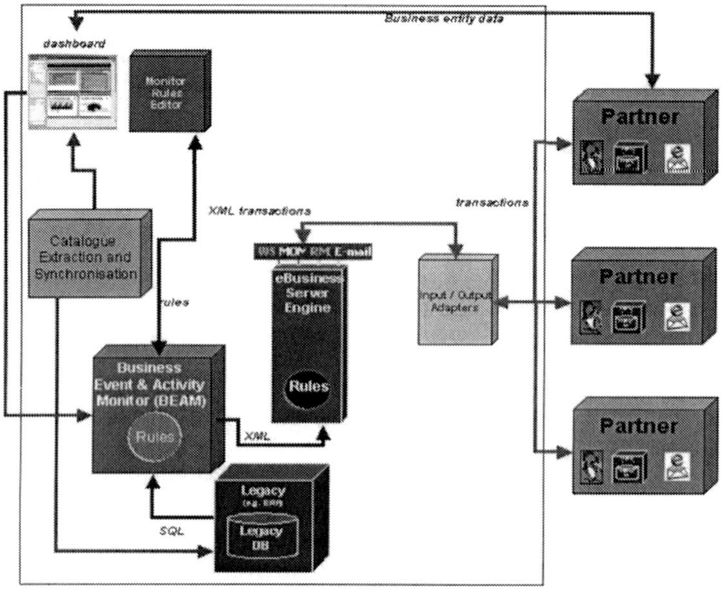

Fig. 2 B2C Architecture

The basic architecture to this module uses the Catalogue Extraction and Synchronisation module to generate a current catalogue in an XML format. This is made using web services technology, and can connect to any ERP of Backend System for which a mapping has been written. The XML product catalogue is used by the Dashboard, which also accesses information from the customers profile after they have logged in. This form allows users to add to their order, automatically updating the orders total for them to view. Once the order is complete the user can preview it, it can be confirmed and paid for by the user. Payment will depend on the customer that has logged in and how they pay will be specified in their customer profile. The generated purchase order is an XML format and as well as product order information, it will include information from the customer profile such as billing information, shipping information and contacts from that customer.

Next we must handle this generated XML purchase Order. Here the eBusiness Server Engine is customisable. Rules can be uploaded to this engine to dictate how any received XML is handled, including this purchase order. Options may include using the web services adaptor to integrate with specific ERP systems; this has maps between the xml formats (purchase order in this case) and the ERP backend, and can write the purchase order data to the ERP. As many ERP systems have specifications for importable files, an XSLT can be applied to the purchase order to convert it to a format of a purchase order that a specific ERP can import. The Purchase order may simply be saved to file (an XSLT can be applied to convert it to any more readable format) and a notification can be generated to inform that an order is awaiting entry. Also the Purchase order may just be emailed to the relevant order entry person.

This Framework works for more than simply the creation of orders and order processing. Using the same methodology we can handle queries from customers, such as quote requests or lead time queries. Similarly to order processing these are all converted into XML and inputted into our eBusiness Processing Engine, again here rules can be uploaded to handle the query in the manner required by this business. Whether we can respond to the query immediately after accessing the backend data source, or we notify a member of staff about the query that needs to be dealt with depends on what rules are running in the eBusiness engine.

2.2.3 Payment Handling

Due to current business operations different payment methods will be handled. Many businesses currently operate allowing credit to customers, and keeping track of their accounts, this functionality is handled in their ERP system, in this case there is little extra to do but to ensure all the required information is in the purchase order when it is being entered.

For non account-holding customers we will provide functionality for handling payments with PayPal or other commercial online payment agencies. PayPal, for example, allows anyone with an e-mail address to make a money transfer to anyone else with an e-mail address, whether the recipient has a PayPal account or not.

There are a number of ways PayPal can accept money from a purchaser:

- The purchaser can create positive balance in his account at PayPal by sending a check, and payments can be deducted from the account
- The purchaser's credit card can be charged
- A checking account can be debited for any payments

If money is received via PayPal, Businesses can request payment from PayPal by check, or can arrange to have PayPal deposit the money into his account. To make a purchase using the PayPal system all that is necessary is provide your name, e-mail address, credit card information, and the billing address for the credit card. To receive a payment less information is required, all that is necessary is account information if payments are to be directly lodged. An attractive feature of PayPal's business and premier accounts is that users have the ability to download Quicken/QuickBooks/Excel-compatible transaction logs, particularly valuable to SMEs, many of which may be using these accounting systems. For business and premier account holders A Web Accept button can be added to a company's website and payment can be accepted there, without traveling to PayPal's website.

2.2.4 Security

Security being key within any business-to-business information exchange, it has to be addressed at the first level of the development model. Here we must ensure the interaction between the end users browser and the Web server must be secure.

To secure the users interaction with the Web Server the widely accepted Secure Sockets Layer (SSL) is used. The SSL protocol, originally developed by Netscape, is the universal standard for authenticating Web sites to users, and for encrypting communications between Internet browsers and servers. All major browsers and servers on the market include SSL. All that is required is for the server to install a digital certificate, or Server ID, to enable SSL capabilities. This is a digital certificate allows Web browsers to verify the site's authenticity and it is used for SSL encryption, which provides a secure information exchange between browser and server.

For secure encrypted SSL communication, all data sent between a browser and a server is encrypted before it is sent and decrypted on receipt. There is also the facility to ensure data has not been tampered with, changed or corrupted, so communications can continue securely and both sides can be confident of privacy and therefore sensitive data, such as credit card numbers, and account information, can be sent safely.

2.2.5 Transmission of messages

Once a company has gone through the tasks of examining its information and data requirements for messaging and formatting this for XML messaging, there are a number of ways in which messages can actually be sent and received.

- As an XML file normally interchanged using the HTTP protocol, or one of its derivatives (e.g. Secure HTTP)
- In the form of a multipart Internet e-mail message (MIME or Secure MIME encoded)
- Internet File Transfer Protocol (FTP)
- Via a HTTP POST message that conforms to Internet's Common Gateway Interface (CGI) specification
- As an EDI message (this would be created by processing the XML file at source using special conversion programs).

2.3 Level 3

2.3.1 Level 3A

The next stage of development is to begin to automate a business's supply chain. Creating closer bonds with partners enables businesses to function more effectively. Being able to receive orders from business customers and deal with them, instantly sending responses, and order confirmation will improve relations, and this automation reduces order lead time and provides essential competitive advantage for both businesses, supplier and customer in the relationship.

2.3.2 Automating B2B

The Architecture for the B2B solution is essentially very similar to that of the B2C Solution (Fig.3).

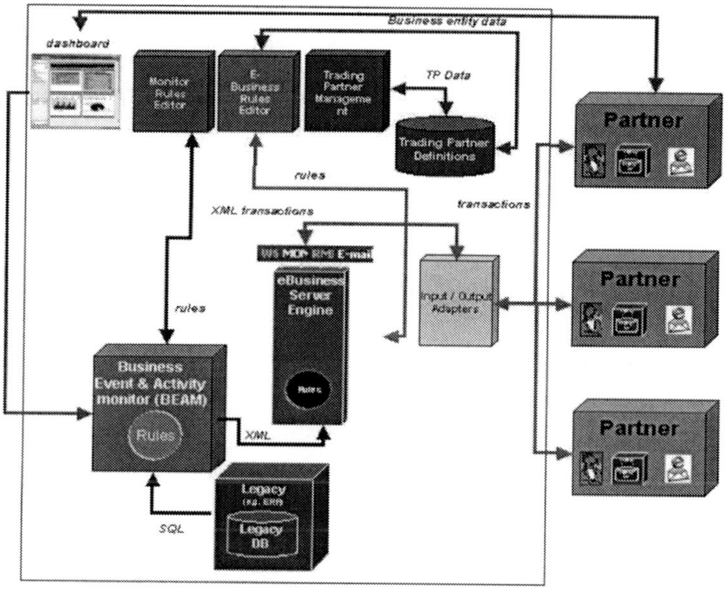

Fig. 3 B2B Architecture

Here the only change is that we now have a set of Trading Partner Definitions so partners can send us orders by whatever means and in whatever format they wish (ensuring not to limit who a business can trade with) and we simply use the profile to select the appropriate XSLT to transform the request to a generic xml format used in our systems. This may involve transforming between different XML formats or EDI to XML, also Email, Flat file, and Word could all be received from partners and are transformed into the correct xml format.

2.3.1 Level 3B

Now we extend this architecture to look at more than incoming messages. To further improve the business processes we must be able to fully leverage our current systems and information and maximize their value. Here we use our BEAM, on the backend systems, to monitor events, particularly relating to our supply chain (Fig.4).

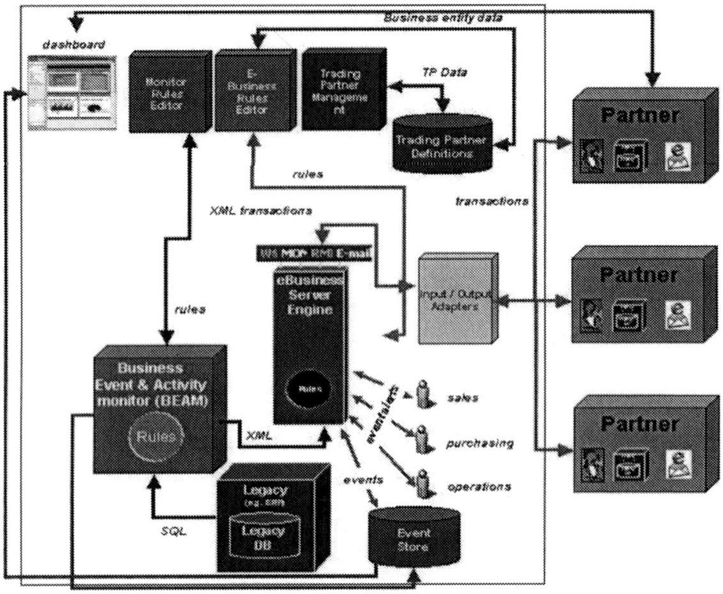

Fig. 4 Event handling

For an example of this, consider monitoring a table of a business's purchase orders. As soon as a new purchase order is entered in an ERP system, this data is pulled out and converted into xml, this is then passed to our eBusiness Server Engine for processing. To complete this example, the eBusiness Server Engine, having access to partner definitions, can select the required XSLT (if any) and can select the message transport technique, it then finally completes and sends the order to the partner. This shows many advantages over the alternatives of printing out orders from and ERP and posting, faxing, ringing or even emailing them to the partner. This can be used in other areas, for example when catalogues from suppliers become outdated, when payments are received from customers, input into the ERP and receipts are to be sent out, when order status changes and many more. Basically this an event driven process where the event can be any change within the ERP system, events logs are also maintained in the event store so partners can log on to the dashboard and see if the status of their orders has changed.

2.4 Level 4

A further step in eBusiness capability development would be to attempt to ease the process of starting to do business with new trading partners. At present before trading partners can begin to trade, large amounts of data have to be exchanged, XSLTs have to be written to convert between formats, and partner and customer profiles have to set up, all in all, a long process that can only alienate potential partners. We address this problem by publishing all the required information to set up the trading partnerships.

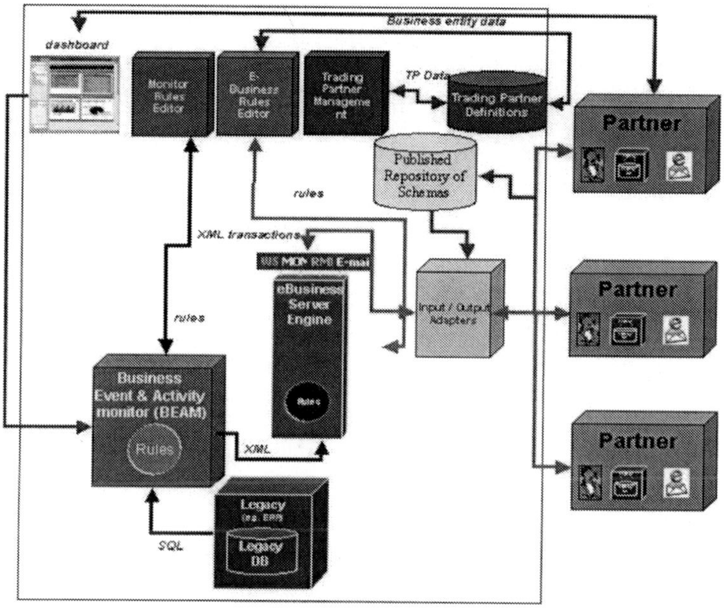

Fig. 5 Publishing Schemas

Any potential partners, suppliers or customers, can view the published documents. These documents should include the XML specifications for the entire set of standard messages for business-to-business trading. Information covering the message flow between parties must also be included (both parties need to know when to expect replies, confirmations, notifications etc). For this purpose it is logical to conform to an xml standard such as ebXML. ebXML is a joint project of OASIS and UN/CEFACT. The aim is to provide an open XML-based framework/infrastructure, focusing on creating a set of standards and specifications that define an interoperability layer, with the purpose of easing global communication and trade. With the knowledge that most large enterprises already have expensive systems in place the specifications are aimed at SMEs. The ebXML effort does not provide any reference implementation of its specifications, but rather encourages the companies involved to implement their own systems using ebXML's specifications.

2.5 Level 5

The final stage of this model looks trying to integrate more fully between an SME and its partners, to improve the overalls supply chains performance, hopefully improve services and products and pleasing the end customer (5). The goals are to reduce costs, reduce time for order fulfilment and dispatch, and eliminate any stock shortfalls. At this phase of eCollaboration partners begin to share information, suppliers can have visibility into levels of their stock within their customers systems, and forecasting information can be passed up the supply chain. With many ERP systems providing forecasting modules sharing sales and requirements forecasts with suppliers could be beneficial to both parties, the suppliers can make sure it has the capabilities top meet future requirements, and the SME ensures it does not fall short.

The sharing of forecasting data is accomplished by using the BEAM, this monitors any ERP forecasts, and Rules within the eBusiness Server Engine can process the output from the BEAM and use partner profile information to send off the forecasts to the relevant suppliers.

Dealing with XML information requests from partners provides visibility into stock levels. This is accomplished using partner profiles to see what information they have privileges to see. The eBusiness engine will then call the web services adapter that will retrieve the required data. The web service adapter again here uses SQL to XML maps to retrieve and format the information before returning it to the eBusiness engine to be dispatched to the requesting partner.

3.0 CONCLUSION

eBusiness is already spreading and increasing across the world. Large enterprises, having the essential investment funds, may be seen to more easily profit from eBusiness projects. However many applications, eSales, eProcurement, and Supply Chain Management applications require the co-operation and uptake of partners and suppliers. Therefore, in many situations SMEs may be under pressure from large enterprise partners to join or integrate with their eBusiness ventures. An increased take up of eBusiness within SMEs, which has been slow to date, is inevitable. The solutions presented in this paper provide an open methodology and technological framework for implementation of an eBusiness strategy. The XML based approach, and the number of possible message transmission techniques, means it is highly configurable, and possible to integrate with many systems.

REFERENCES

1. *Software Engineering Institute*, "The Capability Maturity Model: Guidelines for improving the Software process", Addison-Wesley, 1995.
2. *The US Census Bureau*, http://www.census.gov/mrts/www/current.html
3. *Nordic IT co-operation*, "Barriers to and Opportunities for E-Transactions in Nordic SMEs Analysis based on existing surveys", September, 2003.
4. *Carnegie Mellon University*, "e-Sourcing Capability Model for IT-enabled Service providers" v1.1. October 21, 2002. http://itsqc.srv.cs.cmu.edu/escm/clients/index.html
5. *IBM Retail and CPG Industry Solutions*, "Trading Partner Collaboration, Extending and synchronising the value chain".
6. *The European e-Business Report*, 2003 edition, "A portrait of e-business in 15 sectors of the EU economy", July 2003. http://www.ebusiness-watch.org/marketwatch/resources/SynthRep_2002-03.pdf

The security audit mechanism in the enterprise e-commerce platform

C-Y ZHANG, C-H PIAO, L ZHANG, and **L-J ZHENG**
Computer Department, Shijiazhuang Railway Institute, People's Republic of China

ABSTRACT

The security of the enterprise e-commerce platform has become an ongoing issue that urgently needs to be solved. To enable diversification, complexity and the determination of guilty parties through a network, it is becoming necessary to include an integrated security audit mechanism to protect the e-commerce platform. The checking mechanism in the security audit system records all abnormal or fraudulent behavior and helps the administrator to find the source of such incidents. This article firstly discusses the goals of the security audit mechanism, then goes on to describe these mechanisms for enterprise e-commerce platforms, based on research findings and the peculiarities of such systems. This audit mechanism includes five major areas, namely: system perimeter censorship, strong authentication, privilege management, intrusion alert, audit quoting and so on. The second and the third models described form the basis of this mechanism. The intrusion alert algorithm adopts the analogy reasoning based on grey system theory. In order to enhance the quoting ability of audit clues, we protect the audit data by means of cryptography and long-distance backup.

1 INTRODUCTION

The development of e-commerce is changing the way of human life. The internet is the main physical basis of e-commerce, but the open nature of it leaves many hidden security problems. So, the security of e-commerce systems urgently needs to be solved and has become the subject of recent scientific research. For the diversification, complexity and indetermination of guilty through a network, it has become necessary to include a self-contained secure audit mechanism to ensure the business system security, although some e-commerce systems employ security features such as identity authentication, encryption, integrity and non-repudiation. The clues kept in the secure audit mechanism records and checks all the abnormal or fraudulent behaviors and help the administrator to find the cause of the accident. The quoting quality of audit clue will impulse the unceasing perfecting and development of laws interrelated to e-commerce.

Most of the existing security audit mechanism researches focus on the field of secure operating system. In reference [1], an audit system of secure operating system is defined to record, examine and audit the activity interrelated to security in system. Its main purpose is to examine and prevent illegal users' invasion to the computer system and indicate the legal users' mistaking operation. A design and implement blue print of secure operating system was brought forward in the homemade secure operating system SecLinux by Liu Haifeng in literature [2]. Behavior checkup is an important part in secure audit mechanism. The audit model was brought forward by Dorothy Denning for the first time in reference [3]. This model is made up of three parts, such as event generation machine, behavior mapping and formula set. The abnormal behavior detecting model was sum up into three types, such as statistical model, forecast model and abnormity detecting model based on machine learning and so on. And the last of them is used very diffusely. The secure alert mechanism notifies the behavior detecting result to the administrators in time or takes urgent measure automatically.

Firstly, the function of the secure audit mechanism is brought forward in this article. Then, based on the research fruits in the security audit fields and the peculiarities of e-commerce system, the security audit mechanism for e-commerce system was described in this paper.

2 FUNCTIONS OF SECURITY AUDIT MECHANISM

Primarily, the function of security audit mechanism is to prevent external intruders (whether it is on purpose or not) and internal hackers. The security audit mechanism should establish an effective access control mechanism to store integrated audit records of all behavior to ensure the proof quoting ability after the event. This system holds five functions as follows:

2.1 Check the anti-penetrating ability of the system perimeter

The system perimeter of e-commerce is the exit of the system to Internet. It is the first defence equipment of e-commerce. The technique to construct system perimeter is commonly firewall. Firewall should have fine granularity access control ability. It means that it should not only filter the head of the datagram but also the protocol of application layer and the content of the datagram. There are three basic detecting measures we can take:

(1) Study the filtrating policy of the firewall carefully;
(2) Scan with commercial scanner "out of the firewall", the try to penetrate the firewall;
(3) Intrude into the system with different illegal identities and record the reaction of the system.

2.2 Finding out the inherent leak of the system entities

It can also be called "inner audit". It means appraising the ability of anti-attack of single entity of the system from inner side and protecting them in different grades. This kind of inner audit bases on the two following reasons:

(1) Not all invasive behaviors are from the outer system. Quite a lot of them come from inner network, including destroying on purpose (get secret information) and involuntary mistake operation.

(2) The purpose of the people committing the invasion is to control the important resource, but the safeguard of the important resource is so strange that they sometimes breach the host that has a weak safeguard to make it as the springboard for the next step.

2.3 Checkup the access control policy

Checkup to make sure that all the authorizations and confirmations of e-commerce are accord with the demand of the enterprises. Unauthorized user can not access the database of the enterprise; Authorized user can only access the data that they have right to access. Checkup the control measure mounted for keeping down the audit clue to make sure that it is working effectively [6]. Imitate some business parameters who want to access exceed their authority to see whether the system will reject them.

2.4 Seeing about the preserving ability of audit clue

Audit clue is effective law quoting and main data source of system dynamic defending, so strengthening the preserving ability of audit clue aims at the two kind of behaviors as follow:

(1) Repudiation behavior on purpose;

(2) Steadfast "hacker" behavior.

We won't talk about the repudiation behavior on purpose for it has been talked a lot in other documents. We should take special measure to ensure that the audit data of the system is useable, for the defending device can not stop all the invasive or destructive behaviors (especially inner attacks), and for the invaders always try to wipe out traces of themselves.

2.5 Dynamic detecting and real-time alert

With the ceaseless complicating of the computer information system and the diversification of illegal behaviors, the defending measure of the system has become more and more enormous. The whole of the defending system and the efficiency of the application system is becoming a pair of inconsistency. In order to enhance the efficiency of the system, the system should be able to judge the type of the invasion intelligently and use professional counterattacking (or holding up) tools automatically. When the invasion appears, system will send risk alarm to security officer in a appropriate way, even take some necessary urgent measures (shutting down the system for example).

3 THE DESIGNING MODEL AND IMPLEMENTING POLICY OF SECURITY AUDIT MECHANISM

To meet the function requirements mentioned above, we describe the architecture of every part of secure audit mechanism in e-commerce system and introduce the implementing method of every part.

3.1 Designing thinking

In this system, we propose the two steps of secure audit. Firewall works as a gate keeping system. The gate-keeper can checkup the identity and the access level of the visitor, then decide to let it in or not. Of course, it is necessary to make some records. This is the first step of audit. When the visitor get into the house, he or she may go into a wrong door purposely or not. So, we should do further and more strict identity authentication and privilege validation (employee's card for example) for the important resources (document room for example). As for the very important resources, the outing-passport is needed in addition.

The adoption of two steps audit is propitious to enhance the responding speed and heighten the security of important resources. As for the firewall, we can take some simpler policy. As for the strong authentication and privilege validation, we can use digital certificate.

3.2 Model of secure e-commerce system

In this paper, we take the e-commerce system use security audit mechanism as our secure e-commerce system. The position of security audit center in secure e-commerce system is as shown in Figure 1.

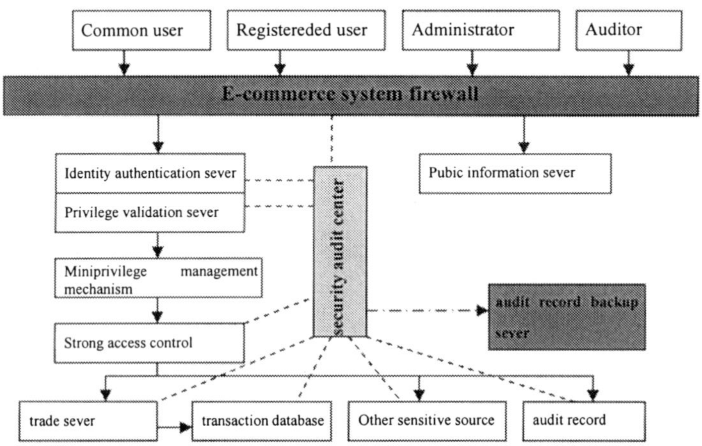

Fig. 1: Secure e-commerce Model

Note:
(1) Audit record backup sever had better be put out of the system (remote point),
(2) ———▸ operating steps,
(3) - - - - - - - setting audit point,
(4) —·—·—▸ remote point backup,
(5) This model has been developed with the help of reference [2].

3.2.1 User access steps

There are four kinds of users in this system, which are common user, registered user, administrator and auditor. Common user can only access the pubic information server via firewall (the other three kinds of users can also access he pubic information sever, but they do not need to pass the strict identity authentication). Administrator and auditor are two special types of registered users. Their access to system should pass a series of checkups, such as identity authentication, privilege validation, minimum privilege management and privilege access controller and so on. After these steps, they access the trade sever, trade database, other sensitive resources and so on according to the privilege they were permitted. Only the auditors have the right to access the audit record and audit backup sever.

3.2.2 Audit point Setting

In this paper, the host-based security audit system is adopted. In this system, the audit software is installed on the key hosts that need to be audited. These key hosts (or severs) are called audit points. As shown in Figure 1, 8 audit points is set in firewall, identity authentication sever, privilege validation sever, strong access control sever, trade sever, transaction database, other sensitive resource storage and audit recording storage point.

The outermost firewall filters all the users simply and only records exceptional behaviour. The identity authentication sever makes strong authentication to users and it requests to record the entering and exiting affairs and the corresponding failure affairs in details. The privilege validation server checkups the rights range and records detailed course information of privilege validation. The strong access control server executes the second-step access control. To special resources, the out passport will be checked. So it should record the way and sequence in which one user access the resources and what resources they have accessed. The trade sever is the center concourse of doing e-commerce activities, so the audit machine will record every transaction of the bargainers. Generally, user accesses the trade database through the trade sever, but the database administrator can access it directly, so the database shall record all the direct or indirect accesses to itself. The other sensitive resource storage audit should record all the accesses to the common secret data. The security auditor is the only legal access user to the audit record storage point and the audit record storage point will record all the access intentions and operating results to the audit records.

3.2.3 Structure of the security audit center

The functions in security audit mechanism are implemented by the security audit center. The structure model of the security audit center is shown as Figure 2. The structure model is composed by five tiers, (1) The first tier is the 8 audits illustrated in preceding context, (2) the second tier is a machine for obtaining audit data, (3) The third tier is four major services of the security audit center, (4) The fourth tier is a data transform to unified connector, (5) The fifth tier is multi-kind of data storage or the data source of other systems.

The machine for obtaining audit data that be placed in the second layer acquire the audit data from the first tier. This is the basic data resource that security audit center implements all the functions. This machine categorizes the primitive data with certain algorithm to prepare for the next step. This tier should follow the given audit data collects the principle to enhance the transplant ability when this audit system is implement on different system platform.

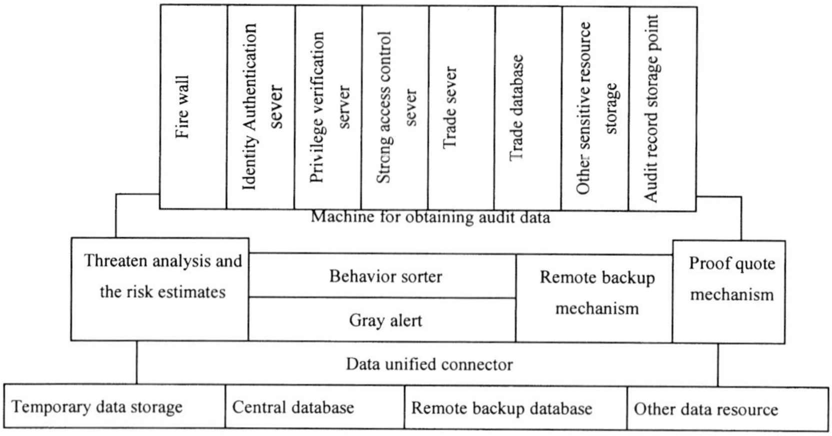

Figure2: The structure model of security audit

As central service layer, the third tier include four main services as follows,

1. The threat analysis and the risk estimates.

   By the periodical analysis on audit data, it estimate the system to find the potential weakness in itself, then make sure the risk level and put forward the platform strengthen suggestion. At the same time, it generates a risk estimating report automatically.

2. Intrusion alarm system.

   On the bases of analyzing audit data by the quantitative analysis of the behavior sorter, sorted the target behaviors checked out by the system intelligently, and make sure whether it have the intrusion characteristic. As the intrusion behavior is detected, the system will adopt an appropriate alert way or an prescriptive urgent measure.

3. Remote backup mechanism.

   Remote backup mechanism is used to enhance proof quoting ability. Under the condition of disaster taking place, this mechanism can reserve the audit clues to assist to investigate the accident reason.

4. Proof quoting mechanism.

   This is an aided mechanism, but it is an important function in this system. This mechanism can give convenience to audit manager to checking the whole audit data, and the data can be quoted visibly and quickly.

The fourth tier provides a unify data connector to separate business logic and the data resources. This model can enhance the transplant ability and the disaster tolerance ability. The fifth tier is special data storage way, such as database, file, other system, etc.. They establish relationships with the system via the fourth tier.

3.2.4 The basic structure of the intrusion alarm system

In this paper, the intrusion alarm system adopts an analogy reasoning method bases on gray system theory. Its technique route illustrate in Figure 3.

(1) Research on human's analogy study mechanism, and consult the running characteristics of the network system, an mathematics model of network system behaviors was established in this paper. The numerical value was used in this model to increase its expressing ability. (2) According to mathematics model, a normal behavior instance database and an abnormal behavior instance database are established. These database are the foundation of analogy reasoning, so they must be extensible based on self-studying. (3) For a particular target behavior, an similar analysis method based on gray system theory is adopt in this model. This method judge the intrusion character in the target behavior in synthesis base on Normal Assumption and Abnormal Assumption. We import in acquaintance degree and decision-making authority into the analogy reasoning method to enhance the efficiency and veracity of the behavior detection. (4) According to the result of the analogy reasoning, implement the real-time alarm or put forward the most appropriate respond suggestion.

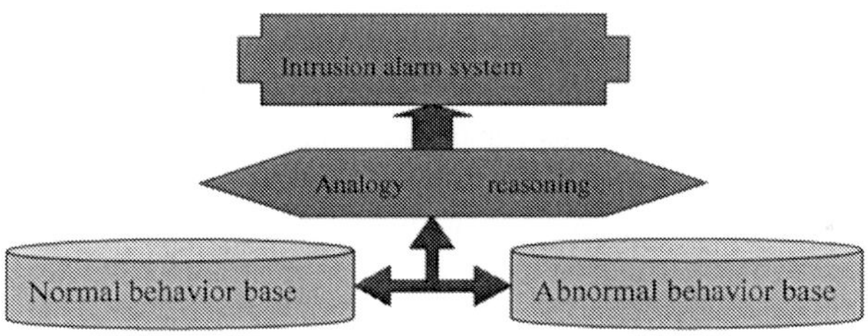

Figure 3 Basic Structure of Intrusion Alarm System

3.2.5 Proof quoting mechanism and visible audit tool

The effect of security audit usually depends on the audit experience of the security auditor. According to the professional judgment, the experienced auditor will discover intrusion from the very slim abnormal clue. Developing an intelligent security audit tool is necessary to accumulate the experience. The security audit tool should have these functions, (1) Collecting the information of audit affairs from each audit point and judging the abnormal type, (2) Compile the audit report automatically to help decision-making and to reduce the audit risk, (3) realize trail audit to visiting entities, (4) alarm to auditor or the bank automatically and adopt the necessary urgent measure, (5) study by itself.

3.3 The operating right of audit system

In order to guarantee the independence of security audit system, the mini-privilege mechanism is adopted here. The security audit task is separated from the administrator's task. System creates three different roles, which are system administrator, security audit manager, security audit operator. To different role, (1) adopting a strong identifier authentication bases on X.509 certificate [7], (2) adopting the privilege management based on attribute certificate [8]. System manager has the right to open the audit function, and the security audit manager has the configure policy and close audit function, and the security audit operator has the right to read and write the audit logs.

3.4 The design strategy of the audit logs

The system adopts the following strategy to raise the integrity and reserve ability of the audit clue. (1) Setting 8 audit points. They are firewall, identity authentication server, privilege validation server, strong access control sever, trade sever, transaction database, other sensitive resource storage and audit recording storage point. (2) The dynamic audit affairs. According to different visitors and different resources level, system can decide which affair should be recorded and how to record it. (3) The management of these records. It includes the record encryption, record integrity checking, affairs signature mechanism, security time stamp, strict access control, segmentation measure of the record, appropriate redundant records and so on. (4) Logs counterfeit. Keep default logs at the default place, but monitor them strictly. (5) Remote backup the audit data. The system makes at least one backup of the audit data outside of the system, to harden the protection of the illegal juggling.

4 CONCLUSION

The security audit mechanism is a necessity supplement of the security technology in e-commerce system. According to policy avowal of the service provider and national laws, the proof quoted in security audit system will have the law force. Under the protection of law, even by dint of the insurance company, the trade risk can be reduced in the e-commerce. With the number increasing of the people to access Internet and the constantly improvement of financial authentication system, many kinds of security protocol was put into use. Then, the e-commerce will have a quiet great progress in China. We should believe that e-commerce will change the traditional business model radically in the near future. The people's life style will be changed greatly. Let us enjoy the security network civilization well.

REFERENCES

1 "A Guide to Understanding Audit in Trusted Systems", NCSC-TG-001, ver2, National Computer Security Center, 1988.

2 Liu H.F., Qing S.H., Liu W.Q., "The Design and Implement of the Audit in Secure Operating System", Computer Research and Development, 2001.10 (38) 1262～1268

3 Denning, D.E., "An intrusion detection model", IEEE Transactions on software engingeering,1987.

4 Zhang J., Gong J., "A Survey of Intrusion Detection Methods", Computer Science, 2003, Vol.30, No.2, pp. 97-99.

5 Deng, Y., Chang, G.C., Wang, X.H.., "The Design and Implement of Network security Monitor and Audit System", Computer Engineering, 2002,12, pp195-197

6 Wei, P.W., "The Main Control and Audit in Electronic Commerce", Journal of Sun Yat-Sen University (Social Sciences edition), 2002, 1 (42), pp103-110

7 Housley,R., Ford, W., Polk, W., Solo. D., "Internet X.509 Public Key Infrastructure Certificate and CRL Profile", 1999.

8 Farrell, S., Housley, R., "An Internet Attribute Certificate Profile for Authorization", 2001.

The electronic business strategy of small enterprises

C-H PIAO, C-Y ZHANG, and **L-J ZHENG**
Computer Department, Shijiazhuang Railway Institute, People's Republic of China

ABSTRACT

There are more than ten million small enterprises in China. How to improve market competency in the information age is the concern of all businesses. Based on analyzing the present situation in China, this paper discusses several problems that should be given attention when starting an e-business as a small enterprise in China. It not only describes step-by-step the way to build a tentative e-business application but also gives an introduction to how to gradually extend the application in depth. Finally, this paper proposes a low cost and easy to deploy security solution that is based on a digital certificate at the appropriate security level.

1 PRESENT SITUATION OF SMALL ENTERPRISES IN CHINA

There are approximately sixteen thousand large enterprises and more than ten million small enterprises in China. Almost all of the large enterprises have used computer information systems that run in different levels. But most small enterprises have no any information infrastructure. Those small enterprises that have purchased computer hardware or have formed LANs have not developed software for applications other than financial management. A few small enterprises have invested in information systems in recent years. They have introduced or developed some management application software and built websites. But only a few have successfully integrated all of their internal applications. The websites are only used as platforms for information distribution or collection.

According to the statistics, the output value of small enterprises makes up 60 percent of the total industry output value in China. The degree of informationalization of small enterprises directly influences whether the whole country's economy can be developed and reasonably sustained. Along with the coming of the information society and sharper market competency, many small enterprises in China have emerging intrinsic demands for informationalization. It has been widely realized that small enterprises must use information technology to improve

G004/053/2004

their competency and increase their economic performance and allow them to proceed to electronic business.

2 THE MAIN PROBLEMS OF ENFORCING E-BUSINESS

Implementing informationalization and developing e-business are restricted by many conditions in small enterprises in a developing country. We can give some main reasons as following.

First, many managers of small enterprises have not fully recognized the importance of informationalization. Although some people have correct understanding of the current situation it is difficult to combine the enterprise's overall goal with informationalization as an organic whole. Secondly, the dominant management software in the market usually costs more than 10,000 US dollars. The government cannot give energetic support to small enterprise in China. Small enterprises find it difficult to afford those expensive e-business solutions. Thirdly, small enterprises are seriously short of professional understanding, not only operations and management but also modern information technology. So it is impossible to enforce a complicated informationalization solution. Fourthly, a lot of small enterprises think that they have achieved informationalization once they have purchased computer hardware, formed a LAN and deployed management software. They also think that setting up a company website means they are conducting e-business. They don't recognize the real connotations of enterprise informationalization and how to reflect advanced management ideas by making use of information technology. Fifthly, the awareness of network security has increased in recent years but it is impossible to deploy expensive and complicated security strategies in small enterprises because of the cost and lack of computer professionals.

3 KEY ESSENTIALS OF STARTING E-BUSINESS IN SMALL ENTERPRISES

3.1 Select appropriate entrance

Small enterprises must try to find a breach has a low cost and is practical according to its requirements and capabilities at the beginning of informationalization. Through enforcing such an information system, the enterprise can feasibly realize the notable effect of informationalization. We can select the breach based on the following principles: The business field of not only being of great significance and fundamental impact on the operation and management but also has considerable quantity of data and demands real time, accuracy and continuity; a new business can directly use information technology in order to avoid reconstructing again; some independent business can also be selected as the breach when the extensibility and integrity are fully considered; some department with higher quality stuff can also first implement informationalization.

Generally speaking , small enterprise can build information system from the following aspects: Implementing fundamental financial management system. Financial software has evolved many years in China. Now most financial applications have transferred from examining and calculating orientation to management orientation. At present, management-oriented financial applications are widely developed and used. Popularizing financial management systems in small enterprises does not run risks because the financial system is quite fixed in China and corresponding software system are very mature.

Implementing office management automation by making use of internal network. The fundamental goal of informationalization is aimed at improving information communication in the enterprise or with the external world. Through building LAN and implementing office automation, information can be rapidly delivered and widely shared.

Building website and exploring e-business step by step. The website is one of indispensable signs of an enterprise in the information age. Because Internet is rapidly popularized in the global, the website can not only present the images of an enterprise but also increase its popularity and contribute to strengthen the interactive communication and relationship between the enterprise and outside world.

3.2 Understand the relationship between advanced and functional technologies

How to give consideration to advanced and functional software developing platforms and tools are very important when a technical solution is planned. Neither concentrating on advanced alone nor functional alone technologies is desirable. According to the practical condition of the enterprise itself both factors should be considered carefully. Enterprises lay stress on the final enforcing effect of the technical solution. For example, some private enterprises have evolved into small group across regions and industries along with the development of market-oriented economic. Further the information communication among a group by making use of information technology is required. Now we will take this as an example to discuss how to plan an appropriate technical solution.

By studying different remote information communicating models such as telephone dialing, ADSL connection and VPN it is obvious that ADSL connection model cost low and functional at the general headquarters. Considering extendibility and usability of the communicating system, advanced Browser/Web Server/Database architecture based on J2EE can be used and JSP as the main developing technology.

Although these small private group enterprises have few computer professionals, some of their staff and workers can skillfully use word or spreadsheet processing software such as Word and Excel. To minimize the cost each branch of group may be equipped with only one microcomputer and one ordinary telephone line. To avoid influence the normal operation of a branch the communicating time between general headquarters and the branch through the line should be as short as possible. So it is not appropriate to fill in forms and then submit data after dialing connection. Therefore the staff of the branch can prepare the data that should be submitted in Word or Excel formats. When the connection is built between two parties by dialing these data files are submitted immediately. This submitting function can be realized by coding corresponding classes uploading data in Word/Excel formats or sending these data in e-mail accessories. Such a way of preparing and submitting data is feasible and simple.

At the general headquarter the server translates accepted data into database records by running corresponding JavaBeans through JDBC. Then any kind of easy and visual requiring and analyzing functions can be further provided.

3.3 Business credit and investment profit

Since existence of current unsound commercial credit system in china, enterprise should pay attention to the risk of its investment. European developed countries trade with other in the way of completely surrounding third-party commercial bank, while business in China mostly just includes buyer and seller. Required by Chinese government, Organizations must trade by transfer accounts or telegraphic money order telegraph, this practice mainly for enhanced the government superintendence, commercial bank assumes no risk of the bargainer, that is to say, there is not any format of protect for behalf. Therefore, small enterprises must decide which first, payment or delivery, depending on the commercial credit degreed and other conditions of its bargainer.

E-business provides chance to the people in the meantime it also occurs new risk. Now, the amount of the people surfing the Internet is very small, the product sold on the Internet is also limited. The chance to be found by the roaming user is very high, this will all influence the investment productivity of the enterprise. According to the analysis of relevant experts, in china, the E-business can save trade cost about 76.59%, in practice it only save 11.61%. Therefore, during the deployment of E-business of small enterprise, in one side we should pay attention to adopt B-B model to engage in operation transaction, in this way, it can not only confirm the identity of the both sides, reduces the risk, but also the amount of the transaction is big, the investment productivity is high. In the other side, it can exert its own flexible advantage, to deploy E-business accordingly.

4 THE REALIZATION OF E-COMMERDE STRATEGY IN SMALL ENTERPRISE

Each enterprise can accord to its own different situation, to realize different E-business strategy. But, generally, you can follow the next phrases to instruct the enterprise E-business application environment.

4.1 The initial phrase

4.1.1The construction of the website

The first step for small enterprise to realize the E-business is to construct a network. Firstly, you must register the domain name of the enterprise. The enterprise can according to its own situations to select paid domain name apply or free domain name apply. Since the security and stability of the free domain name is very bad, so for longtime deployment of the e-business, small enterprise should select the former one. Secondly, you should select the model of network (you can use the DDN, Virtual host and server entrust etc.). Then, the enterprise should create the web pages. On the enterprise network station, it should contain user fed-back forms, email, enterprise member registration, commodity query function etc.

4.1.2Propagate of the website

After having built their own website, medium or small sized enterprise need to propagate widely, as to enhance to the customers' recognizant level to the web site brand .The propagating method chiefly are as follows. Firstly, register the enterprise's web address and product key word at some well-know search engineer website, that is, people could find the enterprise's website easily and go for access, thereby their know the enterprise's general situation, product and service items. In order to enable the website to own more attraction, the enterprise should put attention to modifying the homepage content at any time. Secondly, send the webbing information to the customer from E-mail, or exchange advertisements to build links with other website. Thirdly, from the associated webbing with the other enterprise in the same industry, makes the propagating effect more outstanding. For the medium or small sized enterprise, classifying according to the industry standard 、 then free association ,and associate to hire data special to web can make the propagation more pertinent and effective.

4.1.3Implement the trade

Medium or small sized enterprise's purpose to implement the E-business is to complete the trade. But E-business in China is only at the preliminary stage, so it doesn't exit the condition to make all the trade from the Web, so it needs to base the fact to build a trade system that is adapted to the enterprise's trait.

First, medium or small sized enterprise can collect the customers' information from building Customer Web Forum, preparing for future trade. Secondly, querying on the web and trading offline at present is mainstream. The imperfect Condition such as payment through net, the distribution of material stuff, trade safety, the trade ruler and regulation, it makes the medium or small sized enterprise complete the trade the main query the merchandise from the homepage or in the form of telephone and mailing to aid to complete trade or the two size trade directly.

4.2Go deep into the development of the e-business application

4.2.1 Develop the commerce application and organize market spread

If the enterprise's purpose is to utilize its own website to develop the commerce activity, implement E-business, then it also needs to build application system which can carry through commerce activity at the website. For example a website for merchandise sales, needs to build a special system of merchandise search, stock, payment, order, dealing and customer service .

In addition, it also needs to take market propagating for the website, as to enhance the customers' cognizance develop to the website brand.

4.2.2 Improve commerce operation steps, accelerate the information building of the enterprise

The enterprise owns website, and can complete merchandise sale on the net, it doesn't to say that it complete E-business system building. E-business is the integration and digitalization of implementing the whole trade activity, and is putting electronic tools into the enterprise's production, sale, customer service and the application chain. So, it should accelerate information synthetic construction of the enterprise.

4.2.3The utilization of the external resource

The present drastic competition makes the enterprise when manage the interior resource effectively, at the same time, it must pay more attention to the exterior work, for example tracing the customers' information, developing market, osculating the providers' contact, as to the guarantee the first reflect to the market change, and get the one-up status. Such as implementing provide chain Management and customer relationship management between the enterprises.

4.2.4The fifth step: build e-business community

The e-business community is a multiple alliance, trade and transaction coordinated environment based Internet platform. Full utilization of the information share and real time alternation provided by Internet, completing coordinated commerce operation, reducing medial process can make the trade complete at one time. Now the some electronic communities in the international which focus on the traditional enterprise are forming, for example, General Motors and Ford Motors are trying to build car commerce community.

The steps expatiate before is the necessary process of the E-business building from early days to completion. But to the many enterprises (no matter the scale, fund, benefit) the implement of the course is not an easy thing. Some large enterprises in the developed countries use about 20 ~30 years, but to the Chinese enterprises, it is an arduous task to shorten the gap with the foreign advanced management mode and method in a short time.

5 THE SECURITY STRATEGY OF E-BUSINESS INTRODUCED BY SMALL ENTERPRIZE

The security of e-business is directly connected to the benefit of all sides whose concern the electronic trade, not only relating to technical problems, but also relating to management and legal problems. The technical defense measures of most china's banks apparently can not accommodate the need of large-scale e-trade; the managing standard of e-business has not been finally confirmed; the laws used to cognizing contraband of e-business is still on fumbling phase.

5.1 The function of government on e-business security

The government should create well condition for medium or small sized enterprise to develop e-business. The healthy development of e-business needs corresponding policy, statute, system, and needs government to create security authentication organ and any other e-business environment. The government also needs to institute some preferential policies that will promote the development of China's e-business.

The government should start with three aspects, to create the integrated safeguard system for safe operation of e-business. Firstly, organizing professional to filtrate e-business security technology that accord with the situation of China. For the moment, network security products that we used are mostly "exotic", the problem that we can not avoid is to develop network security products of our own. Secondly, consolidate the e-business security management, standardize dealing between buyer and seller .for the time, the first thing is to constitute formal e-business standard simultaneously, relative management standard should be distributed against time. Thirdly, we should consummate the e-business statute, confirm the law relationship and responsibility among each side of the dealing, sternly hit illegal dealing. Some of those problems are international ones, which should be confronted and solved by the whole world. A feasible solution is to create an independent international entity, whose responsibility is to use a program that is acceptable by the whole world for management and correspondent.

5.2 The security strategy that is the same with small enterprises

The e-business system security technology mainly include: data's security guarantee (data encryption) - the security guarantee for the dealing and the defray. The consummate security resolution project requires a great financing devotion, also needs the complex implement strategy. This is not adapted to the Chinese medium or small sized enterprise. For instance, well known that the SET protocol security is good, it must be the industrial standard of the e-business and the security communication. But the SET protocol's certificate is complex, also cost is high, the payment fashion and authentication beseem using card payment, it is restricted for the other payment fashion. In China, use of the card is far unpopular, plus the costliness expense and the complex deploy mode, it can't be apply to the e-business of the small size corporation.

Here, we only discuss the E-business security scheme that is available to the situation of our country, and has strong availability. This security strategy doesn't need adding any hardware devices. We mainly consider two aspects security: the protection of the user's password and the data's security transfer. To use the MD5 message digest algorithm to realize the protection of the password, we a user log on a network, he send the generated password digest to the server, then the server check the accuracy of the password digest of this user, make sure the corresponding purview of the user, avoid the risk of transferring the password in plaintext.

In the recent two years, with the support of the china government, many provinces build E-business certificate center with authority and equality. The enterprises only need little annual cost to apply the server security certificate to the CA authority center, and gain one-way authority service. Therefore, in the E-business application of the middle-sized and small enterprises, in order to resolve the problem of data's secure transfer, they can adopt SLL technology. SLL technology offers a complete solution scheme on identity authentication, data encryption compression, data integrity, it can not only support one-way authentication(server identity authentication), but also two-way authentication, client-server, by using client browser-server and CA authentication center to offer reliable secure guarantee. SSL security mechanism can be configured on the server, be realized quickly and the encryption procedure is transparent to the user.

REFERENCES

1 Zeng Zi., "Security Analysis on Electronic Business System", *Journal of Guangdong University of Technology,* Vol.18 No.3, pp 22-25

2 Cen, Zhao-yong, Zhai, Jian-she, "Implementation of a J2EE-Based Complex E-Business System", Vol.13 No.3, pp 34-36

3 Chi, Guo-tai, Yang, Bo, Yang, De-li, "The Integrated Management and Control Mode of the e-Commerce", *Proceedings of 2003 International Conference on Management & Engineering,* Vol I, August 15-17, 2003, pp 13-21

4 Shi, Shan-chong, Han, Kuan, Zuo Zhan-ping, "The Systematic Thinking on the development of Chinese E-Business", Proceedings of 2003 International Conference on Management & Engineering, Volume I, August 15-17, 2003, pp 347-353

5 Jin Feng, Cai Jia-mei, Zhang Fan, "Analysis and Description of Basic security problems in e-Commerce System", *Computer Engineering,*Vol.29, No.7, May, 2003, pp 22-25

6 Wang Qian, Yang De-li, "Research on the Secure Techniques & Architecture of Electronic Commerce", *Computer Engineering,* Vol. 29, No.1, January, 2003, pp 72-75

7 Liu Chun-ling. "Discussion on the security strategy for e-commerce system", Liao Ning Taxation College Journal,.Vol.15, No.16, Dec.2003, pp 28-29

8 Ye Hua, Yi Zhi-jian, "Consideration on Implementation of EC for Chinese Enterprises", *China Telecom Development Forum,* pp 3-7.

Computing Infrastructure for e-Engineering

Scalable architecture for multi-device Internet systems

M RAMACHANDRAN
School of Computing, Leeds Metropolitan University, UK
R LEWIS
Volantis Systems Limited, Guildford, UK

ABSTRACT

The aim of this paper is to describe how to design a scalable architecture for a multi-device internet system. The requirements for these applications come from various stakeholders with different browsers and hardware capabilities. The system needs to be designed such a way that it can cater for those requirements and constraints that are imposed by different devices. The role of an architect is to transform those product requirements into realisable functional requirements. We employed a range of architectural design techniques which are a combination of object-oriented architecture and automated design patterns.

1 INTRODUCTION

Internet technology will become ever more pervasive as software design, architecture, and technology evolve to meet the needs of Internet-scale distributed hypermedia applications. Modern WEB and WAP architectures emphasize scalability of component interactions, generality of interfaces, independent deployment of components, and intermediary components to reduce interaction latency, enforce security, and encapsulate legacy systems (1).

Apart from computers, the market for Internet access from various other devices such as mobile phones, PDAs, IDTVs, etc. is increasing rapidly. Demand for Interactive TV and Web access is growing faster than predicted. Many devices are rapidly emerging, for example iDTV, NDS, MSTV, NetBox, Sky Digital, Bush, set-top boxes, etc. It is expected that there are about 7 to 8 million users in the UK for Sky Digital and it is about 50 million iDTV users in the US. This multiplicity of device platforms poses difficult challenges for web application developers. Volantis Systems Ltd, as specialists is multi-device and mobile technology, aimed to develop a system which is capable of deploying Internet applications on any specified device that is known to the system. Figure 1 shows the overview of a multi-device internet system which is capable of generating internet applications automatically for any device of which it has knowledge. The goal in designing such a system is "design once and deploy across multiple devices". One such product known as Mariner provides a solution for pervasive multi-device internet applications. One of the key concepts of Mariner is the separation of content and its presentation. Content (consisting of static and dynamic parts of web site information) is the basic information that a site needs to deliver to a user.

It may be static information, for example a series of paragraphs describing a company. Alternatively, it may be dynamic, for example part of a product catalogue, retrieved from a database in response to a customer enquiry. Layout is the physical placement of content on the display of the user's device. Whereas content is essentially device independent, layout is device dependent. The optimal position for the information being displayed varies depending on the characteristics of the device's display. Thematic information governs the stylistic aspects of a site. We term the totality of this information, across multiple devices, a *theme*. A theme represents the stylistic aspects of the information to be presented and its appearances.

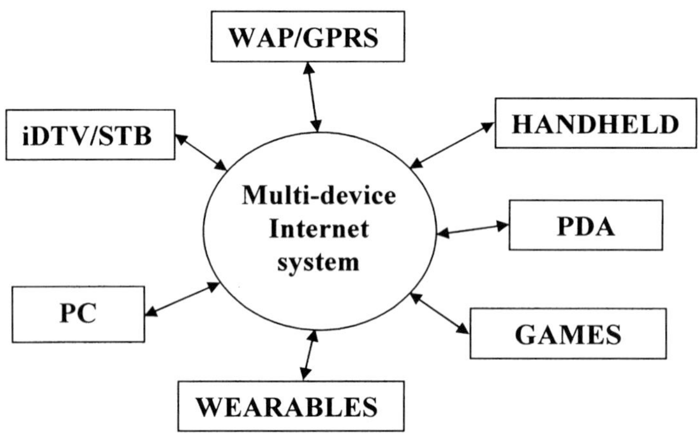

Fig. 1 Multi-device Internet system

2. DESIGN PRINCIPLES

There has been interesting work carried out on architectural design for mobile and distributed internet systems. For example, Bagordia et al (1) has proposed a middleware layers placed between user clients and application servers to perform a variety of functions to support the vision of nomadic computing across varying platforms. In our work, we have designed a number of application-specific protocol layers based on an object-oriented design strategy. We believe this combinational approach provides greater flexibility for reuse and design (2).

For each device there needs to be a variety of knowledge available to identify the device obtain technical information such as browser version, hardware constraints etc. We call this *device knowledge architecture*.

The thematic information for a device is termed a *style*. A style represents device specific information in order to transform the information which is to be presented on that device. So, for example, a corporate brand would be a theme, composed of styles on each device. In the Volantis model, content is authored once, and then tailored to different devices using a combination of layout and style. This contrasts with other schemes in which only content and style are considered., for example the use of style sheets, with HTML or XML. Thematic information provides the overall look and feel to a web site. Themes, in Volantis, are collections of styles. Styles are device-dependent. Themes are device-independent. The requirements for dynamic generation of appropriate pages in response to user requests, obviously demands a significant amount of information about the target device. Other information also needs to be held, for example, thematic information associated with various devices. The individual pieces of information used by a Volantis system when rendering a page are collectively known as policies.

UML (Unified Modelling Language) was used during the design for each customer requirements to scale the architecture for specific needs. UML was also used to analyse software patterns and frameworks.

3. MULTI-DEVICE ARCHITECTURE

The principles of good architectural design will improve scalability, flexibility over a product line. Our previous work on architecture design for such system involves building a number of reusable components based on EJB and other architecture patterns. Figure 2 illustrates a generic architecture developed for the multi-devices internet system. In this architecture a device is identified using JSP query tags which are then passed on to the device knowledge. The system then invokes appropriate themes, styles, presentations, and generates a web page which is suitable for that device. The theme for that device (marked Theme → DeviceTheme in the diagram) is inherited from a general theme and a style for that device. The style (Style → DeviceStyle) is also inherited. It also has the hierarchical knowledge for a range of products, for example, devices such as a mobile phone are manufactured by different manufacturers and each model uses different technologies such as Web, WAP, I-mode, GPRS, etc.

4. EXECUTION ARCHITECTURE

The application server provides a JSP 1.1 compliant container, supporting Java Server Pages (JSP) with custom tag libraries. Within this container, the Volantis services and components execute. When a request is received by the application server, it executes the appropriate Java Server Page. This page has been authored using the Volantis custom tags. As the page is executed, the custom tags are processed to generate the output that will be returned to the browser. Early in the execution of the page, Volantis identifies the device from which the request has been received. This allows it to select the appropriate thematic information and the appropriate layout. Layouts and themes are just part of the information that is stored in the Volantis Repository.

As the Volantis tags execute, they use a range of java classes, together with the information from the repository, to generate the page output. They merge the content from the tags with layout and thematic information to create device specific pages that take account of the different capabilities of the different devices. Application servers often provide additional features for improved reliability or performance. An application that runs in a Volantis system can take advantage of any such facilities. At execution time, the Volantis system behaves just like any other Java Server Page-based application, totality of this information, across multiple devices, a theme (represents the stylistic aspects of the information to be presented and its appearances).

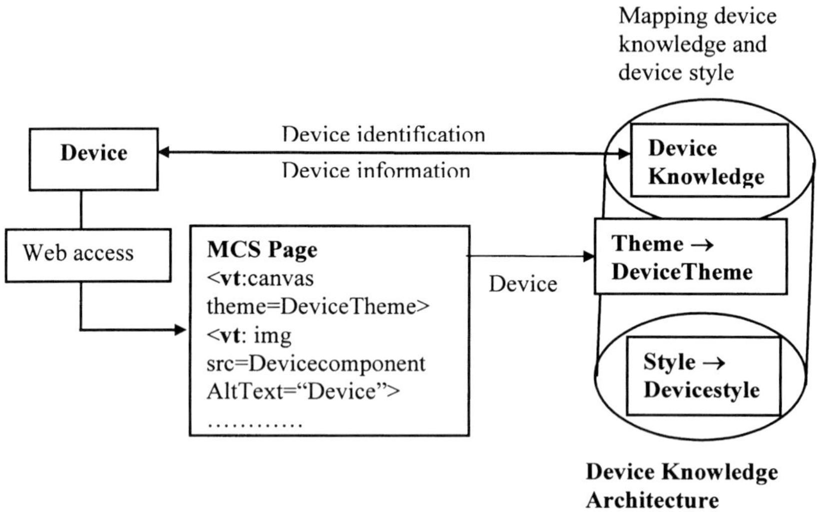

Fig. 2 Scalable Architecture for Multi-Devices

5. LESSONS LEARNED

From our experience in developing this system we identified the following general principles and guidelines:

Identify a development process that is suitable for that application and customise it for your organisation and the nature of the business.

When developing multi-device leading edge applications, initially identify a list of recurring patterns of components and architectures, which will simplify your problem and provides a solution for a range of variations.

Use a mix of architectural approaches. Don't stick with one only because the nature of problems vary at different levels of abstractions

As an Architect, first look at the requirements and their transformation into more realisable notations so that developers at all levels can understand and appreciate the architectural design. Make sure testing is done in parallel with architectural specifications.

6. CONCLUSIONS

A scalable software architecture has been developed that allows a web application to be deployed on any web-enabled device.

The approach is based on the separation of content and presentation, and supported by a knowledge base architecture that allows device independent themes to be transformed into device dependent styles. A combinational approach is used to handle a variety of Internet protocols such as Web, WAP, i-Mode, GPRS, etc

The architecture was developed using an object-oriented design strategy that makes use of software design patterns and frameworks.

ACKNOWLEDGEMENT

The authors would like to thank the Architecture, Development and Requirements Engineering Groups at Volantis Systems Ltd., where the work described was carried out while the first author was working as an Architect.

REFERENCES

1. **Bagrodia, R., Phan, T.** and **Guy, R.** (2003) A Scalable, Distributed Middleware Service Architecture to Support Mobile Internet Applications, Journal of Wireless Networks 9, pp 311–320
2. **Ramachandran, M.** and **Fletcher, M.** (1996) Large scale software reuse: an industrial case study on telecommunication switching system. Proc. Intl. Conf. on Software Reuse, Florida, USA
3. **Stallings, W.** (1999) Integrated Services Architecture: The Next-Generation Internet, Int. J. of Network Management 9, pp 38–43
4. .NET framework, Microsoft Corporation, 2000.
5. **Fielding, T. R.** and **Taylor, N. R.** (2002) Principled Design of the Modern Web Architecture, ACM Transactions on Internet Technology, Vol. 2, No. 2, pp 115–150.
6. **Ramachandran, M** and **Mangano, D.** (2004) Knowledge Based Reasoning for Software Architectural Design Strategies, ACM Software Engineering Notes, March 2004.

A face-based computer login system

S KING, G Y TIAN, and **S WARD**
School of Computing and Engineering, University of Huddersfield, UK

ABSTRACT

With the wide use of computers, it is difficult for end-users to cope with the many usernames and passwords they are exposed to. This paper introduces, using a low cost web-cam and face recognition algorithms, a face-based computer login system. By extending our work on robust face recognition, the system construction is introduced. The system performance is evaluated. Conclusion and further work is also presented.

1 INTRODUCTION

1.1 Biometrics for identification

Increasingly biometrics are being seen as important tools for security and identification, capable of providing enhanced security at sensitive locations or as a replacement for current token password systems. An example of the growth of biometrics for identification can be seen from their adoption by passport agencies worldwide. Of the different biometrics available face recognition is considered the least invasive and because it mimics human behaviour, potentially the most socially acceptable.

The adoption of biometrics for the purpose of logging into a computer has a number of benefits. The primary benefit is one of security; current login procedures based around token-password systems can be used by incorrect personnel and are being increasingly hacked by fraudulent individuals. A person's biometrics however are specific to them and cannot be so easily transferred, copied or stolen. The second benefit is one of convenience; users would no longer be required to remember a number of different passwords and usernames or to carry tokens for identification such as swipe cards. Previously the biometric of choice for such applications has been fingerprints [1] or iris scans [2]. We however argue that face recognition is a more flexible and convenient biometric, allowing a persons identity to be scanned and monitored without any input from the user or for them to even be aware of the process. Face based login systems will have wide applications in granting secure access, particularly applications in e-commerce, on-line collaborative engineering, online learning and access to e-engineering systems.

At the same time as improvements in face recognition and other biometrics have brought them to the attention of security conscious organizations, advances in computer hardware and falling camera prices have made it possible for the average home or office PC to perform real-time face recognition. We therefore now introduce a real-time face recognition system, capable of working in a small office or for a home-user, which is able to run on a typical PC using a low cost USB web cam.

1.2 Face recognition

Approaches to face recognition can be broadly categorised into one of two classes: feature based and image based [3]. Feature based [4] approaches to face recognition use angles and distances between key facial features along with shape to determine a person identity. Feature based approaches are computationally expensive and require accurate localization of the facial features. They have been largely superseded by image based approaches to face recognition. The most popular image based method for face recognition is Eigenface face recognition developed by Turk and Pentland [5]. Eigenfaces are a low dimension representation of a facial image that is, in a last square sense, true to the original image. Eigenfaces make it possible to efficiently store and use a large number of images for face recognition.

Despite the wide acceptance of image based face recognition it is not without its problems. At its heart image based face recognition is a method of image correlation and as such is sensitive to changes in illumination [6], changes in face pose [7] and changes in face alignment [8]. This is because image based approaches are extremely good at recognizing the images of faces they are trained with, but are incapable by themselves of generalizing novel viewing and capture conditions. More advanced image based approaches such as the illumination cone [9] are capable of overcoming many of these problems, however such approaches are computationally expensive and would be unsuitable for real-time use on mid range PC's. We purpose a less computationally expensive approach to face recognition that when combined with a number of appropriate constraints provides an acceptable level of performance.

The rest of this paper is laid out as follows section 2 outlines our system set-up, section 3 discusses our approach to face detection, section 4 details our face recognition approach, section 5 gives details regarding testing whilst section 6 provides conclusions.

2 SYSTEM SETUP

The following set-up is assumed throughout this paper, both in respect to design and testing. It is expected that a single user will be sat comfortably in front of a PC, upon which is mounted a camera. The camera is not expected to be of high resolution; indeed it is assumed that the camera is an inexpensive USB web-cam. The user will be sat facing the front of the PC, it is assumed that the distance between the user and the PC will be such that the user can reach the keyboard at all times and that the users face will be approximately upright at all times. It is assumed that the system will have to work in a typical office environment and therefore the lighting will not be calibrated. Finally it is assumed that the PC will be a mid-range machine, typical of many home and office PC's.

An overview of the face recognition login system is illustrated in figure 1a. A new image is read into the system from the camera. Face detection is then performed on the image, if a face is successfully detected the localised, aligned face image is sent to the face recognition stage. If however a face is not detected in the image then a new image is read in from the camera. The face recognition stage is based on typical PCA face recognition algorithms using Eigenfaces. The PCA algorithms return the closest matching face, and hence matching identity. For a closed system, i.e. one which is only accessed by people authorized to do so, the identity returned from the recognition stage would be used for login, if a more secure login is required the closeness of the match could also be checked against a threshold value, thus if a matched face is not deemed to be a 'good enough' match it can be rejected and a new image read from the camera. To increase the efficiency of the login system a multi-resolution approach is adopted, for face detection low-resolution images are sufficient and this reduces the time taken to localize a face. However for face recognition very low-resolution images are unsuitable and would impair the quality of the results obtained, therefore once a face has been located a higher resolution image of the face area is used for recognition. More details of the face detection and face recognition stages are in sections 3 and 4.

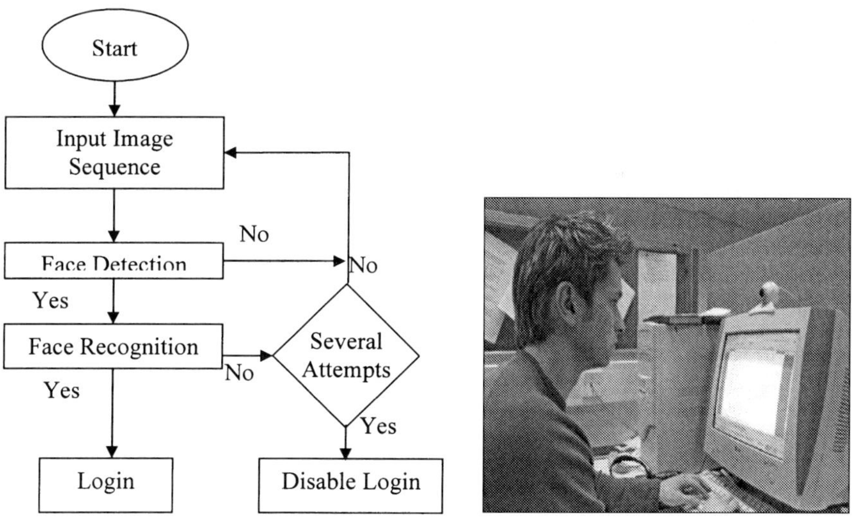

a) b)

Fig. 1 a) Overview of the face login system, b) typical set-up of the face login system

An image of the GUI used for testing the face based login system is shown in figure 2. The functionality of the GUI permits the use of either live video from a web-cam/camera or video from an Avi file to be used for either face recognition or to be stored as training data. The GUI shows in real time the original image, an image showing the detected skin regions, the face detection results and the aligned face image to be stored or used for recognition. When face recognition is being performed the GUI shows in real time the returned ID match and the three closest matching images from the training database. Further options within the GUI allow the results to be stored and/or colour normalization to be used. Though it should be noted that at present the face login system does not utilize the colour normalization functionality.

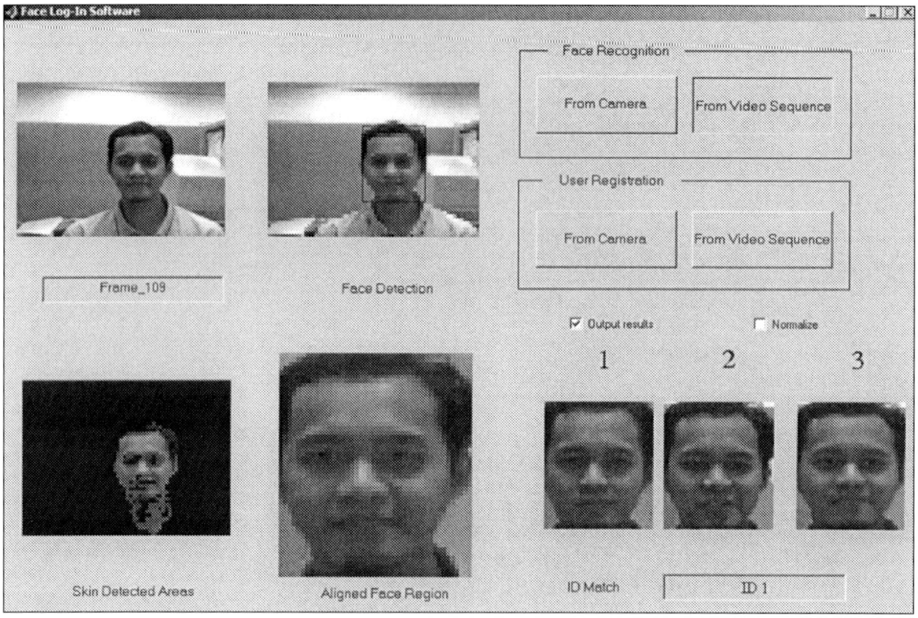

Fig. 2 GUI interface for face login software

3 FACE DETECTION

A number of different approaches to face detection exist [10], Rowley et al. were responsible for producing some of the best face detection results [11] using trained neural networks to detect multiple faces in an image. Such a method however is computationally expensive, a very real concern for a real time system, we instead utilize a skin colour model for face detection. As it is assumed a single user will be present at any terminal the face detector only needs to be capable of localizing a single upright frontal face. Our face detector uses skin colour to identify potential face regions and the shape and size of such a region to verify if it is a face, our approach to face detection is illustrated in figure 3. This approach differs from other approaches, typically the verification that a skin region is a face is accomplished using a face model or by detecting facial features. Our approach is far simpler to implement and is more efficient. As mentioned in section 2 to facilitate the real-time behaviour of our face login system low resolution images are used for face detection, this is possible because of our face detection model.

3.1 Skin colour model

It is commonly recognized that the HSV (hue, saturation and value) colour model is more similar to the human perception of colour than the RGB model. Research has found that different skin colours have similar values for hue and saturation, this makes the HSV colour model very useful for identifying skin coloured regions. The skin colour function in equation 1 is used to determine if a pixel is skin coloured or not, suitable values of hue and saturation were established through testing on a range of different faces and skin colours.

$$\text{SkinColour}(H,S) = \begin{cases} 1 & \text{if}((0 < H < 0.1) \,\|\, (0.9 < H < 1)\,\&\&(0.35 < S < 0.92)) \\ 0 & \text{otherwise} \end{cases} \tag{1}$$

where H = Hue values

S = Saturation values

Figure 4 shows the typical output from the skin colour model given above, small black spots within the skin areas, caused by specular reflections (such as near to the neck areas) or noise from the low quality web cams, are normal and don't affect the login system.

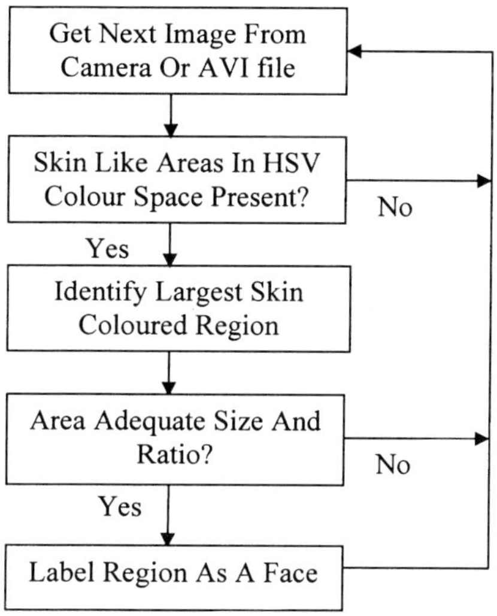

Fig. 3 Face detection process

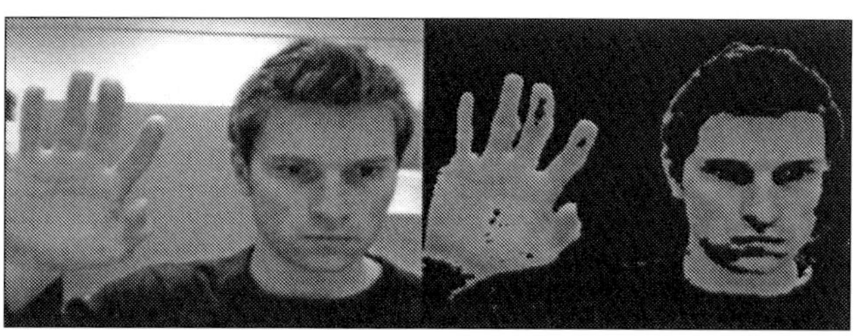

Fig. 4 Typical results of applying our skin colour model to an image

3.2 Verification of face areas

Having identified the areas of an image that contain skin like colours the next task is to determine which areas correspond to a face. Verifying a face is the final component of many face detection algorithms; commonly it is done using a face model [13] or using real face images [14]. As low-resolution images are being used for the face detection using a real face image is not prudent, using a simple face model would be less computationally demanding and would be more suitable when using low-resolution images. To minimize the computational cost used to verify that a skin like region is a face we use a simple and efficient methodology based on relative size and shape. A face is verified as a face using three simple checks, 1) is the skin coloured region the largest skin coloured region in the image, 2) is the skin coloured region at least 150 pixels in size, and finally 3) does the height and the width of the possible face area have the correct aspect ratio. These three checks eliminate skin regions that are not the single largest region in an image, too small for effective face recognition and check that the width and height of the skin region are of a suitable ratio for a face.

4 FACE RECOGNITION

Having successfully detecting a face in a low-resolution image, a higher resolution image is used for face recognition; the high-resolution image is cropped so that the face is centrally aligned within the new image. This is the image that is used for the purposes of face recognition.

A number of approaches to face recognition exist [15]. As has already been mentioned image based approaches to face recognition are currently widely used, this is because they are reasonably efficient and are capable of producing good results. The main drawback of image based approaches to face recognition is their inability to generalize novel viewing conditions. As a consequence of this they are sensitive to changes in pose, expression and lighting. We seek to overcome this problem by utilizing multiple training images for each individual; the multiple images cover a wide range of different pose and lighting conditions. This approach is appropriate for our face login system as it is only aimed at a small user base and training images of an individual can be captured real time using our login system software. Using this approach as opposed to a more complex robust face recognition approach such as illumination cones [9] is considerably more efficient. Eigenfaces based on PCA (principle components analysis) are used for recognition, more specifically we use colour Eigenfaces [16] so that we can take advantage of the extra information contained in colour images. Eigenfaces are used due to their ease and speed, making them well suited to a real time recognition system. The most obvious alternative to Eigenfaces would be Fisherfaces that use LDA (linear discriminant analysis), which despite its additional complexity still has unanswered questions regarding its performance when compare with the more traditional PCA [17].

Details of the Eigenface approach are given below. A face database is made up of M images containing N pixels. Suppose an image Φ_i among a collection of M images, where Ψ is defined as the average image.

$$\Psi = \frac{1}{M}\sum_{i=1}^{M} \Phi_i \tag{2}$$

Each image differs from this mean by a vector $\Gamma_i = \Phi_i - \Psi$. Thus the covariance matrix is defined as:

$$\varphi = \frac{1}{M}\sum_{i=1}^{M}\Gamma_i^T\Gamma_i = AA^T \quad (3)$$

If V_l are the Eigen vectors of $A^T A$ then:

$$A^T AV_i = \mu_i V_i \quad (4)$$

Where, μ_i are the Eigen values.

As AA^T is only an M*M matrix the Eigen vectors, U, can be defined as:

$$U = AV_i = \sum_{K=1}^{M} V_{ik}\Gamma_k \quad (5)$$

K is the number of Eigen vectors to use and that are ordered such that the Eigen values (U) are in decreasing order. This arrangement allows a human face to be encoded as a number of co-efficients; these are weighted values so that the first ten vectors carry far more information than the next ten and so on. Utilizing the first 10 vectors each face is described as a 10*10 matrix. Using a least squared distance it is easy to compare a new face with all the faces in the training database. Colour-eigen faces extend normal Eigenfaces by developing an Eigenface for each colour channel, RGB, and merging the final result.

5 TESTING AND RESULTS

5.1 Testing procedure

5 individuals (labelled as ID 1 to ID 5) were each asked to pose for two video files, the first a short video used for training and the second longer video used for testing. For the first video the individuals were asked to rotate there head left, right and then up and down. The second video file the subjects were asked to pose for was a longer video, in which subjects were asked to look at the screen and move their head, simulating normal behaviour at a computer terminal. The 5 individuals used included examples of different ethnic groups, different age groups and include a user wearing glasses. Figure 5 shows example images of the training data.

Fig. 5 Examples of training data

Using the 5 video files testing was performed at each frame. Testing followed the general outline of the login system illustrated in figure 1a; thus for each frame face detection was attempted, successfully identified face images were then used for face recognition. From this testing process three different sets of results were collected, face detection, face recognition and face recognition using a threshold value.

The first process to test in any face login system is the face detection stage; we identify a successful detection as one that correctly places a box around the face such that the face is centrally located.

Face recognition results are calculated in two different ways, the first and simplest of these examines the closest matching face and determines if this is a correct match. This test is referred to as the first image nearest neighbour recognition rate and is recorded as a percentage of the time the correct individual is logged into the system. The second more secure method of recognition uses a threshold value to determine if the nearest neighbour face match is 'good enough' to use. By altering the value of this threshold value the face login process can be made more or less rigorous. Using this method the false error rates (FER) and false acceptance rates (FAR) can be calculated allowing us to determine the equal error rate (EER). By combining the FER and FAR a receiver operator characteristics curve (ROC) can be deduced, the ROC curve is more useful for determining the security of a system than a simple first image nearest neighbourhood test. Particularly as the ROC shows how the system responds to unknown users. For both face recognition tests the results of the face detection were also used so incorrect face detections will have an affect on face recognition results. In this way the face recognition results are for the complete face login system and not just for ideal captured images.

5.2 Results

Table 1 shows the results of the face detection and the face recognition results performed without a threshold and using purely the nearest neighbour match. The face detector was successful in localising the face 100% of the time; figure 6 shows some images of successful face detections. The face recognition results across the whole system are 95%, which is comparable with the error rate for a text based login systems caused by typing errors or forgotten passwords. This means that the face login software would login the wrong person approximately 5 times per 100 attempts. For a security conscious organization this level of error is unacceptable. By taking advantage of the real time nature of the face login system it is possible to use the nearest neighbour match across multiple frames to improve the face recognition results. Our testing has shown it is able to achieve a 100% correct recognition rate by using multiple frames from real-time captured sequences. In order to improve the correct match ratio, users are required to change their pose in front of the camera slightly and gently.

Fig. 6 Results of applying face detection to 4 images

Table 1 Results of testing the face login system, overall correct login occurred 98% of the time

Video	Login results using first image nearest neighbour recognition rate (%)					
	ID_1	ID_2	ID_3	ID_4	ID_5	Face not detected
ID_1	**98**	0	2	0	0	0
ID_2	0	**100**	0	0	0	0
ID_3	0	1	**95**	1	3	0
ID_4	0	0	0	**100**	0	0
ID_5	0	0.5	2.5	0	**97**	0

By using a nearest neighbour first match recognition classifier the face login system will always login the closest returned match. When the face login system is accessed by users unknown to the system particularly people who shouldn't be permitted login access, this is a problem. By using a threshold value it is possible to stop unknown users logging into the system, this is best demonstrated by an ROC curve. Figure 7 shows the ROC curve for our face login system. The ROC curve shows meaningful information about the level of security our face login system can supply, by showing the crucial trade off between false acceptance rates and false rejection rates. Figure 7 shows that by selecting a suitable threshold value it is possible to eliminate false acceptances at a cost of incurring a false rejection rate of 0.17. In other words to make our system 100% robust, genuine users must accept an approximate 1 in 5 chance that they will not be allowed to log-in each time they attempt to do so. For a real time face based login system this rejection rate is low, as new images are continuously read in and compared resulting in a delay of only a few seconds. For a biometric system involving user input, such as aligning your iris with a scanner, a high number of false rejections would be less acceptable. The equal error rate for the face login system is illustrated using the dotted line and is below 0.02. However, using video sequence for face recognition, the match rate can easily reach 100% with the help of statistics analysis.

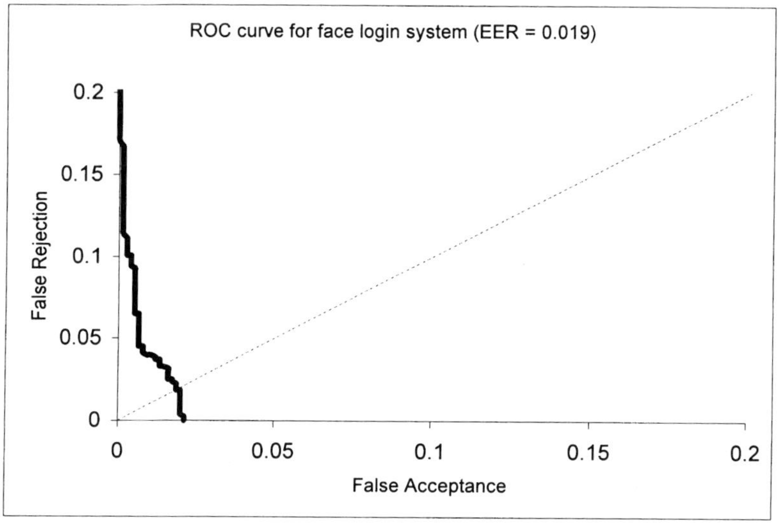

Fig. 7 ROC curve for face login system

6 CONCLUSIONS

We have shown that a real time face login system can produce acceptable levels of accuracy for a small user base. Such a system would have a number of advantages for online secure applications such as e-commerce, on-line collaborative engineering, online learning and restricting access to e-engineering systems.

A number of techniques have been utilized to allow accurate face recognition on a real time system. This has included the use of multiple images for recognition, multi-resolution images for the different stages of the login system, the use of varied training data as the sole means of overcoming lighting and pose changes, the use of different colour spaces and the use of only skin colour and geometry to detect a face.

Future work should investigate how well our face login approach copes when scaled up to a larger sized user-base. Currently results are acceptable for a database of 5 users, increasing the number of users would affect the recognition rate and the speed the system runs. The face recognition component could in the future be expanded to utilize our previous work on illumination normalization [18] and our work on two-step face recognition [19].

REFERENCES

1. **J .K. Lee, S. R. Ryu, K. Y. Yoo,** Fingerprint-based remote user authentication schemes using smart cards, Electronic Letters 38 (12), 2002, pp. 554-555.
2. **J. Daugman,** The importance of being random: statistical principles of iris recognition, Pattern Recognition, Volume 36, Issue 2, February 2003, pp. 279-291.
3. R. Brunelli, T. Poggio, Face Recognition: Features Versus Templates, IEEE Transactions on Pattern Analysis and Machine Intelligence, 1993, pp. 1042-1052.
4. **J. Cox, J. Ghosn, P. N. Yianilos,** Feature-based face recognition using mixture-distance, Proceedings of the International Conference on Computer Vision and Pattern Recognition (CVPR '96), 1996, pp. 209.
5. **M. Turk, A. Pentland,** Eigenfaces for recognition. Journal of Cognitive Neuroscience, 3(1), 1991, pp.71-86.
6. **Y. Adini, Y. Moses, S. Ullman,** Face recognition: the problem of compensating for changes in illumination direction, IEEE Trans. on Pattern Analysis and Machine Intelligence, Vol. 19, No. 7, 1997, pp.721-732.
7. **T. Jebra,** 3D pose estimation and Normalization for face recognition. PhD Thesis, McGill University, Department of Electrical Engineering, 1996.
8. **A.M. Martinez,** Recognizing imprecisely localized, partially occluded and expression variant faces from a single sample per class**,** IEEE Transactions on Pattern Analysis and Machine Intelligence, Vol. 24, No. 6, 2002, pp.748-763.
9. **A.S. Georghiades, D.J. Kriegman,** and **P.N. Belhumeur**, Illumination cones for recognition under variable lighting: Faces. In IEEE Conf. on Computer Vision and Pattern Recognition, 1998.
10. **M. H. Yang, D. Kriegman, and N. Ahuja,** Detecting faces in images: a survey. IEEE Transactions on Pattern Analysis and Machine Intelligence, Vol.24, No. 1, January 2002, pp. 1-25.
11. **H. A. Rowley, S. Baluja, T. Kanade,** Neural network-based face detection. IEEE Transactions on Pattern Analysis and Machine Intelligence, Vol.20, No.1, 1998, pp. 23-38.
12. **Y. Wang, B. Tuan,** A novel approach for human face detection from colour images under complex background, Pattern Recognition, (34), 2001, pp. 1983-1992.
13. **K. Wong, K. Lam** and **W. Siu** , An efficient algorithm for human face detection and facial feature extraction under different conditions, Pattern Recognition, Vol.34, No.10, October 2001, pp. 1993-2004.

14. **K. Wong, K. Lam** and **W. Siu,** A robust scheme for live detection of human faces in colour images, Signal Processing: Image Communication, Vol.18, No.2, February 2003, pp. 103-114.
15. **R. Chellappa, C. L. Wilson** and **S. Sirohey,** Human and machine recognition of faces: A survey. Proceedings of the IEEE, 83(5), 1995, pp.705-740.
16. **L. Torres, J. Reutter, L. Lorente**, The importance of the colour information in face recognition. ICIP (3) 1999 pp. 627-631.
17. **A.M. Martinez,** Recognizing imprecisely localized, partially occluded and expression variant faces from a single sample per class, IEEE Transactions on Pattern Analysis and Machine Intelligence, Vol. 24, No. 6, 2002, pp. 748-763.
18. **S. King, G .Y. Tian, D. Taylor** and **S. Ward,** Cross channel histogram equalisation for colour face recognition. Proceedings of the 4th International Conference on Audio- and Video-based Biometric Person Authentication, 2003, pp. 454-461.
19. **S. King, G .Y. Tian,** A framework for robust face recognition, submitted to the Journal of Computer Vision and Understanding.

Product data tracking, the way to go in an industrial laboratory environment

A HOGENDOORN, G E N KUIPER, P T M VAN LIESHOUT, and **M NOLTE**
Development and Technology, Corus Research, Ijmuiden, The Netherlands

ABSTRACT

In the research laboratories of the steel company Corus many different experiments are done to investigate the various phases of the steel and application production. Series of experiments generate large amounts of measurement and control data, which are stored in files on local PCs or file servers. Within a collection of several ten thousands of data files that are scattered over twenty departments, it can become a time consuming task to find and retrieve the required data.

To avoid an inefficient manual data retrieval process a so-called product tracking system has been developed. As all research equipments make use of data acquisition and control software, it is possible to store the generated data files on a central file server automatically. To ensure the traceability of data files, accompanying metadata is stored in a relational database. Through a user-friendly web interface, all interested users within the Corus research organisation can access the data warehouse. Data files can be downloaded to local PCs for further analyses.

1 CORUS GROUP PLC COMPANY OVERVIEW

In 1999 Corus was formed by the merger of British Steel and Koninklijke Hoogovens to create one of the worlds largest and most capable metal manufacturing and service organisations.

The company aims to be innovative with a strong customer focus, providing metal solutions to an increasingly sophisticated market. The Corus strategy is targeted at creating shareholder value by achieving world-class competitiveness through operating excellence and technological advance.

The emphasis is on achieving leading positions in attractive market sectors where sustainable growth can be achieved.

Corus has manufacturing operations in many countries with major plants located in the UK, The Netherlands, Germany, France, Norway and the USA.
Based on figures from 2002, the company produces around 17 million tons of crude steel annually, which represents approximately 10% of total EU production and positions the company as Europe's fourth largest Steel producer. Corus produces carbon steel by the basic oxygen steelmaking process in the UK and the Netherlands and by the electric arc furnace process in the UK and USA.
It also produces 250 thousand tons pure aluminium metal per year, contributing to a total cast house output of over 300,000 tons per year. The smelters are located in the Netherlands and Germany.
Corus Group employs about 51.000 people worldwide.

1.1 Corus Research, Development & Technology

To develop the Group to the required high technology standards, Corus makes use of its service department Corus Research Development and Technology (further called Corus RD&T). The drive for Corus RD&T is maintaining a research and development focus in areas such as process control and environmental protection, as well as to interface with customer activities in key market sectors. Particular attention is given to engaging with customers at the concept stage of product design in order to ensure that the best product is specified for the application concerned, utilising the most cost effective technology.
Corus RD&T employs almost 900 researchers across Britain and the Netherlands and supplies 'metal solutions' to Corus sites around the world. RD&T work in collaboration with universities and research institutes all over the world as well as with key customers in the automotive, transport, packaging and construction areas.
Within Corus RD&T, large test facilities are available. These contain a lot of experimental equipments as well as full size production installations that were especially adapted for a research environment. Therefore extra measurements and controls are added to the installations. Without losing production performance the installations and equipments are dedicate for research purposes. All equipments produce experimental data on which a part of RD&T knowledge is built.

1.2 Introduction in the Packaging department

One of the departments within de Corus RD&T product applications area is the Packaging department. The research of this department consists of testing newly developed steel grades and the possible application of these materials as well as improving the products and production processes. For instance, making a can wall thinner saves material, but the product quality has to be as good as before. Another hot topic is aiming at material saving to reduce the size of the can closure.

The packaging department has access to a full size industrial can makers line to experimentally imitate beer and beverage cans production. This facility is used to develop new can designs as well as to solve customer's problems on an ad hoc basis.
During the production process of cans different production processes are used and several measurements are done to evaluate the geometrical dimensions and other quality figures.

One of the packaging applications that are investigated is the 2-piece beer and beverage can and the whole production process takes 30 to 50 steps. In the research environment every process step results in one data file with all relevant information about that step for a particular can. Between the production steps extra size measurements are performed. That means that the production of every single can results in 30 to 50 data files. During projects, experimental series up to 100 cans are no exception. Occasionally during projects material grades are changed as well. Summarising all these possibilities, a research project on a 2-piece can may lead to thousands of data files.
Collecting all these files and handling them to come to project results is a major and time-consuming task. Due to the fact that projects could run for months it is sometimes hard to re-locate the data files when required.

2 PROBLEM DEFINITION AND ASSIGNMENT

2.1 Problem definition

As illustrated, research projects in which the effects of several process steps and different materials are studied generate lots of local data files. And although only the packaging department was highlighted, several departments within Corus RD&T are in comparable situations. Until recently, the process and measurement data was mostly stored as separate files on a local PCs or file servers.
To report project results, all related data files have to be collected and relations and correlations have to be derived. That process is done manually and therefore a time consuming job.

2.2 Assignment

Due to the enormous amounts of projects within the different departments and of data files as a result of the projects, in the mid 90's the idea was generated to develop a computer system that could solve the problem of the trace-ability of data. The system had to have the ability to store all relevant information of a product and to track the all the performed production steps necessary to make the required end product. All information had to be easy trace-able, retrievable and available from every desktop PC within the Corus RD&T organisation. The experimental data had to be stored automatically where possible, to prevent manual data storage actions.

3 PRODUCT DATA TRACKING SYSTEM

3.1 Experimental framework within the Product Data Tracking System

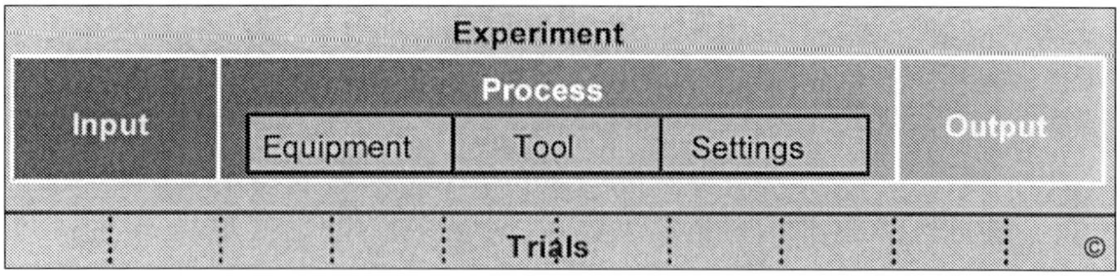

Fig. 1 View of the conceptual approach towards the experimental framework that lies at the basis of the data structure

In figure 3 the concept of the experimental framework is shown. The entity 'experiment' constitutes the basis of the data structure. Experiments make use of one or more input products or materials and a process to come to one or more output products.
The process makes use of machinery (equipment) that is adjusted by a series of settings and in most times the equipment needs one or more tools. One trial is defined as a single execution of an experiment. Therefore a single experiments usually contains multiple trials.
All the mentioned entities are related to each other through the experiment.

3.2 System demands

When the developments were started no commercial Laboratory Information Management System (LIMS) technology was available which met the Corus RD&T requirements. Initially the main demands for the system were the user friendliness and the abilities to access, to implement, to configure and to maintain in an easy way. During the development process extra demands arose, like the scalability of the system. It became clearer that the idea of the Packaging department was an opportunity to structure all the measurement data within Corus RD&T. Therefore the application design had to be more generic, to work within the Corus RD&T operating procedures.
Listing all the demands mentioned and after interviewing several potential users to collect the requirements a choice was made to use web technology to fulfil the needs. Based on web technology however, all requirements together made the whole concept very ambitious, though it seemed the best solution to combine low maintenance costs low with good accessibility.

3.3 Architecture

The entire application was designed according to a 3-tier or 3-layer architecture with a database and fileserver as tier one in the back-end of the system, a web interface as tier three in the front and a number of business components as tier two in between. This kind of architecture results in a scalable application.
Because the use of a database was the most obvious solution, a few other issues could be solved as well. Since within the Packaging department the tools, equipments and materials

information were all kept in different Excel sheets, that information could be stored in the database as well. Consequently, the Excel sheets became redundant and the information was available for every user.

3.4 Software technology

The Corus RD&T web standard is based on Microsoft technology. Initially the database was developed in Microsoft SQL7 and migrated to SQL2000 during the migration of the server park from NT4 to W2000. The interface is based on ASP technology and the business components developed in Visual Basic 6.0.

First prototype html pages were developed to give an impression of the functionality and user interface possibilities. Then the database design was worked out. Although a database for storing tools, equipment and materials data is not difficult, to create a good concept for all experimental procedures that differ from equipment to equipment was more challenging. To guarantee a good performance the choice was made to keep only metadata in the database and to store the accompanying raw data files on a file server. To meet the requirement of scalability a generic database solution was developed. Parallel to the database developments, the web interface was designed and developed to provide the users with the required functionalities. The communication between the database and the web pages was accomplished by the business components. Only the generic applications functions were put in these components, while the equipment related functionalities were programmed in its data acquisition and control (DAQC) software.
The last part that had to be solved was to interface the experimental equipments with the database system. Because a variety of equipments exist, it was decided to communicate by the generic XML-format. One major problem that had to be solved was the communication between the DAQC software and the database on a secured web server. A "gatekeeper" component was developed to give access to the database only for experimental equipments while people were not able to directly access the database environment at all.

3.5 Graphical User Interface

One of the most important parts of an application is the user interface. That has to be intuitive and pleasant to use. For all the different functions, five sections were defined. These sections will be briefly discussed in the next paragraphs.

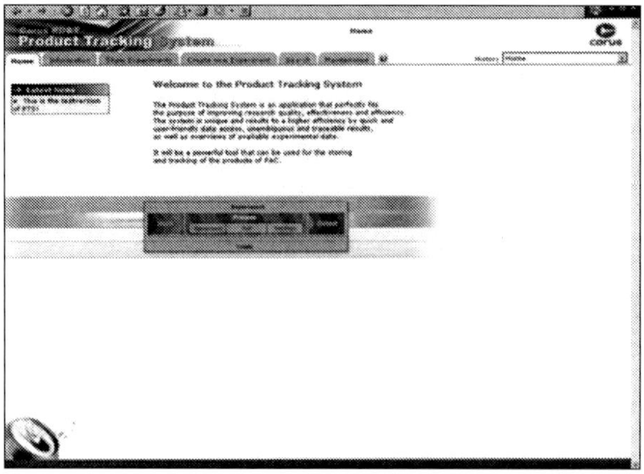

Fig. 2 Home page of the Product Data Tracking System

3.5.1 Information

The 'Information' section replaced the Excel files that where used to store metadata about tools, equipments and materials also called products. In this section it is possible to add or change the entity specifications and to add documents like supplier specifications or calibration reports. The database is also used for the materials stock management.

If a material is produced within Corus RD&T it is possible to track the products' production history from the product information pages. If a part of a certain product in stock is used for further experiments it is possible to track the future production routes as well.

3.5.2 Show Experiments

In the 'Show Experiments' section it is possible to browse through experimental results. An experiment represents one process step. Every time a process step is executed, one trial is performed. Experiments in which 100 cans are processed in the same way the result are 100 trials with their companioning metadata and data files.

The viewed experiments provide the user with information about the used products, tools and equipment and the generated trial files. The mentioned items are directly linked to their own information pages in the application.

It is possible to link extra files to the experiments like reports or process control curve files.

Trial data and data files can be downloaded individually or as a group to the desktop PCs for further data analysis.

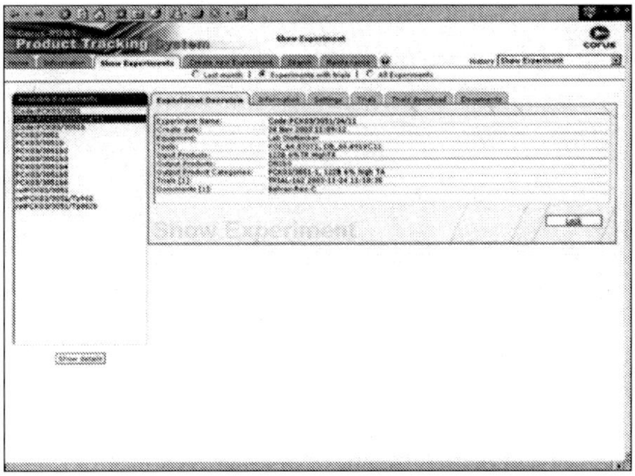

Fig. 3 The experiment overview page

3.5.3 Create new experiments

In the 'Create new experiments' section it is possible to create new experiments from scratch or by reusing existing experiments and make the necessary changes. Within the new experiment, the former results of the reused experiment will be emptied.

The amount of prepare-able experimental information depends ability of the DAQC software to retrieve data from the database. Older equipments are only able to upload results and not to retrieve database information.

3.5.4 Search

With the 'Search' option it is possible to build queries based on every database entity with metadata. In a graphical way complex queries can be created without any knowledge of a querying language.

3.5.5 Maintenance

The Maintenance section is used to compose the data tree structures of the system. It is possible to add or change parameter names and to locate the parameters in the tree structure. Deleting a parameter is only possible when no data is connected to it.

For certain tool and equipment key figures it is possible to add top and bottom limit values which the application uses to check if settings given by the users are within the limits.

3.6 Automated data storage

To prevent manual data handling some kind of automation had to be developed. Due to the fact that most of our experimental equipments run with dedicated in-house developed DAQC software it is possible to 'connect' them to the Product Data Tracking System. Mostly LabVIEW hard- and software is used for in-house DAQC developments. This software had to be extended for direct database communication.

For equipments with software from the supplier, a communication module for storage has to be developed. Most suppliers of software are required to provide data files according to the Corus RD&T standards, the so-called MUS or VHF file types which are comparable with the Microsoft windows INI-file type. The modules extract metadata from the data files and store the data and data files in the system.

Some of the older installations are not computer controlled and new input screens have to be developed. Instead of writing the data in a laboratory journal the data is put directly into the system.

In case of buying new equipments or upgrade the DAQ software to meet new standards, the interaction with the system can be a two-way communication. As discussed the system provides the users with the possibility to prepare experiments at their office. The experimental information is stored in the database and to be used by equipments. When the experiment is executed, the user can retrieve his prepared data directly from the database and use that to control the equipments. This integrated way of working is not obliged and will only be developed when there is a need or an opportunity to do that. These developments are more expensive and hold more compatibility and maintenance risks than the one-way data storage communication.

4 FUTURE DEVELOPMENTS

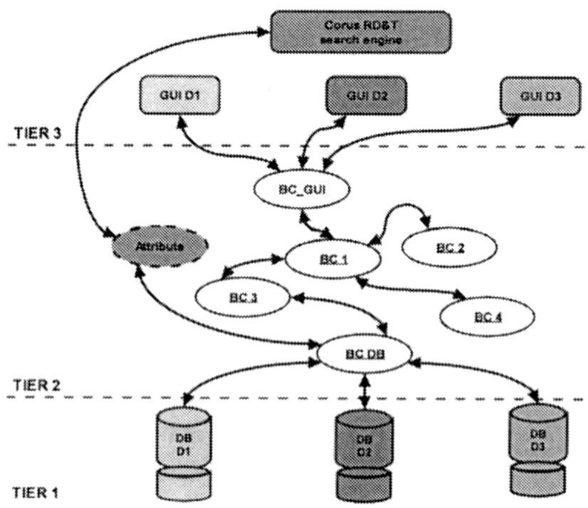

Fig. 4 Overview of the 3-tier software architecture

The Product Data Tracking system was initiated for the Packaging department of Corus RD&T only, however the generic design makes the data storage suited for all experimental equipment. It is also possible to store the results of mathematical modelling.
However some experimental results are confidential and the results may only be known within a single department. To retain data separated from different departments and to keep databases rather small for a good performance and maintainability, it was decided to use one Product Data Tracking System per department.
Mostly data searches are limited to data that has been generated someone's own department. However to have the possibility to find all available data of a certain material a Corus RD&T search engine will be developed to search through all databases. If material data is found in databases of another department then the data itself is not retrieved, but the user receives information about the availability, the location and the expert or owner of the data. That

prevents misinterpretation of data by users who are not experts in the knowledge field of other departments.

Sometimes projects are strategic and only project team members are allowed to browse the data. In the future a security component will be added to be able to shield the database content on a project level from other users. Through a role editor, different types of users will be defined that receive the allowed necessary functionalities and data access. For instance, only the content manager is authorised to maintain the database content and the stock manager has the right to change the stock related metadata of stored materials.

5 CONCLUSIONS

Information is a major part of the core product of RD&T and must be handled with care. Corus RD&T is a ISO 9001/2000 company and the objectives or the certification are trace-ability of information and to be in control of projects. The powerful research tool supports these objectives by improving the accessibility and trace-ability of data and information.
In the first quarter of 2004 the system was rolled out in the packaging department of Corus RD&T and in April to the Aluminium metallurgy research department. These departments are evaluating the system after which the system will be rolled out stepwise to other departments.

The user interface is initiative and user friendly. By making use of web technology a few or the demands are solved directly like the accessibility and the maintainability. Similar, the ability to implement the software within large groups of users is no problem using the Intranet.
Due to the 3-tier architecture the scalability was easy to achieve.
Because every department has its own materials , tools and equipments. That results in a different data structure configuration. Extended maintenance pages provide the database content managers with the ability to configure the data structure to the demand of a particular department.

The Product data Tracking System improves the efficiency of research and development projects. The data is retrievable by making use of easily generated SQL queries. Reliable and complete product history overviews are indispensable for high quality research and the system prevents re-inventing the wheel.
The system saves money by efficient an automated experiment data handling and having all data available from every desktop PC within Corus RD&T. Due to good data management a complete knowledge loss by job-rotation will be reduced.

In the future, all product data tracking systems of the different departments can be combined to a large research information system for products (materials) and product applications.
The product data security is not implemented yet and will remain a point of attention
To the system will be a basis for on-line data analysis. The product data tracking system is a powerful automated, efficient and unambiguous data storage facility.

Improved scheme of double-direction signature in SET protocol

L-J ZHENG, C-H PIAO, C-Y ZHANG, and **Y-F JIA**
Computer Department, Shijiazhuang Railway Institute, People's Republic of China
L-W ZHENG
Department of Food Engineering, HeBei Normal University of Science and Technology, Qinhuangdao, People's Republic of China
J-G QI
TianZheng Jian Li Company, He Bei Water Project Geologic Reconnaissance Institute, Shijiazhuang, People's Republic of China

ABSTRACT
SET is an international standard applied to e-business security. The introduction of a double-direction signature avoids deception between the merchant and the bank, and establishes the authenticity of the customer. However, it doesn't consider the merchant-bank undeniability. Also, the hash algorithm employed is SHA-1 which, with improvements in computer performance and execution speed, is becoming increasingly insecure. In this paper, a new hash function and time stamp are introduced and the undeniability of the sender and receiver is realized. Through the theoretical analysis and test, it is concluded that the scheme has a good measure of security.

1 INTRODUCTION

With the increasing development and penetration of computer networks, more and more trade is being carried out on the Internet. So e-businesses are emerging and beginning to take security seriously. The SET protocol developed by VISA and Master Card is widely employed for Internet transaction security. In the SET protocol, the important innovation of a double-direction signature scheme is introduced. This avoids deception between merchant and bank. However, disappointingly the scheme uses the SHA-1 digital digest algorithm, the security of which, as computer performance and speed improves, is threatened to a certain extent. Also, the double-direction signature scheme doesn't consider the sender and receiver's undeniability in the e-business security, and so it also weakens the SET's security.

This paper puts forward a newly improved scheme based on double-direction signature in the SET. The improvements include: using a new hash function to replace the SHA-1, introducing the concept of a time stamp. Through the theoretical analysis and test, we draw a conclusion that the scheme has a very high security, execution speed, and realizes the sender and receiver's undeniability.

2 DOUBLE-DIRECTION SIGNATURE IN SET PROTOCOL

2.1 Scheme description

The purpose of the double-direction signature in SET protocol is to avoid deception between the merchant and the bank. It guarantees that the merchant can only get the consumer's order information (OI), and the bank can only get the consumer's payment information (PI). Supposing the private key of the consumer is PR_c, the flow of the double-direction signature is as follows [(1)]:

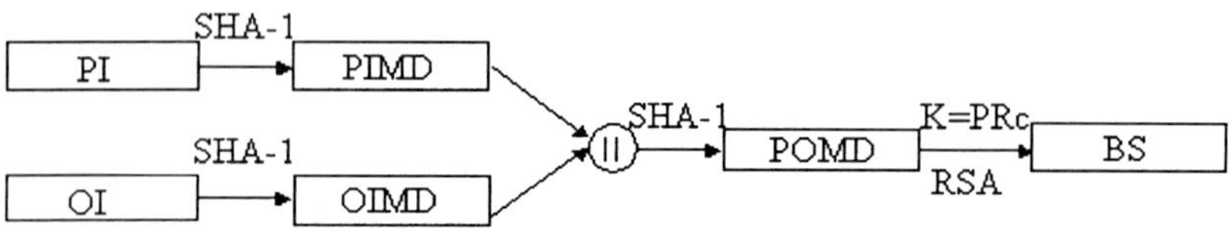

Fig.1 The construction of both-way signature

The consumer calculates the hash data of the payment information (PI) and the order information (OI) by using the SHA-1 algorithm. Then joins the two hash data PIMD and OIMD together, calculates the hash result of the joint data. In the end, the consumer uses his private key to signature the hash result, thus the double-direction signature BS is obtained. The message digest PIMD is conveyed to the merchant, BS and the message digest OIMD is conveyed to the bank. So the merchant and the bank can verify the consumer's signature. It avoids the behavior of revising the order information and the payment information for the interests of the merchant or the bank.

2.2 Potential insecurity of the scheme

Firstly, the double-direction signature uses SHA-1 as the message digest algorithm. The hash length of the SHA-1 is 160 bits. Many researches have shown that, it needs only several days to attack a hash code whose length is 160 bits. It is clear that in the near future, the SHA-1 algorithm will become more and more insecure.

Secondly, the double-direction signature in the SET protocol can ensure the undeniability of the consumer, but it doesn't take the merchant and the bank into consideration. So it is not strong enough to avoid the deception behavior of the merchant or the bank.

3 THE DESCRIPTION OF THE IMPROVED SCHEME

3.1 The construction of a new hash function

The construction of the new hash function is based on the advanced encryption standard Rijndael algorithm, this algorithm is designed by Joan Daemen and Vincent Rijmen[(2)]. For this algorithm, the length of the data block and the key is variable. It is block cipher algorithm. On the whole, the Rijndael algorithm adopts the substitution-permutation network, overlapping many loops, each loop is made up of three layers: non-linear layer, linear-mixed layer, key-plus layer. It has very good difference uniformity, linear deviation, can resist difference cryptanalysis and linear cryptanalysis.

The figure of the constructed hash algorithm is shown as Fig.2, here the length of the block is 128bits, the length of the key is 256 bits, the scheme is described as follows:

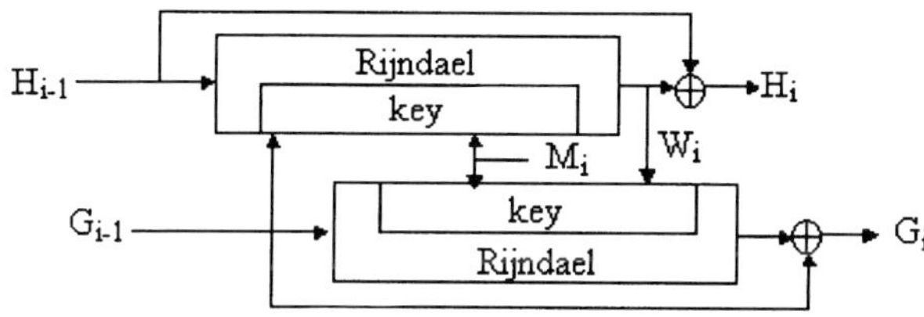

Fig.2 The constructed hash function

$G_0 = I_G, I_G$ is a random initial value,

$H_0 = I_H, I_H$ is a random initial value,

$W_i = Rijndael_{G_{i-1}, M_i}(H_{i-1})$,

$G_i = G_{i-1} \oplus Rijndael_{M_i, W_i}(G_{i-1})$,

$H_i = W_i \oplus H_{i-1}$

The length of H_i, G_i, W_i, I_G, I_H, M_i is 128 bits. M_i represents the ith sub-block of the message. The final hash result is the connection of G_i and H_i. (In the end, we connect G_i and H_i to get the final hash result. Since the length of the G_i and H_i is 128 bits, so the length of the final hash result is 256 bits.)

3.2 Realization of the improved scheme

Before giving the improved scheme, we make the following stipulations:

(1) The sender is consumer C, the receiver is the merchant M and the bank B. (2) T_1, T_2 is a reasonable value arranged by the consumer C and the merchant M.T_1, T_2 is a reasonable value arranged by the consumer C and the bank B. (3) CA is the certification authority, the digital certificate of consumer C is DC_C, the digital certificate of merchant M is DC_M, the digital certificate of bank B is DC_B. (4) The adopted public key algorithm is RSA; the adopted hash algorithm is the newly constructed hash function based on Rijndael, marked as H; E represents the encryption algorithm, D represents the decryption algorithm. (5) PR_C: the consumer's private key. PU_C: the consumer's public key. PR_M: the merchant's private key. PU_M: the

merchant's public key. PR_B: the bank's private key. PU_B: the bank's public key.

The realization of the scheme:

Step1: Consumer C gets the digital certificate DC_M of the merchant and digital certificate DC_B of the bank from the certification authority, and then produces message DS, I_1, I_2. $DS=E(PR_C, H(H(PI)||H(OI)))$, $I_1=E(PU_M, T_1+T_2+DS+OI+PIMD)$□$I_2= E(PU_B, T_3+T_4+DS+PI+OIMD)$.

Step2: Merchant M and bank B respectively get I_1 and I_2 from CA. Merchant M decrypts I_1 by using his own private key: $D(PR_M,I_1)= T_1+T_2+DS+OI+PIMD$; Bank B decrypts I_2 by using his own private key: $D(PR_B,I_2)=T_3+T_4+DS+PI+OIMD$□

Step3: Merchant M and bank B get the public key PU_C of the consumer through the consumer's digital certificate. Merchant M calculates: $D(PU_C, DS)$ and $H(PIMD||H(OI))$, and compares whether they are equal to verify the consumer's signature. Bank B calculates: $D(PU_C, DS)$ and $H(OIMD||H(PI))$, and compares whether they are equal to verify the consumer's signature.

Step4: After the consumer's signature has been verified, I_3 and I_4 are produced during time T_2 and T_4, then they are conveyed to CA. $I_3 = E(PU_c, T_1+T_2+DS+OI+PIMD)$□$I_4= E(PU_c, T_3+T_4+DS+PI+OIMD)$□

Step5: Consumer C decrypts information I_3 and I_4 by using his own private key PR_C, and then calculates $D(PR_C,I_3)$ and $D(PR_C,I_4)$, compares it respectively with $T_1+T_2+DS+OI+PIMD$ and $T_3+T_4+DS+PI+OIMD$, if the comparison result is equal, we can ensure the identity of the merchant and bank.

4 SECURITY ANALYSIS OF THE IMPROVED SCHEME

4.1 Analysis of the undeniability

Since the introduction of the CA in the improved scheme, no one including C,B or M can deny the messages that have been sent to him.

Since only consumer C knows PR_C, nobody can simulate C, so it can ensure the undeniability of consumer C.

Since the existence of the time stamp and CA in the improved scheme, if M or B doesn't receive the information, they can inquire CA's database before T_2 and T4, so it can not only ensure that M or B has indeed received the information, but also ensure the undeniability of the receiver M and B [(3)(4)].

4.2 Resist falsification and replay attack

Supposing that a third side wants to falsify the message transferred between C and M. Firstly he intercepts the message, replaces it with the message encrypted by the falsified PU_C, PU_M. But C and M will find immediately that someone is doing malicious attack. Since he doesn't know PR_C and PR_M, he can't get the meaningful plaintext. In the meantime, the introduction of time stamp can effectively resist replay attack from the enemies.

4.3 Security analysis of the hash function based on Rijndael

4.3.1 Hash performance

Supposing that there are four similar pieces of plaintexts, we calculate their hash values to inspect the hash performance of this algorithm.

Plaintext1="I have a good teacher."
Plaintext2="I have a good teacher!"
Plaintext3="I have a good teacher!!"
Plaintext4="I have a good teacher?"

The hash results are as follows:

Hash(Plaintext1)=dffcc935aa84378d8ae90a73c0a629f4ddfb123fadcab009eacfddf8c9efa6cd;
Hash(Plaintext2)=d88659cef47f07966b481cdcf77bbe628a090f65f702d64eaf00de5fcd67ae67;
Hash(Plaintext3)=e0ffe21cc3e82bcef47bcaca2e481e9de60a0fdefa0754fea3733aaeb0df2299;
Hash(Plaintext4)=9253fd6ab53defb1148accab00937a0653ea377082f77bbe6b4889760defea3
7;

4.3.2 Confusion and diffusion character

Confusion and diffusion are not only two basic techniques to conceal the redundancy in the plaintext, but also are two important factors to measure hash performance.

A hash function must have such characters: the plaintext is irrelevant to he hash result, that is, any slight change of the plaintext can cause the hash result to change at the probability of 50%. Here we define average change bit number and average change probability to measure the confusion and diffusion character.

According to the previous hash result as shown in 4.3.1 and the calculation formula in references (5), the average change bit number and average change probability are respectively 126.54 bits and 48.98%, approaching the ideal value 128bits and 50%.

So we draw a conclusion that the hash function has very good confusion and diffusion character, can effectively resist known-plaintext attack and chosen-plaintext attack.

4.3.3 Calculation speed

The traditional one-way hash algorithms, such as MD5, are complicated methods based on XOR logical calculation. Since the inherent drawbacks, though each step is very simple, the calculation round will be very large. The algorithm constructed in the paper uses Rijndael algorithm, since it has very quick execution speed whether under hardware condition or software condition, so the algorithm constructed has very quick execution speed.

4.3.4 Security performance

The length of the hash code is 256 bits, it can effectively resist brute-force attack. Besides that the Rijndael algorithm itself has very high security, it can resist any attack including difference cryptanalysis and linear cryptanalysis.

5 CONCLUSIONS

The double-direction signature in the SET protocol adopts digital digest algorithm SHA-1, with the improvement of the computer's performance and enhancement of the calculation speed, its security is threatened to a certain extent. In addition, the double-direction signature doesn't consider the sender and receiver's undeniability, it weakens the security of the SET to some degree. So the paper improves the original double-direction signature scheme, introduces a new message digest algorithm and realizes the sender and receiver's undeniability. Through the analysis and verification, we learn that the improved scheme enhances the security of the SET protocol, and it has the possibility to be put into good practice.

REFERENCES

1. Written by William Stallings, translated by YANG Ming, XU Guang-hui, QI Wang-dong, "Cryptography and Network Security: Principles and Practice (second edition)", *Beijing: Publishing House of Electronics Industry*, April, 2001, pp 190-237
2. Lin Mao-qiong, Xiong Kai, Li Min-qiang, "Data Encryption Scheme Based on AES", *Computer Engineering*, Vol.28, No.4, April, 2002, pp 140-142
3. Written by BruceSchneier, translated by WU Shi-zhong, ZHU Zhi-xiong, ZHANG Wen-zheng, *Applied Cryptography*, Beijing: China Machine Press. March, 2003, pp 34-335
4. Li Xian-xian, Huai Jin-peng, "Undeniable Cryptographic Protocol for Both Sender and Receiver and Its Application", *Journal of Beijng University of Aeronautics and Astronautics*, Vol.27, No.2, April, 2001, pp 182-185
5. Zheng Li-juan, Wang Bao-yi, Qi Jian-gang, "A Hash Function Construction based on Rijndael", *Microsoft Development*, Vol.13, No.12, Dec. 2003, pp 119-121

Digital Factory/Enterprise

Web-based integration for virtual factory implementation

A LUNG, R BENHADJ-DJILALI, and **J A GARCIA**
School of Engineering, Kingston University, London, UK

ABSTRACT

The current increasing competition in a global scale enabled by advancements in information and computing technologies (ICT) poses an opportunity as well as a challenge to manufacturing companies worldwide. Important part of this are advancements in CAD/CAM and CAE technologies now grouped in the, still in development phase, concept of Product Lifecycle Management (PLM), that in combination with the convergence of ERP systems and Internet technologies offer the possibility of a fully enabled Supply Chain Management alternative that seems essential to survive in the new century. This paper explores the incorporation of Internet technologies into the manufacturing sector by adopting a web based Virtual Factory (VF) concept model in which PLM is implemented over the Internet to enable an agile aggregation of SMEs.

1 INTRODUCTION

Agility concepts, introduced in the early 1990's as a strategy to face the challenges in the new century, are reaching maturity and proving to be a strategic advantage in current times. Agility enablers such as virtual enterprises, and virtual factories as their application in manufacturing, provide the integration highways and critical information transfer that facilitate the formation of temporary web-based networks. These networks should enable Small and Medium Enterprises (SMEs), in most cases limited in physical resources, to engage in rapid development of complex products in a cost-effective manner since they accommodate heterogeneous platforms, architectures and components regardless of the underlying technology infrastructure and vendors (1). Within these networks, SMEs are able to incorporate new emerging technologies where appropriate, to aggregate their expertise and core competencies in a collaborative and agile approach to respond to a market change or opportunity, which otherwise they would be unable to afford.

The new 'technology led' business environment offers unprecedented channels for wider competition, wider scope in market participation, and diversification in manufacturing operations, which leads to a push for manufacturing enterprises to embrace these newly emerged technologies in order to stay in business. Unfortunately, while the bigger companies may have the resources and expertise to cope with these changes, SMEs don't have the same luck; they are still trying to incorporate latest developments in existing technologies such as Computer Aided Design and Manufacturing (CAD/CAM). Furthermore, they are not even ready to take on any newer developments such as Electronic Data Management (EDM) and Product Data Management (PDM), not to mention Product Lifecycle Management (PLM), which is still in the process of being fully defined. The situation is further complicated by the rapid migration of networking technologies from Local Area Networks (LAN) to Wide Area Networks (WAN), and then to Internet-based networks. Fortunately, because of the simplicity and interoperability of emerging Internet technologies scalable solutions are possible, offering different options of investment and adoption of what seem to be the ultimate tool to be used in the implementation of the concept of agility.

In a short space of time, Internet technologies have evolved from merely distributing static information to allowing dynamic interaction with human users with the latest developments extending this iteration to support communication between highly functional applications. An example of this is the inclusion of actual programming code, rather than just dynamic scripts, in the building blocks of web applications. This not only reduces a layer of communication with remotely located applications but also is a critical factor for their application in the automation of manufacturing operations. In this paper, it is proposed that a virtual factory model can be built using the Internet as the backbone for web-based inter-enterprise integration providing support for the entire product's life cycle from design to distribution. As shown in Fig. 1, the proposed model offers two main phases - the Web Application module A1, and the Manufacturing Services module A2. The Web Application module deals mainly with the management aspects of the model while the Manufacturing Services module represents the required data management for the entire manufacturing process. The Manufacturing Services module, in its own hierarchy, encompasses the Virtual Product/Process Design and the Virtual Enterprise Location/Aggregation service modules A21 and A22 respectively. The subsequent manufacturing Execution and Distribution is included in service modules A23 and A24. Both service modules A23 and A24 include definition aspects as well as control aspects. The VF model as a whole plays the role of a central broker that facilitates data management for product's design definition, partner location and selection, and manufacturing resources management. The partner selection include the identification of the optimal partner for a particular project and the subsequent assignment of responsibilities right from the beginning where parts are ordered and manufactured and all the way to the final product assembly and delivery, all totally monitored over Internet and by the participating companies.

2 THE TOTAL VIRTUAL FACTORY (VF) MODEL

The virtual factory (VF) model, as shown in Figure 1, represents a realistic skeleton for establishing the required information and integration highways that allow the agile formation of temporary interconnected inter-enterprise networks. These virtual networks are made up both of potential registered partners (private domain) and of potential new partners bidding to take part in publicly announced projects (public domain). Although projects can be submitted only by registered members, new members can not only bid for business opened for bidding, but also submit a particular project as their solely intervention in the virtual factory formation. Whichever way, according to the project definition defined in Module A1, the company who raised the original project proposal is regarded as the host enterprise.

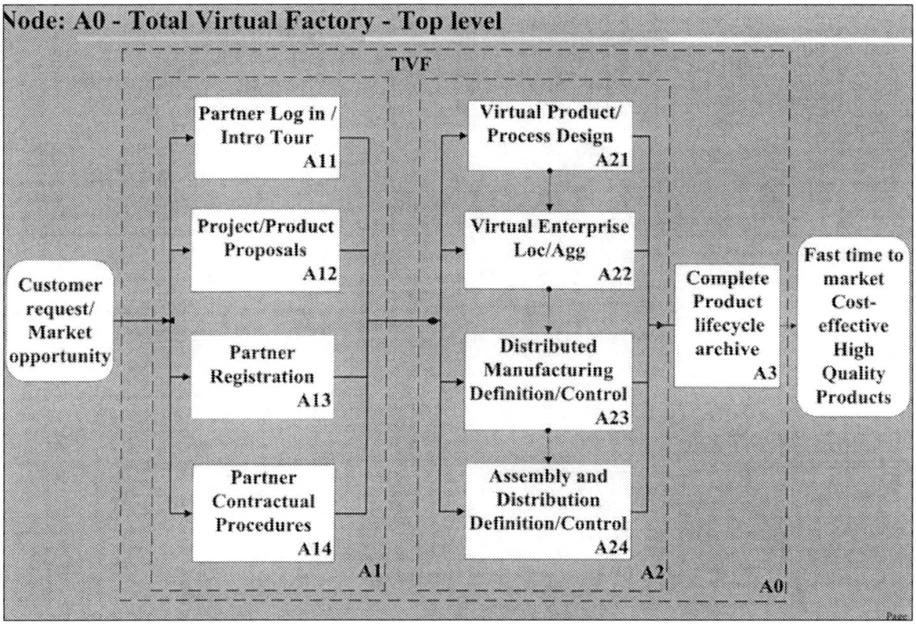

Figure 1 Virtual factory model.

In first instance, the VF will pose as a comprehensive information database for manufacturing enterprises to display their specialities, manufacturing processes, and the core competencies of the various members. Although detailed information can only be provided to active members, the VF would serve as a subject portal for general information for the non-registered manufacturing enterprises. This "visitors" to the VF can browse the different information for the different manufacturing enterprises, execute searches and explore available projects. Members of the VF can in addition, start up or manage their own databases allocating arrays of potential partners focused on product/project-based alliances. Detailed information of manufacturing partners is available to VF members, allowing them to evaluate potential benefits of alliances with other enterprises based on their manufacturing capabilities and core competencies. The VF provides thus a means of communication between two or more enterprises enabling data exchange of all products' lifecycle related information. Detailed information on the manufacturing processes will assist in the selection of the optimal partner from the list of the selected potential partners. Once the initial evaluation stage is complete, online transactions will be then allowed between the involved manufacturing enterprises to establish their detailed agreements in quantities, lead-time, delivery, and other contractual aspects.

Partners, either individuals or enterprises, can specially benefit from the VF model if they are looking for a specific service, which aims to locate a suitable partner for a particular operation. Access is provided over the Internet through a web application interface with executable simulations of the services offered by each of the four modules provided. Appropriate resources are then linked to relevant online executable options that can be run as steps of the whole transaction cycle or in an individual and independent manner. Interaction with modules in both main phases (A1, A2) allows any incoming data to be translated into formats that are compliant to a common standard and that can be readily used by the other potential partners. Past transactions can be retrieved from the VF repository database service either from the user's own past transactions or from other designs with similar characteristics that resulted from transactions completed by other users and partners.

According to the VF model, modules A21 and A22 will not only interpret the incoming request for product design data but also attempt to locate potential partners who have the required resources to achieve the design intend. Modules A23 and A24 are tasked to establish a mechanism for definition and control of the distributed operational aspects. They provide support in the planning and operational stages and control in the execution stage. As a consequence, the two modules 'Distributed Manufacturing' and 'Assembly and Distribution' are further divided in two sub-modules - one for definition and the other one for control. An interactive user/VF product development cycle is maintained throughout all the sub-modules of the Manufacturing Services module in a concurrent manner.

2.1 The Virtual product & process design module.

The first of the manufacturing service modules, 'Virtual Product and Process Design', adopts a fairly traditional design approach defined by Earle (3) as cited by Singh (4) that follows six basic steps:

- Problem Identification.
- Preliminary Ideas.
- Refinement Process.
- Analysis Process.
- Decision Process.
- Implementation Process.

The use of Internet technologies enables the change of these traditional approaches towards more concurrent ones involving simultaneous development of the product life cycle. In this method, the design process is as easily influenced by the design intend as it is by the forthcoming requirements that are due to changes in both the demand and supply sides. Examples of this are price, performance, and functionality in the demand side and material shortage, quality issues, process limitations and over demand of available resources in the supply side (5).

Resulting models associated with each of the steps are made available at the VF throughout all activities in the product development until the design process is finalised. The consequent overlap of the sequential activities then enables a concurrent approach and therefore a reduction in the execution time of the design process and a more effective product life cycle management. An adaptation of Earle's design process is implemented in the VF model in the following manner:

a) Problem definition through user/VF interaction to establish a first draw of requirements and specifications.
b) Preliminary Ideas definition using VF's product database for possible/suggested alternatives.
c) Refinement Process through on-screen simulation of design alternatives that identify detailed product's specifications.
d) Analysis Process with an onscreen representation of the different manufacturing characteristics, different product alternatives, and a first evaluation of these alternatives.
e) Decision Process based on the evaluation of the design alternatives against the original specifications set on the problem definition.
f) Implementation Process with the verification of the completeness of the data model, a virtual manufacturing execution, and the allocation of the model to the VF's database.

This adaptation has been grouped in the different sub-modules as shown in figure 2. As an example, in sub-module A214, after the user's validation, the virtually manufactured product is stored in the repository service of the VF either for future reference, or to be used as the input for module A22 - "virtual enterprise location and aggregation". Preliminary steps to this are the search criteria for the potential partners and the definition and preparation of the relevant documentation (sub-modules A215, A216)

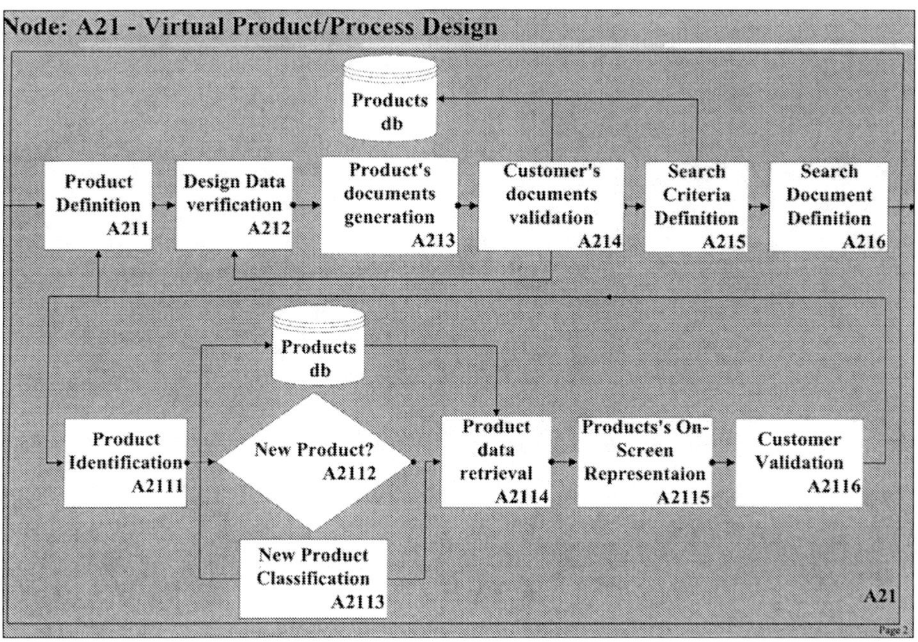

Figure 2 Virtual product/process design.

2.2 The Virtual enterprise location/aggregation module

The "virtual enterprise location and aggregation" module can now be activated by an action request. As illustrated in Figure 3, the request contains the product's specifications used by the data preparation sub-module. This preparation consists of the selection of a suitable format for the 'data collection' sub-module. The suitability of the format depends on the type of agent to be used on the search. Agents can use data definition compressed in formats that range from mark-up language documents to elaborated built-in programs using object oriented languages and specialised in data search over the Internet. At the simple end are the Extensible Mark-up Language (XML) technologies that nevertheless offer to cater for all the requirements in the implementation of the virtual factory communication, from solving the shortcoming of the Hyper Text Mark-up Language (HTML), enabling interoperability for easy initial access and further integration of new members to coordination of factory operations (6). The optimal specifications are placed in a predetermined document that must comply with the preferred format and type of information, as the Document Type Definition (DTF) for the case of XML. These templates can be made publicly available to encourage open access in order to avoid the need for rigid agreements on more complex processes during the initial stages of the virtual factory formation. The data collection can then be achieved as an automated process executed over the Internet, which subsequently provides a group of suitable partners for the virtual factory.

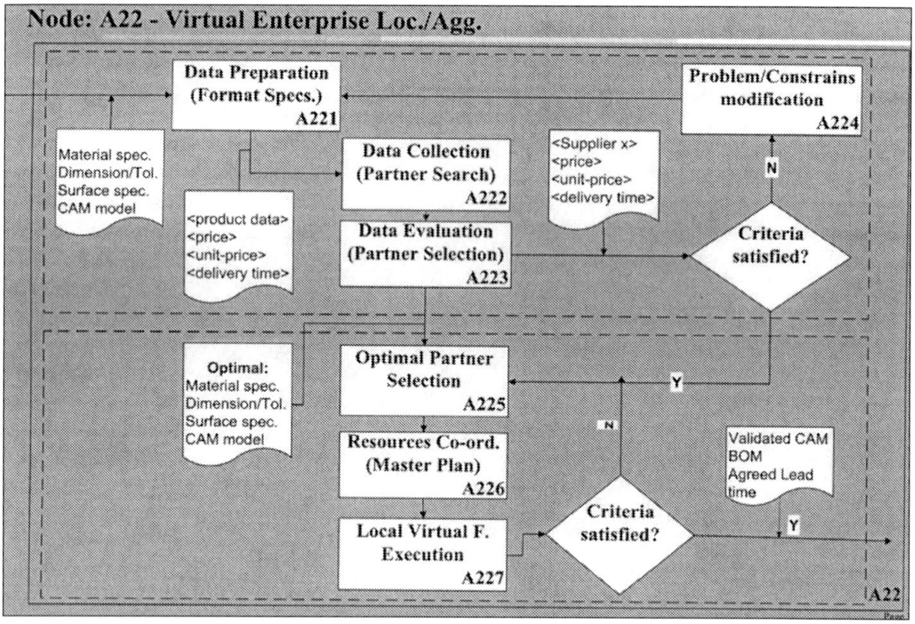

Figure 3 Node A22 Virtual Enterprise Location/Aggregation.

The data preparation, collection, and evaluation in the first phase of the "virtual enterprise location/aggregation" module represent a preliminary partner selection. A more detailed process of verification and selection of the optimal partner(s) is then initiated in the second phase. This phase may involve translation of specific CAD data into compliant generic CAD files so that they can be sent to the selected destinations. Product data can then be distributed to the pre-selected partner enterprises that will provide the execution of a virtual manufacturing of the product at the distributed manufacturing locations. Based on this virtual execution, the user or host enterprise can submit, through the VF interface, a final validation/approval of the optimal product and process, making them readily available for the real execution. This information is stored in the VF repository database as a complete model that contains data related to cost, lead-time, technology required and an optimal partner selection. This data can then be transferred to the subsequent modules in the VF model - the distributed manufacturing module, as well as the assembly and distribution module.

3 CONCLUSIONS AND FUTURE RESEARCH

The proposed total virtual factory model encompasses support for the entire product lifecycle from design to final distribution, all executable via the Internet. This proposed framework represents the use of the latest advancement in Internet technologies to build a realistic Virtual Factory (VF), which in theory does not need to own a vast amount of physical resources. Indeed such a VF can enable a quick response to market changes as it can rely on a vast number of virtually connected beneficiary enterprises to provide required services. The final phase of this research work consists of the implementation of a demonstration model that proves the feasibility of its application in the real manufacturing scenario and in particular in the SME sector. The whole model would allow faster evaluation and verification of the production capability and manufacturing lead-time offered by integration via the VF process. As a consequence an agile product development through rapid Internet-based inter-enterprise integration (web-based integration) can enable a more effective product life cycle management between participating SMEs.

REFERENCES

1 Lung, A; Benhadj-Djilali, R; Garcia, J., "An Internet-based total virtual factory for inter-enterprise integration", *Advances in Manufacturing Technology, XVII*, 2002, pp 377-382. United Kingdom, IMechE.

2 McMahon, C., Browne, J., "*CADCAM: principles, practice and manufacturing management*", United Kingdom, Addison-Wesley, 1998.

3 Earle, J. H., "*Engineering design graphics: AutoCAD release 12*", Wokingham, Addison-Wesley, 1994.

4 Singh, N., "Systems *approach to computer-integrated design and manufacturing*", New York; Chichester, Wiley, 1996.

5 Toussaint, J., Chen K., "Design agility and manufacturing responsiveness on the web", *Integrated manufacturing systems*, MCB UP, 2002, 13/5 pp 328-339.

6 Chryssolouris, G., Makris, S., Xanthakis, V., Mourtzis, D., "Towards the internet based supply chain management for the ship repair industry", *International Journal of Computer Integrated Manufacturing*, United Kingdom, Taylor & Francis, 17/1, Jan.-Feb 2004, pp 45–57.

Development of an agile manufacturing scheduling system for shop floor productions

P VENKAT, K K MATTA, and **Z WANG**
Nevada Manufacturing Research Centre, University of Nevada Las Vegas, USA

ABSTRACT

Manufacturing is the application of physical and chemical processes to alter the properties, geometry, and/or appearance of a given starting material to build parts or products. Manufacturing also includes the assembly of multiple parts, to make products. The process to accomplish manufacturing involves amalgamation of machinery, tools, power and labour. Manufacturing is carried out as a sequence of operations. Each operation brings the material closer to the desired final state [1]. Agile manufacturing can be defined as the capability to thrive and survive in a competitive environment of constant and erratic change by reacting quickly and effectively to varying markets, driven by customer-designed products and services [4].

The ongoing research is focused on developing a finely tuned process, capable of meeting the cost, quality, variability, and time pressures imposed by the marketplace. The primary objectives include the reduction off manufacturing lead-time to the minimum possible, and achievement of a high level of process control.

This research focuses on the increased need for automation in shop floor control systems. It also aims to remove the complex nature of existing software and to create very user-friendly interfaces with easy drag and drop operations. This research paves way to a new scheduling algorithm, which is fast, robust with the help of a database, and is proved efficient. The algorithm supports the real time constraints of a shop floor. The shop floor control system module is a large part of an MRP ii system and difficult implementation procedures soon will be available for testing at small and medium scale industries like Dolphin Machines Inc, which are unable to buy the current software, because of its costs and difficult implementation procedures

1 SCHEDULING ALGORITHM

The main aim of scheduling process is to devise the most optimum way for various activities in a shop floor. When several workstations are required to process a job, the machining and moving of the parts between the workstations can be very time consuming if they are not planned properly [6]. Machine scheduling can take advantage of simultaneously optimising the job schedule and order of operations in each job [7]. Most optimum schedule should satisfy the following conditions:

- There should be no idle machines at any time.
- The batch size should be as large as possible, so that the set up times are greatly reduced.
- The work in progress should be reduced.
- The average time it takes to process one product should be minimized.
- The bottleneck machines are identified, and have to be 100% utilized. On the other hand, the other machines have to be utilized just enough to keep the flow in the entire system, a constant.
- The distance travelled by the material through the different machines should be minimized.

In Shop Floor Control scheduling is an important activity [2]. Despite the numerous developments in scheduling, a common complaint from shop floor personnel is that scheduling techniques and solutions developed are out of touch with the reality of the shop floor. Controlling the production system on the shop floor is extremely difficult because of the complicated manufacturing process and re-entrant characteristics [5]. This is because the actual problems and events in a shop floor cannot be completely depicted in computer simulation or in theory. Hence, a project was conducted at Dolphin Machine Inc. in an attempt to understand the real time scheduling problem. This company does not use any shop floor control software, nor does it use any standard scheduling programs. The schedule is developed on a weekly and sometimes a daily basis using Microsoft Excel (spreadsheet). As a first task the layout of the factory floor has been obtained.

Due to the large number of products and the complexities of the operations involved in creating these products, decision has been made to study the processing of one particular product called the apron hook. The routing sheet for this product was obtained. Then the details of each of these operations were separated and a simple table as below was created.

Step No:	Description	Processing Time (min)
1	Rough machine per roughing program	20
2	Cut coupons 1/2 x 1/2 x 4	10
3	Finish Machine complete per B/P and program. Complete "first part off production" forms for each program	15
4	Deburr in accordance with P707	10
5	Abrasive blast in accordance with P700	20

Table 1 Simplified Route sheet for Apron Hook

Machine	Number of Machines	For Use on Operation
CNC Mill (CM)	16	1 through 3
Deburring Machine	1	4
Grinders	4	5

Table 2 Availability of machines related to Apron Hook

Here the times spent at these operations were assumed. The next table shows the list of machines available at the shop floor and the information about which machines may be used in which processes of this product.

Now the objective is to schedule the given activities and distribute them amongst the available machines in such a way as to finish creating the product in the least amount of time, with minimum waste of resources. There are sixteen machines, which can work on steps 1 to 3. Scheduling the machines using combinations is equal to (16C1 + 16C2 +...16C16) 3. Steps 4 can be done in one way and step 5 can be done in 4 ways only. Therefore, the total number of combinations of creating one product is (16C1 + 16C2 +...16C16) 3 x 4 x 1, which is a very large number. It is impossible to go through each one of these iterations and find the most optimum one. It is also impossible to set up that many constraints as to reduce the solution space to even a few possibilities.

Previous work on this problem has shown that more the number of machines assigned to the first few processes, the faster the process can be complete, so a case where all the possible machines are assigned to each of the steps is to be studied. In the case of the Apron Hook, 16 machines work on the first step, finishing the entire batch of products and then moving on to next step. This also ensures that at any time there are enough of parts of the previous step available for them to process. The work in progress inventory would also be greatly affected. In addition, at no time are any of these machines idle because of lack of raw materials. Again, the advantage seems to be visible only at higher batch sizes as the set up times increase. This is the only disadvantage or scheduling in this manner and can easily be overcome by following the principles of Just-In-Time philosophy.

1.1 Development of the Scheduling Algorithm

In developing the previous model assumptions have been made that all the 16 machines are free to be scheduled at all times and running at infinite capacities. However, the Apron Hook is just one of the large numbers of projects, and some of these machines could already be busy on these other projects. This is a variable that separates the research from the real time model. Thus the real time database is used to store the schedules of each machine, so that the times at which these machines are free could be easily retrieved. However, by including this one variable, the process for finding a schedule that is optimum becomes harder. The solution space is highly reduced, but is broken into too many spaces. The number of constraints on the problem is too high to obtain the optimum. Research software used for scheduling would run continuously for months to calculate the optimum for one scheduling problem. However, our hypothesis that the more number of machines being used in initial stages still holds good. The aim has shifted from finding a path of least time of flow to finding a path of maximum utilization of machines, but the flow time is still being minimized

Considering an example for a product:

Let D be the number of products to be made.
N is the numbers of steps required to complete the product using the same machines.
M is the number of machines available of same capacity and being able to work on any of the above N steps.
t1 be the average set up time for all the steps of a product cycle.
t2 be the average processing time for all the steps of a product cycle.
n1, n2 and n3 … are the number of machines working on each of these steps.
The first step takes (t1 + t2 * D /n1) to be completed.
The second step starts after at least one product is made. That is the remaining (M–n1) machines are idle for (t1 + t2) minutes.
Similarly, the second step takes (t1 + t2 *D/n2)
Hence, M – n1 - n2 machines are idle for 2 * (t1 + t2) minutes.
The total time of processing as a sum of times spent on all the machines:

$$T_{total} = [n1 \times t1 + t2 \times D \div n1 + n2 \times t1 + t2 \times D \div n2 + ...] + [(t1 + t2) + (t1 + t2) \times 2 + ...]$$

$$\Rightarrow T_{total} = [n1 \times t1 + n2 \times t1 + ...] + D \times t2 \times (1/n1 + 1/n2...) + [(t1 + t2) \times (1 + 2 + ..N - 1)]$$

$$\Rightarrow T_{total} = t1 \times (n1 + n2 + ..) + D \times t2 \times (1/n1 + 1/n2...) + [(t1 + t2) \times (N - 1) \times (N - 2) \div 2]$$

Of this above total time for all machines, some machines work at the same time on different steps. Hence, the total processing time is actually much lesser than above. The total flow time can be summarized as:

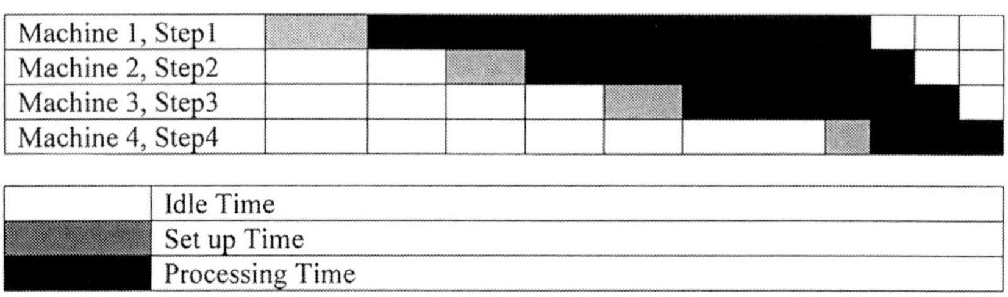

FIG. 1 Total Flow Time For a Product using previous model

$$T_{flow} = [(t1 + t2) \times (N - 1)] + [t1 + D \times t2 \div n_n]$$

Total flow time (Ttotal) = [Idle Time] + [set up time] + [actual processing time]

Among the three components of flow time shown above, the actual processing time cannot be optimised by the system. Using better machines and machining processes can only reduce it. The flow time for products created in this manner is shown next. The 'idle time' component of the total flow time shows only the idle time for the machines involved in the last step. However, except the machines involved in the first step, we can see that all other machines have an idle time. If any of these steps have more machines allocated to them, then these machines will be idle for longer times. This is the idle time component of the total time Ttotal and is greater than the idle time component of the flow time Tflow. The total idle time of Ttotal is calculated as:

$$T_i = [(n2)\times(t1+t2)] + [(n3)\times(2)\times(t1+t2)] + \ldots$$
$$\Rightarrow T_i = (t1+t2)\times[n2 + n3\times 2 + n4\times 3 + \ldots]$$

n1, n2, ... are the distributions of M machines over these N steps. There are several possible combinations here and the idle time increases or decreases according to the combinations. When any step is being processed, the sum of machines is always constant. $n1 + n2 + n3 + \ldots = M$

Consider now the second case, when at the first step,

$n1 = M \Rightarrow n2 = n3 = n4\ldots = 0$

$\Rightarrow T_i = 0$

So during the first step, n1 =M and n2=n3=n4...=0

Similarly during the second step n2=M and n1 = n3=n4 =n5... =0

The total set-up time is increased to

$T_s = M\times t1 + M\times t1 + \ldots$

$\Rightarrow T_s = M\times t1\times N$

The total time as sum of times of operation in all machines is:

$T_{total} = (t1\times M + D\times t2 \div M)\times N$

In addition, the total flow time is:

$T_{flow} = (t1 + D\times t2 \div M)\times N$

The flow time for a product created in this manner is as shown next:

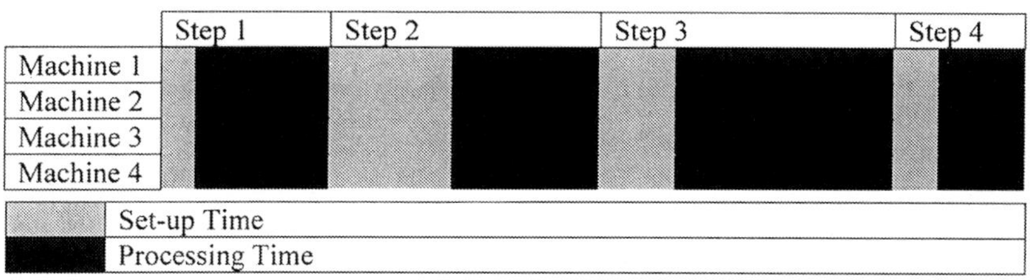

FIG. 2 Shows the Time Flow using the second model

Comparing flow time for the two cases, we can decide which one is better. In first case it is:

$$T_{flow} = [(t1+t2)\times(N-1)] + [t1 + D\times t2 \div n_n]$$
$$T_{flow} = [(t1+t2)\times(N-1)] + [t1 + D\times t2\times N \div M]$$

Assuming that $n_n = M / N$ in a case where the M machines are evenly distributed amongst the N steps.

In the second case:

$T_{flow} = (t1 + D\times t2 \div M)\times N$

Comparing the two we have the $D \times t2 \times N \div M$ in both case which is the total processing time. Usually the set-up times are lesser compared to the actual processing time.
t1 << t2

$$N \times t1 < (N-1) \times (t1 + t2) + t1$$

$$\Leftrightarrow N \times t1 < N \times t1 + (N-1) \times t2$$

Hence, the flow time in the second case is lesser than that in the first case and, the flow time is still optimised. After proving the hypothesis stated earlier, applying the real world constraints of machines being available at certain times only. The scheduling algorithm provides the methodology of applying the proven hypothesis in practice.

1.2 Scheduling algorithm: -

The algorithm follows a fluid-flow type model. The different machines represent different pipes with different diameters and the flow fills as much of the pipes as can be.

FIG. 3 Shows the scheduling problem as similar to a flow problem

The dark areas have already been scheduled, and this data can be obtained from the database. At any time T, if the flow in any two pipes, with areas A1 and A2 need to be synchronous, then the length of flow is the same. Here A1 and A2 are a measure of capacity of the machine as they determine how much flow they allow through it. If N1 and N2 were the number of parts through these machines of capacities C1 and C2 in terms of time per unit product, then Volume / Area or the Length is the same for these machines working simultaneously.
N1 x C1 is the total time spent by the number of parts at this machine.
Hence N1 x C1 = N2 x C2 or (N1/N2 = C2/C1)

This is the case of infinite scheduling occurring at almost the end of the algorithm for each step, when the idle time on any machine is more than sufficient for the products remaining in the batch to be assigned .The products to be scheduled have already been prioritised according to the 'Master Production Schedule'. This algorithm schedules these products in order of priority. For each product, (from the routing sheet) the steps are to be scheduled in order. A unique batch size is first decided based on the total number of products, average time taken in any machine to finish the processing etc.
B = batch size.
R = average time spent in this step.
N = Total Number of products to be created.
Z = number of machine which can simultaneously on this step
B is inversely proportional to R, because the more the time is spent on one product's step, the lesser should be the batch size. At the same time, dividing the number of products to be made into a large number of batches would increase the set-up times largely; hence, B is directly

proportional to N. Also the number of machines, which will work on this step simultaneously, affects the batch size. If many machines are working simultaneously, the batch sizes can be made smaller.

B = k x N / (A x Z)

Where k is the proportionality constant and can be supplied from the statistics or the use of a B-Spleen curve.

From infinite loading, it knows the equipment that can work on this step of the process. The data from infinite scheduling is stored in the table MPS Process Plan Schedule.

Column Name	Data Type	Length
Processed	Char	18
StepNum	Int	4
IDValue	Char	16
IDType	Char	1
Quantity	Bigint	8
Unit Time	Float	8
Position Size	Char	23
Ready For Schedule	Int	4

Table 3 Data from infinite scheduling stored in MPS Process Plan Schedule

If there are two or more machines available simultaneously, processing is done on all of them simultaneously until the batch size is completely assigned. If there are machines of different capacities being used in the same step, the distribution occurs in the ratio of their capacities (In terms of unit time per unit of this product). This ensures that the bottlenecks do not appear and there is a minimum of work in progress. After the first step, the average time for second step is compared with the first. If it is greater, the step can be scheduled any time after the first product from step one is finished. If it is lesser, then their needs to be as many products in step two to be finished after step one completely finishes, as there are machines working on step one. This is done calculating time in the reverse way.

If any step starts at time X and has an average time of a1, the products at this machine are produced at intervals:

X+a1, X+ 2 (a1), X + 3 (a1) …

If the next step has an average time b1, and starts at time Y, the products at this machine are produced at intervals

X+Y+b1, X+Y+ 2 (b1), X+Y+ 3 (b1) …

X is the first steps starting point, which has already been determined; a minimum possible value of Y has to be calculated. Independent of the values of a1, b1, and the value of Y has to be greater than or equal to a1, otherwise there will not be any product of step 1 available for step 2.

1.2.1 Case1:

If a1 $<$ b1 then

For Y = a1

X+ n (a1) $<$ X+ Y + n (b1)

Where n is the number of parts produced.

1.2.2 Case 2:

If a1 $>$ b1 then

There has to be at least as many number of parts in step 2 as there are number of machines (Z) working on step 1, to be scheduled after the completion of step 1.

X+ Y + n (b1) $>=$ X + n (a1) +Z

Y $>=$ n (a1) + Z – n (b1)

If Y amount of time as calculated above is not allowed after the schedule of step1 starts, there will be a time when the number of parts of step1 will fail to provide for step2, resulting again in idle time for machines of step 2.The remaining steps follow the same criteria. They are also scheduled by comparing them with the previous step's schedule. This ensures that the machines are not idle until the entire step is finished and that at any time, there are a product of previous step available for the next step. This goes on for all the products in the MPS.

2 CASE STUDIES ABOUT THE FLOW OF INFORMATION AS IT REACHES THE SFC MODULE ARE AS BELOW:

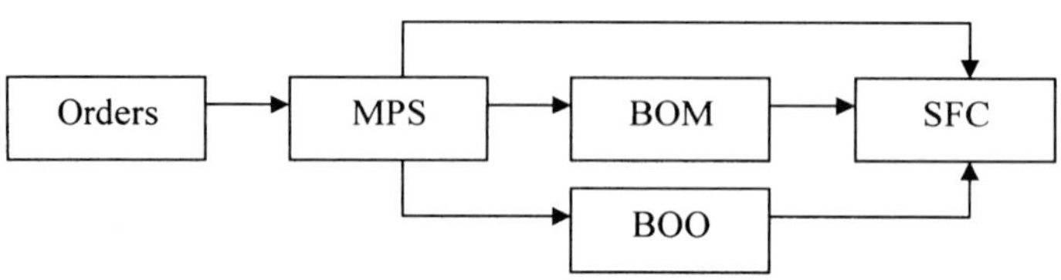

FIG.4 Shows the flow of information to SFC

First, a product ID for the product 'Apron Hook' is generated from the table Product. An order of 750 products is assumed. An order is thus generated from the Order table. The entries in Order table are read into a Master Production Schedule. An MPSID is generated from the table MPS. Since the 'Apron Hook' does not involve an assembly process, the Bill of Materials has only one entry for it. The part 'Apron Hook' is input into the table Parts, and a part ID is generated. In the table MPS Process, the steps in the routing sheet of 'Apron Hook' are translated into step descriptions, possible equipment types which can work on it etc. The current configuration of Apron Hook Project is as follows:

ProductID 10010011 OrderID 200208000008

MPSID 200208080001 PartID 10000002.

The routing sheet is translated to the form below:

Step No:	Description	Machine Type	Processing Time (min)
1	Rough machine per roughing program	CNC Mill	20
2	Cut coupons 1/2 x 1/2 x 4	CNC Mill	10
3	Finish Machine complete per B/P and program. Complete "first part off production" forms for each program	CNC Mill	15
4	Deburr in accordance with P707	Deburring Mc	10
5	Abrasive blast in accordance with P700	Grinder	20

Table 4 BOO for Apron Hook

The Dolphin Machines Inc has 16 CNC mills, 1 Speed Deburring machine and 4 Grinders. These machines can be assigned to the above tasks in any combination. For each such combination, a process layout needs to be developed. In these process layouts, the time taken by each machine needs to be input. Schedule was developed for 6 different combinations. For each of these six combinations, the work in progress, inventory level and the number of finished products are found at one-hour intervals throughout their production time of around 250 hours. At each point, the total work in progress has been calculated and plotted. The plot for the first combination is given below. The total processing time is thus 224 hours.

The different combinations of machines for which the schedule was developed are as follows:

Combination 1		**Combination 2**		**Combination 3**	
Step No	**Number of Machines**	**Step No**	**Number of Machines**	**Step No**	**Number of Machines**
1	3	1	4	1	5
2	3	2	4	2	5
3	3	3	3	3	3
4	1	4	1	4	1
5	2	5	2	5	2
Combination 4		**Combination 5**		**Combination 6**	
Step No	**Number of Machines**	**Step No**	**Number of Machines**	**Step No**	**Number of Machines**
1	6	1	6	1	9
2	6	2	6	2	9
3	3	3	4	3	9
4	1	4	1	4	1
5	2	5	2	5	2

Table 5 Combinations of machines tested

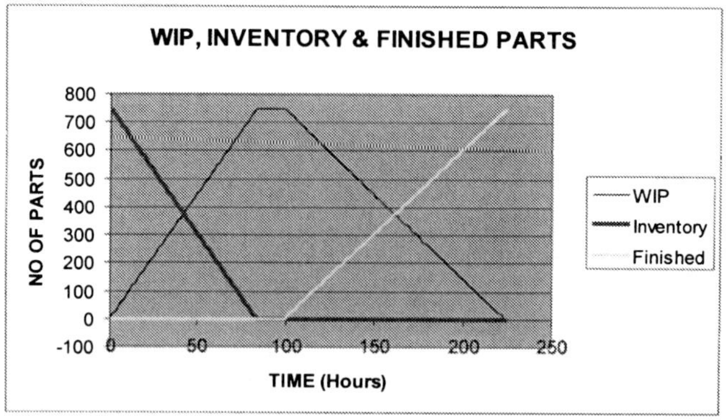

FIG. 5 Shows the WIP, inventory and number of finished parts for combination 1.

The processing times for the six combinations are shown below.

Combination	Processing Time (Hours)
1	224
2	215
3	206
4	197
5	197
6	188

Table 6 Shows the processing time for the six combinations

The above table indicates that there is a decrement in the processing time when more machines are being employed. This decrement is reduced as more machines are added and is not seen when machines used in previous steps are added. These machines are not free to be scheduled until the previous steps are complete. However, a reduction of processing time from 224 hours to 188 hours is quite significant. This implies that orders can be processed 4 days earlier than they were before in case of the Apron Hook's order

3 CONCLUSIONS

The shop floor control is an important and complex task in modern manufacturing. This research concentrates on creating a lean manufacturing system by integrating philosophies of Just-In-Time thinking with the information connectivity of MRP systems. This hybrid model has advanced scheduling algorithm, which optimises the lead-time of a product.

REFERENCES

1. Mikell P.Groover, "*Fundamentals of Modern Manufacturing*", 1996, Prentice Hall Inc
2. Bauer, Bowden, Browne, Duggan, Lyons "*Shop Floor Control Systems*", 1991, Chapman & Hall.
3. http://www.cybor.co.nz/news/matplan.html
4. *International Journal of Production Research*, v 36, n 5, May, 1998, p 1223-1247
5. *The International Journal of Advanced Manufacturing Technology*, Volume 22, Numbers 1-2, pp 75-78
6. *Engineering Costs and Production Economics* ,Volume 21, Issue 1 , February 1991, Pages 21-34
7. *International Journal of Production Research*, Volume 41, Number 13 / 10 September 2003, pp 2957 – 2973.

Research on the strategy and implementation methodology of low-cost moderate automatization semiflexible manufacturing system for leather enterprises

Y ZHANG
Industrial Engineering Institute, Zhejiang University of Science and Technology, Hangzhou, China
X-L WANG
Mechanical Engineering Department, Jiliang University of China, Hangzhou, China

ABSTRACT

In order to enable leather produce enterprises to find their proper positions and deploy their AMT strategies, this paper proposes a new conception – Low-cost Moderate Automatization Semiflexible Manufacturing System (LC-MA-SFMS), and also researches the architecture of the general technology of LC-MA-SFMS. This paper first analyzes the development of manufacturing methodologies, and then discusses the basic economics reasons for LC-.MA-SFMS and the importance of LC-MA-SFMS as an enterprise strategy. Base on these, the author presents that LC-MA-SFMS may be the "principal production mode of 21st century for leather produce enterprises". Finally, this paper provides a general technology framework of LC-MA-SFMS from the point of view of methodology.

1 INTRODUCTION

At present, in manufacturing fields, the new industrial revolution, which revolves round, the advanced manufacturing technology (AMT) is coming [1, 2]. In order to accommodate to new circumstances forwardly and try to capture living and developing space for the cruel market competition, lots of manufacturing enterprise provided a series of new canonical form and philosophic theory for advanced manufacturing system and put them in practice. These new canonical form and philosophic theory include CIMS (Computer Integrated Manufacturing Systems), LP (Lean Production), CE (Concurrent Engineering), BPR (Business Process Reengineering), AM (Agile Manufacturing), VM (Virtual Manufacturing), etc. [3, 4]. In broad sense, each of them is division or class in the complete system of AMT, so they are all predicted "principal production mode of 21st century" [5, 6].

Essentially, AMT is an organic confluent, which is composed of traditional manufacturing system, information technology, automation technology, modernistic management technique and so on. They are interinfiltrated, some are components of one and some is the evolution of another one. It is more important that AMT is the production of information technology, or the offspring of modernistic management thoughtway, or the resultant of well-knit of both.

Today, the research and application of advanced manufacturing technology and advanced manufacturing model are especially important for Chinese enterprises. However, on the other hand, owing to the problems and difficulties caused by the specifically historical context and real-condition, in the course of introducing the advanced manufacturing technology from elsewhere, Chinese enterprises, especially macromedia enterprises are up against tremendous stress and conflict.

How to breach these abhorrent stress and conflict, and then new type manufacturing enterprises can be rebuilt to face 21st century triumphantly by implementing advanced manufacturing technology step by step, Chinese enterprises shoulder heavy responsibilities and embark on a long road.

In order to enable Chinese leather produce enterprises to find their proper positions and deploy their AMT strategies, this paper proposes a new conception – Low-cost Moderate Automatization Semiflexible Manufacturing System (LC-MA-SFMS), and also researches the architecture of the general technology of LC-MA-SFMS.

2 DEVELOPMENT STRATEGY AND GROSS STRUCTURE OF LC-MA-SFMS

Compared with industrial developed countries, Chinese macromedia leather produce enterprises are still in a deeply omnibearing backward state. The uppermost intuitionistic differences are the backwardness of the means and furnishment of manufacturing engineering and the mode and technique of production organizational management. In Chinese leather produce enterprises, the principal facilities of production are still traditional general purpose machine tools and some special purpose machines in addition, and the main mode of organization and management of manufacturing is conventional handcraft and posteriori. The backward status makes the market competitiveness of Chinese leather produce enterprises become weaker badly. In the face of such austere situation, it is the most urgent affairs to develop and innovate in manufacturing mode, which should adapt to the situation of Chinese macromedia leather produce enterprises.

Although some advanced manufacturing automatization monotechnics and advanced manufacturing modes are quite alluring, it should be averted from pursuing the utopian advanced manufacturing technology, which is difficult to be, carried into execution in Chinese macromedia leather produce enterprises now. Based on the present situation of Chinese macromedia leather produce enterprises, there is no doubt that it is impossible and needless to advance the level of manufacturing technology and the automaticity of manufacturing by renewing great lots of technical equipment. On one hand, it is unpractical and infeasible to make the level of manufacturing automation for Chinese leather produce enterprises have a skyrocketing rise because of quick change of global market condition, automaticity only can be improved step by step. But on the other hand, in order to make enterprises exist and develop in the cruel environment of global competition, some low-cost roboticized technical means can be utilized to improve the performance of traditional manufacturing equipment, just like PLC technique, new type digital control technique and so on. Surely the modern mode and technique of production organizational management should be added on coinstantaneously.

In this paper, authors propose a new conception: Low-cost Moderate Automatization Semiflexible Manufacturing System (LC-MA-SFMS). The gross structure of Low-cost Moderate Automatization Semiflexible Manufacturing System is shown in Fig.1. This system is composed of design and manufacture of leather produce equipment, low-cost roboticized technical means improvement for traditional manufacturing equipment, cells of organizational framework of the system and the constituent structure form of the system and the management technique of the system, etc.

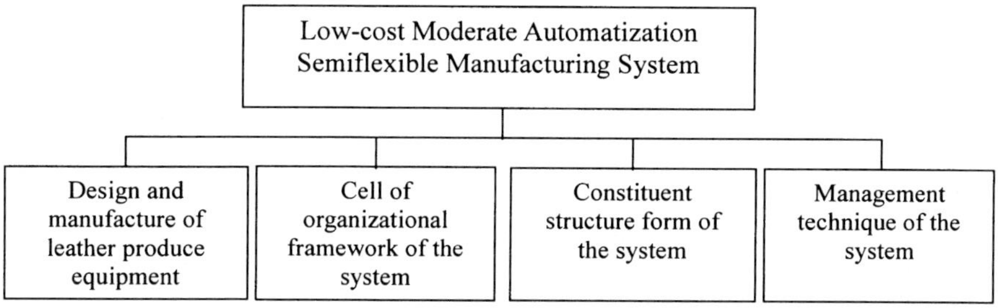

Fig.1 The gross structure of LC-MA-SFMS for leather produce enterprises

3 DESIGN AND MANUFACTURE OF PRODUCE EQUIPMENT FOR LC-MA-SFMS

Leather produce equipment is one of the main foundations for manufacturing cells and system. It must be economical and practical, at the same time, it must also be provided with moderate automatization, and thus, based on the demands, it can be improved step by step. So the basic construction feature of the leather produce equipment should be modular, just as shown in Fig.2. The contents of design and manufacture of produce equipment for LC-MA-SFMS include two: on one hand, the new conception of LC-MA-SFMS should be accepted to improve the performance of traditional leather manufacturing equipment and then make it be roboticized, digitalisation and semiflexible, on the other hand, the new type leather produce equipment should be studied based on the conception of LC-MA-SFMS.

According to the requirement and necessity, the improvement in the control performance of traditional leather manufacturing equipment should be done step by step. Some mature and low-cost roboticized technical means, which met the case of modularization, standardization and serialization, can be utilized to improve the control performance of traditional leather manufacturing equipment, just like PLC technique, new type digital control technique, digital display technique, IPC control technique and so on.

The new type leather produce equipment based on the new conception of LC-MA-SFMS should be studied to meet with the basic structure of LC-MA-SFMS. The characteristics of the new type leather produce equipment, which studied by Object-Oriented design method, should include modularization, standardization, serialization, digitalization, moderate automatization and semiflexibility, etc. Thus, the new type leather produce equipment can satisfy the users not only their current demands but also their escalatory need in the future.

4 CELLS OF ORGANIZATIONAL FRAMEWORK OF LC-MA-SFMS

Manufacturing cell of LC-MA-SFMS is the logic entity with certain relative unattached machining functions. It is the basic integrant of the manufacturing system. It is composed of produce equipment, information, human and other basic elements. The proportion of the position and action of these basic elements acting in the manufacturing cell is determined by the requirements of the function and performance of the respective cell. The difference proportion relation always results in the difference performance of each manufacturing cell. As the basic element for LC-MA-SFMS, manufacturing cell must have some essential properties, just like automatic rhythmicity, collaboration, semiflexibility, etc. In LC-MA-SFMS, any one of the manufacturing cells will be in contact with other cells about materials, information and human. Therefore each manufacturing cell must be provided with basic interfaces for materials, information and human, etc.

According the above, the manufacturing cell of LC-MA-SFMS is composed of three basic elements, viz. produce equipment, information and human. It also has three essential properties, viz. automatic rhythmicity, collaboration and semiflexibility. Finally, the manufacturing cell must be provided with three basic interfaces, viz. materials interface, information interface and human interface. The basic structure of the manufacturing cell of LC-MA-SFMS is shown in Fig.3.

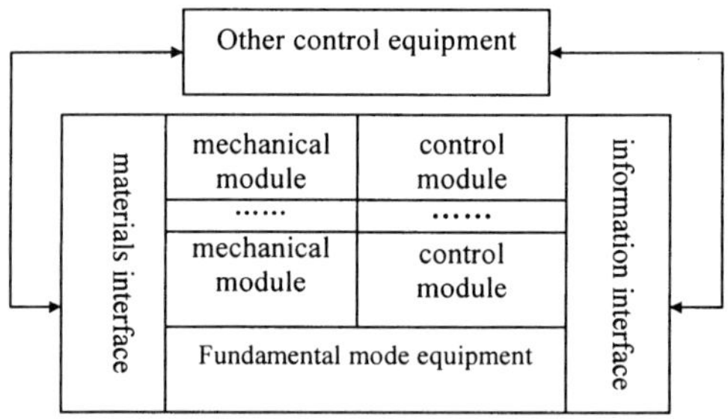

Fig.2 The basic construction feature of the produce equipment for LC-MA-SFMS

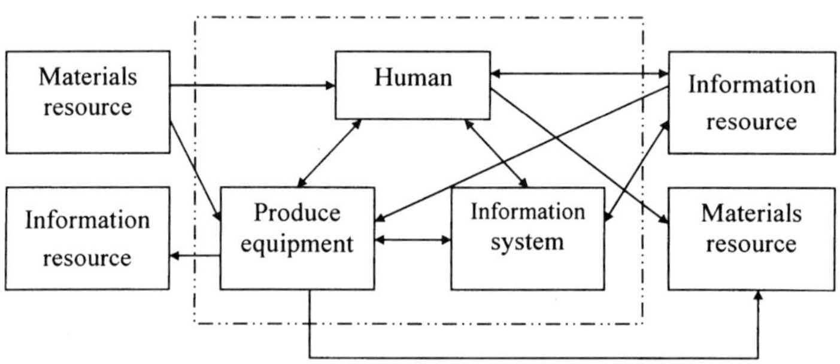

Fig.3 The basic structure of the manufacturing cell of LC-MA-SFMS

5 CONSTITUENT STRUCTURE FORM OF LC-MA-SFMS

LC-MA-SFMS is composed of several manufacturing cells, its structure of constituent form is shown in Fig.4. From the figure, it can be seen that the constituent form structure of LC-MA-SFMS is nonlinear. The relationships among the substrate cells that belong to the same manufacturing cell are determined by interior collaborative rules, on the other hand, the relationships among the manufacturing cells are determined by exterior collaborative rules. The constituent forms for the whole system or each cell on every layer are dynamic. According to the dynamic change of each interface of the system, these forms can be corrected on real time. Then these forms should be redesigned in term of the global optimization strategy for the overall system. Thus, LC-MA-SFMS can respond to the dynamic change of the global market rapidly.

6 MANAGEMENT TECHNIQUES OF LC-MA-SFMS

The management mode of LC-MA-SFMS combines independent collaboration management in each cell with negotiable collaboration management among cells. In this case, the cell can manage its internal affairs independently by any entente with others, and its exterior affairs must be managed by negotiating with other independent cells. So each cell is a controller, as well as, it is a performer. The relationship among the cells is shown as Fig.5.

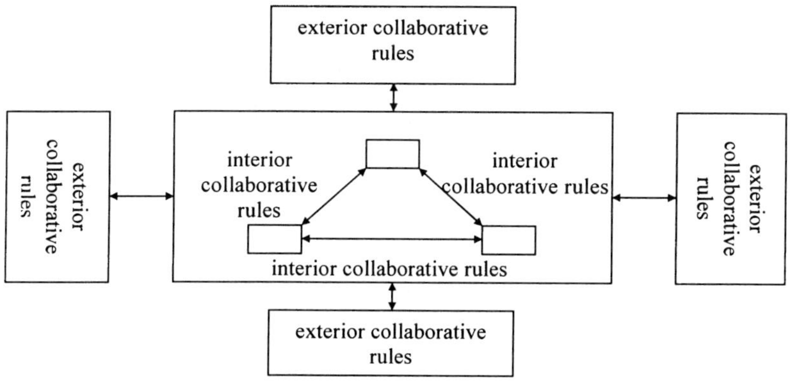

Fig.4 The structure of constituent form of manufacturing cell of LC-MA-SFMS

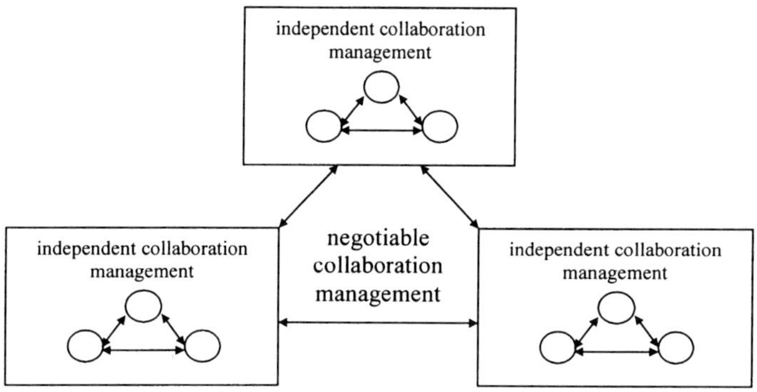

Fig.5 The relationship among the cells of LC-MA-SFMS

7 CONCLUSIONS

In this paper, in order to enable Chinese leather produce enterprises to find their proper positions and deploy their AMT strategies, the author provides a new conception: Low-cost Moderate Automatization Semiflexible Manufacturing System (LC-MA-SFMS), and also research the architecture of the general technology of LC-MA-SFMS. This new conception predicts that after Chinese enterprises have acquired some kinds of core competition advantage and research & development competencies, enterprises can join the world's specialization and cooperation system by LC-MA-SFMS based on the proper industrial organizational structure. On the other hand, China government should adjust enterprise resources deployment and competent advantages by economics rules. Then Chinese leather produce enterprises can compete with transnational corporations, and form a kind of resource dynamic optimizes mechanism.

REFERENCES

1 Shaw, M. J., "Information-Based Manufacturing with the Web", *International Journal of Flexible Manufacturing Systems*, 2000(12): pp 115-129

2 Cheng, K., Pan, P.Y. and Harrison, D.K., "Internet as a tool with application to agile manufacturing: a web-based engineering approach and its implementation issues", *International Journal of Production Research*, 38(12), 2000, pp 2743-2759.

3 Gek Woo Tan, Hayes, C.C. and Shaw, M. J., "An Intelligent-Agent Framework for Concurrent Product Design and Planning", *IEEE Transactions on Engineering Management*, 1996, 4(3): pp 801~806

4 Molina, A., et al., "Modeling manufacturing resources, processes and strategies to support concurrent engineering", *International Conference on Concurrent Engineering: Research and Applications*, 1994, Pittsburgh, PA: pp 41-50

5 Shi, H. M., "Research on Multi-Agent System With Application To Distributed and Networked Manufacturing", Keynote Paper at the *5th International Conference on Mechatronics and Machine Vision in Practice*, Proceedings, Nanjing, 1998, China: pp 9-14

6 Graves, R. J., et al., 'Estimating tools to support multipath agility in electronics', *Manufacturing Technology Part C: Manufacturing*, 1996: pp 48-56

On-line Education and Industrial Training

Agents for operator training

A CRISPIN and **P CLAY**
School of Technology, Leeds Metropolitan University, UK
T BAYES
SATRA Limited, Kettering, UK
D REEDMAN
R & T Mechatronics Limited, Melton Mowbray, UK

ABSTRACT

The paper discusses how agents can be used for operator training. It has been shown that agents can be used for task guidance in both simulator-based training and real world operation of complex industrial equipment such as robots. The paper proposes an agent framework to help leather cutters undertake part placement tasks in a simulator based training environment. Footwear manufacture involves cutting components from a hide to maximise material utilisation. A training simulator allows operators to be trained without wasting leather by using digitally scanned parts and hides. Integrating agents into a training simulator allows user actions to be monitored and feedback to be provided on leather cutting practices. Issues relating thread agent development and the delivery of information using client/server or distributed architectures are considered.

Keywords: Agents, Operator Training, Simulator for Leather Cutters, Design Consistency Checking, Microsoft .NET Framework, Client/Server Architecture

1 INTRODUCTION

Cutting of footwear components involves placing component parts onto a hide to balance a number of optimisation criteria including minimising material waste, ensuring that the directions of certain components are related to anisotropic axes and avoiding regions of unacceptable quality (Heistermann, J. and Lengauer, T. 1995). A significant proportion of the manufacturing cost is invested in the cutting of the hide and so the efficient utilisation of this resource is of prime importance. The difference to the costs incurred between a good and average cutter can be as much as 5% of the total manufacturing cost. Although techniques are being researched for automating the cutting process (see Crispin et. al. 2003) most leather cutting is undertaken manually by operators who are awarded accreditation when they are capable of meeting required standards and frequently rewarded to maintain those standards. The research outlined in this paper addresses the design of an interactive graphical simulator for training leather cutters which is capable of monitoring user actions to provide feedback on nesting methodologies and making an assessment on material utilisation.

The advantage of a simulator is that an operator can be trained within a simulated environment without wasting leather by using digitally scanned hides and component parts. Interactive graphics allow hands-on practice in part placement allowing an operator to develop spatial awareness skills. Software agents are being integrated into the simulator to assist the user in placement decisions. The agents monitor user actions and provide feedback on leather cutting practices. Agents are capable of autonomous actions and can make decisions (i.e. can reason) so that they can respond to a sensed environmental change. For example, an overlap agent would detect if the user moves a part so that it partially covered another part and take appropriate action. The paper discusses the software design of the agent-based training simulator for leather cutters.

2 NESTING SIMULATOR

2.1 Graphical user interface

A lay-plan simulator has been developed for reading, manipulating and displaying polygon file formats (.dig and .rat files) specific to the shoe industry as shown in Fig. 1. Hide files contain a list of boundary points with millimetre resolution and closed polygons point lists relating to grade regions marked on the hide. In Fig.1 the darker regions of the hide represent poorer quality grade material. Shoe component part files contain point co-ordinates defining the boundary of the part and have been scanned with a resolution of one thousand of an inch (mil). Part data has to be scaled to correspond to hide dimensions. The required numbers of shoe components to be cut from the hide are displayed in the parts requirement toolbar. Components can be dragged from the toolbar and dropped on the hide for placement. Each component shape can be moved and rotated so that optimal nesting placement can be obtained. The software has been designed with a multiple document interface (MDI) so that components can be placed across a number of hides. The percentage coverage of components placed on the hide is displayed in real-time.

2.2 .NET Framework

The lay-plan simulator has been developed using the Microsoft .NET framework. The .NET framework is an environment for building and running software including web-based applications. It contains a huge number of classes that encapsulate many of the tasks required by the developer when designing software applications. For example, graphics classes are provided for 2D-vector drawing and the socket class library simplifies the design of a TCP server. Visual Studio.NET ($C^{\#}$ programming language) has been used for rapid application development.

The .NET framework contains classes for graphics programming using GDI+ (Graphics Device Interface), the Windows DLL responsible for displaying information on screens and printers. Important for 2D vector graphics are the graphics path and matrix classes. A graphics path object can be defined so that complex shapes (components and hides) can be drawn as a single unit that can be manipulated using transform matrices. The .NET framework matrix class encapsulates methods such as RotateAt(θ, P_{xy}) and Translate(Δx, Δy) that are used to perform affine matrix operations where θ is the rotation angle, P_{xy} the (x, y) co-ordinate rotation point, Δx the x co-ordinate translation and Δy the y co-ordinate translation. A single

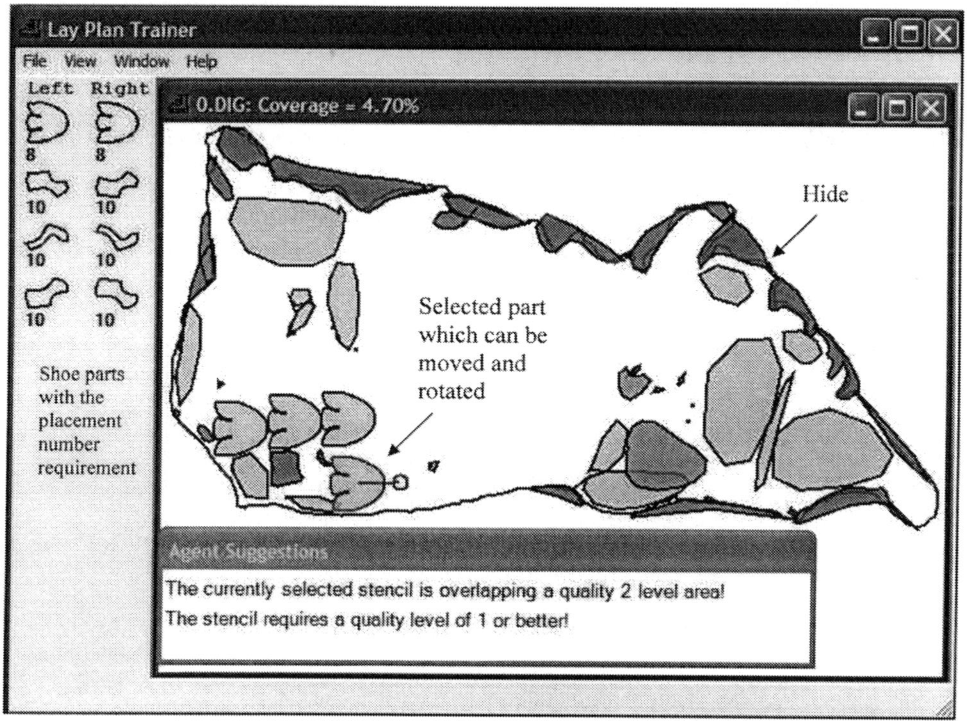

Fig.1. Agent based lay-plan simulator

transformation matrix can be formed by using these methods which can then be applied to geometric data in a single operation. The order in which the methods are applied is important as it affects the resulting transformation matrix.

Using the matrix class the simulator allows component parts to be rotated around their centroid point. Hide vector objects can be rotated around the centre point of the hide bounding box. A matrix transformation is applied to convert a graphics path with hide and shape data placed at world coordinates (measured in mm) to screen coordinates (measured in pixels) depending on the current viewing parameters (i.e. zoom and rotation). Any interaction with the screen using the mouse (e.g. shape selection and placement) can be converted to world coordinates by applying the inverse of this transformation matrix. The accuracy of positioning is dependant on the scale of the current view. A 1mm square lattice space results with the zoom set to 100%.

3 AGENT BASED SYSTEMS

An agent can be defined as an autonomous software module that can sense and act upon its environment in a rational manner. Autonomous behaviour means that the agent can act without the direct intervention of users (or other agents) so that it has control over its own actions. Agents are independent reasoning systems specified in terms of logic. Russel and Norvig (1995) describe how to build and implement agent programs and the book is a standard text in the field. It considers four types of agent program, namely; reflex agents, reflex agents with internal state, goal based agents and utility agents. The main pitfalls of agent orientated development are discussed by Wooldridge and Jennings (1998). The paper makes claims about bad practice in multi-agent system development to prevent developers falling into the same traps. It is based on a decade of experience in developing multi-agent systems in a range of applications.

Related work is that of St-Cyr et. al (2000) who have investigated using agents to improve the usability of CAD applications. The work investigated using an agent to help in the design of office layout. The agent checks the legality of actions requested by the user before performing them. For example, if a designer tries to place a computer on a heater the agent reports this as an illegal request and suggests alternative actions.

Van Breemen (2002) discusses integrating agents into software applications. The paper discusses the problem of whether to use a dedicated agent platform or not. The decision depends on the requirements of the application, retaining copyright, and the reason why agents are to be used in the first place. Implementing agents from scratch can have greater flexibility and allows the developer to decide which components should be modelled as agents although the downside can be increased development time. It is the route chosen for this work as the goal is to develop a simulator from the ground up with integral agent functionality. The Van Breemen paper discusses well-known software design patterns for implementing sensing, effectors and agent communication. Grosso et. al. (2003) discusses agent development with the .NET framework.

3.1 Integrating agents into the graphics viewer

The integration of agents into the lay-plan simulator needs to consider the issues of agent architecture, user interaction and the encapsulation of knowledge which will be used to guide the operator making decisions on component placement. The placement of components on a hide to maximise material utilisation is a difficult problem. This is because both the hide and shoe component shapes are highly irregular and specific constraints need to be considered. No two hides are the same and parts have to placed so that they do not overlap other parts or the edge of the hide and might have angle constraints so that they are aligned to specific stretch axis directions. Parts can not be placed in regions of unacceptable quality. This knowledge has to be integrated into the agents so that they can provide operator guidance for part placement.

The paper proposes the use of design rule checking agents to provide assistance to operators to monitor the legality of placement positions made by the operator. The work is investigating the use of reflex agents for providing assistance to the operator relating to grade region placement, the overlap of the current component with another shape or the border of the hide and direction angle constraint violations. Fig. 2 shows a schematic diagram of the structure of the reflex agents.

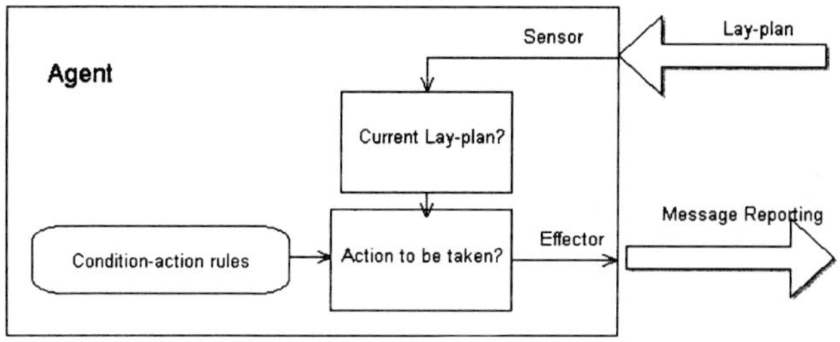

Fig.2. Schematic diagram of the reflex agents

An agent senses the lay-plan environment (i.e. the position and rotation of the current shape being placed by the operator on the hide) and finds a rule whose condition matches the current situation and performs the action associated with that rule. With a reflex agent the knowledge is hard coded using set condition-action rules that are if-then relationships such as:

if (current component overlaps hide boundary) **then** (initiate boundary overlap message)

A reflex agent responds immediately to its current perception of the lay-plan and provides assistance by reporting constraint violations.

A multi-agent architecture (i.e. a system composed of multiple interacting agents, where each agent is an independent computational unit in it own right) is being developed to create a training simulator for leather cutters. The work is investigating the use of thread agents that run autonomously in a separate processor thread. Specifically the research is developing the following agents:

(i) Overlap agent
(ii) Region placement agent
(iii) Direction angle agent
(iv) Supervisor agent

The purpose of the overlap agent is to check the border of the current shape for overlap with either the hide border or the border of another shape.

The region placement agent is responsible for checking that a component part is not placed in a region on the hide of unacceptable quality. The digitised hide files contain grade region areas that have associated grade number values ranging from 1 to 5. Each component part is assigned a region grade number (set when imported into the viewer) which represents the lowest grade region in which the part can be placed. The region placement agent checks the border of the current shape for overlap with the grade regions mapped onto the hide using the grade number values.

The direction angle agent is responsible for ensuring that parts are orientated to meet stretch axis constraints imposed by the hide. Leather nesting requires that certain parts (e.g. shoe upper vamps) are orientated to follow the spine direction in order to maximise elasticity properties. To approximate alignment constraints it has been assumed (for implementation purposes) that the only area with significant directionality is near to, and parallel with, the spine of the hide. Shapes have desirable directionality deviation properties associated with them. Those without any recommended directional constraints are set to 90°. Any shape with significant directionality constraints has the deviation angle set at some small number of degrees (typically 5°). If the shape possesses significant alignment requirements the direction agent will suggest that the shape be placed within the area of high directionality. If the shape is within this area the direction agent calculates the deviation of the angle that the shape is placed at from the direction of the hide alignment property. The direction angle agent will detect if this value is outside the required shape alignment deviation angle

The supervisor agent is responsible for agent communication and displaying the outputs generated by the other agents. The work is investigating a simple string based communication protocol to allow messages generated by the agents to be displayed in a message dialog box. String messaging minimises the overheads of managing agent communications.

4. CLIENT/SERVER ARCHITECTURE

In distributed computing several computers are used to complete a task. Some applications are inherently distributed (e.g. automated transactions between two or more computer systems) while other types of application are distributed for performance reasons (e.g. solving large-scale scientific problems). It may also be important to have some aspects of the application running on a server for centralised monitoring purposes or to retain intellectual property rights. In this case, a client/server assessment agent is being developed to allow centralised monitoring and to retain expert knowledge of hide assessment at the server end. A balance must be found between the amount of computation saved on the client and the amount of information transferred across the communication link.

The assessment of leather nesters in the world wide shoe industry is necessary to award accreditation for those who meet required standards. The advantages of locating the assessment software on a server in a traditional client/server architecture is that it allows local assessment management and software agent improvements to be implemented on the server. However, multiple platform development and world wide distribution of client software must be included in supply costs. An alternative method of delivery is to develop software to be compatible with standard web server/browser protocols or at least compatible with easily accessible browser plug-ins. The advantages of this type of web delivered application are development for a single format, complete control of software running on server and local software management. However with this approach, software traffic needs to be considered as this is a system where a guaranteed response time is desired.

In a client/server-computing environment sockets are used to establish a connection between two programs across a network using the TCP/IP protocol. At one end of the network is the server that creates a communication end point identified with the IP address of the computer running the server and a port number. The server waits passively for connection requests. At the other end is the client (e.g. computer running the lay-plan trainer software) that creates a communication endpoint and actively connects to the server by specifying the IP address and port number of the server. Once communications is established read and write calls can be used to send data between the client and server. Security issues relate to keeping out unwanted traffic and allowing genuine client requests. One approach is to implement a login procedure and to encrypt data at both the client and server ends.

The work is investigating a TCP server using the .NET framework socket class libraries which is installed as a service to evaluate the quality of a lay-plan based on percentage coverage. The socket and threading class libraries simplify server code development by providing high level functionality. The server code creates a socket, associates a defined port number (arbitrarily chosen as 1999) with the socket, establishes a queue for connection requests (listens), waits for and accepts a connection, gets data from the client, returns a response to the client and finally closes the connection. The client code is built into the training simulator. The client algorithm creates a socket, connects to the server, sends a request, gets and displays the response and finally closes the connection. A detailed assessment of a lay-plan needs to consider factors such as hide grade, the degree of continuous stack nesting of components, component rotation constraint violations and component directionality throughout the lay.

5. CONCLUSIONS

A prototype agent-based training simulator for leather cutters is being researched which is based on using vector graphics techniques. Design rule checking agents have been proposed for part placement tasks. Initial studies have focused on reflex agents and encapsulating design constraints as condition-action rules. Thread agent based coding methodologies are being studied whereby each agent is coded as a separate unit (class) and operates independently. An assessment process agent is also being investigated based on client/server architecture to fulfil an e-engineering philosophy whereby aspects of the training simulator are distributed and offered as a service. Future work will investigate graphical visualisation techniques and other forms of agent architecture.

REFERENCES

1. Heistermann, J. and Lengauer, T., "The nesting problem in the leather manufacturing industry", *Annals of Operations Research.* 1995, Vol. 57, pp. 147-173,
2. Crispin, A. J. Clay, P. Taylor, G. E. Bayes, T. Reedman, D., "Genetic algorithms applied to leather lay-plan utilisation", *Journal of Engineering Manufacture, I Mech E Part B, Short Communications in Manufacture and Design,* 2003, Vol. 217, pp 1753-1756.
3. Crispin, A. J. Clay, P. Taylor, G.E. Bayes, T. Hackney, R. Reedman, D., "Genetic algorithms applied to material utilisation in leather shoe manufacture", *International Journal for Manufacturing Science and Production,* 2003, Vol. 5, No 1-2, pp17 – 21, ISSN 0793-6648
4. Russel, S. and Norvig, P., "Artificial Intelligence (A modern approach)", Prentice Hall, New Jersey, 1995. ISBN 0-13-360124-2.
5. Wooldridge, M. and Jennings, N.R., "Pitfalls of Agent-Oriented Development", *Proceedings of 2nd International Conference on Autonomous Agents.* Pp. 385-392, Minneapolis, USA, 1998.
6. Van Breemen A., "Integrating Agents in Software Applications", *NetObjectsDays Conference,* Erfurt, Germany. 2002.
   Also http://citeseer.nj.nec.com/vanbreemen02integrating.html
7. Grosso, A. Gozzi, A. Coccoli, M. and Boccalatte A., "An Agent Programming Framework Based on the C# Language and CLI", 1st International Workshop on C# and .NET Technologies on Algorithms, Computer Graphics, Visualisation, Distributed and Web Computing, pp 13-20, University of West Bohemia, Czech Republic, 2003. ISBN 80-903100-3-6.
8. St-Cyr, O. Lesperance, Y. Stuerzlinger, W., "An Intelligent Assistant for Computer-Aided Design", Smart Graphics: AAAI 2000 Spring Symposium, pp. 24-28, ISBN 157735110-X, March 2000.

Panoramic imaging based e-laboratory construction

D GLEDHILL and **G Y TIAN**
School of Computing and Engineering, University of Huddersfield, UK
J LIAO
College of Manufacturing Science and Technology, Sichuan University, People's Republic of China
D CLARKE
Rotography Limited

ABSTRACT

The design and development of digital 3D work is a labour consuming but attractive task, particularly construction of realistic digital laboratories which has wide applications for engineering applications e.g. e-learning and practical training. This paper introduces a novel method for construction of an e-laboratory or digital laboratory using panoramic imaging. By extending our recent work on 3D panoramic imaging, the paper reports we use a structured light based panoramic imaging system for construction of our metrology laboratory. Within the panoramic environment, 3D objects can be selected and visualised in 3D for inspection and learning. After a discussion of the experimental work, conclusion and further work are presented.

1 INTRODUCTION

The automatic 3D reconstruction of real world objects and environments for computer editing and visualisation has attracted much attention from both research and commercial institutions. Automatic reconstruction of objects and environments can be achieved much quicker than traditional manual processes. Accurate texture information can also be included from the automatic processes. The two automatic methods covered in this paper are stereo vision and structured light, combined with panoramic imaging to introduce the reconstruction of scenes.

Stereo vision is a method for recovering depth information from two or more images taken from different viewpoints. Hemayed *et al* [1] classified stereo imaging into five methods, point matching, scan line matching, curved-segment matching, line-segment matching and edges with region matching. 3D reconstruction from stereo images is covered in many papers, and the reader is directed to [1-3] for further reading. A typical stereo system includes two cameras, aligned in parallel set apart by a distance called the baseline. Figure 1 shows a typical stereo vision setup.

Panoramic imaging uses images to create a panorama of an environment. There are different ways of constructing a panorama, from multiple shots, or using special mirrors for a single shot system. By capturing more than one panorama at a location it is possible to reconstruct a 3D scene [4]. Optical systems which can be used for stereo panoramic imaging have been researched by Pritch *et al* [5] and Svoboda *et al* [6], who also looked at the epipolar geometry needed for stereo panoramic imaging [7]. There are five possible panoramic imaging methods which can be used to capture the images for stereo processing detailed as follows [4].

The first of the five methods mosaics images from a single 'off the shelf' style camera to create panoramic images. The images are joined together using a mosaic technique. Usually the camera is rotated about its nodal point, but this does not enable 3D structure reconstruction. There are many commercial packages able to mosaic images like these. Peleg and Ben-Ezra [8] moved the point about which the camera rotates to a point behind the camera. Two panoramic images are then constructed from left and right halves of the captured images. This enables stereo viewing for a human, but no depth can be calculated. Ishiguro *et al* [9] rotated a standard camera around a circular path. Two images from different parts of the path can be compared and depth computed. Unfortunately it takes about 10 minutes to capture a panorama in this way. Peer *et al* [10] proposed a method of extracting 3D information from a mosaiced panorama.

The second method uses a camera with moving parts, for example Benosman *et al* [11] mounted 2 line scan camera systems above each other. The line scanning cameras are then rotated about a vertical axis. The distance between the cameras is the baseline. Because the system is rotated it means it can not be used for real time situations. Petty *et al* [12] also used two line scan cameras to construct and test a stereo panoramic system.

The third method uses a fisheye lens, and is used widely in the commercial environment for panoramic imaging, such as the Ipix [13] system. The wide field of view is ideal for capturing the minimum number of images possible for creating panoramic images. A camera with a wide angle lens, such as a fisheye lens, can be calibrated and used for 3D reconstruction if the camera model is extended to include distortion parameters.

The fourth method uses a camera with a single mirror. Using a single mirror on a camera usually means mounting the mirror above the camera in order to capture as much of the scene as possible. An example would be a commercial parabolic mirror from 0-360.com [14]. Nayer *et al* [15] presented prototypes of panoramic cameras which use a parabolic mirror. Geb [16] proposed a spherical mirror for robot navigation. Hicks and Bajcsy [17] proposed a range of reflective surfaces which provide a large field of view. Chahl and Srinivasan [18] and Yagi *et* al [19] proposed a conical mirror. Yamazawa *et al* [20] and Svoboda *et* al [21] proposed a system with a hyperbolic mirror.

The fifth method is to use a single camera with multiple mirrors. A single camera with two planar mirrors is equivalent to a two camera stereo system. Gluckman *et al* [22], Arnspang *et al* [23], Gothasby *et al* [24] and Nene and Nayar [25] all used two planar mirrors in front of a single camera to compute depth. Southwell *et al* [26] proposed a single camera with a double lobed mirror.

The most important requirement of stereo panoramic imaging for 3D object and scene reconstruction is correspondence, which is used to determine the disparity. Many applications of 2D image registration, or correspondence, are covered in Odone and Fusiello's paper [29]. Brown [30] also gave an in-depth review of image registration techniques. There are two main methods of detecting and matching features in the images, area based methods and feature based methods.

Area based techniques use correlation of intensity patterns of a pixel with the intensity pattern around the corresponding pixel in another image. First a point of interest is chosen in one image. Then cross-correlation is used to search for the corresponding pixel in the other image. Area based methods suffer because they use the intensity values at each pixel directly, and are therefore sensitive to changes in viewing position, absolute intensity, contrast and illumination. Also occlusions can give erroneous depth information. Zitnick and Kanade [31] use a co-operative stereo algorithm for stereo matching, which also takes into account occlusion detection.

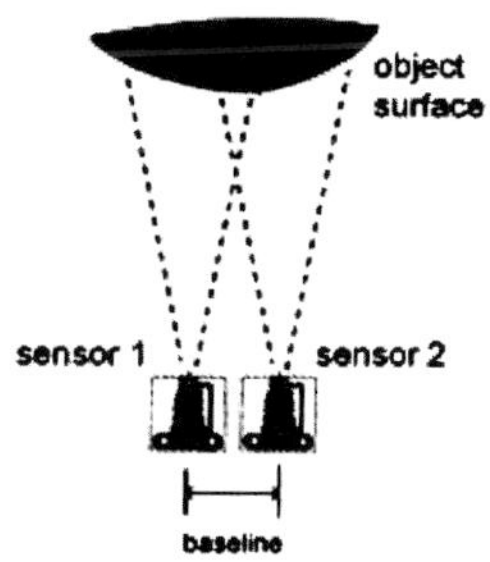

Figure 1. A typical stereo vision setup

Feature based methods are based on features in the images for correspondence detection, rather than image intensities themselves. There are two features that are most commonly used, edge based and interest points or corners.

Edge detectors attempt to recover the discontinuities in the photometrical, geometrical and physical characteristics of objects in the images. This information creates variations in the grey-level image. There are three main ways to use the variation information, discontinuities, local extrema and the 2D features formed where at least two edges meet. Bao and Xu [32] use edge-preserving visual perception modelling to build mosaics. Kim *et al* [33] use edges in their graph matching algorithm and use camera focal length to improve stitching reliability, which finds corresponding edges for the matching. Edge detection is used in Bourque's *et al* [34] work for determining interesting points from where a robot should capture a spherical panorama.

Interest point detectors attempt to find locations in the image where the signal changes two-dimensionally, including locations such as corners, T-junctions or significant changes in texture. Many interest point detectors exist, as reviewed by Schmid *et al* [35]. Interest points, as local invariant features [36], are robust to changes in lighting, viewpoint, rotation, translation and scale [37] and are used in many areas including image retrieval and indexing [38], object recognition [39], mosaicing [36,] and colour correction [41].

Structured light techniques involve the projection of a known light pattern onto an object and measuring the deformations in the pattern to determine depth. Different patterns can be used, point-wise, stripe-wise, or a more complex pattern. When the position of the sensor (e.g. camera) is known with respect to the emitter (e.g. LCD projector) triangulation can be used to determine depth. Figure 2 shows a typical structured light setup [42] where the 3D positions of the sampled points are computed by triangulation given the sampled point projection P(a) on the sensor plane and the known relative position/orientation of the emitter and the sensor. Valkenburg [43] used structured light to measure 3D.

In this paper an area based method is used to determine correspondence. Correspondence methods achieve more accurate results in high texture environments. In low texture environments the correspondence methods produce more errors. This paper uses a random binary dot structured light pattern to overcome this limitation. The structure light pattern is directed at areas of low texture.

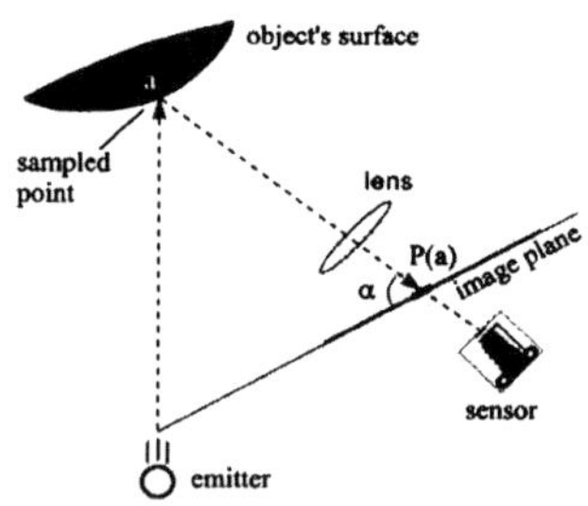

Figure 2. Typical structured light setup

The rest of the paper is organised as follows, Section 2 introduces the stereo panoramic system for visualisation of an environment and selection of objects for visualisation in 3D. The binary dot pattern, the correspondence algorithm and the pattern removal algorithm are also introduced in Section 2. Section 3 introduces the experimental results and section 4 presents the conclusion and further work.

2 STEREO PANORAMIC BASED SYSTEM USING STRUCTURED LIGHT

Panoramic imaging creates 2½D visualisations. Although the images are 2D, they are viewed in such a way as to perceive 3D. Stereo vision however uses multiple 2D images to create 3D objects. This paper proposes the combination of stereo vision and panoramic imaging to create a virtual environment with photorealistic rendering and the possibility to select an object of interest for visualisation and inspection in 3D. The process is outlined in Figure 3. The panoramic images can be captured using any panoramic imaging technology, for example multiple shots mosaiced into a single image, 2 fisheye images, or a single shot parabolic mirror based system. The system proposed in this paper uses a single shot mirror based system to capture the panoramas. The panorama is visualised in an internet browser, with 'hotspots' over selectable objects. The user is able to select an object to view in 3D.

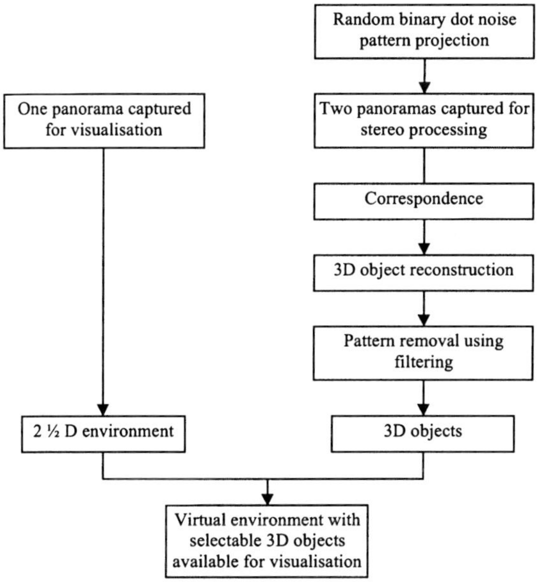

Figure 3. Outline of process for capture and creation of virtual environment with selectable 3D objects

2.1 Random binary dot pattern creation

Reconstruction from stereo images, either panoramic or from standard cameras, requires accurate correspondence information. Correspondence techniques are prone to errors in areas of low texture. To overcome this limitation it is proposed that a random binary dot noise pattern is projected onto the areas of low texture in the environment. The system generates a random pattern until a threshold is reached, determining the coverage of black pixels. Figure 4 shows the random binary pattern. To be able to remove the pattern later for visualisation the noise pattern requires unique points which can be easily removed. To ensure this a constraint is imposed on the random pattern generator so that no two black pixels are adjacent to each other. An array P is created with all the values set to 1. For each random location $P(x, y)$, if the sum of the surrounding pixels b = 8 then the point $P(x, y)$ is made to equal 0. This is shown in Eq. (1).

$$P(x,y) = \begin{cases} 1 & b = 8 \\ 0 & otherwise \end{cases} \quad (1)$$

where b is the sum of the pixels surrounding $P(x, y)$ and

$$x = rand(1, width) \quad (2)$$

$$y = rand(1, height) \quad (3)$$

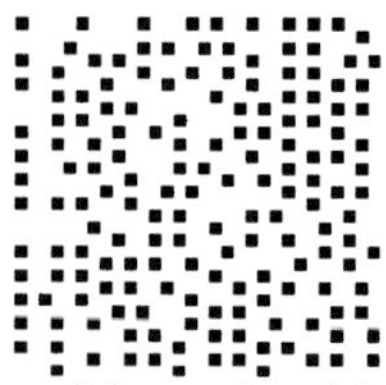

Figure 4. An example crop of the random binary distribution pattern

2.2 Correspondence

The proposed system uses an area based method for the correspondence search in the panoramas. The area based method calculates correspondence by minimising the sum of errors from the feature difference where the features are extracted from different orientation filters as illustrated in Figure 5.

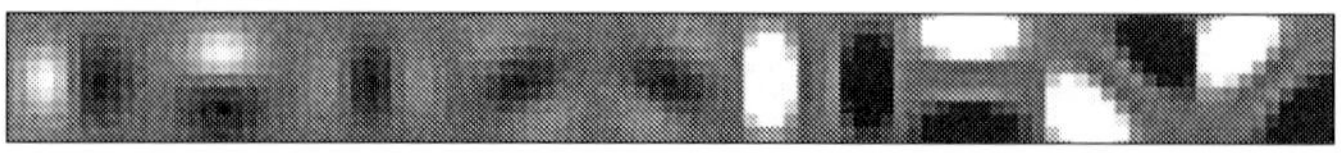

Figure 5. The Gaussian filters used for the 2D convolution

As illustrated in **Error! Reference source not found.** a 'window' N moves through the image data in the left image for each point $P(x,y)$. A similar 'window' M takes image data from the right image from areas within a defined distance of the left window location, at points $R(x+n,y)$, where n is the search range $\left[-\frac{s}{2},\frac{s}{2}\right]$ and s is the search number in the second image. The feature vectors at the two search windows N and M can be calculated by the convolution in Eq. (4) and Eq. (5).

$$c_{xyi} = N_{xy} \oplus f_i \qquad (4)$$

$$d_{(x+n)yi} = M_{(x+n)y} \oplus f_i \qquad (5)$$

Where f is the filter and $i = [1, g]$, g is the number of filters, and x and y are the current pixel location. For each window location $M(x+n,y)$ an error is calculated, $err(x+n,y)$ using Eq. (6).

$$err(x+n,y) = \sqrt{\sum_{i,n=1}^{g} (c_{xyi} - d_{(x+n)yi})^2} \qquad (6)$$

The minimum value of $err(x+n,y)$ is determined to be the best match, which produces the correspondence information for which $R(x+n,y)$ matches $P(x,y)$. The correspondence information is used with Eq. (4) to determine depth.

2.3 3D reconstruction

To enable reconstruction of objects within the scene, stereo panoramic imaging can be used. When using a parabolic mirror system, if two panoramic images are captured in vertical orientation, then the epipolar lines are vertical. By using this information the search for corresponding points within the panoramas becomes a 1D problem. Figure 6.a shows the epipolar line transformation when converting from the captured image, in polar coordinates, to the final image, in rectangular coordinates. Figure 6.b shows a diagram explaining the reconstruction used for extracting 3D.

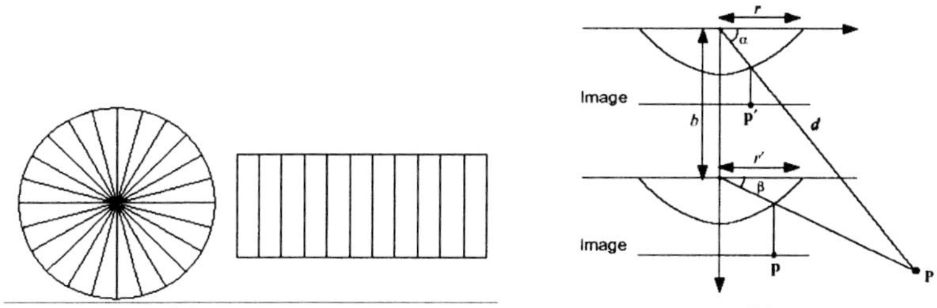

(a) (b)

Figure 6. (a) The epipolar lines from the panoramas and the (b)maths used for reconstruction from vertically mounted stereo panoramic image capture

As shown in Figure 6, each corresponding image point is the orthographic projection of a point on the parabola, which can be defined by the angles α and β. From the law of sines, depth d can be calculated using Eq. (7).

$$d = \frac{\cos(\beta)}{\sin(\alpha - \beta)} b \qquad (7)$$

where b is the baseline of the stereo system. If p and p' are the corresponding image points of the scene point P, and x and x' are the distances of the image points from the centre of the parabolas then the angles α and β are determined by Eq. (8) and Eq. (9).

$$\tan(\alpha) = \frac{r^2 - x^2}{2rx} \qquad (8)$$

$$\tan(\beta) = \frac{r'^2 - x'^2}{2r'x'} \qquad (9)$$

where r and r' are the radii of the parabolas. If correspondence is calculated once the images have been converted to rectangular coordinates, and x and x' are the heights of corresponding points in the top and bottom panoramas, disparity $x - x'$ is inversely related to depth as shown in Eq. (10).

$$\left(\frac{x - x'}{\sqrt{R^2 + x^2}}\right) b = \frac{1}{d} \qquad (10)$$

where R is the radius of the cylinder.

Using the calculations above for reconstruction of objects in a scene means fewer images need to be captured. Alternatively each object in the scene could be reconstructed individually using stereo image techniques, or using commercial software such as Canon's Software Object Modeller (SOM) software. Canon's SOM software uses silhouette techniques for modelling 3D objects, which has the disadvantage of not being able to model indentations in objects.

2.4 Pattern removal

The texture used for visualisation is the same image used for reconstruction, therefore the noise pattern must be removed from the image for a visually accurate and pleasing result. This paper uses filtering techniques to remove the noise. Initially a median filter was used to remove the noise. Although the pattern was removed the filter also softened the image, an expected but undesirable result. A new filtering technique has been developed which attempts to remove the noise without affecting the sharpness of the image. The proposed filtering method adapts adaptive median filter, which considers the ratio of a pixel with its surrounding pixels. This ratio determines whether or not a pixel is affected by noise, and if so the affected pixel is adjusted based on nearest neighbour interpolation

For each pixel in the image a ratio is calculated, shown in Eq. (11).

$$r = \frac{a}{b} \tag{11}$$

where a = intensity value for the pixel of interest and b equals:

$$b = \frac{c + d + e + f + g + h + i + j}{8} \tag{12}$$

where c to j are the surrounding pixels, shown in Figure 7. A threshold is then considered and the pixel of interest changed as necessary, as shown in Eq. (13).

$$a = \begin{cases} b & if(r > thresh) \\ a & otherwise \end{cases} \tag{13}$$

c	d	e
j	a	f
i	h	g

Figure 7. The pixel distribution for equations (11), (12) and (13)

Figure 8 shows the filtering technique used on an image of a product box. The original captured image (a) is shown with a median filtered image (b) and the new filtered image (c). The median filter has removed the noise, but softened the image. The new filter has removed the noise like the median filter, but kept more of the sharpness of the original image.

(a) (b) (c)

Figure 8. (a) Showing the structured light pattern in an image, (b) after it has been removed using median filtering and (c) after removal using the new filter

3 EXPERIMENTAL STUDIES

Figure 9 and Figure 10 show the results from captures of the CMM (Coordinate Measuring Machine) laboratory. Figure 9 shows the panorama capture, with a screen shot from a viewpoint looking at the CMM machine. Figure 10 shows the head of the CMM machine in 3D, created from multiple photos with Canon's SOM software.

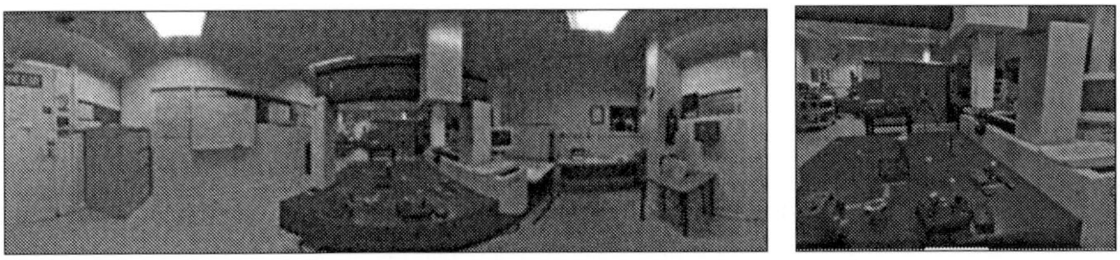

(a) (b)

Figure 9. (a) Example panorama from the CMM laboratory and (b) a single viewpoint from a panorama viewer

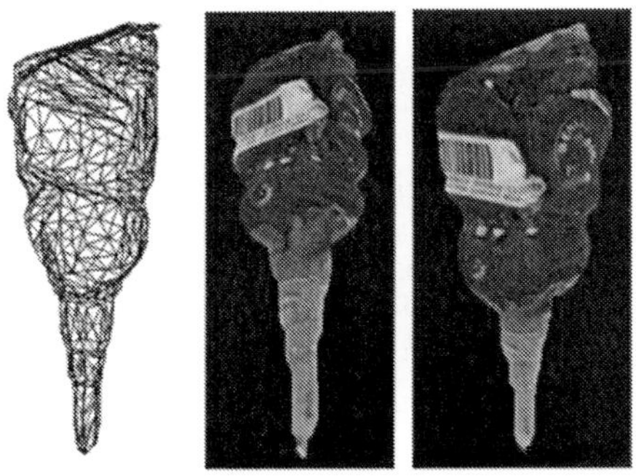

(a) (b) (c)

Figure 10. CMM head from CMM machine in laboratory, created with SOM. (a) The wire frame model, and (b and c) a textured model from two different viewpoints

4 CONCLUSION AND FURTHER WORK

An integration approach by using panoramic imaging and stereo vision has been presented and developed for image based 3D virtual environment reconstruction. The experimental studies have illustrated the approach is fast and low cost. It is low cost because it uses standard cameras, plus a single shot mirror, which is much cheaper than the cost of a CMM machine. The system is also fast because the panoramic image can be converted quickly using standard image processing algorithms. The 3D reconstruction is faster and more accurate than traditional 3D model generation. Further work will involve more accurate automatic model generation and the combination of panoramas and 3D using VRML for complex environment reconstruction.

ACKNOWLEDGMENT

The authors would like to thank the EPSRC, Rotography Ltd and The Royal Society for funding the work.

REFERENCES

1. Hemayed, E. E., Sandbek, A., Wassasl, A. G., Farag, A. A., *"Investigation of Stereo-Based 3D Surface Reconstruction"*, Proc. SPIE, Vol. 3023, pp. 191-202, Feb. 1997
2. J. Park and S. Inoue, *"Hierarchical Depth Mapping from Multiple Cameras"*, Proc. ICIAP'97, vol.1, pp.685-692, Florence, Italy, Sept. 1997
3. H. M. Hamden, E. E. Hemayed and A. A. Farag, *"A Fast 3D Object Reconstruction Using Trinocular Vision and Structured Light"*, *Proc. of SPIE Conf. 3522*, Nov. 1998
4. Gledhill, D., Tian, G. Y., Taylor, D., Clarke, D.: *"Panoramic Imaging – A Review"*, Computers and Graphics 27(3), pp. 435-445, 2003
5. Pritch, Y., Ben-Ezra, M., Peleg, S.: *"Optics for Omnistereo Imaging"*, Foundations of Image Understanding, Kluwer Academic, July 2001, pp. 447-467
6. Svoboda, T., Pajdla, T.: *"Panoramic Cameras for 3D Computation"*, Czech Pattern Recognition Workshop, Perslak, Czech Republic, February 2000
7. Svoboda, T., Pajdla, T., Hlavac, V.: *"Epipolar Geometry for Panoramic Cameras"*, Fifth European Conference on Computer Vision, Freiburg, Germany, June 1998
8. Peleg, S., Ben-Ezra, M.: *"Stereo Panorama with a Single Camera"*, Proc. Computer Vision and Pattern Recognition Conf., pp. 395--401, 1999
9. Ishiguro, H., Yamamoto, M., Tsuji, S.: *"Omni-directional Stereo"*, IEEE Transactions on Pattern Analysis and Machine Intelligence, 14(2), pp. 257-262, February 1992
10. Peer, P., Solina, F.: *"Mosaic-Based Panoramic Depth Imaging with a Single Standard Camera"*, Proceedings of the IEEE workshop on Stereo and Multi-Baseline Vision 2001, 75-84, 2001
11. Benosman, R., Maniere, T., Devars, J.: *"Panoramic stereovision sensor"*, 14th International conference on Pattern Recognition, Brisbane, Australia, pp. 767-769. IEEE Computer Society Press, August 1998
12. Petty, R., Robinson, M., Evans, J.: *"3D Measurement using rotating line-scan sensors"*, Measurement Science and Technology, 9(3), pp. 336-346, March 1998
13. IPIX, www.ipix.com
14. 0-360.com, www.0-360.com
15. Nayar, S. K.: *"Catadioptric Omnidirectional Camera"*, International Conference on Computer Vision and Pattern Recognition, Puerto Rico, USA, pp. 482-488, IEEE Computer Society Press, June 1997
16. Geb, T.: *"Real-time Panospheric Image De-warping and Presentation for Remote Mobile Robot Control"*, IEEE Transactions on Robotics and Automation, 1998
17. Hicks, A. R., Bajscy, R.: *"Reflective Surfaces as Computational Sensors"*, The Second IEEE Workshop on Perception for Mobile Agents, June 1999
18. Chahl, J., Srinivasan, M.: *"Range Estimation with a Panoramic Visual Sensor"*, Journal of the Optical Society of America, 14(9), pp. 2144-2151, September 1997
19. Yagi, Y., Nishizawa, Y., Yachida, M.: *"Map-based Navigation for a Mobile Robot with Omnidirectional Images Sensor COPIS"*, IEEE Transactions on Robotics and Automation, 11(5), pp. 634-648, October 1995
20. Yamazawa, K., Yagi, Y., Yachida, M.: *"Obstacle Detection with Omnidirectional Image Sensor Hyperomni Vision"*, IEEE International Conference on Robotics and Automation, pp. 1062-1067, 1995
21. Svoboda, T., Pajdla, T., Hlavac, V.: *"Epipolar Geometry for Panoramic Cameras"*, 5th European Conference on Computer Vision, Germany, Vol. 1406 of Lecture Notes in Computer Science, pp. 218-232, June 1998
22. Gluckman, J., Nayar, S.: *"Planar Catadioptric Stereo: Geometry and Calibration"*, Conference on Computer Vision and Pattern Recognition, Colorado, Vol. 1, pp. 22-28, IEEE Computer Society Press, June 1999
23. Arnspang, J., Nielsen, H., Chritensen, M., Henriksen, K.: *"Using Mirror Cameras for Estimating Depth"*, 6th Conference on Computer Analysis of Images and Patterns, Czech Republic, 970 in Lecture Notes in Computer Science, pp. 711-716, September 1995
24. Goshtasby, A., Grover, W.: *"Design of a Single-Lens Stereo Camera System"*, Pattern Recognition, 26:913-937, 1993

25. Nene, S. A., Nayar, S. K.: *"Stereo with Mirrors"*, 6th Conference on Computer Vision, India, pp. 1087-1094, January 1998
26. Southwell, D., Basu, A., Fiala, M., Reyda, J.: *"Panoramic Stereo"*, 13th International Conference no Pattern Recognition, Austria, Vol. A, pp. 378-382, IEEE Computer Society Press, August 1996
27. Gluckman, J., Nayar, S. K., Thoresz, K. J.: *"Real-Time Omnidirectional and Panoramic Stereo"*, Proceedings of the Image Understanding Workshop, Monterey, California, November 1998. DARPA
28. Zhu, Z., Rajesekar, K. D., Riseman, E. M., Hanson, A. R.: *"Panoramic Virtual Stereo Vision of Cooperative Mobile Robots for Localizing 3D Moving Objects"*, IEEE Workshop on Omnidirectional Vision, Hilton Head Island, June 12, 2000, pp. 29-36
29. Odone, F., Fusiello, A.: *"Applications of 2D Image Registration"*, Research Memorandum RM/99/15, Heriot-Watt University
30. Brown, L. G.: *"A Survey of Image Registration Techniques"*, ACM Computing Surveys Vol.24, No. 4 (Dec. 1992), pp. 325-376
31. Zitnick, C. L., Kanade, T.: *"A Cooperative Algorithm for Stereo Matching and Occlusion Detection"*, IEEE Transactions on Pattern Analysis and Machine Intelligence,. Volume: 22 Issue: 7, July 2000, pp 675-684
32. Bao, P., Xu, D.: "*Complex Wavelet-Based Image Mosaics using Edge-Preserving Visual Perception Modelling*", Computers and Graphics 23(1999) 309-321
33. Kim, H. S., Kim, H. C., Lee, W. K., Kim, C. H.: "*Stitching reliability for estimating camera focal length in panoramic image mosaicing*", 15th International Conference on Pattern recognition, 2000, pp 596-599
34. Bourque, E. Dudek, G.: "*Automated Creation of Image Based Virtual Reality*", Proceedings of SPIE, "Sensor Fusion and Decentralised Control in Autonomous Robotic Systems" volume 3209, 1997, pp 292-301
35. Schmid, C., Mohr, R., Bauckhage, C.: *"Evaluation of Interest Point Detectors"*, International Journal of Computer Vision, 37(2):151-172, 2000
36. Zoghlami, I., Faugeras, O., Deriche, R.: *"Using Geometric Corners to Build a 2D Mosaic from a Set of Images"*, In Proceedings of the International Conference on Computer Vision and Pattern Recognition, Puerto Rico, June 1997
37. Mikolajczyk, K., Schmid, C.: *"Indexing Based Scale Invariant Interest Points"*, In International Conference on Computer Vision, 525-531, July 2001
38. Schmid, C., Mohr, R.: *"Image Retrieval using Local Characterisation"*, Proceedings of the 3rd International Conference on Communicating by Image and Multimedia, pp. 45-50, May 1996
39. Lowe, D. G.: "*Object Recognition from Local Scale Invariant Features*", Proc. Of the International Conference on Computer Vision, Corfu (Sept.1999)
40. Tian, G. Y., Gledhill, D., Taylor, D.: *"Comprehensive Interest Points Based Imaging Mosaic"*, Pattern Recognition Letters 29(9-10), pp. 1171-1179, 2003
41. Tian G. Y., D. Gledhill D., Taylor D. and Clarke D., *"Colour correction for panoramic imaging, Proceeding"*, IV2002 (IEEE), pp 483-488.
42. Rocchini, C., Cignoni, P., Montani, C., Pingi, P., Scopigno, R., *"A Low Cost 3D Scanner Based on Structured Light"*, Eurographics 20 (3), pp. 299-308, 2001
43. Valkenburg, R. J., *"Accurate 3D Measurement using a Structured Light System"*, Image and Vision Computing (16), No. 2, pp. 99-110, 1998

Developing web-based java applets for on-line learning of geometric tools

X L YOU
Consultant, Bournemouth, UK

ABSTRACT

Geometric tools are widely used in industries such as computer graphics, CAD and virtual reality. Web-based learning environment can explain concepts, theories and methods of geometric tools vividly. In order to develop on-line learning of this subject, a user interface is designed and Web-based Java applets are written in the platform independent object-oriented Java programming language. The World Wide Web (WWW) language is used to create Web pages. The developed Java applets are embedded in the Web pages. Learners can execute these applets via the internet, and obtain interactive documents of this subject according to their needs and preferences.

Keywords: on-line learning, Web-based Java applets, geometric tools

1 INTRODUCTION

Due to the development and increasing availability of the internet, Web-based education is becoming more and more popular. There have been a lot of publications investigating Web-based education. These publications can be classified into three groups. The first group investigated general theories and methods of Web-based education. The second group studied the development of Web-based teaching system of some subjects. And the third group discussed how to develop Web-based learning system of these subjects.

A lot of work investigated general theories and methods of Web-based education. Granlund et al. discussed some basic properties of learning using Web-based simulation with the focus on different types of learning goals (instructional goals) and on proper instructional strategies (pedagogical strategies) for Web-based simulation [1]. Bouras et al. presented the data components of an open and distance learning (ODL) system that access the information stored in a database and the file system, their underlying technology, their interaction with the network services, and features regarding the ways they address issues faced in an open vendor-independent distance learning environment and outlined the system's overall architecture [2]. Sheremetov and Arenas developed a Web-based learning environment called

Virtual Learning Spaces which is composed of knowledge, collaboration, consulting and experimentation spaces as a collection of agents and conventional software components working over the knowledge domains [3]. Pahl presented his approach to the design and maintenance of educational systems in rapidly evolving environments and illustrated the consequences of evolution and change for these systems and for the educators and developers responsible for their implementation and deployment [4]. Rafaeli et al. described a Web-based and distributed system named QSIA (Questions Sharing and Interactive Assignments) that serves as an environment for learning, assessing and knowledge sharing [5].

In addition to the investigations into general theories and methods of Web-based education, there are some publications which aim to develop Web-based teaching systems of some subjects. For example, in the work of Papaspyrou et al., the existing course material was supplemented by Web-based courseware and integrated into a Web-based novel networked learning environment for Web-based teaching of software engineering [6]. In order to create a more cost effective way to deliver complex and rapidly changing curriculum content and to enhance the learning experience for the student and to encourage the development of self motivated learning skills, Bailey designed a series of Web-based tutorials for teaching digital electronics to help weaker students and stretch the stronger students. The tutorials consist of a series of HTML Web pages with embedded Java applets. The applet either is a complete working program that dynamically illustrates the concept or provides a working environment for the student to experiment and work through their solution [7]. Roy and Lee presented some of the developments of interactive activities which can be built into Web pages for computing in an engineering degree [8]. Kofke and Mihalick demonstrated an "object-oriented" view of molecular simulation that can help conceptualize and organize its elements, and discussed software development tools that they have constructed to enable non-experts in simulation or graphical programming to create instructive molecular simulations [9]. Kasyanov and Kasianova introduced their Web-based systems intended to support computer science teaching and learning [10]. Zaidel et al. developed a multifunctional interactive radiology teaching environment accessible for medical students, residents, physicians, and researchers to present teaching components related to clinical studies [11]. Hamam presented a multimedia teaching and research tool for optometrists and for the students of ophthalmology which uses animated Java applets to help users visualize the behavior of complex optical systems [12]. Henderson et al. proposed a Web-based approach that provides an easy-to-use, low-cost, and effective solution that can be added onto most picture archiving and communication systems (PACS) allowing for easy storage, classification, viewing, editing, and sharing of teaching cases [13].

The development of Web-based learning systems of some subjects is also very important for Web-based education. Hui and Cheung investigated an approach for developing a web-based learning environment for building energy efficiency and solar design in Hong Kong [14]. Wilkins and Barrett introduced their virtual building and construction environment which demonstrates the capabilities of the World Wide Web as a means of delivering a virtual environment using multimedia for teaching/learning in construction engineering and technology [15]. Hough and Marlin presented Web-based interactive learning modules for undergraduate process control through three categories of questions: Checking Your Reading, Study Questions and Thought Questions [16]. Chalk made a survey of Web-based learning environments such as diagramming tools and program animators supporting the teaching of some standard topics in a software engineering course and proposed an approach to classification based on how the student explores or expresses the model represented in the software [17]. Butler and Brockman developed a self-paced, Web-based learning tool that introduces computer concepts by simulating the Lego Mindstorm's RCX programmable brick and can be used as a supplement to an introduction to engineering systems course [18]. Sung and Ou discussed a Web-based learning system by applying the principle of constructivism and virtual reality (VR) technologies to the computer-aided design (CAD) curriculum and integrating networking, CAD and VR into a Web-based learning environment [19]. Kuo et al. designed an intelligent Web-based interactive language learning (IWiLL) system to support English learning on the Internet which has an interactive English writing environment and an authentic conversation learning environment [20]. Jung et al. presented an instruction model called Web-based instruction model for information communication technology (ICT) which emphasizes learners' autonomy, and helps them decide the learning objectives and strategies for themselves, acquire information communication ethics such as the utilization of sound information, network communication etiquette, and copyright protection, and develop their creativity and problem solving abilities [21]. Yen and Li designed a Web-based learning and instructional system for pneumatics including course material, remote data acquisition modules, and a pneumatic laboratory set which can be used for pneumatics instruction, computer aided sequential control design and pneumatic laboratory practice [22]. Using Macromedia Flash for animations, Design Science WebEQ to dynamically present equations and Input Control integrated with Question Mark for preparing mathematical questions and reporting on assessments, Humar et al. developed an e-learning system which supports dynamic educational materials with mathematical equation derivation and other multimedia elements [23]. Freasier et al. presented a Web-based interactive homework quiz and tutorial software package to improve the learning of first-year undergraduate chemistry students [24]. Sitthiworachart and Joy introduced the design and implementation of a Web-based peer assessment system which can help students to develop their understanding of computer programming [25]. Yuan and Teng presented a Web-based package CALSB for computer-aided learning of structural behavior which includes an analysis engine based on the matrix stiffness method, structural games and paradoxes, and provides a virtual laboratory environment for users to build and test two-dimensional skeletal structures [26].

Geometric tools are used to create geometric primitives. Nowadays, their applications have ranged from computer graphics [27], scientific visualization, medical image analysis, engineering simulation, to virtual worlds. Visualization on and interaction through a computer are the two key features of this subject. The World Wide Web (WWW), Java language and Hypertext provide a more effective means to present this subject than the traditional education based on textbooks and blackboards. Therefore, it is especially suitable for learners to study this subject through Internet.

2 USER INTERFACE

In this section, we first introduce the basic knowledge of Internet and World Wide Web, then Web learning and its advantages, and finally a user interface for Web-based learning of geometric tools.
The Internet is a worldwide system of computer networks by which users at any one computer can get information from any other computer if they obtain permission to do so. Besides the

World Wide Web (WWW) introduced below, the Internet has many other parts such as email, ftp, and usenet.

The World Wide Web is a subset and the most widely used part of the Internet. It can be viewed in three parts. The first is the server that is hosting (storing/connecting to the Internet) HTML documents. The second is the surfer or end user viewing HTML documents. The third is the protocol making the bidirectional communication possible. Using the Web, people can get access to millions of pages of information through a Web browser, such as Netscape Navigator or Microsoft Internet Explorer.

HyperText Transfer Protocol (HTTP) provides a simple way for users to request and receive hypertext documents over the Internet. Unlike HTTP, which defines how information is sent or received, HyperText Markup Language (HTML) is a coded format language used for creating hypertext documents on the World Wide Web and controlling how Web pages appear. The location of each object on the World Wide Web can be described with a Uniform Resource Locator (URL). A URL is an address in a specific format that can uniquely identify an individual object on the Web.

With the benefits of platform independence and global accessibility, the World Wide Web is creating tremendous opportunities to provide a learning environment that is available to all learners. Web-based learning means the activities occurring in such an environment.

Compared to the classroom-centered learning method, Web-based learning has its own features. For example, it enables learners to access to resources in any location, to study in their own environment, to take full advantage of their time, to interact with the simulation, to visualize the models on a computer screen, to share information with other learners, to communicate their ideas through internet, and to control their own learning progress.

In order for learners to execute the Java applets of geometric tools, we design a user interface with which learners can view the contents of geometric tools and execute the required Java applets. It can be also used as a main frame for the further development of online learning system of this subject. As shown in Fig. 1, the user interface consists of two regions: Navigation Region and Content Region.

Navigation, appearing with hypertext and popularized with the development of the World Wide Web, is the process through which learners experience the domain content, i. e., decide where they are, where everything else is, and how to get to particular objects and places. In the Navigation Region, the content list of geometric tools is presented. The list includes chapters, sections, subsections and so on in a tree structure. For example, from the Chapter of Vector Algebra, learners can view any section of Vector basics, Vector Space, Affine Spaces, Affine transformations and Barycentric Coordinates and Simplexes. From the sections, learners can view all the subsections and select one of the sections or subsections to acquire the knowledge. All the Java applets are listed at the end of the content list. Such a treatment is suitable for the users who have had the knowledge of the subject or its some contents and only require these applets for the purpose of applications such as for their assignments. From these listed Java applets, the users can directly execute the required Java applets.

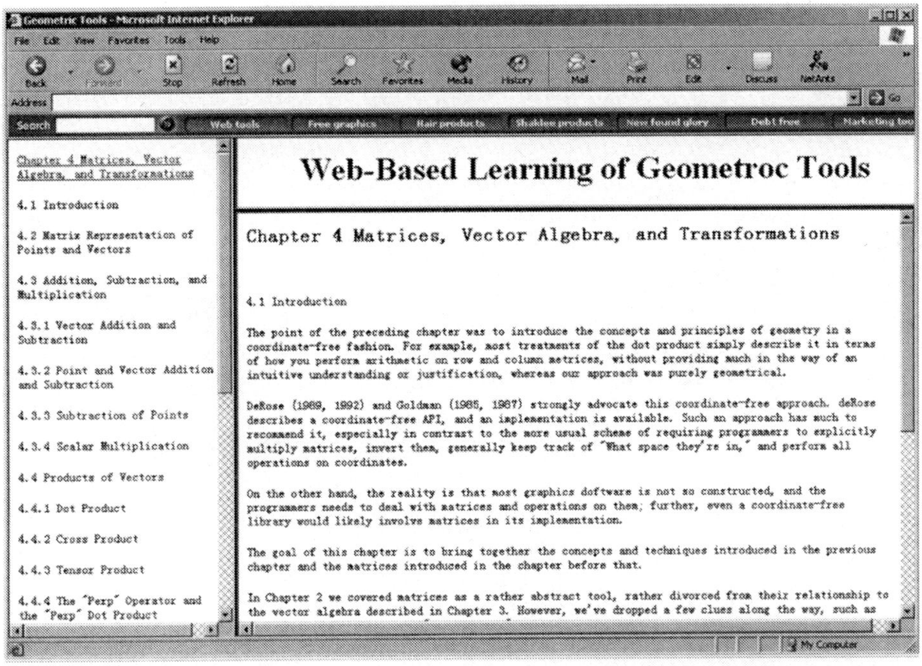

Fig. 1. User interface for online learning of geometric tools

Content Region is divided into top and bottom parts. The top part gives the title "Web-Based Learning of Geometric Tools". The bottom part gives the contents of the chosen chapter, section or subsection. Learners can move the scroll bar on the right of the computer window to browse the total contents or move to a specified page. The Java applets are also embedded in the corresponding contents. The users can run these Java applets when they are learning the relevant knowledge.

3 JAVA APPLETS

The contents of geometric tools include: matrices, linear algebra, vector algebra, transformations, 2D geometric primitives, distance in 2D, intersection in 2D, 3D geometric primitives, distance in 3D and intersection in 3D etc.

In order to implement these contents on the Web, some existing technologies are available such as COM/ActiveX, Active Server Pages (ASP), Dynamic HTML (DHTML) and Java programming language. For the subject of geometric tools, a lot of mathematical calculations are involved. ASP and DHTML are difficult and time-consuming when used for such a case. In contrast, ActiveX can use Visual Basic or Visual C++ to develop active Web contents and be easily deployed over the Internet. However, there are still two key shortcomings. One is the machine dependent because it is usually used on the Windows platform. The other is the security problem. The client's hardware and file systems can be accessed directly by ActiveX applications. In order to ensure security, a global and bureaucratic registration is required.

Among them, Java programming language is an ideal tool to develop Web applications. This is because Java programming language has the following advantages. Firstly, it is an object-oriented programming language similar to advanced programming languages such as C++. Secondly, it is platform independent. Java applications, called Java applets when embedded in Web pages can be executed through a Web browser. Therefore, the combination of Java and World Wide Web provides a powerful and effective means for online learning.

Due to the above advantages, we will use Java programming language to develop some Java applets of geometric tools. Here, we briefly introduce 4 representative Java applets.

3.1 Matrix multiplication

Matrix multiplication is to determine a new matrix from two or more existing matrices through a multiplication algorithm. It is an important content of matrix operations. Matrix operation includes addition, subtraction, multiplication, transposition and inverse of matrices.

As shown in Fig. 2, when a matrix multiplication is to be carried out, the users should input the order of the left two matrices and the values of the elements of these two matrices. After the Run button is chosen, the Java applet performs the operation and gives the calculated values in the rightmost matrix.

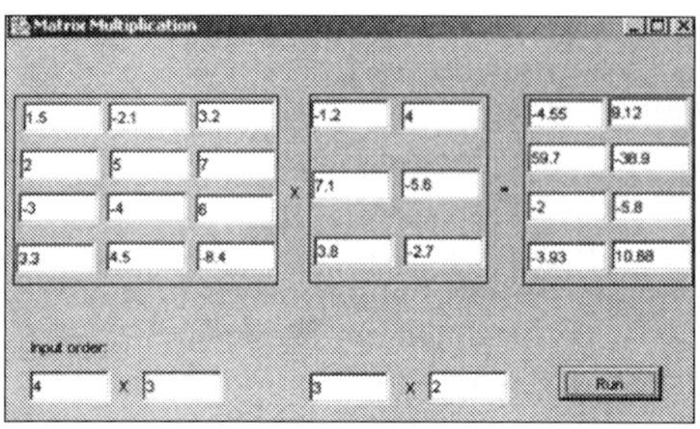

Fig. 2. Java applet of matrix multiplication

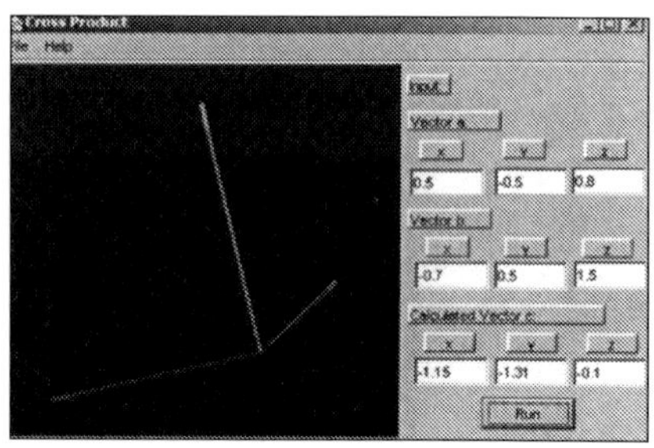

Fig. 3. Java applet of cross production

3.2 Cross product

Cross product belongs to vector algebra. The contents of vector algebra have addition, subtraction, scalling, dot product, cross product and coordinate transgormation.

Cross product operation is to produce a new vector from the two existing vectors. This newly generated vector is perpendicular to the plane decided by the two old vectors and its direction is determined by the right-hand rule.

With the Java applet of cross product operation, the users input the coordinate values of the two existing vectors, i. e., vector **a** and vector **b**, from the upper part of the right panel of the

Java applet interface shown in Fig. 3. After the Java applet obtains these data, it draws these two vectors in the left canvas and carries out the cross product operation. The calculated coordinate values of the new vector **c** are displayed in the bottom part of the right panel and the new vector is drawn in the left canvas. In the figure, the vector in red, green and blue is vector $\mathbf{a}$, $\mathbf{b}$ and $\mathbf{c}$, respectively.

3.3 Two-dimensional B-spline curves

Two-dimensional (2D) B-spline curves are a key point of two-dimensional geometric primitives. In addition to B-spline curves, lines, triangles, rectangles, polylines, polygons, circles, ellipses, parabolas, Bézier curves, and NURBS curves are also belonged to geometric primitives. These two-dimensional geometric primitives, like three-dimensional ones, are widely used in computer graphics, computer aided design and other graphical applications.

A two-dimensional B-spline curve of degree j can be constructed from a set of points P_i and a monotone set of parameters t_i through the mathematical equation of

$$\mathbf{X}(t) = \sum_{i=0}^{n} \mathbf{B}_{i,j}(t)\mathbf{P}_i \tag{1}$$

In the equation, the points P_i are called control points, the parameters t_i are called knots which have the property of $t_i \le t_{i+1}$ ($0 \le i \le n$), its vector form is called a knot vector, $t \in [t_0, t_n]$, j in the above equation is defined as $1 \le j \le n$, and the polynomials $\mathbf{B}_{i,j}(t)$ are defined by the Cox-de Boor recursion formulas

$$\mathbf{B}_{i,j}(t) = \frac{(t - t_i)\mathbf{B}_{i,j-1}(t)}{t_{i+j-1} - t_i} + \frac{(t_{i+j} - t)\mathbf{B}_{i+1,j-1}(t)}{t_{i+j} - t_{i+1}} \tag{2}$$

where

$$\mathbf{B}_{i,1}(t) = \begin{cases} 1, & t_i \le t \le t_{i+1} \\ 0, & \text{otherwise} \end{cases} \tag{3}$$

Here, we consider cubic B-spline curves which are most widely used. For this case, the subscript j in the above equations takes the value of 4 and each B-spline curve segment is defined by 4 control points.

Implementing these equations into a Java applet, we can draw an arbitrary B-spline curve. As shown in Fig. 4, when users click once on the canvas, a control point is input. After 4 control points are input, the first B-spline curve segment is drawn on the canvas. Then, the input of one more contorl point will lead to one more B-spline curve segment is drawn on the canvas.

3.4 Three-dimensional B-spline surfaces

Like two-dimensional B-spline curves, three-dimensional B-spline surfaces are one of the most important elements of three-dimensional (3D) geometric primitives. Three-dimensional geometric primitives consist of planes, platonic solids, quadric surfaces, 3D curves and 3D surfaces.

A B-spline surface can be constructed from a set of control points $\mathbf{P}_{i_0,i_1}$ ($0 \le i_0 \le n_0$, $0 \le i_1 \le n_1$) and two monotone sets of knots $\{s_i\}$ ($0 \le i \le n_0$) and $\{t_i\}$ ($0 \le i \le n_1$) through the equation of

$$\mathbf{X}(s,t) = \sum_{i_0=0}^{n_0} \sum_{i_1=0}^{n_1} \mathbf{B}_{i_0,j_0}(s)\mathbf{B}_{i_1,j_1}(t)\mathbf{P}_{i_0,i_1} \tag{4}$$

where

$$\mathbf{B}_{i,j}(r) = \frac{(r - r_i)\mathbf{B}_{i,j-1}(r)}{r_{i+j-1} - r_i} + \frac{(r_{i+j} - r)\mathbf{B}_{i+1,j-1}(r)}{r_{i+j} - r_{i+1}}$$

$$\mathbf{B}_{i,1}(r) = \begin{cases} 1, & r_i \le r \le r_{i+1} \\ 0, & \text{otherwise} \end{cases} \tag{5}$$

$(r = s,t)$

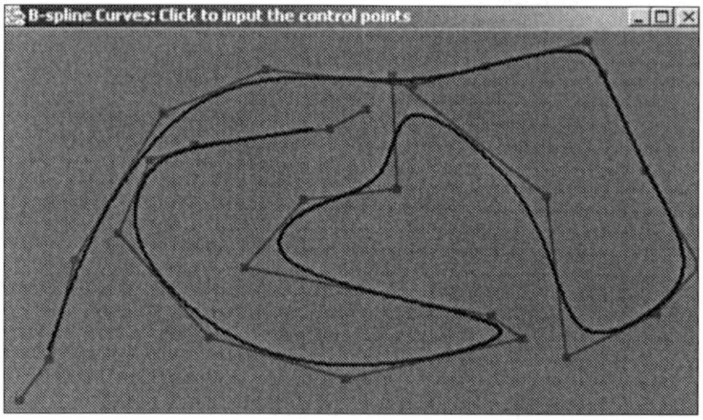

Fig. 4. Java applet of B-spline curves

Here we consider most widely used bicubic B-spline surfaces. The subscripts j_0 and j_1 in the above equations take the value of 4 and each surface patch are defined by 16 control points.

The user interface of the Java applet implemented from the above equations is indicated in Fig. 5. In order to generate a B-spline surface patch, users are asked to input the coordinate values of 16 control points from the panel on the right of the user interface. After the Java applet obtains there data, it uses Eq. (4) to calculate the coordinate values of the points on the B-spline surface patch and draws it on the left canvas.

4 CONCLUSIONS

In this paper, we have introduced the implementation of Web-based Java applets for online learning of geometric tools. A user interface was designed which not only provides an access to Java applets but also presents a main frame able to used for the further development of online learning system of this subject. Four represevtative examples of Java applets were presented to indicate their applications in Web-based learning of geometric tools.
Because the Java programming language is platform independent and the Java applets can be implemented on the World Wide Web, they are widely accessible and easy to use. Therefore, they are especially suitable for online learning of the subject on geometric tools.

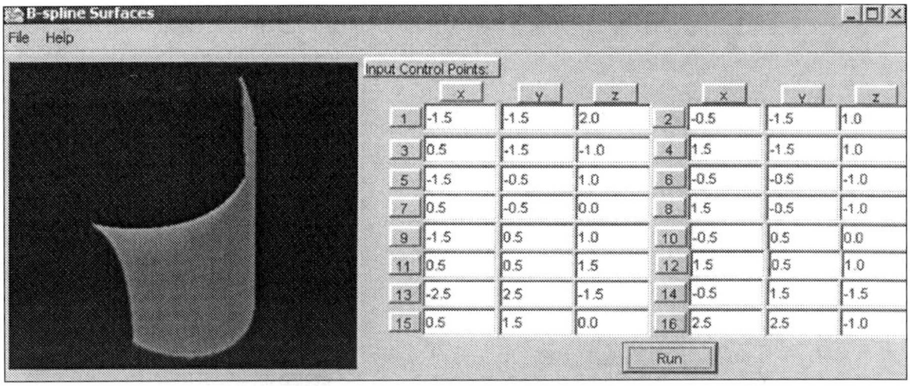

Fig. 5. Java applet of B-spline surfaces

REFERENCES

1. Granlund, R., Berglund, E. and Eriksson, H. (2000) "Designing Web-based simulation for learning", *Future Generation Computer Systems*, 17(2), 171-185.
2. Bouras, Ch., Destounis, P., Garofalakis, J., Gkamas, A., Sakalis, G., Sakkopoulos, E., Tsaknakis, J. and Tsiatsos, Th. (2000) "Efficient Web-based open and distance learning services", *Telematics and Informatics*, 17(3), 213-237.
3. Sheremetov, L. and Arenas, A. G. (2002) "EVA: an interactive Web-based collaborative learning environment", *Computers & Education*, 39(2), 161-182.
4. Pahl, C. (2003) "Managing evolution and change in Web-based teaching and learning environments", *Computers & Education*, 40(2), 99-114.
5. Rafaeli, S., Barak, M., Yuval Dan-Gur and Toch E. (2004) "QSIA – a Web-based environment for learning, assessing and knowledge sharing in communities", *Computers & Education*, available online at www.sciencedirect.com.
6. Papaspyrou, N., Retalis, S., Efremidis, S., Barlas, G. and Skordalakis, E. (1999) "Web-based teaching in software engineering". *Advances in Engineering Software*, 30(12), 901-906.
7. Bailey, D. (2000) "Interactive Web based tutorials for teaching digital electronics", *Proceedings of the SPIE* - The International Society for Optical Engineering, 4228, 212-218.
8. Roy, G. G. and Lee, P. L. (1999) "Interactive Web-based teaching for computing in an engineering degree", *International Journal of Engineering Education*, 15(5), 358-364.
9. Kofke, D. A. and Mihalick, B. C. (2002) "Web-based technologies for teaching and using molecular simulation", *Fluid Phase Equilibria*, 194-197, 327-335.
10. Kasyanov, V. N. and Kasianova, E. V. (2002) "Web-based systems for supporting computer-science teaching and learning", *SIGCSE Bulletin*, 34(3), 238.
11. Zaidel, M., Hopper, K. and Iyriboz, T. (1999) "Interactive Web-based radiology teaching file" *Journal of Digital Imaging*, 12, 203-204.
12. Hamam, H. (2003) "An interactive Web-based tool for teaching optics and vision", *Optics & Photonics News*, 14(8), 30-34.

13. Henderson, B. T., Camorlinga, S. and DeGagne, J. C. (2003) "A cost effective Web-based teaching file system", *Journal of Digital Imaging*, 16, 36-37.
14. Hui, S. C. M. and Cheung, K. P. (1999) "Developing a Web-Based Learning Environment for Building Energy Efficiency and Solar Design in Hong Kong", *Solar Energy*, 67(1-3), 151-159.
15. Wilkins, B. and Barrett, J. (2000) "The virtual construction site: a Web-based teaching/learning environment in construction technology", *Automation in Construction*, 10(1), 169-179.
16. Hough, M. and Marlin, T. (2000) "Web-based interactive learning modules for process control", *Computers & Chemical Engineering*, 24(2-7), 1485-1490.
17. Chalk, P. (2000) "Webworlds-Web-based modeling environments for learning software engineering". *Computer Science Education*, 10(1), 39-56.
18. Butler, J. E. and Brockman, J. B. (2001) "A Web-based learning tool that simulates a simple computer architecture", *SIGCSE Bulletin*, 33(2), 47-50.
19. Sung, W. T. and OU, S. C. (2002) "Web-based learning in the computer-aided design curriculum", *Journal of Computer Assisted Learning*, 18(2), 175-187.
20. Kuo, C.-H., Wible, D., Chen, M.-C., Sung, L.-C., Tsao, N.-L. and Chio, C.-L. (2002) "The design of an intelligent Web-based interactive language learning system", *Journal of Educational Computing Research*, 27(3), 229-248.
21. Jung, H., Jun, W., Gruenwald, L. and Hong, S.-K. (2002) "The design and implementation of a Web-based teaching-learning model for information communication technology application education", *Lecture Notes in Computer Science*, 2436, 56-68.
22. Yen, C. and Li, W.-J. (2003) "Web-based learning and instruction support system for pneumatics", *Computers & Education*, 41(2), 107-120.
23. Humar, I., Pustisek, M. and Bester, J. (2003) "Developing dynamic educational material with integrated mathematical notation for Web-based E-learning system", Conference of Engineering as a Human Endeavor: Partnering Community, Academia, Government, and Industry, Nov. 5-8 2003, Westminster, CO, USA, *Proceedings of Frontiers in Education Conference*, Vol. 1, pp. T3F19-T3F24, published by Institute of Electrical and Electronics Engineers Inc., USA.
24. Freasier, B., Collins, G. and Newitt, P. (2003) "A Web-based interactive homework quiz and tutorial package to motivate undergraduate chemistry students and improve learning", *Journal of Chemical Education*, 80(11), 1344-1347.
25. Sitthiworachart, J. and Joy, M. (2003) "Web-based peer assessment in learning computer programming", *Proceedings of 3rd IEEE International Conference on Advanced Learning Technologies*, 9-11 July 2003, Athens, Greece, pp. 180-184, published by IEEE Comput. Soc, Los Alamitos, CA, USA.
26. Yuan, X. F. and Teng, J. G. (2003) "Interactive Web-based package for computer-aided learning of structural behavior", *Computer Applications in Engineering Education*, 10(3), 121-136.
27. Schneider, P. J. and Eberly, D. H. (2003) "Geometric Tools for Computer Graphics", London: Morgan Kaufmann Publishers, Elsevier Science, USA.

Authors' Index